Digital Innovations in Architecture, Engineering and Construction

The Architecture, Engineering and Construction (AEC) industry is experiencing an unprecedented transformation from conventional labor-intensive activities to automation using innovative digital technologies and processes. This new paradigm also requires systemic changes focused on social, economic and sustainability aspects. Within the scope of Industry 4.0, digital technologies are a key factor in interconnecting information between the physical built environment and the digital virtual ecosystem. The most advanced virtual ecosystems allow to simulate the built to enable a real-time data-driven decision-making. This Book Series promotes and expedites the dissemination of recent research, advances, and applications in the field of digital innovations in the AEC industry. Topics of interest include but are not limited to:

- Industrialization: digital fabrication, modularization, cobotics, lean.
- Material innovations: bio-inspired, nano and recycled materials.
- Reality capture: computer vision, photogrammetry, laser scanning, drones.
- Extended reality: augmented, virtual and mixed reality.
- Sustainability and circular building economy.
- Interoperability: building/city information modeling.
- Interactive and adaptive architecture.
- Computational design: data-driven, generative and performance-based design.
- Simulation and analysis: digital twins, virtual cities.
- Data analytics: artificial intelligence, machine/deep learning.
- Health and safety: mobile and wearable devices, QR codes, RFID.
- Big data: GIS, IoT, sensors, cloud computing.
- Smart transactions, cybersecurity, gamification, blockchain.
- Quality and project management, business models, legal prospective.
- Risk and disaster management.

Xinyan Huang · Wai Cheong Tam
Editors

Intelligent Building Fire Safety and Smart Firefighting

Springer

Editors
Xinyan Huang
Department of Building Environment
and Energy Engineering
Hong Kong Polytechnic University
Kowloon, Hong Kong

Wai Cheong Tam
Fire Research Division
National Institute of Standards
and Technology
Gaithersburg, MD, USA

ISSN 2731-7269 ISSN 2731-7277 (electronic)
Digital Innovations in Architecture, Engineering and Construction
ISBN 978-3-031-48160-4 ISBN 978-3-031-48161-1 (eBook)
https://doi.org/10.1007/978-3-031-48161-1

This Springer imprint is published by the registered company Springer Nature Switzerland AG
The registered company address is: Gewerbestrasse 11, 6330 Cham, Switzerland

Paper in this product is recyclable.

Preface

Driven by globalization and fast urbanization, the density, scale, and complexity of the urban infrastructure have dramatically increased in the past few decades. More innovative architectural design concepts and new construction materials are proposed to satisfy the development of urbanization and the demand for mega-cities. The 2017 Grenfell Tower fire in London resulted in 72 fatalities, revealing many problems, e.g., the usage of combustible facades and lack of experience in fighting facade fires. The massive use of Li-ion batteries has caused many electric vehicle fires, and essentially, any innovation and evolution in the energy system will face new fire problems. Today, the estimated cost of fire on society is 1-2% of annual GDP, including all fire losses, public fire services, and fire protection systems.

Driven by climate change, the vast growth of wildland-urban interface (WUI) and disastrous wildfires pose extreme challenges to fire safety design, firefighting, emergency response, and recovery. For example, the 2018 California Camp Fire caused at least 85 fatalities and destroyed more than 18,000 structures. Moreover, unexpected fire incidents, from the 9-11 WTC fire and collapse to the 2020 di Beirut blast, continue challenging urban fire safety and resilience. High-density urban areas rely on complex interconnected infrastructure, e.g., high-rise buildings tunnels, tunnels, underground spaces, and emerging energy systems. However, the high-density populations stretch the capacity of infrastructure resilience in the event of an urban fire disaster requiring extremely efficient emergency management. These infrastructures are, therefore, highly vulnerable to major disruptive fire events (e.g., façade fires and WUI fires) and derived hazards (e.g., the large-scale urban fire after earthquakes) and require more innovative solutions to achieve fire safety and resilience.

Before fire safety became a formal discipline, it was often a side job of architects, civil engineers, and mechanic engineers when they considered the safety of structures and products. From the beginning, fire safety science and engineering are multidisciplinary in nature, mainly influenced by the subject of combustion physics, structural engineering, safety science, etc. Historically, the development of building fire safety is also driven by new scientific understandings and innovations in other areas, as illustrated in Fig. 1. Since the 1950s, new findings in heat transfer, fluid mechanics, and combustion have helped establish the foundations of fire dynamics.

Many new concepts have been proposed, such as compartment fire, piloted ignition, fire spread, and plume. The understanding of compartment fire development also introduces the standard fire curves for structural fire tests. These early efforts eventually established the fire safety scientific community, professional society, and engineering education in most countries around the 1970s. Thus, its history of fire safety science and engineering is much shorter than other engineering disciplines, not to mention thousands of years of the professional firefighting history.

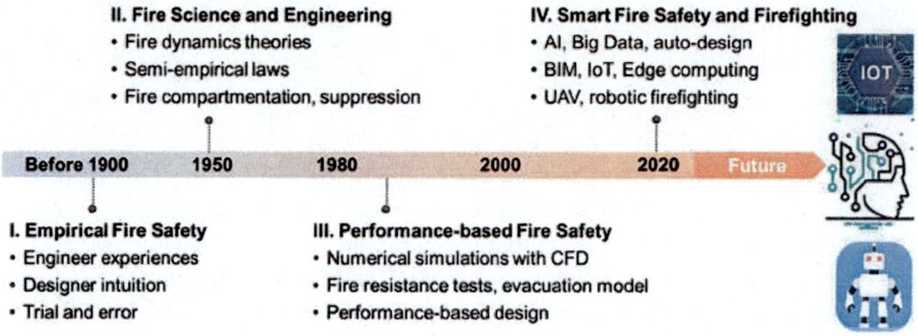

Fig. 1 The historical development of fire safety engineering and research

Along with the development of numerical simulation and the improvement of computational speed from the 1990s, the computational fluid dynamics (CFD) simulation is quickly applied in modelling fire and smoke motions. Together with other computational methods (e.g., evacuation modelling and structural fire finite element analysis), the framework of performance-based fire engineering design is established, as another milestone in the fire safety discipline. Today, the use of fire and evacuation modelling tools has become a routine or paradigm for most fire engineers' daily practices.

Over the past decade, new smart technologies, e.g., artificial intelligence (AI), Internet of things (IoT) sensor networks, cloud and edge computing, augmented reality (AR) and virtual reality (VR), 5G communication, UVAs, and robotics, have significantly changed our daily life and moved forward many research frontiers. Today, most smartphones have the function of AI-driven facial recognition to unlock, and large AI-power language model like ChatGPT has fundamentally changed the ways of working. Nevertheless, more advanced AI technologies like self-driving cars still need more time before massive application. A new AI-driven revolution is coming that will significantly improve our understanding of fire safety and fundamentally change the ways of firefighting practices (Fig. 2). Many researchers have explored these smart concepts over the last decade from different aspects and made important discoveries with demonstrations, which will be emphasized in this book.

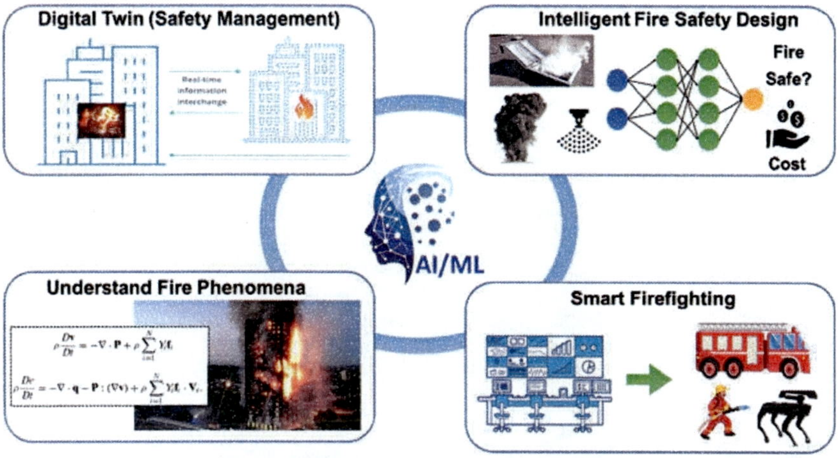

Fig. 2 Potential applications of AI on building fire research, safety management and firefighting

Arguably, the concept of Smart Firefighting originated from a workshop, *Smart Firefighting: Where Big Data and Fire Service Unite*, held on March 24–25, 2014, in Arlington, Virginia. Since then, the concept of Smart Firefighting has changed dramatically due to the rapid development of new technologies. Today, the usage of AI methods has become a key sign of smart firefighting technology, such as fire image recognition using deep learning and AI-driven fire detection and forecast.

On the other hand, the terminology of Smart Firefighting has been over-used in both the industry and research community, causing confusion and misconception. For example, many so-called "Smart Fighting" systems only connect the addressable fire alarm control panel to the computer and create a vivid user interface. Others assume that "Smart Fighting" is to use the existing surveillance camera network to monitor the fire and evacuation. Although these technologies improve building fire safety, they are still conventional methods that simply provide more information from the fire scene information. Therefore, it is important to first clarify the concept and level of "smart" before introducing potential applications of Smart Firefighting and discussing the technical challenges.

Fig. 3 provides a reference hierarchy of Smart Firefighting, ranked by the usage of AI and the level of intelligence. As a baseline, the past and current building fire safety and firefighting practices can be defined as Level 0, where no AI and machine learning are used. Then, Level 1 will use AI-assisted Fire Detection and Building Fire Safety Design. Level 2 will introduce the Smart Fire Digital Twin (or Fire Metaverse) for better fire safety management. Level 3 is AI-assisted Firefighting, where building fire protection systems will be controlled by AI, and semi-autonomous UAVs and firefighting robots will support firefighters. Level 4 will be Autonomous Robotic Firefighting, where human-free robots will be the leading fire service, replacing most firefighters' work. We have to acknowledge that today's technology and application of Smart Firefighting are still at the very early stage, where most of the research focuses on Levels 1 and 2. This book, with 20 chapters, will cover all four levels of Smart Firefighting, driven by AI, IoT, AR, robotics, and other new technologies, introduce the latest research progress, and prospect new research frontiers.

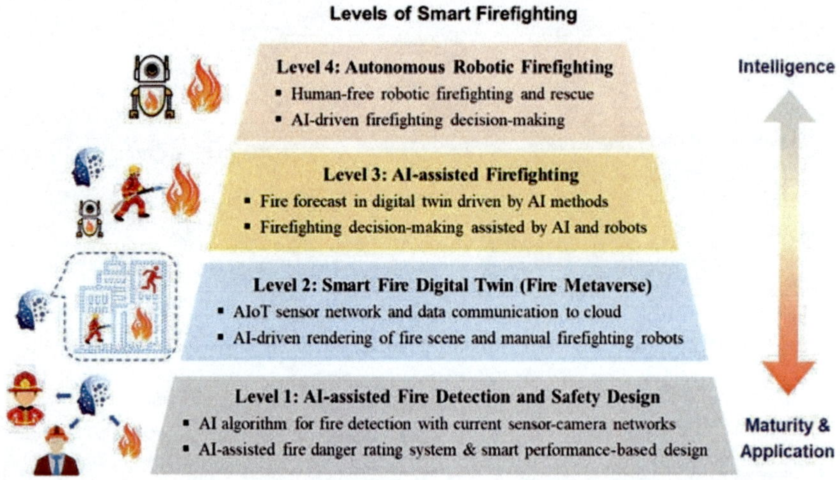

Fig. 3 Concepts and levels of smart firefighting and fire safety driven by artificial intelligence

This book has four sections, starting from **Introduction to Fire Safety and Smart Building**. In Chapter 1, Nils Johansson et al. present an overview of key concepts and fundamental works in *Building Fire Dynamics and Safety*. Then, we enter Chapter 2 by Craig Weinschenk, to discuss knowledge gaps and recent research achievements in *Fire Hazards and Firefighting in the Residential Environment* for pre-, during-, and post-emergency. In Chapter 3, *Smart Buildings: State-of-the-Art Methods and Data-Driven Applications*, Cheng Fan et al. showcase how Internet-of-Things and Artificial intelligence can work together to enable large-scale forecasting in commercial building settings. Yuanyuan Wang and his colleagues provide an *Introduction of Artificial Intelligence* in Chapter 4 and take a deeper dive into basic concepts, mathematics, applicability and example cases of common machine learning algorithms and novel deep learning methods for fire hazard prevention.

From **Intelligent Fire Safety Design**, Yanfu Zeng and Xinyan Huang first present *Artificial Intelligence Powered Building Fire Safety Design Analysis* in Chapter 5 and deliver an AI-powered software IFETool that can accelerate design, review and approval processes for performance-based building fire safety analysis. In Chapter 6, Xiao Chen showcases *Fire Safety Design of Confined Building Environment with Mechanical Ventilation* and presents the latest scientific advances for safety designs in special confined spaces. Then, in Chapter 7, *Automation in Structural Fire Resilience Design*, Mhd Anwar Orabi and his colleagues revisit the concepts of prescriptive design and performance-based design in structural fire engineering. It is in this chapter that automation concepts are applied to streamline thermomechanical analyses to facilitate the lengthy building design process. Jeanette Cobian-Iñiguez and her colleagues provide a thorough literature review on *Advanced Fire Safety in Wildland-Urban Interface* in Chapter 8. This chapter details the concepts of home hardening, land management, and fire-adapted communities and illustrates the adoption of AI to tackle fire safety problems in WUI. Andre Thompson's Chapter 9, *New Fire Protection Materials*, delivers a review of advanced chemical fire-retardant additives to reduce residential upholstered furniture flammability – a potential solution to prevent flashover and to increase the time available for egress. Chapter 10 brings us to *Smart Safety Design for Firefighting, Evacuation, and Rescue*, where Yuxin Zhang and Xinyan Huang revisit ASET and RSET and present a new concept, Safe Firefighting Time (SET), that fills the knowledge gaps from existing approaches.

Advanced Fire Identification includes Chapters 11–15. In Chapter 11, Tianhang Zhang et al. stress the importance of *Fire Database and Cybersecuirty* and introduce the framework of a fire database for a wide range of data and IoT-based blockchain technology to reduce cyber vulnerabilities for future smart fire safety systems. Chapter 12 is titled *Sensors and Modeling Driven Real-time Fire Forecast*, where Wolfram Jahn delivers well-established numerical approaches for fire event forecasting in enclosed building environments. Yoon Ko and her co-workers overview the use of machine learning paradigms, including traditional machine learning and deep-learning approaches, for *Fire and Smoke Image Recognition* in Chapter 13. Then, in Chapter 14, Xiaoning Zhang and his colleagues present a promising technology using the *Internet of Things and Digital Twin in Fire Safety Management* and deliver demonstrations in room fire forecasting and fire drill evacuation. Chapter 15, led by Qi Tong, offers a deeper look into *Applying Machine Learning to Evaluate the Performance of Thin-Walled Steel Members in Fire* where thermal effects on structural components are considered to identify fatigue regions.

Smart Firefighting includes the last five chapters. In Chapter 16 *Building Fire Hazard Predictions Using Machine Learning*, Eugene Fu et al. present deep-learning models to forecast the onset of flashover in residential building settings. In Chapter 17, Christopher Brown et al. deliver a *Data-Driven Wireless Fire Hose Flow Rate Apparatus* and discuss its implementation to provide actionable information to firefighters. Then, in Chapter 18, Ruggiero Lovreglio and his colleagues showcase the use of AR and VR together with machine learning to facilitate the development of smart evacuation systems in *Digital Technologies for Fire Evacuations*. Dillon Dzikowicz et al. bring us to look into the number-one cause of U.S. firefighters' deaths through *Cardiovascular Function and Deleterious Adaptations among Firefighters: Implications for Smart Firefighting* in Chapter 19. Finally, Chapter 20 is titled *Robotic Firefighting: A Review and Future Perspective*, where Meng Wang et al. present the current firefighting robot functionalities and capabilities and discuss the need for research efforts in future transition from human-based firefighting to autonomous robotic firefighting in extreme hazardous emergency conditions.

Kowloon, Hong Kong — Xinyan Huang
Gaithersburg, USA — Wai Cheong Tam

Contents

Abbreviations

3D	Three-dimensional
ACC	Automated compliance checking
AEC	Architecture, engineering, and construction
AHJ	Authority having jurisdiction
AHRR	Average heat release rate
AI	Artificial intelligence
ANN	Artificial neural network
API	Application programming interface
APIs	Application programming interfaces
ASET	Available safe egress time
BAS	Building automation system
BEMS	Building energy management systems
BI	Bottom ignition
BID	Burning item database
BIM	Building information modelling
BRI	Building research institute
CAD	Computer-aided design
CAM	Class activation map
CCTV	Closed-circuit television
CFD	Computational fluid dynamics
CFDB	Corsican fire database
CMMS	Computerized maintenance management system
CNN	Convolutional neural network
CPS	Cyber-physical systems
CPU	Central processing unit
CV	Computer vision
DAE	Denoising autoencoder
DBE	Data block engine
DC-CNN	Dual-channel convolutional neural network
DINS	Damage inspection (dataset)
DL	Deep learning

E	Young modulus
EHC	Effective heat of combustion
FAC	Fire adapted communities
FBs	Fire barriers
FCD	Fire calorimetry database
FDD	Fault detection and diagnosis
FDS	Fire dynamics simulator
FEM	Finite element model
FPUF	Flexible polyurethane foam
FRs	Flame retardants
GAN	Generative adversarial network
GNN	Graph neural network
GPS	Global positioning system
GPU	Graphics processing unit
GRU	Gated recurrent unit
HF	Heat flux
HRR	Heat release rate
HVAC	Heating, ventilation and air conditioning
IAFSS	International association for fire safety science
IC	Incident commander
IDE	Integrated development environment
IEQ	Indoor environmental quality
IMU	Inertial measurement unit
IoT	Internet of things
IoU	Intersection over union
IP	Image processing
IPFS	Interplanetary file system
IR	Infrared ray
ISO	International standards organization
LBP	Local binary pattern
LBPV	Local binary pattern variance
LSTM	Long short-term memory
LTB	Lateral-torsional buckling
LWIR	Long-wave infrared ray
MaCFP	Measurement and computation of fire phenomena
MaP	Materials and products database
ML	Machine learning
MLP	Multilayer perceptron
MLR	Mass loss rate
MODIS	Moderate resolution imaging spectroradiometer
MPC	Model predictive control
MQH-method	Method to calculate gas layer temperature developed by Mccaffrey, Quintiere and Harkleroad
MWIR	Mid-wave infrared ray
NFID	National fire information database

NFPA	National fire protection association
NIST	National institute of standards and technology
NWP	Numerical weather prediction
OSI	Open systems interconnection
PCA	Principal component analysis
PHRR	Peak heat release rate
PIR	Pyroelectric infrared
PLM	Product lifecycle management
PoA	Proof of authority
PoS	Proof of stake
PoW	Proof of work
PR	Polynomial regression
PSPNet	Pyramid scene parsing network
PTZ	Pan tilt zoom
R-CNN	Region-based convolutional neural network
RF	Random forest
RFID	Radio-frequency identification
RGB	Red-green-blue
RNN	Recurrent neural networks
ROM	Reduced order model
RUF	Residential upholstered furniture
SVM	Support vector machine
SVR	Support vector regressor
TCNN	Transpose convolutional neural network
TDMA	Time-division multiple access
THR	Total heat release
ToF	Time-of-flight
UAS	Unmanned aerial systems
UAV	Unmanned aerial vehicle
UI	User interface
UV	Ultraviolet
VID	Video image detection
VID_SF	Video image detection for smoke and fire
VIIRS	Visible infrared imaging radiometer suite
VIS	Visible
WebGL	Web graphics library
WSN	Wireless sensor network
WUI	Wildland-urban-interface
XGB	Extreme gradient boosting

Symbols

A_o Enclosure opening area (m^2)
A_T The total internal enclosure surface area, excluding the opening area (m^2)
b Flange width (m)
f_y Yield strength (Pa)
H_D Height to the hot gas layer interface (m)
H_N Height to the neutral plane (m)
H_o The height of the enclosure opening(m)
h_k The effective heat transfer coefficient (kW/m^2K)
h Height (m)
$k\rho c$ The thermal inertia (kW^2s/m^4K^2)
L Column length (m)
M Moment of structure element (N-m)
$\dot{m}_a$ Mass flow of ambient air entering an enclosure (kg/s)
$\dot{m}_g$ Mass flow of gas leaving an enclosure (kg/s)
N Load (N)
$\dot{Q}$ Heat release rate (kW or MW)
T Temperature (°C or K)
t Time (s)
t_f Flange thickness (m)
t_w Web thickness (m)

Greeks

α Growth rate (kW/s^2)
θ Temperature of the steel member (K)
ρ Density o (kg/m^3)
ψ Ratio between end-moments applied on the beam member

Introduction to Fire Safety and Smart Building

Building Fire Dynamics and Safety

Nils Johansson, Björn Karlsson, and Stefan Svensson

1 Introduction

A fire can be initiated and developed in various ways, and it is impossible to describe and predict every possible type of fire and its development in a building. However, it is possible to provide a general understanding of how fires in buildings develops and spreads.

Oxygen, fuel, and energy are all needed for combustion, and these three basic components are often illustrated as the fire triangle (see Fig. 1). If one of these components is not available, it cannot burn. "Oxygen" is in the air, "Fuel" can be any type of combustible material such as furniture, clothing, or waste. For ignition to occur "Energy" needs to be provided, this energy is needed to gasify the fuel to such an extent that a flammable mixture will be present. Examples of ignition sources that can initiate a fire are open flames or faults in electrical products.

Once ignition has taken place, the spread of the fire will depend on how big the ignition source is and on how well or badly the ignited object burns. The size of a fire in flammable liquids (like gasoline or ethanol) and plastics increases rapidly until the entire liquid surface or object is involved in the fire, while other products such as wood can burn more slowly.

Another factor that will govern the size of the fire is the amount of oxygen available. A fire out in an open environment such as a controlled bonfire (see Fig. 2) or

N. Johansson (✉)
Division of Fire Safety Engineering, Lund University, Lund, Sweden
e-mail: nils.johansson@brand.lth.se

B. Karlsson
Faculty of Civil and Environmental Engineering, University of Iceland, Reykjavík, Iceland

S. Svensson
Swedish Civil Contingencies Agency, Revinge, Sweden

X. Huang and W. C. Tam (eds.), *Intelligent Building Fire Safety and Smart Firefighting*,
Digital Innovations in Architecture, Engineering and Construction,
https://doi.org/10.1007/978-3-031-48161-1_1

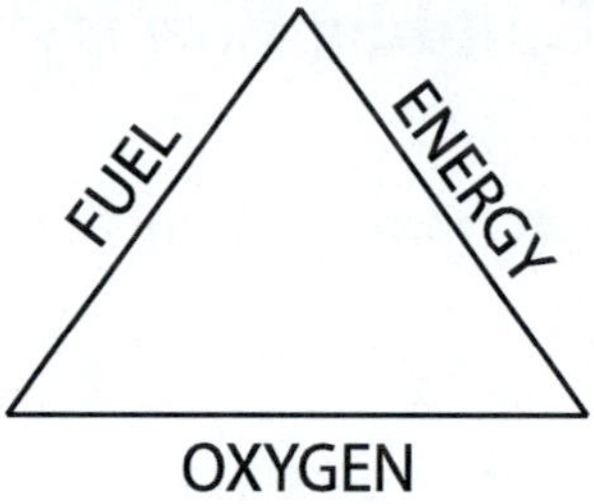

Fig. 1 The fire triangle (*Illustration* by N. Johansson)

an uncontrolled wildfire has free access to oxygen, while a fire in an enclosure can have limited air supply.

Most of the textbooks on building fires concentrate on fires in furniture or other items in rooms. However, intelligent fire safety designs and smart firefighting could possibly be applied to all types of fires in buildings, the presentation of building fire

Fig. 2 Bonfire in the open with free access to oxygen (*Photo* by N. Johansson)

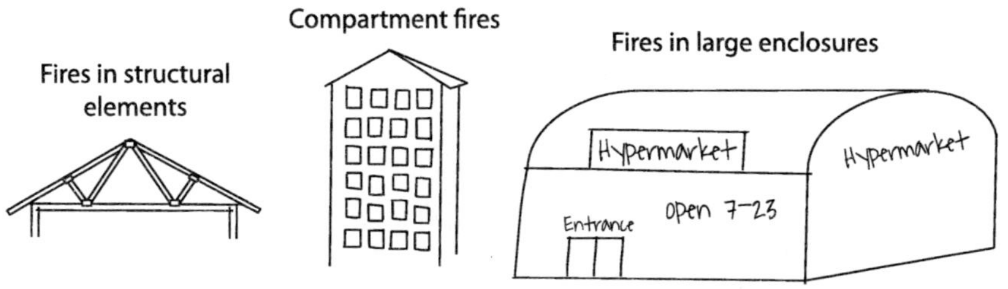

Fig. 3 The three overarching categories of building fires which are based on the characteristics of the building and the space in which the fire is contained (*Illustration* by N. Johansson)

dynamics here will therefore be based on three different overarching categories of building fires. These categories are based on the characteristics of the building and the space in which the fire is contained (see Fig. 3). The first category "Fires in structural elements" involves fires that start in, or spread to, a structural element e.g., inside walls, attic space, facade, or roof. The second category is the "Compartment fire", and it includes fires that start in, or spread to, enclosures that are not large in relation to the fire. That is enclosures where a hot gas layer forms, the temperature will be rather uniform throughout the layer, flashover can occur, and the fire can transition into a fully developed fire. The third category "Fires in large enclosures" involves fires that are small compared to the size of the enclosure, resulting in substantial temperature differences within the hot gas layer, thus it is not possible to assume uniform temperature conditions throughout the compartment.

The terminology might create some confusion since "Fires in large enclosures" are also fires in compartments; however, the "Compartment fire" is an established concept that refers to fires in small- and medium-sized enclosures (e.g. residential rooms, cellular offices, or small industrial units) and therefore, that terminology is used here.

Different models and assumptions are needed when studying these different kinds of categories of fires, and an understanding of fundamental fire dynamics and building fire safety is essential to develop and use methods and tools for fire safety designs and smart firefighting. If the user is not familiar with how building fires develop and spread, errors and mistakes might be introduced when these kind of methods and tools are applied. This chapter provides a general description of building fire dynamics and important features of the three mentioned categories of fires.

2 Fires in Structural Elements

"Fires in structural elements" is not clearly defined in the literature and based on its name, it can be perceived differently by different people. In a combined interview study and case study of actual fires in structural elements in Sweden [1], "Fires in structural elements" were found to be constituted by two sub-categories of fires namely concealed (or hidden) fires and fires in building materials. These more or less corresponds to the terms “internal structural fires” and “external structural fires” that has been used by Svensson [2].

There are examples of case studies of real fires which shows that concealed fires and fires in building materials, especially fires in connection to facades and roofs (external fires), pose a great risk for large fire damage. Figure 5 gives a schematic presentation of the results in a Canadian study [3] where 120 fires in residential buildings were categorized based on where the fire started and the final damage of the fire. Figure 5 illustrates that most compartment fires cause minor fire damage, while a larger proportion of hidden and especially external fires lead to more serious fire damage.

2.1 *Concealed Fires*

Concealed fires refer to fires in smaller spaces that are often difficult to access for the fire service in order to extinguish, e.g. in suspended ceiling spaces, attics and installation attics (see example in Fig. 4). Concealed fires also include fires in composite structures such as in walls (partition walls and partition walls) and joists. In a composite wall with an air cavity, the so-called chimney effect can occur if there is some ventilation present [4]. This type of fire is distinguished by the fact that the enclosure volume is smaller, and the oxygen supply is limited compared to “Compartment fires”. A building fire that occurs in a compartment can initially burn with flames, while a concealed fire generally smolders, and if air flows to the fire for some reasons (e.g., if parts of a roof collapse) the fire can quickly flare up.

Figure 6 illustates how a concealed fire can progress. A fire starts in a concealed space (point 1), e.g., by heat transfer from a hot duct, from overheating of electrical installations, or by spread from a room fire to the concealed space. The space is often

Fig. 4 A fire has been hidden inside an attic space in a villa. The photo shows how the flames started to penetrate through ventilation openings between the outer wall and roof (*Photo* by S. Svensson)

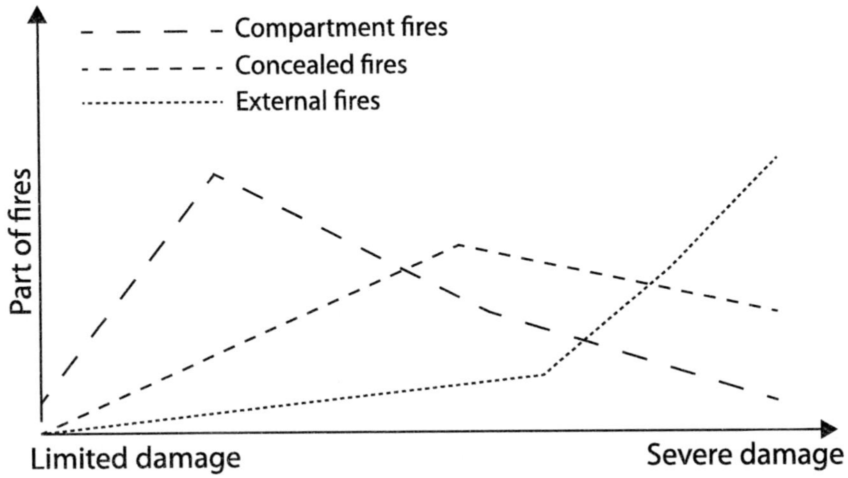

Fig. 5 Schematic representation of the results from a study of 120 residential fires [3] (*Illustration by N. Johansson*)

small and there is limited air supply, which means that the fire will be ventilation-controlled. The extent of the damage increases with time, but not as quickly as for a fire with a good supply of oxygen. If the containment breaks in an early stage (e.g., if fire is in a small attic space and there is no fire rating on the supporting roof structure) or if openings to the fire are created by the fire services, air can flow in and the fire flares up (point 2). This can lead to a chimney effect in the construction and spread of fire to more spaces in the building, and the extent of the fire damage increases. If the concealed space is intact, the fire services can carefully intervene by applying extinguishing agents, e.g., with a cutting extinguisher or fog spike, without the fire flaring up (point 3). If the fire is allowed to continue, containment can eventually give way and more air will flow to the fire which will mean that the extent of the damage will increase (point 4).

Fig. 6 Schematic representation of some typical behaviour in concealed fires (*Illustration by N. Johansson*)

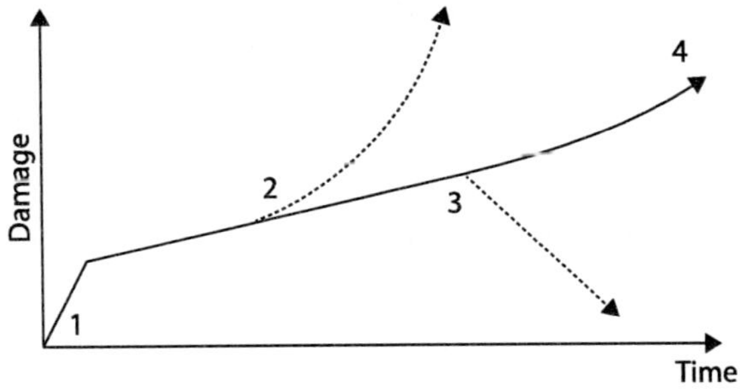

2.2 *Fires in Building Materials*

The second sub-category, Fires in building materials, involves fires on building surfaces and materials inside and outside the building, i.e., fires on walls, ceilings, and floors as well as on facades and external roofs. The fire behaviour in building materials inside occupied spaces (e.g. fire in combustible interior walls) are considered to a large extent be influenced by the enclosure and this type of fires are therefore not considered to be covered by what is here referred to as "Fires in structural elements". Instead, this type of fire is considered to belong to one of the other two overarching categories of building fires, "Compartment fires" or "Fires in large enclosures" (see Sects. 3 and 4).

In the case of fires outside the building, e.g., a roof fire or facade fire, there is free access to air, and the fire is affected by external conditions (e.g. wind speed and wind direction). Therefore, the course of the external fire cannot be equated with fires in compartments, nor with concealed fires. There have been several examples of fires in tall buildings where the exterior vertical fire spread has been very fast along a combustible façade [5, 6] and the fire in Grenfell Tower [7] is one of the more noticeable. In a comprehensive study and literature review in 2015 [8] funded by the *Fire Protection Research Foundation*, twenty facade fires in high-rise buildings were studied. In the study it was concluded that:

- Extensive facade fires often lead to extensive property damage but less often to deaths.
- The incidents often occur in countries with insufficient or no regulations on combustible material in facades at the time the building was constructed.
- Interior fires ("Compartment fires") that spread to the outside of the building are the most common starting scenario for exterior fires.
- Falling burning materials can pose a great danger and contribute to fire spreading downwards.
- U-shaped facades (e.g. when glazed balconies create a corridor in a vertical direction along the facade) mean a risk of faster fire spread than flat facades.

However, the fire in Grenfell Tower shows that fatal facade fires can occur even in countries where comprehensive building regulations exist. New innovative building products and materials are constantly being introduced, which can be a challenge for the regulatory system and the different test methods for façade systems.

3 Compartment Fire

The "Compartment fire" is, compared to the other categories of building fires, well researched, and there are several textbooks that cover it [9, 10]. One of the first comprehensive studies on the subject was done by Kawagoe in the 1950s [11] at the *Building Research Institute* (BRI) in Japan. Kawagoe studied fully developed

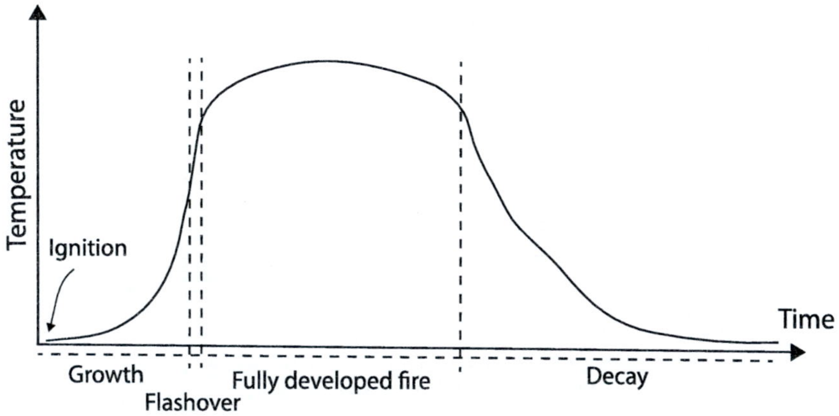

Fig. 7 Typical temperature development in a compartment fire curve (*Illustration* by N. Johansson)

fires and summarized a total of ten experiments in rooms with dimensions and fire sizes that resulted in rather uniform temperatures in the rooms. The work was later extended at BRI [12], and pioneering work was also conducted at the *Fire Research Station* in the UK [13, 14].

The main characteristic of the "Compartment fire" is that the hot gas layer created by the fire has a uniform temperature and gas composition. If a fire occurs in a small or medium-sized enclosure, the fire and fire plume will cause turbulence and stirring of the hot gases, which results in that the hot gas layer will be well-mixed. A compartment fire progresses through a specific sequence: growth, flashover, fully developed fire, and decay. Figure 7 depicts this sequence, but it is important to note that the time axis of the figure is not to scale. The growth period typically lasts a few minutes, while flashover usually occurs within a few seconds. The fully developed period can vary depending on the amount of fuel and ventilation conditions, but it usually lasts between 15 and 60 min for a single room. The decay period is more difficult to estimate, but it generally lasts between 1 to 3 h. Figure 8 shows a growing fire that soon will transition to a fully developed fire.

3.1 The Growth Period

After ignition, the fire will increase in size, and this is often called the growth period or the pre-flashover stage of the compartment fire. Depending on how ignition has occurred the fire can smolder before transitioning to flaming combustion. Smoke is produced, which means that smoke detectors normally can detect fires even at this early stage. Once flaming combustion is present, the fire will most likely increase in intensity unless someone intervenes, an extinguishing system is activated or if there is very little combustible material (fuel) present. A fire plume is created above the

Fig. 8 The photo shows how a fire has started in a room and spread to some furniture (*Photo* by S. Svensson)

flames, and air is entrained in the plume as it rises upwards. When the plume hits the ceiling, the flow will turn and vertical buoyant force in the plume will create a horizontal flow underneath the ceiling. This flow is called a ceiling jet. The ceiling jet will spread underneath the ceiling until it hits an obstacle, like a wall. The flow will turn, smoke will accumulate under the ceiling, and a hot gas layer will form in the room (see Fig. 9).

Beneath the hot gas layer, the air is rather unaffected since there is a flow of ambient air into the compartment through from adjacent spaces or the ambient. These so-called stratified conditions are often pictured as two distinctly different zones, and the gas properties in each zone are regarded as uniform (see Fig. 10 and red and blue box in Fig. 11). The height to the interface between these two zones is often labeled H_D and the height to the neutral plane (at which the pressure difference between the compartment and the ambient is zero) is labeled H_N. The smoke that

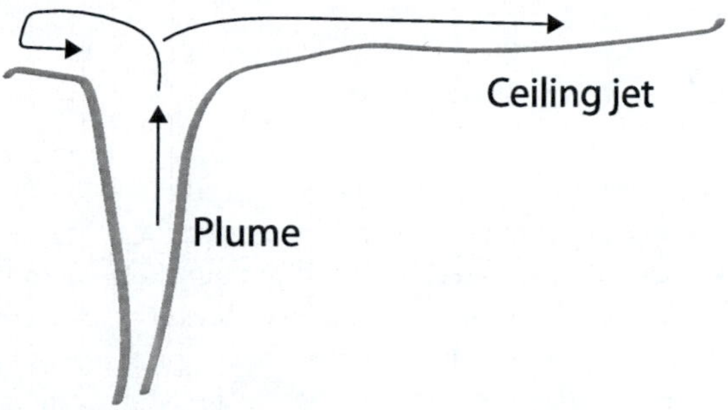

Fig. 9 A fire plume and ceiling jet in an enclosure (*Illustration* by N. Johansson)

Fig. 10 A distinct smoke layer that has developed in an enclosure fire (*Photo* by S. Svensson)

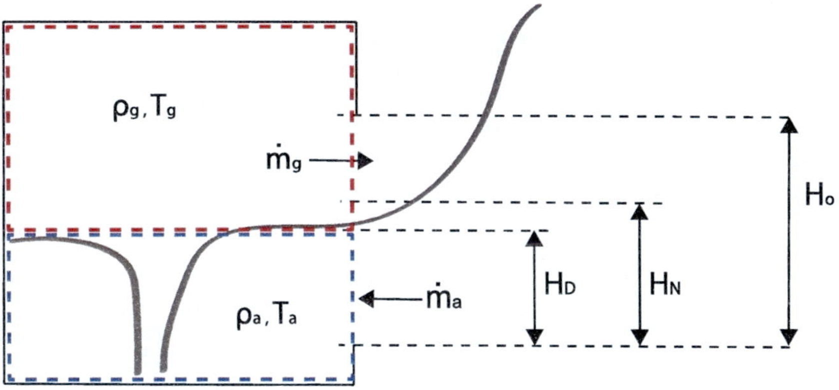

Fig. 11 The stratified fire with a hot gas layer and ambient lower layer. Hot gases flows out of the compartment above H_N and into the compartment below H_N (*Illustration* by N. Johansson)

Table 1 Rough values of fire heat release rates from burning different fuels

Burning wastepaper basket	50–100 kW
Wood chair with cushion	100–250 kW
A burning 1 m^2 pool of gasoline	2.5 MW
Two-seat sofa	1–3 MW
Burning wood pallets stacked to 3 m height	7 MW

accumulates in the room can spread through doors or other openings (e.g., windows or ventilation ducts) to other rooms in the building.

During the early stages of the fire, the fire is usually fuel-controlled, meaning that it is the amount of fuel involved in the fire that determines the size of the fire. It is common to use the heat release rate (HRR), to describe the size and power of a fire. It is because the rate of energy or heat released gives a measure of how destructive and dangerous the fire can be. Thus, the HRR is an important parameter and often something that needs to be estimated for any calculations of the consequences of a fire. In Table 1, some rough estimates of HRR from various fuels are presented.

A simple, and often rather reasonable, way to describe the growth of the fire during this stage, is to assume that the HRR increases exponentially time. By multiplying time squared with a constant α, , various fire growths can be modeled. With this model the HRR ($\dot{Q}$) as a function of time is expressed (see Eq. 1).

$$\dot{Q} = \alpha \cdot t \quad (1)$$

where α is the growth factor in kW/s^2 and t is the time, in seconds. The start of this exponential growth depends both on the material that is burning and the ignition source. Equation 1 has been found to correspond well with the growth rates exhibited by various combustibles and typical α-values for different types of fuels can be found in different textbooks [9, 15].

In many cases, information about the building contents might not be readily available when doing analyses, then the α-value can be assigned based on the building type and occupancy. There are some building codes and standards that have given recommendations for growth rate values for various types of occupancies, and these have been summarised by Karlsson and Quintiere [9], see Table 2. Even though some recommended values exist (like in Table 2) it is important to use engineering judgment and carry out a sensitivity analysis to check the credibility of a calculation.

Table 2 Typical fire growth values for different occupancies [9]

Type of occupancy	Fire growth (kW/s^2)
Dwellings	0.012 (medium fire growth)
Hotels, nursing homes	0.012 (medium fire growth)
Shopping centers, entertainment centers	0.047 (fast fire growth)
Schools, offices	0.012 (medium fire growth)

When the HRR has been established it is possible to study the consequences of the fire. Karlsson and Quintiere [4] gives a good overview of simple hand-calculation methods for ventilation flows, gas temperature and smoke filling in the pre-flashover fire. An example of such a simple but yet powerful and useful method is the so-called MQH-method developed by McCaffrey, Quintiere and Harkleroad [16]. The MQH-method is used to calculate the excess temperature in the hot gas layer, ΔT, during the pre-flashover fire. The MQH-method as presented in Eq. 2 is based on experimental work conducted in the early 1980s.

$$\Delta T = 6.85\left(\frac{\dot{Q}^2}{A_o\sqrt{H_o}h_k A_T}\right)^{1/3} \quad (2)$$

where $\dot{Q}$ is the HRR in kW, A_o is the opening area in m^2, H_o is the height of the opening in meters, h_k is the effective heat transfer coefficient in $kW/(m^2\ K)$ and A_T is the total internal enclosure surface area (excluding the opening area) in m^2. The effective heat transfer coefficient, h_k, is calculated as $h_k = \sqrt{k\rho c/t}$, where $k\rho c$ is the thermal inertia in $kW^2s/(m^4K^2)$ of the bounding material. A calculation example is provided in Example 1 to give the reader an Insight into the use of Eq. 2.

Example 1

There is a 500 kW fire in a compartment that is 5 m long, 4 m wide and 2.5 m high. The compartment has a door opening to the outside that is 2 m high and 1 m wide. The ambient temperature is 20 °C and the compartment is made of concrete with a thermal inertia of 2 kW^2s/m^4K^2. Calculate the temperature in the hot gas layer after 5 min

Solution: The area of the opening is 2 m^2 and the total internal surface area excluding opening is $2 \cdot (5 \cdot 4) + 2 \cdot (5 \cdot 2.5) + 2 \cdot (4 \cdot 2.5) - 2 = 83\ m^2$. The effective heat transfer coefficient is $\sqrt{2/300} = 0.0816\ kW/(m^2\ K)$. The excess temperature can now be calculated with Eq. 2

$$\Delta T = 6.85\left(\frac{500^2}{2\sqrt{2}\cdot 0.0816\cdot 83}\right)^{1/3} = 161\ K$$

The temperature in the hot gas layer is then 161 + 20 = 181 °C

3.2 *Flashover*

If it gets hot enough in the room during the growth period, flashover can occur, meaning that every combustible in the room starts burning. The *International Standards Organization* (ISO) [17] defines flashover as "transition to a state of total surface involvement in a fire of combustible materials within an enclosure" which means that flashover is not a protracted phenomenon but rather a transitional phase between the growing and the fully developed fire (see Fig. 7). If and when flashover occurs depends on several factors, such as fuel quantity, fuel properties, fuel location, surrounding geometry and the oxygen availability.

Flashover is often characterized by a hot gas layer temperature that exceeds 500 °C and a heat flux from the hot gas layer to the floor that exceeds 20 kW/m^2. These levels are far beyond what an unprotected human can survive; therefore, evacuation must be completed before flashover occurs. The fire growth and temperature development in a compartment just prior to flashover can be rapid due to the escalating HRR in the room, which is a consequence of the increasing temperature in the hot gas layer.

There are simple semi-empirical correlations that can be used to estimate what HRR is needed in a compartment to reach temperatures that will cause flashover. One such method is based on the MQH-method (Eq. 2). By setting $\Delta T = 500$ °C and re-arranging the equation, it can be used to find the energy release rate required for flashover $\dot{Q}_{FO}$ in a compartment (see Eq. 3).

$$\dot{Q}_{FO} = 610\left(h_k A_T A_o \sqrt{H_o}\right)^{1/2} \tag{3}$$

Example 2

Estimate the heat release rate required to reach flashover in the compartment presented in *Example 1*

Solution: The values for the compartment geometry and the effective heat transfer coefficient from *Example 1* is used in Eq. 3

$$\dot{Q}_{FO} = 610\left(0.0816 \cdot 83 \cdot 2\sqrt{2}\right)^{1/2} = 2{,}670 \text{ kW}$$

The heat release rate required to reach flashover in the compartment is 2,670 kW. In this case it has been assumed that $t = 300$ s, which might not be the case if the fire has a slow growth rate

3.3 *Fully Developed Fire*

As everything is involved in the fire at flashover the HRR increases rapidly. At flashover the fire goes from being fuel-controlled to ventilation-controlled, this means that it is not the amount of fuel that governs the rate of energy release instead it is the size of the openings. In ventilation-controlled fires, there is not enough oxygen in the room to combust all the fuel that is produced in the compartment. This means that there will be excess fuel that flows out of openings and can combust outside the compartment where there is abundant of oxygen available (see Fig. 12). The terms regime I and regime II [18] are sometimes used to distinguish between ventilation-controlled and the fuel controlled-burning, respectively.

Fig. 12 A fully developed fires where combustion occurs outside a window due to the ventilation-controlled conditions in the compartment (*Photo* by S. Svensson)

The increased turbulence and the transition from a localized fire to that the room is totally involved in the fire results in that the stratified conditions that were present during the growth phase are replaced by a single well-mixed gas volume, meaning that the smoke now fills up the entire room (see red box in Fig. 13). Hot gases will flow out of the compartment opening above the neutral plane and the fire will be supplied with ambient air below the neutral plane. If the opening size is kept constant during the fully developed fire, the HRR will be rather constant and the temperature in the hot gas layer will be stable. The temperature in a fully develop fire is normally in the range of 800–1200 °C.

The heat release rate in the compartment in a ventilation-controlled fire depends on the size of the opening since it is the flow of oxygen into the compartment that regulates the HRR. As proposed by Kawagoe [11], the correlation in Eq. 4 can be used to find the maximum possible HRR (in MW) in a compartment that an opening will allow in a ventilation-controlled fully developed compartment fire.

$$\dot{Q}_{vent} = 1.518 \cdot A_o \sqrt{H_o} \tag{4}$$

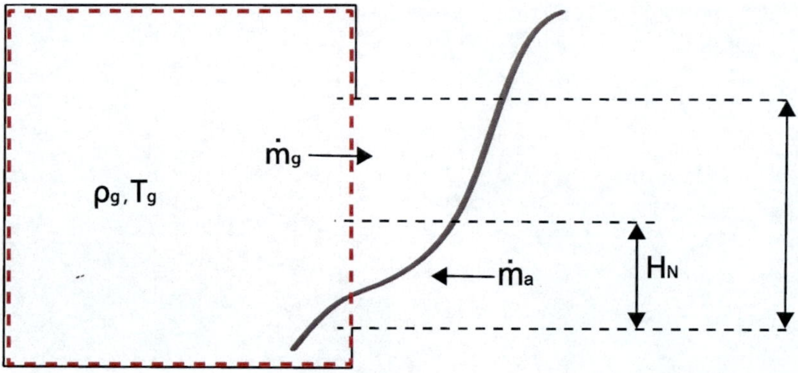

Fig. 13 The fully developed fire with a well-mixed hot gas layer (*Illustration* by N. Johansson)

Example 3

Examine if the openings in the compartment described in *Example 1* are large enough that flashover can occur, i.e., that the HRR calculated in *Example 2* can be reached

Solution: The opening had an area of 2 m^2 and height of 2 m, with these values $\dot{Q}_{vent}$ can be calculated with Eq. 4

$\dot{Q}_{vent} = 1.518 \cdot 2\sqrt{2} = 4.29$ MW

The calculated HRR is larger than the HRR required for flashover. This means that flashover can occur in the compartment and if it occurs all combustibles in the compartment will be involved in the fire and the HRR in the compartment will increase to the ventilation limiting value of 4,290 kW

The high temperatures during the fully developed fire will put a stress on fire separations and the load-bearing systems in the building. When building products or elements are tested to be classified according to a certain fire resistance class the classification test is performed according to a standard. The product that is tested is placed in a furnace where the temperature exposure follows a so-called standard fire curve (see Fig. 14).

Standard fire test curves are fictitious heat exposures developed with the purpose to just classify different building products and do not need to match more realistic fires. Thus, if a product has passed a standardized test for a certain time, that time does not correspond to the duration of a more realistic fire. Instead, it should be seen as the classified product can, regarding some criteria (related to e.g. integrity or insulation), withstand a certain thermal load.

In performance-based design of fire protection design of load-bearing and separating structures, it is important to study realistic fires. There are different analytical methods to calculate temperatures during fully developed fires. In these methods, the temperature is assumed to be uniform throughout the compartment and controlled

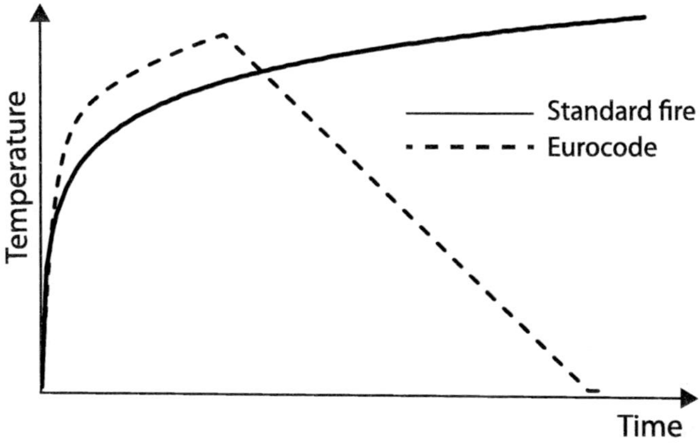

Fig. 14 A standard fire curve and example of a natural fire according to Eurocode (*Illustration* by N. Johansson)

by the size of the openings, the insulation of the boundaries and the fire load, i.e., the amount of energy contained in combustible material in the room. The Eurocodes [19] contains a method to calculate time–temperature curves for so-called natural fires (see Fig. 14). The term "natural fire" is misleading because the time–temperature curves in the Eurocode are still strong simplifications of the temperature development in fully developed compartment fires.

Even in the case of a fully developed fire, it can be of interest to study life safety and evacuation. In such cases, the analysis usually concerns an adjacent part of the building where evacuation still can be in progress, and to which smoke and heat are transported from the nearby fully developed compartment fire.

3.4 Decay Phase

If there is no intervention of the fire service the decay period will begin when about 70–80% of the combustible material in the room has been consumed by the fire [20]. If the fire has been ventilation-controlled, it may revert to being fuel-controlled as the HRR decreases. Due to the thermal inertia in materials and building products it can be very important to study the entire course of the fire, until the fuel runs out completely. The heat wave will propagate slowly through a structural element with low thermal inertia which means that the maximum temperature in a material can be reached during the decay phase (see Fig. 15), long after the maximum temperature in the compartment was reached.

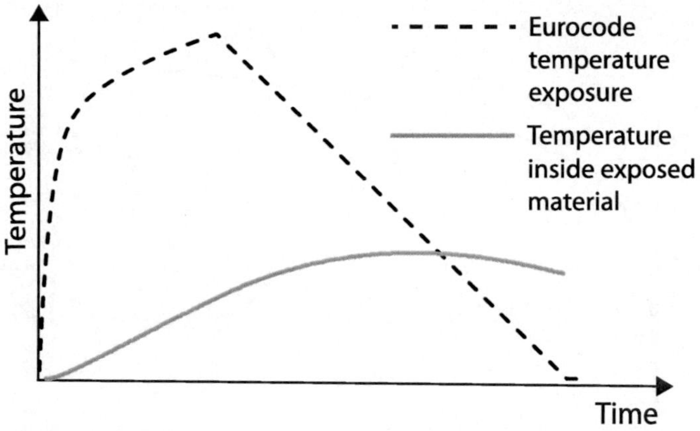

Fig. 15 Example of a natural fire exposure according to Eurocode and the temperature inside an exposed material. The maximum temperature in the exposed material occurs later than the maximum fire temperature (*Illustration* by N. Johansson)

3.5 Under-Ventilated Compartment Fires

As the fire grows larger and the amount of oxygen is limited, the fire can transit into a so-called ventilation control. If the enclosure has no openings or only small leakage areas, the hot gas layer will soon descend toward the flame region and eventually cover the flame. The air entrained into the combustion zone now contains little oxygen and the fire may die out due to oxygen starvation. The dashed line in Fig. 16 shows that a fire reaches a point (1) in Fig. 16 prior to flashover when oxygen depletion occurs and the HRR, as well as the gas temperature, decreases. Even though the energy release rate decreases, the pyrolysis may continue at a relatively high rate, causing the accumulation of unburnt gases in the enclosure. If the ventilation conditions do not change the fire will eventually go out (2). However, if a window breaks, or if the fire service creates an opening (3), the hot gases will flow out through the top of the opening and cold and fresh air will flow in through its lower part. This may reduce the thermal load in the enclosure, but the fresh air may also cause an increase in the HRR. The fire may then resume and grow toward flashover (4).

However, other more devastating phenomena can occur as a result of an under-ventilated fire. When a door or window is opened the inflowing air may mix with the unburnt pyrolysis products from the oxygen-starved fire. Any ignition sources, such as a glowing ember, can ignite the resulting flammable mixture. This leads to an explosive or very rapid burning of the gases. Expansion due to the heat created by the combustion will eject the burning gases out through the opening causing a so-called backdraft. Usually, a backdraft will only last for a very short time, in the order of seconds. The backdraft can then be followed by a fully developed fire.

Another possible consequence of an under-ventilated fire is a smoke gas explosion. A smoke gas explosion occurs when unburnt gases from an under-ventilated fire leaks

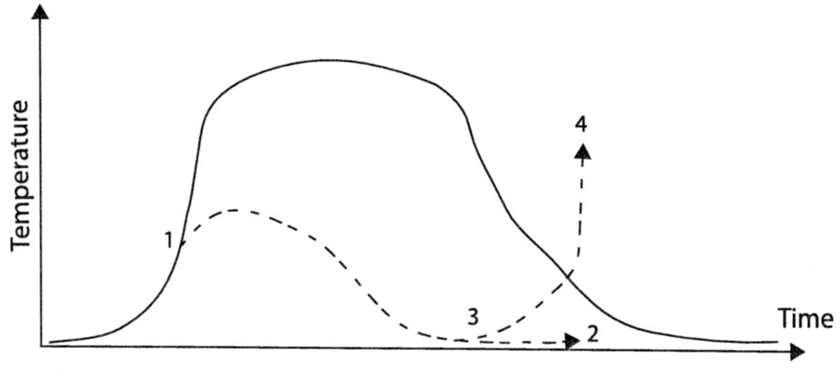

Fig. 16 Possible temperature development in a under-ventilated compartment fire (*Illustration* by N. Johansson)

into a closed space adjacent to the fire room. The unburnt gases mix with the air in the adjacent spaces and form a combustible gas mixture. A small spark can then be enough to cause an ignition of the gas, resulting in an explosion. This will result in very high pressures, and there can be very serious consequences to the building and the people within it. Bengtsson [21] gives a detailed description of the phenomena.

3.6 The Validity of the Compartment Fire Concept

As the size of an enclosure increases, the fundamental assumption of the "Compartment fire" concept (i.e. that the hot gas layer is uniformly mixed) becomes more uncertain. This is because the turbulence generated by the fire and plume is not strong enough to effectively mix the layer in larger enclosures, resulting in non-uniform conditions. As a consequence, the usefulness of the compartment fire framework decreases in larger enclosures.

There is no clear definition when the compartment fire framework should or should not be applied. However, the ISO has published some guidance on the use of zone models [22], which gives some hints of the possible enclosure dimension limits of the compartment fire framework, see Table 3.

Another example of limits that can be mentioned is related to thc Eurocode method, presented in Sect. 3.3. It is namely written in the Eurocode standard that

Table 3 Acceptable dimension and heat release rate/volume ratios [22]

Enclosure dimension ratios	HRR/volume
Length/width ≤ 5	>0.1 kW/m^3
Lenght/height ≤ 5	<1 MW/m^3

the parametric fire method is valid for enclosures up to 500 m^2 and 4 m room height [19].

Furthermore, it should be noted that the models discussed in this section are applicable only to single rooms, whereas in the case of a building fire, multiple rooms are typically interconnected. As a result, numerical models, such as two-zone models, are required to calculate the fire dynamics for multiple connected rooms.

4 Fires in Large Enclosures

The uniform temperature assumption, which is the backbone of the "Compartment fire" concept, can as mentioned in Sect. 3.6 be questioned for small fires in large enclosures. In a small- and medium-sized enclosure, the fire will mix the smoke and the temperature will be rather uniform, but this effect is reduced as the size of the enclosure increases. The temperature will generally decrease with an increasing lateral distance from the fire, and there will be greater vertical distribution in the hot gas layer temperature as well. Since there are few analytical methods for studying small fires in large enclosures, the most reasonable approach to study gas temperatures in large enclosures is to use an appropriate Computational Fluid Dynamics (CFD) model. There have also been attempts to study the conditions in large enclosures with so-called zone models which compared to CFD are much less computationally intensive [23].

A fire in an atrium is an example of a fire in a large enclosure. To address this situation, simple hand-calculation methods exist for calculating smoke filling and designing natural and mechanical ventilation [9]. However, it is important to note that these methods are often based on the same assumption as used for "Compartment fires," which assume that the hot gas layer created by the fire has a uniform temperature and gas composition. As a result, it is crucial to view the results obtained from these methods as strong estimates.

It does not make any sense to use the terms pre- and post-flashover fires when studying fires in very large enclosures. Flashover in the sense that was defined in Sect. 3.2 will not occur in enclosures when there is no hot gas layer with a uniform temperature. This does not mean that the flashover cannot occur locally in a part of the enclosure. For example, high shelves or partitions might result in parts of the enclosure that experience conditions where the compartment framework can be applicable. It should also be stressed that the fact that flashover does not occur in a large enclosure does not necessarily mean that the risk to people and property are reduced.

It has been argued that fires in large spaces are not likely to become ventilation-controlled since the availability of air will be high due to the larger presence of openings and leakages than in smaller enclosures. This means that the opening will not limit the size of the fire and it is instead important to study how the fire progresses from fuel package to fuel package. This type of progressive burning has been seen in experimental studies [24] and in actual events, and it is sometimes referred to as

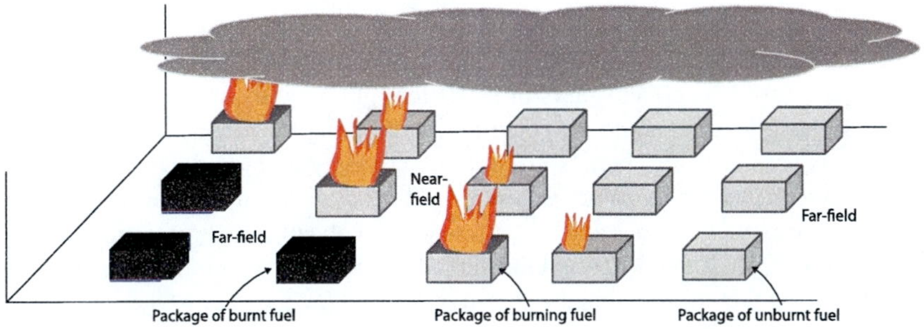

Fig. 17 Schematic description of the travelling fire concept in a large enclosure (*Illustration* by N. Johansson)

traveling fires [25] which is a concept that was developed to calculate the temperature exposure to structural elements in large enclosures. In the travelling fires concept, the thermal environment is split into the near and far field. The near-field is in the region close to the fire and the far-field is more remote from the flames where the structure is exposed to the hot gas layer. The regions are not static but move as the fire progresses to new fuel in the enclosure. This means that there is a large range of possible scenarios depending on the distribution of combustibles and where the fire starts. Figure 17 illustrates the concept of the travelling fire.

5 Role of Fire Dynamics Theory in Smart Firefighting

In this chapter, examples of different engineering tools to model or calculate the consequences of fires are presented. It should be obvious to the reader that the compartment fire framework presented in Sect. 3 in this chapter is more developed and holds more analytical methods. But, there are also different theoretical frameworks related to the other categories of buildings that can be applied to support and improve fire service activities. There have been studies that have shown that there is a great potential to use this type of tool before and during a fire service operation [1].

The HRR is a parameter that is of great importance to the potential damage of a fire and if it is not determined with some accuracy, calculation tools will not give any meaningful results. One way to determine the HRR is with video analysis of the fire, and another way is to use so-called inverse fire modelling [26].

In inverse fire modelling signals and consequences from the fire are interpreted, and engineering models are used to calculate backwards to estimate e.g., the HRR. It has been shown how computer models can be used to, based on data from measurements and theoretically based estimates, also predict the future development of fire in a building [27]. The fire development forecast can be updated with new data from e.g. thermal cameras or temperature sensors in the building.

One of the challenges of implementing smart firefighting is to process and analyze the collected data using simple and accurate models that can generate useful outputs for fire services. The effectiveness of smart firefighting will largely depend on the timescale involved. It should be noted that firefighting operations in the majority of cases is time-critical. There is rarely time to do any more in-depth calculations or simulations and the implementation and outcome of such operations is based on real-time assessments. Consequently, personnel involved in such operations, especially fire officers, should have well-established knowledge and understanding on fire theory and factors affecting the spread of fire and fire gases as well as on how to mitigate and manage fires.

In this regard, it can be argued that the potential of smart firefighting is greater in fire prevention and investigation than in firefighting operations, as these areas typically involve longer timeframes. However, when it comes to events that occur less frequently and with a longer duration than the more common residential fires, such as fires in larger and more complex buildings, there is clear use of smart firefighting in the operational area. The fire service may also be in a greater need for more automated support and information for decision-making in larger and more complex buildings. However, as it was seen in the previous sections, there are no or few simple engineering methods available for larger enclosures which means that specific digital models might need to be developed for individual buildings.

The application of fire dynamics theory has the potential to improve fire service operations across various domains. However, to leverage its full potential, it is necessary to systematically map the appropriate fire theory/model for a given situation and identify the necessary data needed. This will require a comprehensive understanding of the underlying principles of fire dynamics and its practical implications.

References

1. N. Johansson, S. Svensson, Review of the use of fire dynamics theory in fire service activities. Fire Technol. **55**(1), 81–103 (2019). https://doi.org/10.1007/s10694-018-0774-3
2. S. Svensson, *Fire ventilation*, 2nd edn. (Swedish Civil Contingencies Agency, 2020). ISBN: 978-91-7927-036-0
3. P.L. Senez, K. D. Calder, and H.H. Li, Fire loss statistical consideration in relating failure and building damage to the building code objectives, in *Presented at the 13th International Fire Science and Engineering Conference, 2013* (2013)
4. K. Livkiss, *'Fires in Narrow Construction Cavities: Fire Dynamics and Material Fire Performance*, Doctoral Thesis (compilation). Lund University, Division of Fire Safety Engineering (2020)
5. K.T. Nguyen, P. Weerasinghe, P. Mendis, T. Ngo, Performance of modern building façades in fire: a comprehensive review. Electron. J. Struct. Eng. **16**, 69–87 (2016). https://doi.org/10.56748/ejse.16212
6. A.C.Y. Yuen et al., Evaluating the fire risk associated with cladding panels: an overview of fire incidents, policies, and future perspective in fire standards. Fire Mater. **45**(5), 663–689 (2021). https://doi.org/10.1002/fam.2973

7. S.T. McKenna et al., Fire behaviour of modern façade materials – understanding the Grenfell Tower fire. J. Hazard. Mater. **368**, 115–123 (2019). https://doi.org/10.1016/j.jhazmat.2018.12.077
8. N. White, M. Delichatsios, Fire hazards of exterior wall assemblies containing combustible components, in *SpringerBriefs in Fire* (Springer, New York, NY, 2015). https://doi.org/10.1007/978-1-4939-2898-9
9. B. Karlsson, J. G. Quintiere, *Enclosure Fire Dynamics*, Second Edition. CRC Press (2022). https://doi.org/10.1201/b22214
10. D. Drysdale, *An Introduction to Fire Dynamics*, 3rd ed. John Wiley & Sons, Ltd, 2011. https://doi.org/10.1002/9781119975465
11. K. Kawagoe, *Fire Behaviour in Rooms*, Building Research Institute, Tokyo, Japan, 17
12. K. Kawagoe, Estimation of fire temperature-time curve in rooms (Part III), in *Transactions of the Architectural Institute of Japan*, vol. 140, no. 0, pp. 63–70,73 (1967), https://doi.org/10.3130/aijsaxx.140.0_63
13. P. Thomas H., *Theoretical considerations of the growth to flashover of compartment fires*, Fire Research Station, Boreham Wood, UK, 633 (1967). Accessed: Jan. 20, 2023. [Online]. Available: https://publications.iafss.org/publications/frn/663/-1
14. A. Heselden, *Some fires in a single compartment with independent variation of fuel surface area and thickness*, Fire Research Station, Boreham Wood, UK, 469 (1961)
15. M.J. Hurley et al., (eds.) *SFPE Handbook of Fire Protection Engineering* (Springer, New York, NY, 2016). https://doi.org/10.1007/978-1-4939-2565-0
16. B.J. McCaffrey, J.G. Quintiere, M.F. Harkleroad, Estimating room temperatures and the likelihood of flashover using fire test data correlations. Fire Technol. **17**(2), 98–119 (1981). https://doi.org/10.1007/BF02479583
17. ISO, *Fire safety — Vocabulary*, ISO 13943:2017 (2017)
18. J.L. Torero, M. A.h, A.-E. C, and C. A, Revisiting the Compartment Fire. Fire Saf. Sci. **11**, 28–45 (2014)
19. ISO, 'Standard—Eurocode 1: Actions on structures - Part 1–2: General actions—Actions on structures exposed to fire', Brussels, Belgium, EN 1991-1.2:2002, (2002)
20. G. V. Hadjisophocleous, J. R. Mehaffey, 'Fire Scenarios', in *SFPE Handbook of Fire Protection Engineering*, ed. by M. J. Hurley, D. Gottuk, J. R. Hall, K. Harada, E. Kuligowski, M. Puchovsky, J. Torero, J. M. Watts, and C. Wieczorek. (Springer, New York, NY, 2016), pp. 1262–1288. https://doi.org/10.1007/978-1-4939-2565-0_38
21. L.-G. Bengtsson, *Enclosure Fire*. Swedish Rescue Services Agency (2001)
22. ISO, 'Fire safety engineering—Guidance for use of fire zone models', ISO/TS 13447:2013, 2013. Accessed: Jan. 31, 2023. [Online]. https://www.iso.org/standard/53799.html
23. W.J. Bong, *Limitations of Zone Models and CFD Models for Natural Smoke Filling in Large Spaces* (2012). https://doi.org/10.26021/2161
24. E. Rackauskaite et al., Fire experiment inside a very large and open-plan compartment: x-ONE. Fire Technol. **58**(2), 905–939 (2022). https://doi.org/10.1007/s10694-021-01162-6
25. J. Stern-Gottfried, G. Rein, Travelling fires for structural design-Part II: design methodology. Fire Saf. J. **54**, 96–112 (2012). https://doi.org/10.1016/j.firesaf.2012.06.011
26. K.J. Overholt, O.A. Ezekoye, Characterizing heat release rates using an inverse fire modeling technique. Fire Technol. **48**(4), 893–909 (2012). https://doi.org/10.1007/s10694-011-0250-9
27. W. Jahn, *Inverse Modelling to Forecast Enclosure Fire Dynamics* (2010). [Online]. https://era.ed.ac.uk/handle/1842/3418. Accessed 11 Jan 2023

Fire Hazards and Firefighting in the Residential Environment

Craig Weinschenk

Abstract Traditionally, the fire service has improved by following a model based on experience [1]. Introduced by Brunacini, this is an iterative model that describes the experience-based approach to improvement and how fire service operations evolve over time. At the core of this approach are a department's standard operating guidelines (SOGs) which drive how activities will be performed (i.e., strategy and tactics) at an emergency. Firefighters train to department SOGs and subsequently work to execute based on their training at emergencies. Post-incident, firefighters assess their training/SOGs based on the outcome. These assessments can lead to revisions to the SOGs and corresponding training.

1 Introduction

Traditionally, the fire service has improved by following a model based on experience [1]. Introduced by Brunacini, this is an iterative model that describes the experience-based approach to improvement and how fire service operations evolve over time. At the core of this approach are a department's standard operating guidelines (SOGs) which drive how activities will be performed (i.e., strategy and tactics) at an emergency. Firefighters train to department SOGs and subsequently work to execute based on their training at emergencies. Post-incident, firefighters assess their training/SOGs based on the outcome. These assessments can lead to revisions to the SOGs and corresponding training. Ultimately, the more incidents a fire department responds to, the more the SOGs can be refined. This process can be visualized by the top loop in Fig. 1.

A limitation of an approach that relies primarily on individual experience and self-assessment, is that it is rare for an individual to have a complete picture of what everyone else experienced on the fireground. Further, how an individual experience's fire behavior may lead to a change of their behavior or the department's SOGs. How-

C. Weinschenk (✉)
Fire Safety Research Institute, UL Research Institutes, Columbia, MD 21045, USA
e-mail: Craig.Weinschenk@ul.org

X. Huang and W. C. Tam (eds.), *Intelligent Building Fire Safety and Smart Firefighting*, Digital Innovations in Architecture, Engineering and Construction,
https://doi.org/10.1007/978-3-031-48161-1_2

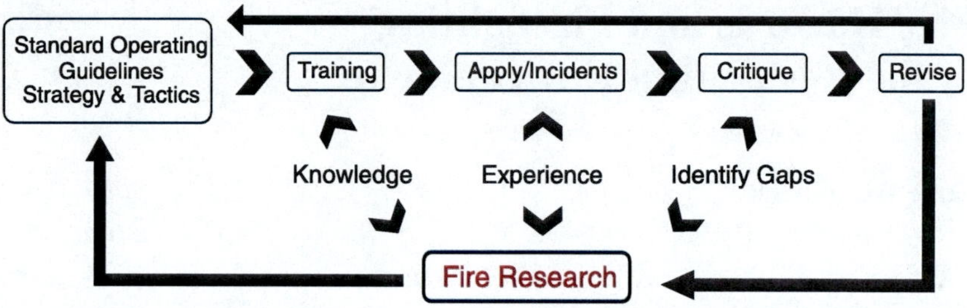

Fig. 1 An enhanced fire service improvement model highlighting the importance of fire dynamics knowledge and fire service research [2]

ever, this experience can be based on an incomplete understanding of the factor(s) that led to those outcomes.

The number of fires that occur in the United States have decreased by 3.2% from 2010 to 2019 [3], however the annual fire deaths during the same time period have increased by 24.1% [3]. The majority of these fires and fire fatalities occur in residential structures; between 2014 and 2018, "69% of the reported residential fires were in one- or two-family structures, causing 85% of the residential fire deaths [4]." In 2021, United States fire departments responded to approximately 1.35 million fires [5]. These fires caused 3,800 civilian fire deaths and 14,700 reported civilian fire injuries and nearly $16 billion in property damage [5]. Although there are many factors that can contribute fire fatalities and injuries, the fire service improvement model (Fig. 1) can be used to incorporate fire research into incidents and in their post incident reviews.

The bottom loop of Fig. 1, proposed by Kerber [2], highlights the role that research plays within fire service improvement system to fill in gaps highlighted above. The modifications proposed by Kerber suggest that the improvement model should be based on a foundation of fire dynamics, which in turn enables the incorporation of fire research. In the simplest sense, fire dynamics is the study of how fluid mechanics, thermodynamics, chemistry, and heat transfer interact to influence fire behavior. Fire service research includes the intersection between how fires start, spread, and develop and how tactics employed by firefighters are used for suppression. Essentially, without knowledge of fire dynamics, a firefighter could be misinterpreting the fire dynamics and propose SOGs changes in the wrong direction. Further complicating this situation is that fireground variables continue to evolve over time, precipitated by new technologies that have impacted both construction practices and the fuel loads within the structures. Johansson and Svensson confirm the importance of fire dynamics, stating that "solid knowledge about fire dynamics is fundamental in any type of fire service activities associated with fire safety in buildings [6]."

When assessing the fire hazards and corresponding firefighting response from changes in the residential built environment, it is important to consider the relative position along the timeline of an emergency: pre-emergency, emergency response, and post-emergency. Prior to a fire (pre-emergency), recognition of best practices can be utilized to minimize the chances of fire as well as minimize the impact of the fire on potential occupants and structures. If a fire does occur (emergency response), fire fighters can leverage tactics supported by science-based research to minimize the impact to potential occupants and involved structure(s). Finally, after the fire is suppressed (post-emergency), steps can be taken to limit the effects of exposure on the firefighters, fire investigators and other responding personnel. Johansson and Svensson further highlighted the importance of fire dynamics in all three phases of a fire response. In particular, fire dynamics theories are currently most used pre- and post-emergency compared to during a response [6]. This is due to a combination of the operational time frame of an emergency response and established tactical policies set by a department. The pair concluded that throughout all phases, there is room for the fire service to incorporate additional fire dynamics knowledge [6]. This chapter will explore the intersection of these three phases of a fire response and how fire service research has been used to provide recommendations to improve the safety of firefighters and the civilians they protect.

2 Pre-emergency

In a paper from 2012, Kerber highlighted how the residential fire environment has changed over time [7]. Several experiments were conducted to compare the impact of changing contents and construction methodology on fire development and structural integrity in residential houses. The changes examined included home size, size of open interior spaces (e.g. open concept floor plans), prevalence of synthetic fuels, and construction materials/methods. When looking at furniture in a typical residential home within the last 50 years, those furniture items are generally constructed of synthetic materials, such as polyester, polyurethane and polystyrene. Research shows that these materials tend to spread faster, release more energy, and generate more products of incomplete combustion than the natural materials, such as cotton, wool and solid wood, out of which older furnishings are made. Three pairs of experiments showed that for living room fires started with a small flaming ignition on the sofa, the time to flashover was less than 5 min when the sofa was made with synthetic materials compared to 30 min when the sofa was made with natural materials.

Floor system collapse time for different construction approaches exposed to a post-flashover fire were also compared [8]. In these experiments, a basement level was built from 0.61 m (2 ft) wide by 0.61 m (2 ft) tall by 1.22 m (4 ft) long interlocking concrete blocks. The interior dimensions of the basement were 6.1 m (20 ft) wide by 11 m (36 ft) long by 2.44 m (8 ft) tall. The 6.1 m (20 ft) span floor systems were built on top of a wood sill plate that was attached to the top edge of the block walls. Lightweight engineered wood floor assemblies included engineered I-Joists (depth 0.3 m (11 7/8 in.), flange thickness 0.04 m (1 1/2 in.), flange width 0.06 m (2 1/2 in.), web depth 0.23 m (8 7/8 in.), and web width 0.01 m (3/8 in.) and parallel chord metal plate connected wood trusses (depth 0.36 m (14 in.) and width 0.9 m (3 1/2 in.)). The legacy construction was dimensional lumber (width 0.04 m (1 1/2 in.)

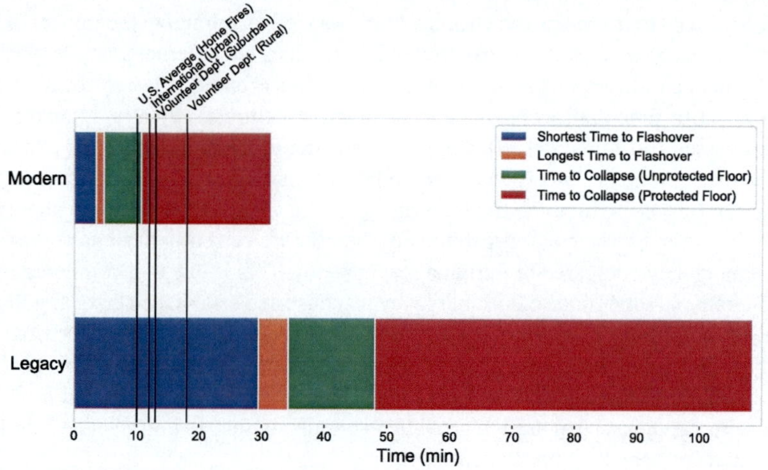

Fig. 2 Fire service arrival times versus fire development and collapse times adapted from [7]

and depth 0.29 m (11 1/4 in.)). The experiments showed the potential to collapse was sooner for the lightweight construction compared to the solid wood joists (legacy construction) [8]. Additional experiments demonstrated that the failure time of wall linings, windows, and interior doors have also decreased over time. These earlier failure times allow for more content fires to become structure fires which results in faster fire propagation and subsequently shorter times to collapse [7]. Figure 2 shows the time to flashover and the time to collapse for modern and legacy construction types overlaid with typical United States fire service response times.

Continued evolution of the built environment necessitates firefighters to understand the conditions of the structure to which they arrive. Fire conditions can change rapidly when firefighters begin to attack an under-ventilated fire compartment, and floor systems can collapse quickly and with little warning. Operating conditions need to be understood beginning with a proper size-up so the tactics used (ventilation, search, and suppression) can be implemented in an effective manner. Ultimately, as the fire environment continues to change, training and tactics need to correspondingly adjust to have the greatest opportunity to be effective. With recognition of the evolving response timeline and corresponding fire behavior, it is important to consider mitigation options that can exist prior to a fire emergency.

2.1 *Impact of Isolation*

Several UL FSRI research studies have shown the positive impact of a closed door on occupant tenability during a residential fire [9–16]. In particular, results from a firefighter suppression study [9], which incorporated the use of instrumented occupant tenability packages [17], led to a fire service tactical consideration pertaining to search operations and closed doors that in part states: "A victim located in a bedroom during a search with a closed door between them and the fire has a much higher likelihood of [survival] than a victim with an open bedroom door [9]."

The impact of a closed door was further explored in a study focused on search and rescue tactics in residential fires [18]. This study built upon prior research by conducting 21 experiments in two identical purpose-built single-story, single-family residential structures to quantify the impact of search and rescue tactics coupled with ventilation and suppression actions and timing. Each fully furnished structure included four bedrooms, two bathrooms and an open-floor kitchen and living room. A subset of these experiments—where fires were ignited in the bedroom (N= 6), kitchen (N= 2) and living room (N= 1)—had the doorway to one of the four bedrooms closed prior to ignition. As part of the search operations, that bedroom door was opened and closed to simulate a firefighting crew searching the space. Prior to the door being opened, thermal conditions in the hallway outside of the closed bedroom reached peak ceiling temperatures above 700 °C (1292 °F) and heat fluxes to the floor greater than 14 kW/m^2. Over this subset of experiments, gas concentrations 3 ft above the floor in the hallway reached a minimum concentration of 1.2% O_2 and maximum measured concentration of 5% (50,000 ppm) CO (the upper measurement limit of the CO analyzer). For an unprotected occupant, these thermal and toxic conditions would exceed typical tenability limits. On the bedroom side of the closed door, the minimum O_2 concentration and maximum CO concentration (measured 3 ft above the floor on the bed) was 19.1% and 0.2% (2000 ppm), respectively. The corresponding temperature and heat flux measured on the bed did not exceed 30 °C (86 °F) and 1 kW/m^2.

As an example, consider the pair of images in Fig. 3 from bedroom fire as part of the search and rescue tactics in residential structures study [18]. The isolated bedroom shows soot deposition around the door jam, due to the transport through gaps around the door and door frame (Fig. 3a). For the closed door bedroom, the minimum measured O_2 was 20%, and the maximum measured CO_2 and CO were 0.9% and 0.1% respectively. Comparatively, the non-isolated bedroom shows soot deposition throughout the ceiling and upper portion of the walls (Fig. 3b. As a result, for the open door bedroom, the minimum measured O_2 was 11.2%, and the maximum measured CO_2 and CO were 9.2% and 1.6% respectively. Additionally, damage to the lamp shade in Fig. 3b is an indication of the differences in thermal exposures. The 3 ft temperature peaked at 30 °C (86 °F) in the isolated bedroom compared to a peak 3 ft temperature of at 135 °C (275 °F) in the non-isolated bedroom.

Data from the search and rescue study [18] corroborates the previous research studies which showed the effectiveness of a door that was closed prior to ignition for exposure reduction. Communication of these empirical findings regarding the effectiveness of a closed door is also a critical component to pre-emergency hazard mitigation. Dedicated public-safety websites, such as *closeyourdoor.org* which is designed for both the general public and first responders are a good resource [20].

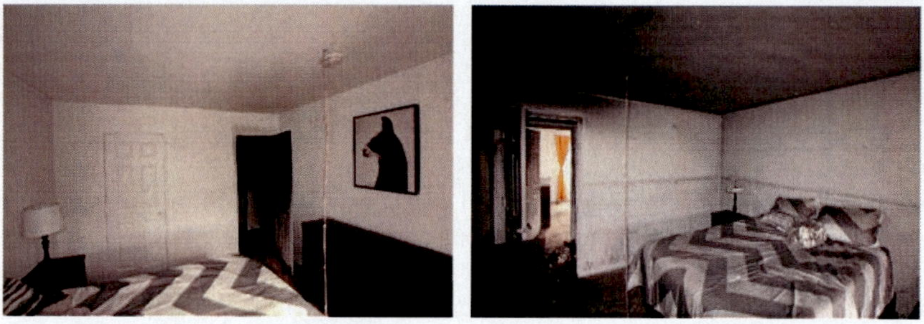

(a) Bedroom with Closed Door (b) Bedroom with Open Door

Fig. 3 Photographs showing the differences between bedrooms with a closed interior door versus an open interior door [19]

2.2 Training

As discussed in relation to Fig. 1, a critical component to the fire service improvement cycle is training. For the training to be effective, it is important that firefighters incorporate science-based recommendations into training to compliment their experiential-based core. In particular, research has helped the fire service use fire dynamics knowledge to (1) select of appropriate fuel packages for live-fire training and (2) develop a fundamental understanding of hose-stream mechanics.

2.2.1 Live-Fire Training

NFPA 1403: Standard on Live Fire Training outlines minimum requirements for conducting live fire training in fixed-facility and acquired structures [21]. One of the key requirements is that only wood-based products with known or well characterized burning behavior should be used as part of the training fuel package.

A common criticism from firefighters using wood-based training fuels, is that the fires produced by these fuels are not necessarily representative of the fires seen in contemporary homes filled with synthetic furnishings [22]. To help quantify NFPA 1403-compliant training fuel packages relative to synthetic furniture items, experiments were conducted in UL Solutions' calorimetry laboratory in Northbrook, IL [23]. Twenty-one NFPA 1403-compliant training fuel packages (composed of varying combinations of wood pallets, straw, excelsior, oriented strand board (OSB), medium density fiberboard (MDF), and dimensional lumber) and 12 furniture items (upholstered sofas, upholstered chairs, and a queen-sized bed) were ignited in a free burn test configuration under an oxygen consumption calorimeter.

Figure 4 shows the distribution of training fuel packages and furniture items on a spectrum of measured peak heat release rates (HRR). Training fuel packages ranging from a single bale of straw (15.4 kg) to seven pallets with one bale of straw (135 kg)

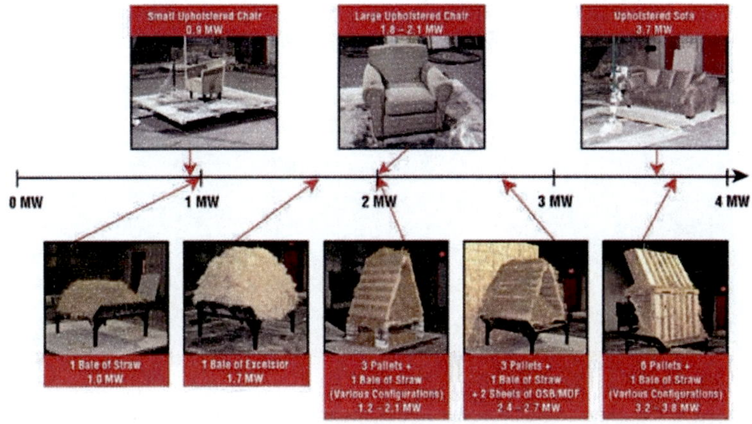

Fig. 4 Comparison of peak heat release rates of NFPA 1403 compliant fuel packages with upholstered furniture items [24]

had peak HRRs that varied from 1.0 MW to 3.8 MW. The peak HRR of the furniture items that weighed between 8.6 kg and 105 kg ranged from 0.9 MW to 3.7 MW.

The data from Regan showed that wood-based training fuel packages have peak HRRs that are comparable to single furniture items [23]. Although it is uncommon in a residential setting to find only one furniture item in a room, this work establishes some empirically-generated reference points firefighters can use to maximize potential impact when selecting NFPA 1403 compliant fuel packages for live-fire training.

2.2.2 Hose-Stream Mechanics Training

Over the past 10 years, fire service research has emphasized the importance of applying water to the fire as quickly as possible from the safest available location [9–12, 15, 25, 26]. To accomplish this task, it is important for firefighters to train on skills such as: water application to the fire compartment from the interior/exterior of the structure; selection of hose stream type, placement, and movement; minimizing/maximizing air entrainment; and identifying the position of flow paths within the structure. To increase the fundamental understanding of these fireground skills, two series of water-flow experiments were conducted [27, 28].

A set of 83 water distribution experiments were conducted to quantify the impact of changing the nozzle type, flow/pressure, or flow position on water dispersion within a compartment [27]. The experiments were conducted in a 3.66 m (12 ft) wide 5.48 m (18 ft) long room built on top of 48 water collection bins (which served at the floor of the room) to measure how water was distributed within the structure after impacting ceilings, walls, and door/window frames. Figure 5 shows the exterior and interior of this test prop.

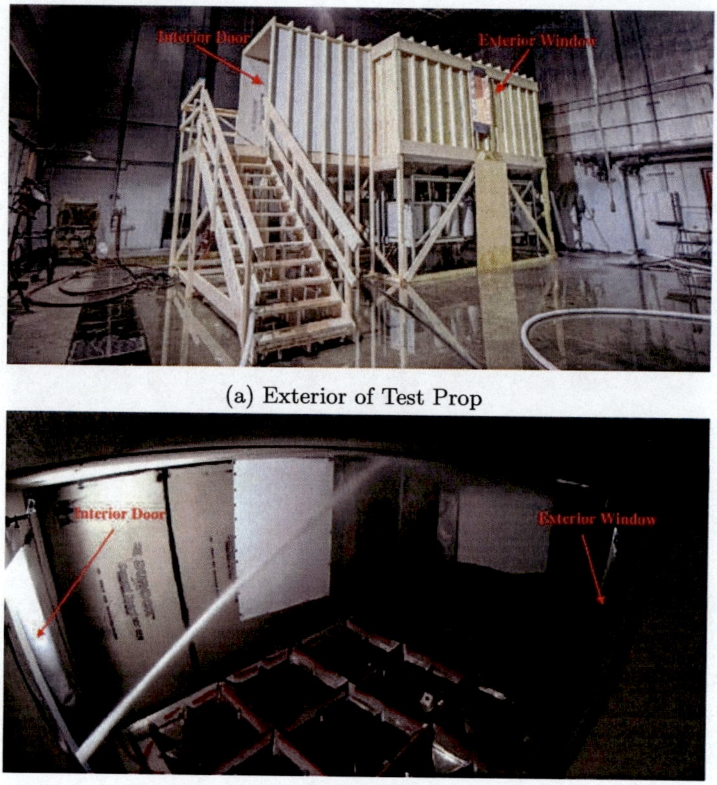

(a) Exterior of Test Prop

(b) Interior of Test Prop

Fig. 5 Photographs showing an exterior view of the water prop and interior view of the prop with a straight stream hitting the ceiling. Note, to measure water distribution, the floor of the room was composed of 48 water collection bins [27]

The results of the study showed the momentum from fire service hose streams prevents the stream from bouncing off the ceiling when flowed at a shallow angle, allowing the water to flow across the ceiling until a wall is encountered. As a result most of the water collects on the floor in the room at a location opposite the nozzle. Using a steep angle directed at the ceiling of a compartment breaks up the stream which results in improved distribution of water over more wall area of the room. As the angle decreased, the amount of surface coating decreased. This was consistent regardless of the nozzle type, pattern, or even the direction the water was applied from.

Thirty air entrainment experiments were conducted in a two-story, purpose-built structure designed to reflect a typical residential home [28]. The data showed that to increase entrainment, a firefighter should create a more broken stream or fog pattern, move the nozzle rapidly, and/or provide the largest stream length. If the intent is to

Fig. 6 Photograph of training prop specifically designed for firefighters to learn about water mapping and air entrainment [29]

limit air entrainment, the nozzle firefighter should use a fixed nozzle position on a smooth bore or straight stream and limit the stream length where possible.

Due to the complexity of the experimental structures used in these studies, teaching firefighters to apply this knowledge in hands-on manner using typical training structures was not feasible. Therefore, a prop (Fig. 6) that can be built by a fire department was designed with the distinct purpose of training on the suppression tactics of water mapping and air entrainment. Dimensioned drawings, material cut lists, and construction directions along with written and video lesson plans are freely available to any interested parties: Hose Stream Mechanics Toolkit [29].

3 Emergency Response

During an emergency, fire research can be incorporated as part of an on-going size-up and continues through a coordinated fire attack (suppression and ventilation), search, and when applicable—rescue.

3.1 Size-Up

NFPA 1006/1700 [30, 31] define size-up as "the ongoing observation and evaluation of factors that are used to develop strategic goals and tactical objectives." Fire service tactical considerations from studies on suppression [9], basement fires [15], fires in acquired single-family and multi-family structures [10, 25], identify size-up as a critical component to an emergency response that must occur on every fireground. Sizing-up an incident is not only about identifying problems but also identifying opportunities for engagement and mitigation [32]. Rapidly identifying the occupant status and determining the likelihood that someone may be trapped can help identify tactical priorities. That decision can become even more critical if the first arriving crew will be operating for several minutes before another crew arrives.

One recurring point of emphasis in the literature is the need to conduct a proper 360 °C size-up of the structure in question. Visualizing all sides of the structure is important because it can help the first-arriving officer determine what the problem is, where the problem is, and develop a course of action that can be relayed to incoming units. Taking the time to size-up a structure properly can lead to safer operations conducted in a timely manner [33]. During a 360 °C size-up, potential victim locations, building features (construction and layout), and other hazards such as exterior propane cylinders can also be identified [34]. With modern construction practices focused on energy efficiency through a tighter building envelope, utilizing a thermal imaging camera during the 360 °C size-up may help identify hotspots or potential areas of fire involvement [35].

There are several factors that build a good size-up with priorities varying based on the situation. A size-up should include fire extension and location [31], smoke presentation [36], building construction [37], building layout [38], and occupancy type [38]. Weather is another important factor in size-up, especially the wind direction and speed. Research has shown that winds speeds as low as 4 m/s can create untenable conditions remote from the seat of the fire [39, 40].

3.2 Coordinated Fire Attack

Although not comprehensive, firefighter research in residential structures to date was historically built by isolating tactics. In the simplest sense, ventilation tactics of horizontal [11], vertical [12], and positive pressure [13] and suppression tactics [9, 27, 28] were independently studied in purpose-built structures to provide recommendations to the fire service. These tactics were combined as part of *The Study of Coordinated Fire Attack in Acquired Structures*. Using contemporary fuels in real structures (single-family homes, multi-family structures, and commercial strip mall units), this study examined the effectiveness of tactics as they relate to changing fireground timelines.

A coordinated fire attack should be thought of as the systematic approach to the implementation of suppression and ventilation tactics. These tactics can occur sequentially or simultaneously with proper communication to minimize the time lag between them. To properly execute a coordinated attack, one of the key takeaways was for all firefighters to have a strong understanding of how their respective tasks on the fireground impact fire dynamics.

To understand how ventilation effects fire dynamics, is to understand the influence of ventilation on the gas flows within a structure. The acquired structure experiments confirmed findings from the ventilation experiments on the importance of identifying and controlling flow paths within a structure (i.e., use of a hoseline, closed doors, fans, etc.).[1] The increase in available oxygen due to ventilation led to an increase in the heat release of the fire. These experiments confirmed the importance of (1) identifying the fire location, (2) identifying existing intake and exhaust locations (i.e., where the air is coming from and smoke is going to), and (3) identifying points of firefighter entry relative to the fire location [10, 25, 26].

Across these experiments, a common theme emerged with respect to fireground coordination. When ventilation was provided by opening the front door for access, or by conducting horizontal or vertical ventilation, and the suppression was delayed, the ventilation led to an increase in the oxygen available for combustion. Additional oxygen provided to a vent-limited fire, resulted in an increase in the heat release rate of room and contents fires. As the time difference between ventilation and suppression lengthened, there was an increase in severity of the thermal and toxic hazard for potentially trapped occupants within the structure. When the time for water application following an opening being made was decreased, the removal of heat prior to the renewed source of oxygen reaching the combustibles that were still pyrolyzing reduced the temperatures and combustion gas production within the structure.

3.3 *Search and Rescue*

Several full-scale research studies, both in laboratory structures and acquired structures, have shown that there are survival spaces upon fire department arrival [9, 10, 14, 25]. To determine tenability, these studies assessed both thermal and toxic gas exposures. In particular, the laboratory fire experiments led to the following finding:

> Two things impact the survivability of a given space in the structure: the proximity to the fire and the elevation in the space. The proximity to the fire compartment relates to the thermal exposure where the elevation in the space relates to the exposure to toxic gases and thermal exposure [9].

To build on those studies, twenty-one experiments were conducted to provide more specific guidance for search and rescue to be applied to the fire service improve

[1] Note: a flow path is the interior volume between a higher-pressure source and a lower-pressure space within the structure and/or atmospheric pressure exterior vent.

Fig. 7 Photograph of firefighters taking bedroom windows to initiate search ahead of suppression [18]

model [18]. These experiments were conducted in fully-furnished, 4 bedroom, 2 bathroom structures with an open-concept living room/kitchen and included a heating, ventilation, and air conditioning system to accurately represent a ranch-style home, the most common single-family structure type in the United States [41, 42]. In these experiments, the fire dynamics and interior conditions were examined as firefighters created exterior ventilation openings (Fig. 7) and opened/closed interior doors in relation to suppression timing (e.g., pre-suppression, during-suppression, and post-suppression).

For searching firefighters, it was found that if the fire compartment door could be closed (e.g., close the bedroom door for a bedroom fire and close the front door for a kitchen/living room fire), flame spread in the compartment and thermal exposure risk for those freighters could be reduced. Cutting off adjacent spaces within the structure from the fire compartment, also reduced toxic and thermal exposure for potentially trapped occupants. Additionally, for the scenarios where fire department created ventilation preceded suppression as part of search operations (i.e., breaking a window for entry such as that shown in Fig. 7), isolation of the space of entry from the fire compartment, typically via a closed door, was shown to be effective at reducing toxic and thermal exposure rates compared to spaces that were not isolated.

If an occupant is found, these experiments also provided guidance to reduce exposures during removal. The data showed that occupant removal at the lowest functional elevation was effective at reducing exposures compared to higher elevations, however there were additional factors that could further improve conditions. Prior to suppression, removal of an already isolated occupant could lead to higher exposures compared to remaining in that isolated space until after suppression, particularly if the removal pathway required passing the fire compartment. Suppression, either from the interior and exterior, was effective at reducing the rate of thermal exposure. The rate of toxic exposures also decreased following suppression due to the reduced production of combustion gases, however, to minimize toxic exposures, suppression should be coordinated with additional structure ventilation.

4 Post-emergency

It has been well documented that today's structure fires can produce high levels of airborne particulate and gases/vapors including certain aldehydes, polycyclic aromatic hydrocarbons (PAHs), and volatile organic compounds (VOCs) [43]. While most of the exposure literature is based on data during fire responses or training fires, it does not mean that the exposure hazards end with suppression. Two post-emergency areas of concern include exposure to and accumulation of particulates to responding firefighters and fire investigators.

4.1 Firefighter Exposures

Cardiovascular events and cancer are two leading causes of firefighter fatalities. Horn et al. conducted a series of controlled live-fire experiments at the Illinois Fire Service Institute to provide data to inform firefighters on how to minimize cardiovascular strain and chemical exposures [44]. During the experiments the structures were instrumented to quantify the exposure fire dynamics and firefighters were instrumented to measure cardiovascular function (hemostatic balance, electrocardiogram, autonomic function) and toxic exposure (metabolites in blood, breath, and urine) before, immediately post-firefighting activity, and after 2 or 12 h of recovery.

To decrease the risk of a sudden cardiac event, Horn et al., recommend among that all firefighters have received a proper medical evaluation consistent with NFPA 1582 guidelines [45], firefighters utilize incident scene rehabilitation to provide rest, hydration and cooling; and medically monitor firefighters for signs and symptoms of heat-related illness or cardiovascular events [44].

The study generated several recommendations to limit chemical exposures, beginning with the usage of SCBA during knockdown, overhaul, and other firefighting activities where exposure to combustion byproducts is likely. Once gear is contaminated it is important for firefighters to perform a preliminary exposure reduction process (e.g., use of soap and water as shown in Fig. 8) and to remove gear before entering scene rehabilitation areas or entering apparatus. Additionally, firefighters should avoid storing contaminated gear in apparatus, personal vehicles, or living areas and it should be thoroughly cleaned before reuse. Finally, after removing gear,

Fig. 8 Photograph of firefighters performing fire service contamination reduction procedures

firefighters should wash hands and neck skin immediately and shower as soon as possible after a fire response [46].

4.2 Fire Investigation

Post fire, conducting an investigation is an important step in the fire service improvement model—the critique block in Fig. 1. Accurate investigations are important to determining cause and origin, understanding the fire dynamics of incidents, and to better understand cause and effect relationships between expected and actual outcomes. Sometimes additional expertise is needed to fully understand and critique an incident [47]. Further, computational tools like NIST's Fire Dynamics Simulator [48] can be used to better understand incidents such as near-misses, so that proper revisions to the improvement cycle can take place [49, 50].

However, a concern is that elevated and hazardous levels of air-borne particulate may be encountered during all phases of the post-fire scene investigation and exposure can depend on the activities of the fire investigator [51]. In conjunction with the residential search and rescue study [18], air concentrations of contaminants were measured during post-fire investigation of residential fires with furnishings common to bedroom, kitchen, and living room fires. Prior to each fire, baseline pre-fire area air samples were collected over 60 min. in the room where the fire was ignited. Post-fire samples were collected in several time windows: after the fire was extinguished but

Fig. 9 Post fire photograph following kitchen fire in a purpose-built residential structure. Gas and particulate sampling hardware in foreground

during overhaul, after overhaul was completed, and 1, 3, and 5 d post-fire. Figure 9 shows an example of the post-fire gas sampling infrastructure utilized in this study.

The results of the gas sampling indicated that high levels of particulate were present when drywall was removed from the structure, which is of specific importance to fire investigators who may be on site any time this activity is conducted. While airborne concentrations of benzene and naphthalene are often reported in highest concentrations prior to suppression, formaldehyde concentrations appeared to be the most abundant compounds relative to its occupational exposure limit in investigation timeframes [51]. Further, Horn et al., found that formaldehyde concentration could exceed recommended exposure limits in several phases of the investigation, including ceiling limits following suppression and the 1-day post fire [51]. These findings demonstration the importance of respiratory protection during-investigation and post-investigation decontamination.

5 Continuing Forward

The fire service has had to continually adapt to changes in residential built environment—over the past several decades there has been increasing use of lightweight construction and use of synthetic materials in particular. Firefighters will need to continue to adapt to address the evolving hazards associated with the growing shift to electrification. With the increased prevalence and speed of evolution of new hazards such as photovoltaic systems, energy storage systems, electric transportation (scooters, bikes, vehicles, etc.) as well as future unknown hazards, firefighters may not be able to develop training or SOG's rapidly enough to sufficiently address them. In such cases, the fire service may need to rely on research to provide their baseline [52]. Further, the fire service may also need to utilize emerging technologies such as machine learning and artificial intelligence throughout the incident response timeline to help identify ignition/incident timelines, firefighter location/health, and building performance among other hazards. Ultimately, the fire service will need to leverage all aspects of their improvement cycle to continue to safely respond to emergencies and serve the citizens of their jurisdictions.

References

1. A. Brunacini, A ready/aim/fire change plan, in *Fire Engineering*, vol. 162 (2009)
2. S. Kerber, *Utilizing Research to Enhance Fire Service Knowledge* (2020)
3. U.S. Fire Administration. U.S. Fire Statistics (2019)
4. S. Ahrens, R. Maheshwari, Home structure fires, in *Technical report, National Fire Protection Association, Quincy, MA* (2020)
5. S. Hall, B. Evarts, Fire loss in the United States during 2021, in *Research Report, National Fire Protection Agency, Quincy, MA* (2022)
6. N. Johansson, S. Svensson, Review of the use of fire dynamics theory in fire service activities. Fire Technol. **55**, 81–103 (2019)
7. S. Kerber, Analysis of changing residential fire dynamics and its implications on firefighter operational timeframes. Fire Technol. **48**, 865–891 (2012)
8. S. Kerber, D. Madrzykowski, J. Dalton, R. Backstrom, *Improving Fire Safety by Understanding the Fire Performance of Engineered Floor Systems and Providing the Fire Service with Information for Tactical Decision Making* (Underwriters Laboratories, Northbrook, Illinois, 2012)
9. R. Zevotek, K. Stakes, J. Willi, *Impact of Fire Attack Utilizing Interior and Exterior Streams on Firefighter Safety and Occupant Survival: Full-Scale Experiments* (UL Firefighter Safety Research Institute, Columbia, Maryland, 2018)
10. K. Stakes, J. Bryant, N. Dow, J. Regan, C. Weinschenk, *Analysis of the Coordination of Suppression and Ventilation in Multi-Family Dwellings* (UL Firefighter Safety Research Institute, Columbia, Maryland, 2020)
11. S. Kerber, *Impact of Ventilation on Fire Behavior in Legacy and Contemporary Residential Construction* (Underwriters Laboratories, Northbrook, Illinois, 2010)
12. S. Kerber, *Study of the Effectiveness of Fire Service Vertical Ventilation and Suppression Tactics in Single Family Homes* (UL Firefighter Safety Research Institute, Northbrook, Illinois, 2013)
13. R. Zevotek, S. Kerber, *Study of the Effectiveness of Fire Service Positive Pressure Ventilation During Fire Attack in Single Family Homes Incorporating Modern Construction Practices* (UL Firefighter Safety Research Institute, Columbia, MD, 2016)
14. N. Traina, G. Horn, S. Kerber, D. Kyritsis, Occupant tenability in single family homes: part I-impact of structure type, fire location and interior doors prior to fire department arrival. Fire Technol. **53**, 1589–1610 (2017)
15. D. Madrzykowski, C. Weinschenk, *Understanding and Fighting Basement Fires* (UL Firefighter Safety Research Institute, Columbia, Maryland, 2018)
16. S. Kerber, J. Regan, G. Horn, K. Fent, D. Smith, Effect of firefighting intervention on occupant tenability during a residential fire. Fire Technol. **55**, 2289–2316 (2019)
17. N. Traina, R. Kesler, S. Kerber, R. Zevotek, T. Lee, G. Horn, Ex-vivo porcine skin model for estimation of trapped occupant burn risk in pre-and post-suppression fire environments. Fire Technol. **55**, 2465–2489 (2019)

18. C. Weinschenk, K. Stakes, *Analysis of Search and Rescue Tactics in Single-Story Single-Family Homes Part III: Tactical Considerations* (2022)
19. C. Weinschenk, *Analysis of Search and Rescue Tactics in Single-Story Single-Family Homes Part I: Bedroom Fires* (2022)
20. Close Before You Doze. http://closeyourdoor.org. Accessed 12 Dec 2022
21. NFPA 1403. *Standard on Live Fire Training Evolutions* (National Fire Protection Association, Quincy, MA, 2018)
22. G. Horn, J. Stewart, R. Kesler, J. DeBlois, S. Kerber, K. Fent, W. Scott, B. Fernhall, D. Smith, Firefighter and fire instructor's physiological responses and safety in various training fire environments. Saf. Sci. **116**, 287–294 (2019)
23. J. Regan, Heat release rate characterization of NFPA 1403 compliant training fuels. Fire Technol. (2021)
24. J. Regan, *Research Corner: Bridging the Gap Between Training Fuels and Furniture*. https://www.firehouse.com/operations-training/article/21204434/research-corner-bridging-the-gap-between-training-fuels-and-furniture (2021). Accessed 12 Dec 2022
25. J. Regan, J. Bryant, C. Weinschenk, *Analysis of the Coordination of Suppression and Ventilation in Single-Family Homes* (UL Firefighter Safety Research Institute, Columbia, Maryland, 2020)
26. C. Weinschenk, R. Zevotek, *Exploratory Analysis of the Impact of Ventilation on Strip Mall Fires* (UL Firefighter Safety Research Institute, Columbia, Maryland, 2020)
27. C. Weinschenk, K. Stakes, R. Zevotek, *Impact of Fire Attack Utilizing Interior and Exterior Streams on Firefighter Safety and Occupant Survival: Water Mapping* (UL Firefighter Safety Research Institute, Columbia, Maryland, 2017)
28. C. Weinschenk, K. Stakes, R. Zevotek, *Impact of Fire Attack Utilizing Interior and Exterior Streams on Firefighter Safety and Occupant Survival: Air Entrainment* (UL Firefighter Safety Research Institute, Columbia, Maryland, 2017)
29. K. Stakes, D. Ey, Hose stream mechanics (2022). https://fsri.org/programs/hose-stream-mechanics. Accessed 12 Dec 2022
30. National Fire Protection Association, Quincy, MA. *NFPA 1006, Standard for Technical Rescue Personnel Professional Qualifications* (2021)
31. National Fire Protection Association, Quincy, MA. *NFPA 1700, Guide for Structural Fire Fighting* (2021)
32. J. Norman, *Fire Officer's Handbook of Tactics*, 5th edn. (PennWell Corporation, Tulsa, Oklahoma, 2019)
33. J. Mittendorf, *Truck Company Operations*, 2nd edn. (PennWell Corporation, Tulsa, Oklahoma, 2011)
34. M. Dugan, Conducting a proper size-up. *Firehouse* (1998)
35. D. Stratton, The basics of enhancing your size-up. Fire Eng. **171**(6) (2018)
36. D. Dodson, The art of reading smoke—FDIC (2014). https://www.fdic.com/articles/2014/06/david-dodson-the-art-of-reading-smoke.html
37. C. Clausing, J. Fortney, *Essentials of Fire Fighting*, 7th edn. (Fire Protection Publications, Stillwater, Oklahoma, 2018)
38. J. Coleman, *Searching Smarter* (PennWell Corporation, Tulsa, Oklahoma, 2011)
39. D. Madrzykowski, S. Kerber, *Fire Fighting Tactics Under Wind Driven Conditions: Laboratory Experiments. NIST Technical Note*, vol. 1618 (National Institute of Standards and Technology, Fire Research Division Engineering Laboratory, 2009)
40. S. Kerber, D. Madrzykowski, *Fire Fighting Tactics Under Wind Driven Fire Conditions: 7-Story Building Experiments. NIST Technical Note*, vol. 1629 (National Institute of Standards and Technology, Fire Research Division Engineering Laboratory, 2009)
41. American Housing Survey 2019. Washington, DC (2021)
42. Trulia's Guide To The Most Popular Home Styles in America (2017). https://www.trulia.com/blog/guide-to-most-popular-home-styles. Accessed: 15 Jan 2023
43. K. Fent, D. Evans, K. Babik, C. Striley, S. Bertke, S. Kerber, D. Smith, G. Horn, Airborne contaminants during controlled residential fires. J. Occup. Environ. Hyg. **15**(5), 399–412 (2018)

44. G. Horn, S. Kerber, K. Fent, B. Fernhall, D. Smith, *Cardiovascular and Chemical Exposure Risks in Modern Firefighting: Interim Report* (2016)
45. NFPA 1582, *Standard on Comprehensive Occupational Medical Program for Fire Departments* (National Fire Protection Association, Quincy, MA, 2018)
46. K. Fent, B. Alexander, J. Roberts, S. Robertson, C. Toennis, D. Sammons, S. Bertke, S. Kerber, D. Smith, G. Horn, Contamination of firefighter personal protective equipment and skin and the effectiveness of decontamination procedures. J. Occup. Environ. Hyg. **14**(10), 801–814 (2017)
47. M.B. McKinnon, S. DeCrane, S. Kerber, *Four Firefighters Injured in Lithium-Ion Battery Energy Storage System Explosion—Arizona* (UL Firefighter Safety Research Institute, Columbia, Maryland, 2020)
48. K. McGrattan, S. Hostikka, J. Floyd, R. McDermott, M. Vanella, E. Mueller, *Fire Dynamics Simulator, User's Guide*, 6th edn. (National Institute of Standards and Technology, Gaithersburg, Maryland, USA, and VTT Technical Research Centre of Finland, Espoo, Finland, 2023)
49. C.G. Weinschenk, K.J. Overholt, D. Madrzykowski, *Simulation of a Wind Driven Residential Basement Fire—Riverdale Heights, MD. NIST Technical Note* (National Institute of Standards and Technology, Gaithersburg, MD, 2015), p. 1870
50. M.B. McKinnon, D. Madrzykowski, *Four Firefighters Burned in Residential House Fire—Georgia* (UL Firefighter Safety Research Institute, Columbia, Maryland, 2022)
51. G. Horn, D. Madrzykowski, D. Neumann, A. Mayer, K. Fent, Airborne contamination during post-fire investigations: hot, warm and cold scenes. J. Occup. Environ. Hyg. **19**(1), 35–49 (2022)
52. Fire Safety Academy (2023). https://www.training.fsri.org. Accessed 15 Jan 2023

Smart Buildings: State-Of-The-Art Methods and Data-Driven Applications

Cheng Fan, Fu Xiao, and Huilong Wang

1 Introduction

The rapid development of IoT technologies has enabled the collection of massive amounts of data during building operations, making it increasingly promising and feasible to devise data-driven methods for smart building management. As an example, comprehensive sets of sensors and actuators are being equipped in the heating, ventilation and air-conditioning system (HVAC) to achieve real-time monitoring and controls over different components, such as chillers, water pumps, cooling towers and air handling units. Such information is typically collected with high frequencies (e.g., 30-s or 1-min), which can be of significant value to describe the complicated operating patterns and interactions among building services systems. At present, extensive research interests have been drawn from both the building industry and academia to utilize such information to devise data-driven solutions. Compared with conventional management methods which rely heavily on domain expertise and human labors, data-driven approaches are highly promising to enhance the efficiency and effectiveness of various building management tasks.

This chapter serves as an overview of the state-of-the-art research and applications for smart building management, using building energy system as an example. More specifically, Sect. 2 introduces main characteristics of information collected

C. Fan · H. Wang
Key Laboratory for Resilient Infrastructures of Coastal Cities, Ministry of Education, Shenzhen University, Shenzhen, China

Sino-Australia Joint Research Center in BIM and Smart Construction, Shenzhen University, Shenzhen, China

F. Xiao (✉)
Department of Building Services Engineering, The Hong Kong Polytechnic University, Hong Kong SAR, China
e-mail: linda.xiao@polyu.edu.hk

X. Huang and W. C. Tam (eds.), *Intelligent Building Fire Safety and Smart Firefighting*, Digital Innovations in Architecture, Engineering and Construction,
https://doi.org/10.1007/978-3-031-48161-1_3

during building operations, while emphasizing its compatibility with different kinds of data analytics. Section 3 introduces data preprocessing techniques for enhancing building operational data quality. Sections 4, 5 and 6 describe representative data-driven applications in various building management tasks, i.e., fault detection and diagnosis, occupancy analysis and optimal controls. The state-of-the-art data-driven methods and associated challenges in practice have been discussed in-depth. It is believed that the contents of this chapter are helpful for grasping the status-quo and major research trends in the smart building field.

2 Major Types and Characteristics of Building Operation Information

Buildings possess intrinsic geometric, parametric and physics-related information. In general, there are three types of information available for analysis in large-scale public buildings. The first type is image and video data collected through closed circuit television (CCTV) systems. Such information has unstructured formats and is typically of large amounts, which imposes great challenges for knowledge extraction and applications. Thanks to the development in deep learning and computer vision technologies, existing studies have developed advanced tools to utilize such information for object detection, image segmentation and classification tasks. For instance, Lee et al. developed deep learning models to identify human postures and behaviors in buildings, which in turns facilitated optimal controls of building services systems [1]. Ayala et al. proposed a deep learning approach for automatic fire alarms, which helps to achieve intelligent building safety management [2]. Another type of unstructured information collected during building operations is the textual data and it is typically system maintenance records generated by building management staffs. Existing studies mainly adopted text mining techniques to extract useful knowledge from such information. Gunay et al. puts forward a text-mining method to extract information about failure patterns in building systems and components from computerized maintenance management systems (CMMS) databases [3]. Ding et al. proposes a text mining-based methodology to gain insights from relevant literature on building energy saving [4].

Besides the abovementioned unstructured information, structured building operational data, which describe the operating conditions of building services systems, are also available and serve as the major resources for building data analytics. Building operational data are typically collected by Building Automation Systems (BAS) or Building Energy Management Systems (BEMS) through IoT technologies and stored in multi-relational databases. As an example, Fig. 1 presents an example two-dimensional data table for storing building operational data, where each column represents a data variable and the row represents data collected at a time step. The data collection intervals may range from seconds to hours, providing detailed electronic descriptions of building dynamic operations at various temporal scales. The

Time	Energy consumption data	Physical operating parameters	Environmental conditions	Miscellaneous data
10:00	$\{X_{11},X_{12},\ldots,X_{1m}\}$	$\{Y_{11},Y_{12},\ldots,Y_{1n}\}$	$\{Z_{11},Z_{12},\ldots,Z_{1p}\}$	$\{L_{11},L_{12},\ldots,L_{1q}\}$
11:00	$\{X_{21},X_{22},\ldots,X_{2m}\}$	$\{Y_{21},Y_{22},\ldots,Y_{2n}\}$	$\{Z_{21},Z_{22},\ldots,Z_{2p}\}$	$\{L_{21},L_{22},\ldots,L_{2q}\}$
12:00	$\{X_{31},X_{32},\ldots,X_{3m}\}$	$\{Y_{31},Y_{32},\ldots,Y_{3n}\}$	$\{Z_{31},Z_{32},\ldots,Z_{3p}\}$	$\{L_{31},L_{32},\ldots,L_{3q}\}$
13:00	$\{X_{41},X_{42},\ldots,X_{4m}\}$	$\{Y_{41},Y_{42},\ldots,Y_{4n}\}$	$\{Z_{41},Z_{42},\ldots,Z_{4p}\}$	$\{L_{41},L_{42},\ldots,L_{4q}\}$
14:00	$\{X_{51},X_{52},\ldots,X_{5m}\}$	$\{Y_{51},Y_{52},\ldots,Y_{5n}\}$	$\{Z_{51},Z_{52},\ldots,Z_{5p}\}$	$\{L_{51},L_{52},\ldots,L_{5q}\}$
15:00	$\{X_{61},X_{62},\ldots,X_{6m}\}$	$\{Y_{61},Y_{62},\ldots,Y_{6n}\}$	$\{Z_{61},Z_{62},\ldots,Z_{6p}\}$	$\{L_{61},L_{62},\ldots,L_{6q}\}$

Fig. 1 A typical example of structured building operational data

data collected can be further divided into four major categories, i.e., energy consumptions, physical operating parameters, environmental conditions, and miscellaneous data.

(1) Energy consumption data: Energy meters are used to measure energy consumptions of different building systems and components, such as the electricity power consumptions of chillers, pumps and fans, and the gas consumptions of boilers or heaters. Such data are highly valuable for building energy performance assessment.

(2) Physical operating parameters: The operating conditions of building services components can be described through real-time measurements and control signals. As an example, the actual operating conditions of chillers can be described using physical measurements, such as chilled water temperatures, chilled water flow rates, compressor operating frequencies, on/off status, and etc. Such variables can be numerical or categorical, such as *ON* and *OFF* status. Numerical variables may present different scales and variations, which make data normalization or standardization a must before most predictive modeling tasks. In addition, some numerical variables can be represented as categorical variables due to the operating characteristics of different components. For instance, measurements on water flow rates are numerical values, yet their varying ranges may be quite limited, specifically when a constant-speed pump is used.

(3) Environmental conditions: Building operations are subject to both indoor and outdoor environmental conditions, such as the indoor and outdoor temperature and humidity, as well as the extreme fire environments. Such variables are valuable for building performance assessment and controls, as they are closely related to indoor comforts and energy efficiency issues. At present, a wide variety of outdoor environment variables are publicly available and can be used to facilitate smart building management. For instance, outdoor illuminance data can be used for daylighting and renewable energy controls, CO_2 concentrations can be used to formulate optimal ventilation strategies, and smoke visibility can be used for fire emergency response.

(4) Miscellaneous data: Thanks to the rapid development in IoT and sensing technologies, more detailed building and occupant data are available in smart buildings. For instance, occupancy data of different spaces can be readily collected using smart entrance security or camera-aided technologies. Such information can be used to optimize HVAC control logics with the aim of enhancing indoor environment quality. Personal belongings, such as smart phones, smart bracelets and laptops, also provide information for smart controls. Smart phones or smart bracelets typically contain inertia measurement units, which can capture high-frequency and multi-axis acceleration data for human gesture identification. Laptops are typically connected with wireless networks and such information can be used to infer the relative locations of building occupants. It can be imagined that more detailed and fine-grained information will be available during operations, which further opens the opportunities for smart building controls and optimization.

Building operational data have intrinsic high complexity due to interactions among various services components. Meanwhile, the relative low-quality issue makes it more challenging to ensure the reliability and validity of building data analytics. The quality of building operational data is restricted by the accuracy of sensors and suffers from deterioration due to measurement errors and transmission problems. The raw data collected during building operations may contain large amounts of missing values, outliers, drifting and stagnant measurements. In practice, it is a must to apply customized data preprocessing methods before the use of any machine learning and artificial intelligence algorithms.

3 Data Preprocessing Tasks for Building Operational Data

The general process of building data analytics has three key stages, i.e., data preprocessing, knowledge discovery, and knowledge applications. Data preprocessing is an indispensable step considering possible deficiencies in building data quality. There are five major tasks for preprocessing building operational data, i.e., data cleaning, data reduction, data scaling, data transformation and data partitioning [5] (Fig. 2).

Data cleaning addresses missing value and outliers, which may stem from errors in sensors, transmission, and storage systems. Missing values can be imputed through model-free and model-based methods. Model-free methods are easy to implement, which may use means or medians as estimate s for missing values. It should be noted that such methods are not effective when imputing missing values lasting for long time periods, as it cannot reflect dynamics in system operations. Model-based methods, such as regression-based methods, are more suitable for estimating continuous missing values, yet more efforts are needed for model development. Outliers in building operational data are typically identified through either univariate or multivariate approaches. The univariate approach assumes that a variable follows a certain distribution, based on the outliers that are identified from a statistical point

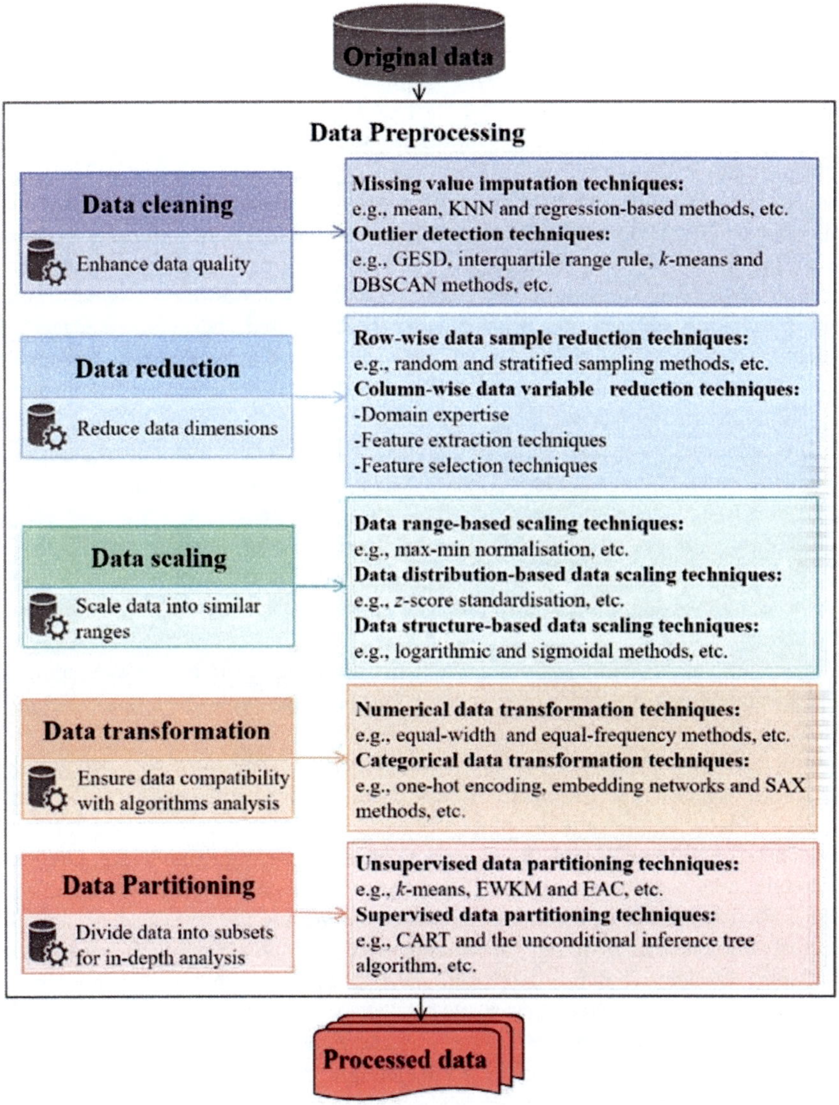

Fig. 2 Typical data preprocessing tasks for building operational data analysis [5]

of view, such as the 3-sigma method for normal distribution data. Such approach can only identify outliers with extreme values. By contrast, the multivariate approach considers possible correlations among multiple data variables for outlier identification. Existing studies mainly adopted unsupervised clustering analysis algorithms, such as k-means and density-based clustering methods, for applications.

The second task is data reduction, which aims to reduce data analysis complexity through row-wise or column-wise reduction methods. The row-wise data reduction methods focus on reducing the total number of data samples. As an example, random sampling can be applied to create a subset for analysis. Stratified sampling, which considers the proportions of data classes, is more suitable when class proportions are desired to be the same. For instance, the energy profiles of commercial buildings may present huge differences between weekdays and weekends. Random sampling may not strictly preserve the proportions between weekdays and weekends, while stratified sampling can be applied to maintain such information between the original and sampled data. The column-wise data reduction methods aim to select useful variables for in-depth analysis. Both domain expertise-based, statistical-based feature selection (e.g., correlation-based filters [6]) and feature extraction (e.g., principal component analysis [7] or autoencoders [8]) methods, can be applied for this purpose.

The third task is data scaling, which transforms variables with different scales to similar ranges for predictive modeling. The max–min normalization and z-score standardization methods are most used in the building field. The former transforms a variable to a range between zero and one, while the latter creates a variable with a mean of zero and a standard deviation of one. In practice, the latter may provide more flexibility as it is less sensitive to extreme values. Additionally, data structure-based methods, such as logarithmic methods, can be applied when variable distributions are highly skewed.

The fourth task is data transformation, which deals with mapping between numerical and categorical variables to ensure the compatibility with learning algorithms. Most building operational data are numerical variables. However, some learning algorithms, such as the A-priori algorithm for rule mining, only work with categorical variables. In such a case, the equal-width or equal-frequency methods can be applied to transform numerical variables into categorical ones. Meanwhile, categorical variables need to be transformed into numerical ones for predictive modeling. One-hot encoding is the most common way to transform a k-level categorical variable into $k - 1$ encoding vectors. When k is too large, embedding networks can be applied to create numerical embeddings for efficient representations. It should be mentioned that some variables in building operational data, even though stored as numerics, should be treated as categorical variables. As an example, the *Month* or *Daily Hour* are typically stored using numerical values, yet they should be treated as categorical variables as their intrinsic orders are not meaningful for predictive modeling.

The fifth task is data partitioning, which aims to divide building operational data into subsets to ensure the sensitivity and reliability of data analysis results. For instance, unsupervised learning methods, such as clustering analysis, can be applied to identify typical operating conditions of chiller systems [9]. Such clustering results can be used for data partitioning, based on which more fine-grained knowledge can be discovered across various operating conditions. Supervised methods, especially decision tree-based methods, can also be applied for data partitions. The main idea is to select a variable as outputs, while using operating schedule-related variables (e.g., *Year*, *Month*, and *Day type*) as inputs for predictive modeling [10]. In such a case, the resulting decision tree splitting rules can be used for data partitioning.

All the above-mentioned data preprocessing methods can be integrated for enhancing the quality of building operational data and ensuring the reliability of data analysis results. In the later sections, various learning techniques which can be applied to extract useful knowledge for building management tasks, such as building services system modeling, fault detection and diagnosis, occupancy identification, and etc., are discussed.

4 Data-Driven Fault Detection and Diagnosis Methods for Building Systems

4.1 Basic Unsupervised and Supervised Learning-Based FDD Methods

From the data perspectives, operating faults in building systems can be regarded as data outliers or anomalies. Both supervised and unsupervised learning methods have been developed for fault detection and diagnosis (FDD) tasks. As an example, scholars have used the unsupervised principal component analysis method to decompose and reconstruct the original operational data, and then established diagnostic indicators based on the reconstruction error for fault identification [11]. The other popular approach is to use clustering analysis for fault identification. Clustering analysis-based methods assume that the operation data under similar working conditions will form a relatively close cluster in the multi-dimensional space. Therefore, possible faults can be identified by calculating the relative distance or density distributions between each data sample and cluster centers [12].

Unsupervised learning-based methods, to some extent, can avoid possible decision-making malpractices caused by the over-dependence on subjective experience. However, the analysis results from the unsupervised algorithms are usually inexplicable and redundant. As an alternative, supervised learning is a more direct and effective approach, which in essence tackles the fault diagnosis task as multi-classification problems. It requires the data to be labeled and such labels are used to describe the actual conditions of different data samples (such as *Normal*, *Valve leakage fault*, *Fan failure fault*). The most widely used approach is to use machine learning algorithms to capture the complicated relationships between input and output variables at the same time step. Once developed, such models can be applied for real-time fault identification. Existing studies have shown that ensemble methods, such as random forest, light gradient boosting machine (LightGBM) and extreme gradient boosting (XGB), typically result in the best fault diagnosis performance [13]. Considering that building systems have intrinsic dynamics, time-series modeling approaches have also been explored to enhance the fault diagnosis performance. On the one hand, users can introduce additional input variables to represent measurements at previous few time steps for more accurate fault modeling. In such a case, the learning algorithm stays unchanged. On the other hand, users may directly apply advanced temporal modeling algorithms for analyzing temporal building operational data. The main techniques used in this field are recurrent neural networks, such as the gated recurrent unit (GRU)-based and the long short-term memory (LSTM)-based models [14].

4.2 *Transfer Learning-Based FDD Methods*

Supervised learning-based FDD methods are straightforward to use, yet the high requirement in labeled data makes them less feasible in practice. At present, researchers have proposed various approaches to handle the potential data quality problem for reliable fault modeling. Transfer learning, which has gained great success in computer vision, has provided a promising approach to tackling the data shortage problem. As shown in Fig. 3, the main goal of transfer learning is to migrate the knowledge learned from the source domain to solve new tasks in the target domain. In terms of building FDD problems, the main intuition is to integrate operational data collected from information-rich source buildings to facilitate the fault modeling in the information-poor target building.

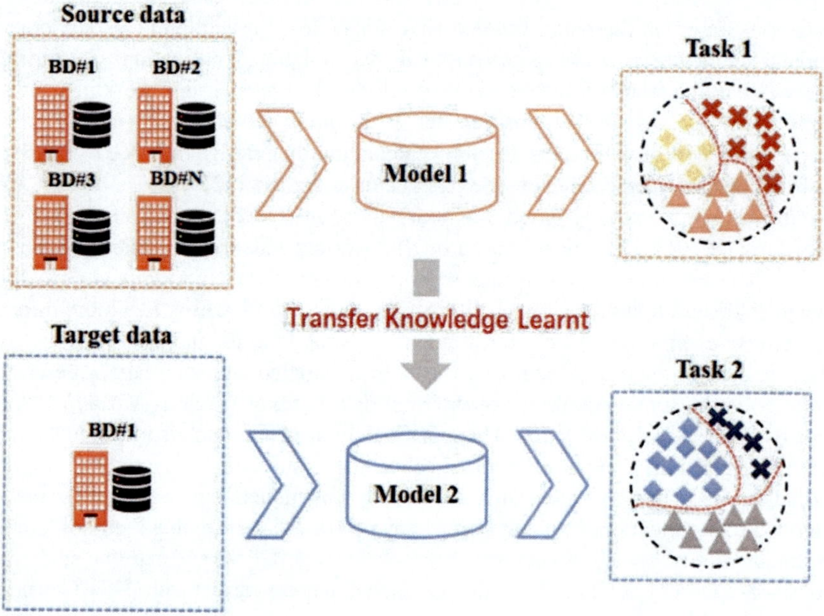

Fig. 3 Schematic diagram of transfer learning process

As an example, Liu et al. achieved knowledge transfer between two chiller systems using convolutional neural networks (CNNs) [15]. A temporal dimension was manually inserted to convert tabular building operational data into 2D image data. CNNs were developed based on the rich operational data collected from the source chiller, and customized to the target chiller using limited operational data. Experimental results showed that transfer learning helped to enhance the fault diagnosis accuracy by up to 8.18 ~ 12.63%. In practice, different building systems or components may collect completely different data variables. Advanced solutions are therefore needed to tackle the heterogeneous data challenge for source data integration and unification. Fan et al. proposed a novel image-based transfer learning framework which provides a format-compatible, information-expanding data basis for building operational data analysis [16]. The main idea is to apply the t-SNE algorithm to obtain the 2D coordinates of each variable, based on which an image can be generated for each data sampling using their measurements as pixel values. In this study, two strategies have been explored for effective knowledge transfer. One is called the freezing strategy, which utilizes the pre-trained model for image feature extraction. In such case, some or all convolutional layers for feature extractions using the source domain data are frozen, while the rest are trained using the target domain data. The other is the fine-tuning strategy, which adopts the pre-trained model for weight initialization and all the model weights are fine-tuned using target domain data. The research results indicated that the fine-tuning strategy was more efficient for cross-domain FDD knowledge transfer in the building field.

4.3 Semi-Supervised Learning-Based FDD Methods

Transfer learning-based methods assume the availability of source domain data for pre-trained model development, which is a rather heavyweight solution for individual buildings. A natural question arises as if there is no such source domain data, what approaches should the user take to enhance the reliability of data-driven models. It turns out that individual buildings typically have sufficient amounts of operational data, yet only a small subset of them are labeled and compatible with supervised learning algorithms. As a solution, semi-supervised learning methods, which aims to explore and utilize the information hidden in unlabeled data for predictive modeling, have gained increasing interests in the building field.

In general, there are four approaches for semi-supervised classification, i.e., generative model-based, low-density separation-based, graph-based, and self-labeling approaches [17]. The first two types have specific assumptions about the distributions of unlabeled data and decision boundaries, which may limit their potential in analyzing real-world data. As an example, Li et al. proposed a semi-supervised data-driven fault diagnosis method for chiller systems based on semi-generative adversarial networks [18]. It used semi-generative adversarial networks to learn the information on unlabeled data distributions, based on which the quality of supervised fault classification models was enhanced.

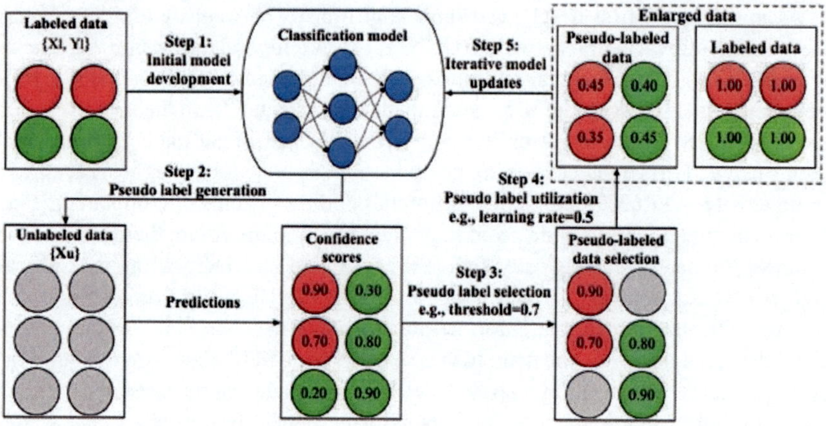

Fig. 4 The self-training strategy for semi-supervised neural network development [19]

The third approach relies on the graph theories for semi-supervised learning. It is mainly used to analyze unstructured data (e.g., images or text) and usually requires larger amounts of computational resources. By contrast, the fourth approach, i.e., the self-labeled approach, is more popular for its simplicity and flexibility in practical implementation. The main idea is to firstly adopt the supervised learning paradigm to train a basic model using labeled data alone, and iteratively update model parameters using high-quality pseudo-labels generated from unlabeled data. Figure 4 illustrates the self-training process for developing semi-supervised neural networks for fault diagnosis. It consists of five major steps, i.e., initial fault diagnosis model development using labeled data alone, pseudo label generation from unlabeled data, customized pseudo label data selection, and iterative updates of the fault diagnosis model for improved performance [19]. Such a learning framework is compatible with various supervised learning algorithms, making it a promising candidate for analyzing building operational data. The research results validated the value of semi-supervised learning in FDD tasks, indicating that the utilization of unlabeled data can lead to an accuracy increase of over 20% given limited labeled data.

4.4 Self-Supervised Learning-Based FDD Methods

Semi-supervised learning enables the integration between unlabeled and labeled data for predictive modeling. One possible limitation is that semi-supervised learning may suffer from overfitting and poor generalization performance given extremely labeled data. Taking the self-labeled approaches described in Sect. 4.3 as an example, the base model is developed through the supervised learning protocol using labeled data alone. Given limited labeled data, such a model may have rather poor performance, which may severely affect the quality of pseudo labeled data generated from unlabeled data, making it an ineffective semi-supervised learning process. To tackle the labeled data challenges associated, existing studies have resorted to self-supervised learning to unlock the potential of large amounts of unlabeled building operational data. In general, self-supervised learning consists of two key steps. The first is to develop pre-trained models based on user-defined tasks (also known as pretext tasks) to learn useful data representations from unlabeled data, and the second is to customize pre-trained models for specific predictive modeling tasks through transfer learning protocols.

Self-prediction and contrastive learning are two major approaches of self-supervised learning. The former focuses on reconstructing a data sample or predicting one part of the data sample given other parts, while the latter aims at learning data invariant embeddings given different data transformations. Self-supervised learning has been widely used in natural language processing and computer vision fields [20]. Existing self-supervised learning frameworks are mainly designed for analyzing sequential or image data [21]. Considering that building operational data are in essence tabular data, customized solutions are needed to implement self-supervised learning methods in the building field. As shown in Fig. 5, the data reconstruction-based self-prediction methods are the most direct route to self-supervised tabular data learning. The main idea is to develop denoising autoencoders (i.e., DAEs) to reconstruct unlabeled building operational data from their noisy versions. A mask matrix, which has the same dimension as the unlabeled data, is defined through random 0–1 sampling to indicate the position of added data noise. Such data noise can be generated through a certain noise injection method, such as Gaussian noise, swap-noise, and etc. Besides data reconstruction, pre-trained models can be developed for both data reconstruction and noise location prediction. If trained properly, these encoder models should learn meaningful embeddings which are robust to data noises and can be transferred to downstream predictive modeling tasks [22]. Contrastive self-supervised learning methods can also be applied for analyzing tabular building operational data. As shown in Fig. 6, one possible approach is to use data augmentation techniques to generate positive and negative pairs for each data sample, and then develop Siamese encoder models to learn meaningful latent embeddings to classify whether two data samples are similar or dissimilar to each other.

In previous studies [23], a lightweight transformer architecture has been proposed to analyze tabular building operational data and data experiments have been conducted to investigate the value of self-supervised learning in FDD tasks. The experimental results demonstrated the value of the self-supervised learning in fault classification, especially when limited labeled data are available for fault modeling. The results indicate that self-prediction pretext tasks with 10–20% swap-noises levels are more suitable for tabular building operational data analysis. By contrast, contrastive learning is more sensitive to hyperparameter settings, making it less feasible for practical applications in the building field.

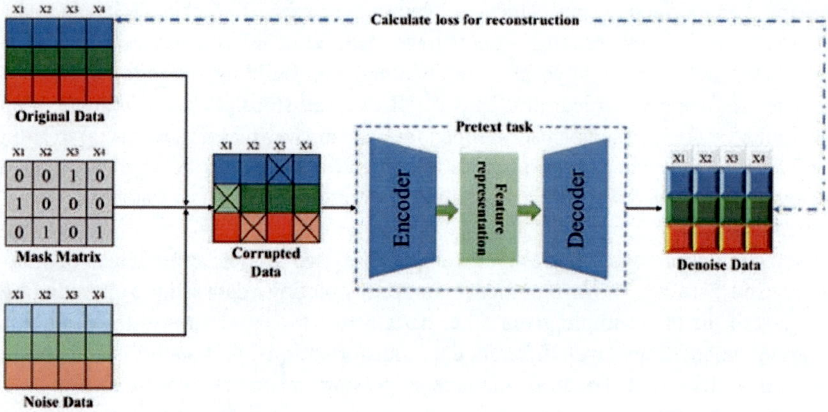

Fig. 5 Illustration of self-prediction pretext tasks for tabular building operational data

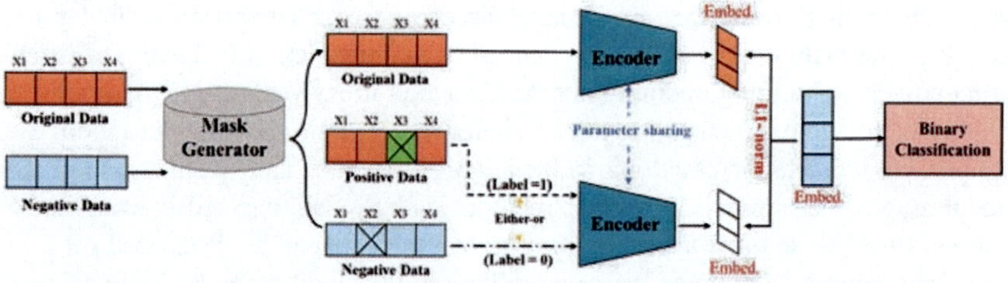

Fig. 6 Illustration of contrastive learning pretext tasks for tabular building operational data

5 Data-Driven Human Behavior Identification for Building Systems

Building is designed to provide a comfortable indoor environment for building occupants and ensure their safety in case of an emergency. For example, the accurate and real-time identification of human behavior is essential for preventing unnecessary energy uses while maintaining high levels of indoor comfort. Data-driven methods have been developed to facilitate human behavior identification tasks. More specifically, existing studies differ in their data sources, machine learning techniques used, and occupant-centric applications. To summarize, four types of data have been adopted for human occupant behavior studies, i.e., environmental data, smart meter data, video camera data, and physiological data. Environmental data are measurements on indoor and outdoor environments, such as air temperature, humidity, carbon dioxide concentration, luminosity, and noise level. Previous studies have utilized temperature, relative humidity and carbon dioxide concentration data as inputs for forecasting the number of occupants in buildings [24]. Smart meters, which track energy consumptions of different building systems, such as HVAC systems, lighting systems, and socket systems, can also be used as proxies for occupant modeling [25]. Video cameras can collect massive amounts of image data on human activities and serve as the most straightforward approach for occupant behavior identifications. The video data from Kinect have been adopted for detecting various human discomfort poses, such as wiping sweat, shaking T-shirt, and shoulder shaking [26]. Despite its high accuracy, such approach requires large amounts of computing resources and may violate occupant privac y. By contrast, the physiological data collected through wearable devices have gained increasing interests in the past few years. Such data cover a wide range of physiological indices, such as the heart rate, blood oxygen, skin temperature, and movement acceleration. Previous studies have utilized such data to predict thermal comfort and thereby, enabling the detailed description of individual-level thermal preferences across different spaces[27].

In terms of the human behavior modeling methods, existing studies have used both supervised learning and unsupervised learning algorithms for analysis. Supervised learning-based methods focus on developing regression or classification models on indoor occupancy presence [28], occupant number [29], occupant activities [30], thermal comfort levels [31]. Unsupervised learning-based methods, such as clustering analysis, which have been developed with the aim of differentiating thermal preferences based on longitudinal environmental and physiological data [32], while principal components analysis (PCA) were frequently adopted to reduce data dimensionality for further analysis [33].

From the application perspectives, existing studies can be divided into three types, i.e., occupancy detection, occupants number prediction, and human thermal comfort identification. Occupancy detection determines whether someone is in the room and serves as the basis of occupant-centric control logics, e.g., whether the lighting and HVAC systems in a space should be turned on. Occupant number prediction focuses on estimating the occupant numbers. It can be regarded as a regression problem and the predictions can be integrated with control logics for more energy-efficient operations, e.g., determine the fresh air volume for air conditioning systems [34]. Thermal comfort identification aims to identify the thermal preferences of indoor occupants given different HVAC operating conditions. Data-driven models have been developed for such purposes using the air temperature, relative humidity, and air velocity as model inputs [35].

To further illustrate the usefulness of data-driven methods in human behavior identification, two examples on occupant detection and thermal preference identification have been provided in below figures. The first is to utilize multi-source environmental data to achieve non-invasive occupancy detection [36]. Figure 7 presents the pairwise plot between environmental and time variables collected in a meeting room. The blue dots signify that the room is being occupied, while the red dots signify that the meeting room is vacant. When there is a discernible distinction between the red dots and the blue dots in the row and column of an input variable, it demonstrates that this input variable has a great potential in detecting occupancy. Conversely, this input variable has limited efficacy in indoor occupancy detection. For example, a marked divergence between the red dots and the blue dots can be observed in the row and column of the indoor light intensity. It indicates that the indoor light intensity has the capability to distinguish the indoor occupancy. After initial data exploration, machine learning models can be developed for occupancy detection and occupant number prediction tasks. As shown in Fig. 8, a fairly high occupancy detection accuracy of 97.09% can be achieved using the indoor temperature, outdoor temperature, indoor relative humidity, indoor carbon dioxide concentration, indoor noise level, indoor light intensity and hour as inputs to extreme gradient boosting (XGB) model.

As another example, the high-frequency Inertial Measurement Unit (IMU) data collected through wearable devices can be utilized to infer human thermal comfort. The main idea is that humans may perform certain activities when feeling either hot or cold, and the resulting IMU data can be used to identify thermal comfort-related activities. As shown in Fig. 9, in our previous studies, a framework for analyzing human IMU data has been proposed for human thermal comfort identification. Firstly, a prototype IMU dataset was created containing 30 prototype activities, i.e., 10 cold-related (e.g., rubbing hands, wearing coat), 10 hot-related (e.g., wiping sweat from face, taking coat off), and 10 neutral (e.g., using phone, typing) activities. Secondly, data preprocessing steps on features extraction and variable selection are conducted to prepare the IMU data into suitable inputs for predictive modeling. Finally, an online thermal comfort activity recognition method was developed based on machine learning algorithms. Figure 10 presents the resulting confusion matrix on thermal comfort-related activity identification. No. 1 to 10 are cold-related actions, No. 11 to 20 are hot-related actions, and the others are neutral actions. The results indicate relatively high identification performance with an accuracy of 86.2%. Such prediction results can be integrated with HVAC system control logics to achieve customized and energy-efficient operations.

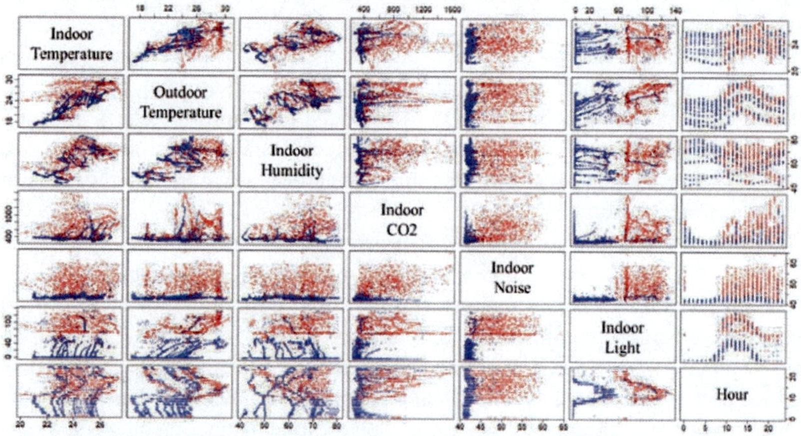

Fig. 7 The pairwise plot of environmental and time variables given different occupancy states [36]

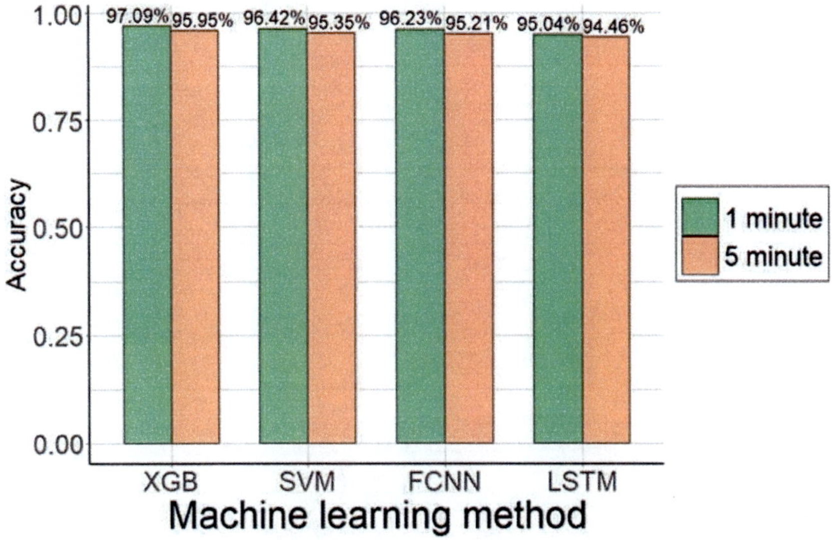

Fig. 8 Accuracy of different classification methods with different data collection frequencies [36]

6 Data-Driven Optimal Control Methods for Building Systems

Data-driven technologies have been used in three main types of optimal control methods in the building field, i.e., model predictive control (MPC), rule-based control, and reinforcement learning control.

MPC is a constrained optimal control strategy that can calculate the optimal control action by maximizing/minimizing a given objective function based on the prediction of the controlled object's future behavior. For example, the objective of building control could be to maximize the thermal comfort and indoor environmental quality (IEQ) for occupants considering temperature, humidity, CO_2 concentrations, illuminance, etc. In addition, minimizing monetary costs under time-varying energy prices is also a popular objective because energy markets are undergoing massive changes to increase their flexibility by demand side management.

The model predictive control framework mainly includes three parts, model prediction, rolling optimization, and model correction [37], as shown in Fig. 11. The data-driven technology can be used for training the prediction model. Here is a typical example. A data-driven model is used to predict building cooling load based on disturbances (e.g., weather conditions). Then according to the cooling load, the rolling optimization can be conducted to determine the optimal chilled water outlet temperature setpoint for achieving minimal monetary costs. In fact, this prediction model could also be other types of models, such as the white or gray box models. However, compared with the white box model, the computation load of the data-driven model is relatively low which is more proper to be used for online optimal

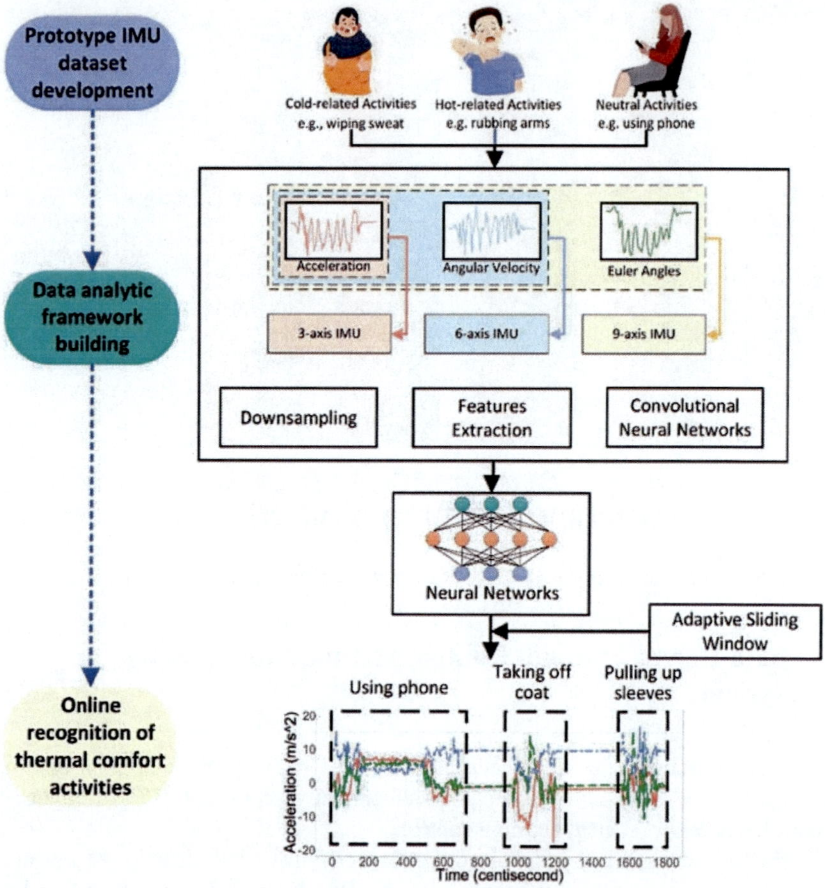

Fig. 9 A framework of thermal comfort inference using human IMU data

control. Compared with gray box models, the prediction accuracy of the data-driven model could be more satisfied with the development of many advanced machine learning technologies. Although data-driven models normally require a large amount of training data, it becomes less problematic with the rapid development of IoT and sensing technologies. It should be mentioned that many advanced machine learning technologies that can obtain high accuracy are non-linear, which then can cause non-convex problems in the rolling optimization process. Although this problem can be solved by heuristic numerical optimization, it still faces the risk of divergence [38].

For rule-based control, data-driven technology can be used to extract rules from the results of other advanced control strategies [39], e.g., the MPC mentioned above. Therefore, the control action (u shown in Fig. 11) can be obtained directly according to the disturbances ($\widehat{d}$ shown in Fig. 11). One typical example is that it can directly find the relationship between weather conditions and the optimal chilled water outlet

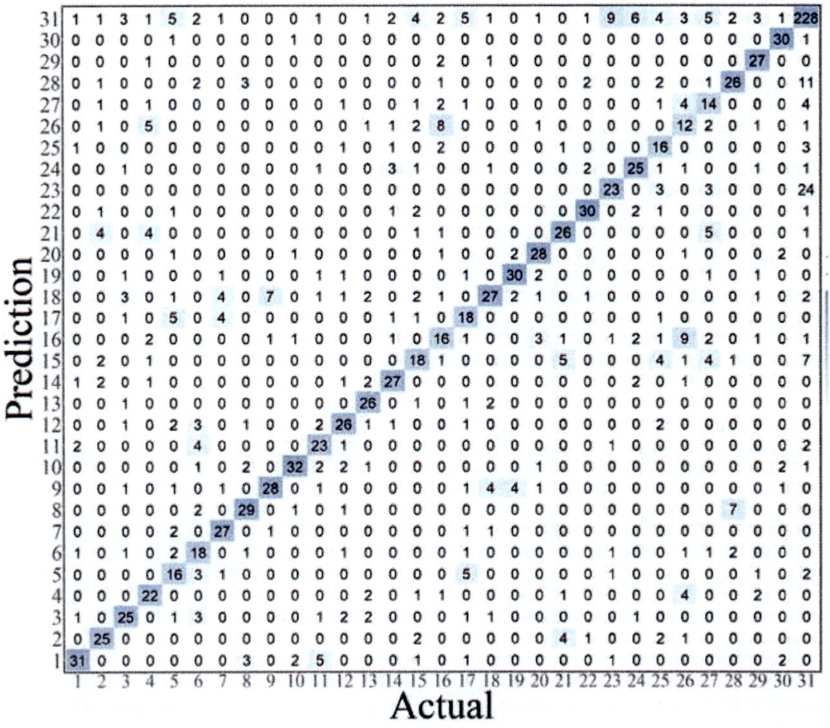

Fig. 10 Confusion matrix of thermal comfort activities recognition

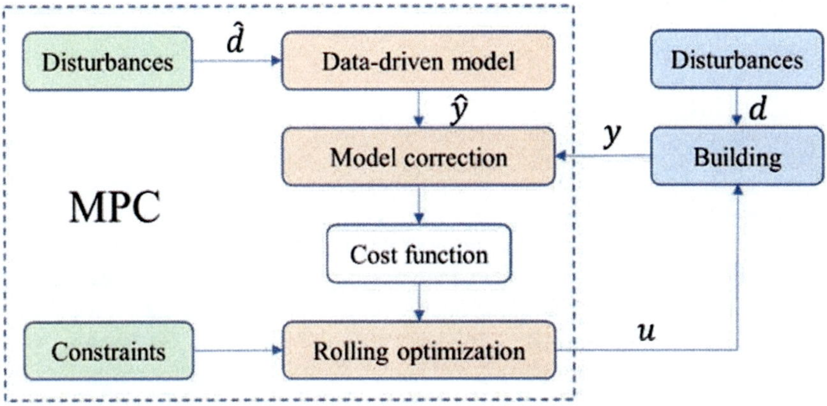

Fig. 11 Data-driven model for model predictive control (MPC)

temperature setpoint based on the control actions of MPC in the training stage. In this way, the process of cooling load prediction and online optimization is not required in the operation stage, which can save the cost of IT infrastructure both in terms of hardware and software. On the other hand, rule-based control can only obtain near-optimal control, which means that it can only mimic MPC but not be able to obtain the same results as MPC. In addition, after these rules are extracted and used in the operation stage, it is normally difficult for them to be amended, as the training dataset is fixed. It indicates that the control performance will be degraded when the practical cases are outside of the predefined conditions. Therefore, the rule-based control has relatively low adaptability.

Reinforcement learning control can also map disturbances ($\widehat{d}$ shown in Fig. 11) to control actions (u shown in Fig. 11). However, different from the rule-based control, the learning process is based on a trial-and-error procedure [40]. It means that the reinforcement learning control agent will try different chilled water outlet temperature setpoints and calculate the reward (monetary cost) under different weather conditions. Then gradually, this agent can find the optimal chilled water outlet temperature setpoint according to the weather conditions for obtaining the most reward (minimal monetary cost). Compared with rule-based control, it has high adaptability as it can be trained and updated at every step in the operation stage. On the other hand, safe operation (especially in the initial phase) should be noted in real applications because of the trial-and-error procedure mentioned above. This could be accomplished by a suitable modification of the trial procedure. For example, instead of making use of heuristic trial methods that are blind to the risk of actions, the initial learning phase could be guided by the available prior knowledge of the problem [41]. In addition, the long training periods could also be an obstacle to solve before the real application of this control method.

7 Conclusive Remarks

Buildings are becoming not only energy-intensive, but also information-intensive. The massive amounts of building operational data have provided an ideal platform to devise data-driven solutions for smart building management. This chapter presents the overall picture of building data analytics together with their applications for building energy management. Focusing on the structured tabular data collected by Building Automation Systems (BAS), data analytics have been designed for data preprocessing, knowledge discovery and knowledge applications. More specifically, data preprocessing techniques have been developed with the aim of enhancing original data quality while ensuring the validity and sensitivity of data analysis results. Data-driven applications have been developed to facilitate typical building operation management tasks, such as fault detection and diagnosis, occupancy recognition, and optimal controls.

It is realized that the main obstacle in practice is the data quality problem, which may severely affect the applicability of learning algorithms and validity of data analysis results. As a prominent example, an essential premise of supervised learning algorithms is that sufficient labeled data are available for model training. However, the acquisition of data labels can be time-consuming and labor-intensive. As a result, most building operational data are unlabeled and not compatible with advanced machine learning algorithms. Therefore, to further enhance the applicability and feasibility of data-driven methods, advanced learning paradigms should be adopted in the building field to tackle potential data shortage problems, such as transfer learning, semi-supervised learning, generative modeling and self-supervised learning. At present, the research in these fields is still at their preliminary stages and more research efforts for comprehensive analysis are needed. Despite possible challenges, it can be foreseen that with the advances in building data analytics, data-driven innovations will become more prevalent in the building to facilitate building energy saving and smart energy management.

Acknowledgements The authors gratefully acknowledge the support of this research by the National Natural Science Foundation of China (No. 52278117), the Philosophical and Social Science Program of Guangdong Province, China (GD22XGL20) and the Shenzhen Science and Technology Program (No. 20220531101800001 and 20220810160221001).

References

1. M.F.R. Lee, Y.C. Chen, C.Y. Tsai, Deep learning-based human body posture recognition and tracking for unmanned aerial vehicles. Processes **10**(11), 2295 (2022)
2. A. Ayala, B. Fernandes, F. Cruz, D. Macêdo, A. L. Oliveira, Zanchettin, C., Kutralnet: A portable deep learning model for fire recognition, in *2020 International Joint Conference on Neural Networks*. (IEEE, 2020), pp. 1–8
3. H.B. Gunay, W. Shen, C. Yang, Text-mining building maintenance work orders for component fault frequency. Build. Res. Inf. **47**(5), 518–533 (2019)
4. Z. Ding, Z. Li, C. Fan, Building energy savings: Analysis of research trends based on text mining. Autom. Constr. **96**, 398–410 (2018)
5. C. Fan, M. Chen, X. Wang, J. Wang, B. Huang, A review on data preprocessing techniques toward efficient and reliable knowledge discovery from building operational data. Front. Energy Res. **9**, 652801 (2021)
6. L. Zhu, D. J. Hill, C. Lu, Auto-Starting Semi-Supervised Learning-Based Identification of Synchrophasor Data Anomalies. IEEE Internet Things J. (2022)
7. F. Xiao, C. Fan, Building information modeling and building automation systems data integration and big data analytics for building energy managementm in *Research Companion to Building Information Modeling* (2022), pp. 525–549
8. C. Fan, F. Xiao, Y. Zhao, J. Wang, Analytical investigation of autoencoder-based methods for unsupervised anomaly detection in building energy data. Appl. Energy **211**, 1123–1135 (2018)
9. C.W. Chen, C.C. Li, C.Y. Lin, Combine clustering and machine learning for enhancing the efficiency of energy baseline of chiller system. Energies **13**(17), 4368 (2020)
10. C. Fan, Y.T. Lei, Y.J. Sun, M.S. Piscitelli, R. Chiosa, A. Capozzoli, Data-centric or algorithm centric: exploiting the performance of transfer learning for improving building energy predictions in data-scarce context. Energy **240**, 122775 (2022)

11. X. Xu, F. Xiao, S. Wang, Enhanced chiller sensor fault detection, diagnosis and estimation using wavelet analysis and principal component analysis methods. Appl. Therm. Eng. **28**(2–3), 226–237 (2008)
12. X.J. Luo, K.F. Fong, Y.J. Sun, M.K.H. Leung, Development of clustering-based sensor fault detection and diagnosis strategy for chilled water system. Energy Build. **186**, 17–36 (2019)
13. W. Yao, D. Li, L. Gao, Fault detection and diagnosis using tree-based ensemble learning methods and multivariate control charts for centrifugal chillers. J. Build. Eng. **51**, 104243 (2022)
14. S. Taheri, B. Talebjedi, T. Laukkanen, Electricity demand time series forecasting based on empirical mode decomposition and long short-term memory. Energy Eng. **118**(6), 1577–1594 (2021)
15. J. Liu, Q. Zhang, X. Li, G. Li, Z. Liu, Y. Xie, B. Liu, Transfer learning-based strategies for fault diagnosis in building energy systems. Energy Build. **250**, 111256 (2021)
16. C. Fan, W. He, Y. Liu, P. Xue, Y. Zhao, A novel image-based transfer learning framework for cross-domain HVAC fault diagnosis: from multi-source data integration to knowledge sharing strategies. Energy Build. **262**, 111995 (2022)
17. X. Yang, Z. Song, I. King, Z. Xu, A survey on deep semi-supervised learning. IEEE Trans. Knowl. Data Eng. (2022)
18. B. Li, F. Cheng, X. Zhang, C. Cui, W. Cai, A novel semi-supervised data-driven method for chiller fault diagnosis with unlabeled data. Appl. Energy **285**, 116459 (2021)
19. C. Fan, Y. Liu, X. Liu, Y. Sun, J. Wang, A study on semi-supervised learning in enhancing performance of AHU unseen fault detection with limited labeled data. Sustain. Cities Soc. **70**, 102874 (2021)
20. K. Ohri, M. Kumar, Review on self-supervised image recognition using deep neural networks. Knowl.-Based Syst. **224**, 107090 (2021)
21. G. Li, J. Wu, C. Deng, M. Wei, X. Xu, Self-supervised learning for intelligent fault diagnosis of rotating machinery with limited labeled data. Appl. Acoust. **191**, 108663 (2022)
22. J. Yoon, Y. Zhang, J. Jordon, M. van der Schaar, Vime: Extending the success of self-and semi-supervised learning to tabular domain. Adv. Neural. Inf. Process. Syst. **33**, 11033–11043 (2020)
23. C. Fan, Y.T. Lei, Y.J. Sun, L.K. Mo, Novel transformer-based self-supervised learning methods for improved HVAC fault diagnosis performance with limited labeled data. Energy **278**, 127972 (2023)
24. L.M. Candanedo, V. Feldheim, Accurate occupancy detection of an office room from light, temperature, humidity and CO2 measurements using statistical learning models. Energy Build. **112**, 28–39 (2016)
25. Y.R. Yoon, Y.R. Lee, S.H. Kim, J.W. Kim, H.J. Moon, A non-intrusive data-driven model for detailed occupants' activities classification in residential buildings using environmental and energy usage data. Energy Build. **256**, 111699 (2022)
26. B. Yang, X. Cheng, D. Dai, T. Olofsson, H. Li, A. Meier, Real-time and contactless measurements of thermal discomfort based on human poses for energy efficient control of buildings. Build. Environ. **162**, 106284 (2019)
27. H. Park, D.Y. Park, Prediction of individual thermal comfort based on ensemble transfer learning method using wearable and environmental sensors. Build. Environ. **207**, 108492 (2022)
28. S. Zemouri, Y. Gkoufas, & J. Murphy, A machine learning approach to indoor occupancy detection using non-intrusive environmental sensor data. in *Proceedings of the 3rd international conference on big data and internet of things*, (2019), pp. 70–74
29. O. Ardakanian, A. Bhattacharya, D. Culler, Non-intrusive occupancy monitoring for energy conservation in commercial buildings. Energy Build. **179**, 311–323 (2018)
30. A. Meier, W. Dyer, C. Graham, Using human gestures to control a building's heating and cooling System. in *Proceedings of the 9th International Conference on Energy Efficiency in Domestic Appliances and Lighting* (2017), pp. 627–635
31. A. Ghahramani, G. Castro, B. Becerik-Gerber, X. Yu, Infrared thermography of human face for monitoring thermoregulation performance and estimating personal thermal comfort. Build. Environ. **109**, 1–11 (2016)

32. P. Jayathissa, M. Quintana, M. Abdelrahman, C. Miller, Humans-as-a-sensor for buildings—intensive longitudinal indoor comfort models. Buildings **10**(10), 174 (2020)
33. G. Song, Z. Ai, G. Zhang, Y. Peng, W. Wang, Y. Yan, Using machine learning algorithms to multidimensional analysis of subjective thermal comfort in a library. Build. Environ. **212**, 108790 (2022)
34. M. Khalil, S. McGough, Z. Pourmirza, M. Pazhoohesh, S. Walker, Transfer learning approach for occupancy prediction in smart buildings. in *2021 12th International Renewable Engineering Conference*. (IEEE, 2021), pp. 1–6
35. B. Brik, M. Esseghir, L. Merghem-Boulahia, H. Snoussi, An IoT-based deep learning approach to analyse indoor thermal comfort of disabled people. Build. Environ. **203**, 108056 (2021)
36. W. He, Y. Ma, C. Fan, X. Wang, B. Huang, Non-intrusive Indoor Occupancy Detection Methods Based on Machine Learning Techniques, in *International Symposium on Advancement of Construction Management and Real Estate*. (Springer, Singapore, 2022), pp.1186–1201
37. R. Tang, C. Fan, F. Zeng, W. Feng, Data-driven model predictive control for power demand management and fast demand response of commercial buildings using support vector regression. In *Building simulation*, Vol. 15, No. 3, pp. 317–331 (Tsinghua University Press, 2022)
38. J. Drgoňa, J. Arroyo, I.C. Figueroa, D. Blum, K. Arendt, D. Kim, L. Helsen, All you need to know about model predictive control for buildings. Annu. Rev. Control. **50**, 190–232 (2020)
39. A. Domahidi, F. Ullmann, M. Morari, C.N. Jones, Learning decision rules for energy efficient building control. J. Process Control **24**(6), 763–772 (2014)
40. Q. Fu, Z. Han, J. Chen, Y. Lu, H. Wu, Y. Wang, Applications of reinforcement learning for building energy efficiency control: A review. J. Build. Eng. **50**, 104165 (2022)
41. E.T. Maddalena, Y. Lian, C.N. Jones, Data-driven methods for building control—A review and promising future directions. Control. Eng. Pract. **95**, 104211 (2020)

Introduction of Artificial Intelligence

Yuanyuan Wang, Eugene Yujun Fu, Xinwei Zhai, Chunxi Yang, and Fengchun Pei

1 Introduction

Artificial intelligence (AI) is referred to as the intelligence developed by machines with mathematical modeling. In particular, AI is manifested by machine's ability to effectively perceive, process, and reason diverse kinds of signals, and make satisfactory responses. It assists human in the tasks of recognition, analysis, and decision-making to increase efficiency while reducing errors.

Today, AI has been receiving increasing attention and developed to encompass a diverse range of applications including smart city, digital healthcare, intelligent manufacturing, intelligent education, intelligent transportation and so on, exhibiting significant impact on human life and society. The boosting of these AI applications largely benefited from the development of computing techniques and devices, and internet facilities. Among them, machine learning plays one of the most essential roles. It is an inevitable trend to add and develop "smart" features for best practices in attacking real-life problems including smart building and fire fighting.

In building fire safety, AI can be applied to intelligently sense and model time-series (e.g., temperature signals) and image/video (e.g., satellite image) data to detect signs of fire, such as smoke, flames, or temperature changes, to achieve early fire detection and immediate response. Moreover, AI builds machine learning models based on historical fire incident data, weather conditions, building structures, and other relevant factors to predict fire hazards in specific locations. This facilitates in proactive planning, resource allocation, and risk mitigation. AI can also assist in

Y. Wang · X. Zhai · F. Pei
Faculty of Education, Shenzhen University, Shenzhen, China

E. Y. Fu (✉) · C. Yang
Department of Rehabilitation Sciences, The Hong Kong Polytechnic University, Hong Kong, China
e-mail: eugene.fu@polyu.edu.hk

X. Huang and W. C. Tam (eds.), *Intelligent Building Fire Safety and Smart Firefighting*, Digital Innovations in Architecture, Engineering and Construction,
https://doi.org/10.1007/978-3-031-48161-1_4

assessing fire hazards by analyzing data patterns and identifying potential risks based on factors such as materials used, emergency escape routes, and occupancy type.

Focusing on the general principles and technologies of ML, this chapter provides a brief introduction of AI. It illustrates, with examples of applications about how fire fighting can be benefited from these technologies. The chapter is organized as follows. Section 2 delineates the overview of training, evaluating, and applying ML models, and describes the basic concept, mathematics, and the applicability of the classic ML algorithms. The chapter is followed by the introduction of deep learning approaches in Sect. 3, including the commonly adopted deep neural network architectures. Section 4 demonstrates the example cases of applying ML in image/video and time series data analysis. Section 5 briefly presents the idea of transformer and few-shot learning, which are the current research interests in the area of AI and ML. Finally, this chapter is concluded in Sect. 6.

2 Machine Learning Techniques

As the amount of data continues to accumulate at an exponential rate in every aspect of human society, the task of developing efficient computer algorithms to analyze and interpret this data becomes increasingly challenging. Machine learning (ML) is regarded as an urgent response to this critical need of such an era, which is growing tremendously and garnering global attention in recent years. Starting from the overall procedure of training, evaluating, and running ML models, this section introduces the basic concept and some classic algorithms of ML.

2.1 Procedure for Machine Learning

ML broadly refers to technologies that enable computers to learn automatically from the given data without being explicitly programmed [1]. Its primary focus is algorithms that generate models based on data to support for pattern recognition and decision making. The optimal machine learning architecture and algorithm may be different, depending on the nature of the data (e.g., image, video, text data with or without labelling) and task (e.g., binary, multiclass classification). However, the overall workflow is relatively the same. In the classic machine learning process, a series of algorithms is applied to a set of profiled data instances. They first clean and extract features from the data. Based on the features, a selected machine learning algorithm learns the data distribution and relations, and fits a hyperplane in multi-dimensional space, to separate the data instances. As depicted in Fig. 1, there are four primary steps for it, which are described as follows:

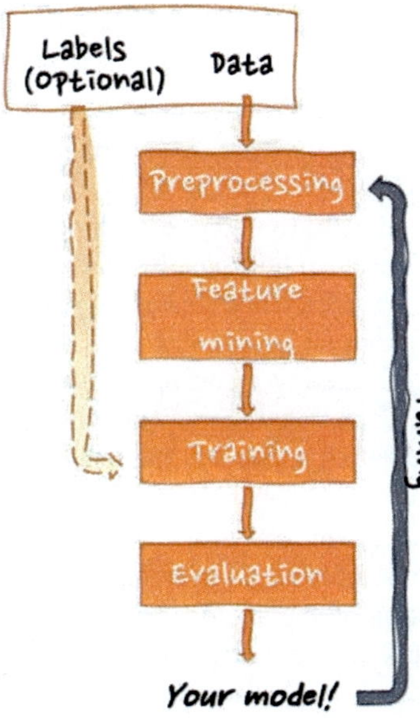

Fig. 4.1 General procedure of machine learning

- Data preprocessing: the data is cleaned and prepared for further modelling, involving the handling of missing values, outliers, noise, data format, and normalization;
- Feature mining: features are engineered and extracted from the cleaned data, the most relevant and informative ones are mined and selected to improve model performance;
- Model training: the data is split into training, validation and test set. A proper machine learning algorithm is selected, which is trained on the training set and adjusted using validation set; and
- Model evaluation: the model performance is evaluated using the test set with common performance metrics.

ML process often requires iterations between these steps to achieve the optimal setting for enhancing the model performance. There are two common subsets of ML, namely unsupervised learning and supervised learning, depending on the nature of the data and training task [2]. They are mainly different in the model training step. In unsupervised learning, the machine learning algorithm is trained on unlabeled training samples to understand data patterns and cluster them without any prior guidance. While in supervised learning, the algorithms are trained on labeled training samples to learn the association between data and labels (i.e., with guidance). Their details and common algorithms are presented in the following subsections.

2.2 Unsupervised Learning

In unsupervised learning, machine learning algorithms learn the intrinsic properties and patterns of the data, from unlabeled training samples, i.e., learning without any prior guidance. Clustering is the most extensively studied unsupervised learning task. It refers to the process of grouping a set of data samples into different meaningful clusters, so that samples from the same cluster are more similar to each other than those from other clusters [3]. The grouping is conducted by clustering algorithms without the prior knowledge of what the groups should look like.

Clustering algorithms have been applied in a wide range of applications, from healthcare and bioinformatics [4], marketing analysis [5], information security [6], to fire detection [7]. There are majorly five types of clustering methods: partitional clustering, density-based clustering, hierarchical clustering, grid-based clustering, and model-based clustering [8]. This section particularly introduces three most commonly used clustering algorithms, namely k-means, density-based spatial clustering of applications with noise (DBSCAN), and Gaussian mixed model (GMM). This section particularly introduces three most commonly used clustering algorithms, namely k-means, density-based spatial clustering of applications with noise (DBSCAN), and Gaussian mixed model (GMM).

K-means is the most commonly used partitional clustering algorithm. It partitions data samples into a predetermined number (K) of clusters. It aims to minimize the distance between data samples within the same cluster while maximizing the distance between data samples across different clusters. Specifically, there are four major steps of K-means algorithm: (1) randomly select K centroids for initialization; (2) assign each data sample to the closest centroid to achieve K clusters; (3) compute the new centroid for each cluster based on the data samples assigned to it; and (4) repeat the steps of 2 and 3 until convergence (i.e., the clusters remain stable throughout the iterations). Figure 2 illustrates an example of clustering by K-means algorithm.

Parameters, such as the number of clusters (K), stop criteria (e.g., the number of iteration), and distance metric, may have significant impacts on the clustering performance. They thus should be carefully determined based on the nature of the data, before running the K-means algorithm. Taking the distance metric as an example, though Euclidean distance (Eq. 1) is commonly used to compute the distance in K-means, other distance metrics may particularly help when dealing with some specific data. For instance, Manhattan distance (Eq. 2) may be more effective in handling categorical and ordinal data. It is often suggested to test with different combinations of parameters and compare their performance to find the optimal setting.

$$d(x, y) = \sqrt{(x_1 - y_1)^2 + (x_2 - y_2)^2 + \cdots + (x_n - y_n)^2} \tag{1}$$

$$d_{12} = \sum_{k=1}^{n} |x_{1k} - x_{2k}| \tag{2}$$

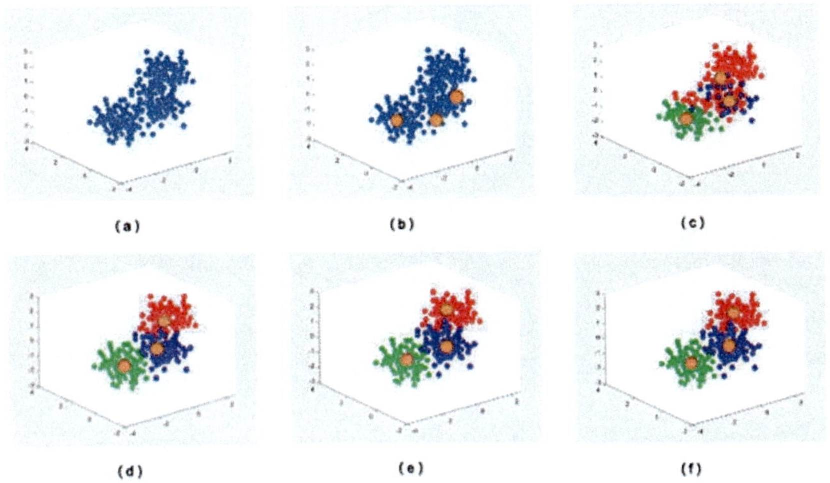

Fig. 2 The demonstration of an example K-means ($K = 3$) clustering process: **a** the given dataset; **b** the initialized K centroids; **c** assign data samples; **d** update clusters and compute new centroids; and **e** to **f** iterate the data assigning and cluster updating steps until convergence

The major strengths of K-means algorithm include simplicity, ease of implementation, low computational complexity, and comprehensibility. However, it is limited by requiring predetermined cluster number, and is sensitive to the initialization and outliers. Moreover, it is often not efficient when handling high dimensional data, and non-spherical or irregularly shaped clusters.

Density-Based Spatial Clustering of Applications with Noise (DBSCAN) is a commonly used density-based clustering algorithm, which works without making assumptions about the specific form of the underlying data density. It shows the significant effectiveness in handling datasets with irregular shapes and varying densities in multiple applications [9].

In particular, DBSCAN defines a neighborhood around each data point in the dataset, and group densely populated neighborhood points together to form clusters. After that, points that do not fall within any dense neighborhood (i.e., clusters) are considered as outliers or noise [10]. Figure 3 illustrates an example of clustering by DBSCAN algorithm. It starts with categorizing each data point in the dataset into "core point", "border point", and "noise point" based on two parameters, namely "Eps" (epsilon) and "$MinPts$" (minimum number of points). Specifically, a point that has at least $MinPts$ number of Eps reachable neighbors (points locate within its Eps radius) is categorized as core point. A core point's neighbor point that has fewer than $MinPts$ reachable neighbors is categorized as border point. And finally, points other than core and border points are noise points.

In summary, DBSCAN can be conducted in the following steps: (1) randomly select a core point p, , which is not assigned to any clusters; (2) assign p to a new cluster and expand the cluster by adding p's neighbor core points and their reachable

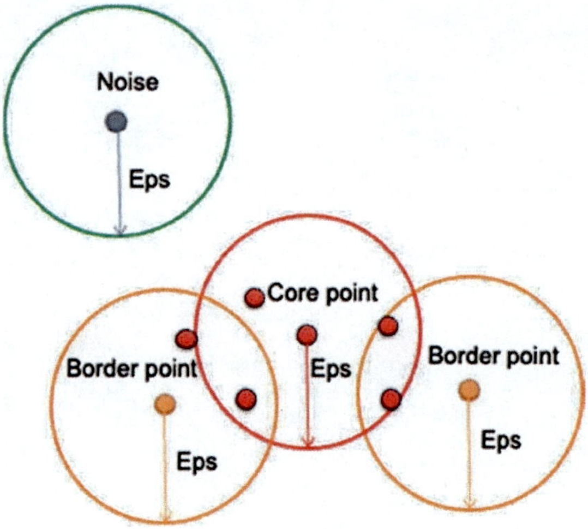

Fig. 3 An example of how DBSCAN works

neighbor core points; (3) assign all border points that are within the reachable radius of the cluster; and (4) repeat the first three steps until all the core and border points are processed, and assigned to the clusters. Any other points that do not fall within any clusters are considered as outliers or noise. $MinPts$ and Eps could have significant impacts on the clustering performance. Again, it is often suggested to test with different values for the parameters to find the optimal combination.

Compared with the traditional K-Means algorithm, the biggest difference between DBSCAN and K-means is that DBSCAN does not require the input category number k. Its biggest advantage is that it can find clusters of arbitrary shapes, unlike K-Means, which is generally used only for clustering convex sample sets, and it can also find outliers while clustering. However, DBSCAN struggles with datasets where the density of clusters varies significantly or in high-dimensional spaces, as it becomes challenging to define suitable values for the parameters. Moreover, DBSCAN's computational complexity increases with the size of the dataset, making it less efficient for large-scale datasets.

Gaussian Mixed Model (GMM) is a commonly used probability function-based model for clustering, density estimation, and dimension reduction tasks [11]. Theoretically, GMM is especially efficient in handling data that contains several distinct distributions, empowering accurate modelling of complex data patterns. It assumes that the given data is generated from a mixture of multiple gauss distributions with different means and standard deviations, as shown in Fig. 4. GMM describes these distributions in a multidimensional probability density function with linear weighted combination as depicted in Eqs. (3) and (4):

$$P(X|\lambda) = \sum_{i=1}^{K} \omega_i f_i(X) \tag{3}$$

$$f_i(X) = \frac{1}{2\pi^{D/2}|\sum_i|^{1/2}}\exp[-\frac{1}{2}(X - u_i)^t \sum_i^{-1}(X - u_i)] \tag{4}$$

where, ω_i denotes the weight of the *i*th gauss component showing its contribution to the mixture distribution, u_i and $\sum_i$ denotes its mean and the covariance matrix respectively. A gauss distribution is expressed by its weight, mean, and covariance matrix as $[\omega_i, u_i, \sum_i]$, and a GMM characterizes a collection of K distributions with this representation (i.e., the parameters of a GMM can be represented as $\lambda = \{[\omega_1, u_2, \sum_3], \ldots, [\omega_i, u_i, \sum_i], \ldots, [\omega_K, u_K, \sum_K]\}$). The clustering is conducted by determining the gauss components and data points belonging to them.

The specific steps of GMM algorithm are summarized as follows: (1) randomly initialize the mean, variance, mixing coefficients of K Gaussian distributions, and initialize the likelihood function value as 0; (2) calculate the probability of data point belonging to each Gaussian distribution based on density function and Bayes' theorem (known as "E-step"); (3) update the mean, variance, and mixing coefficients of the K gaussian distributions based on the results in E-step with weighted averaging strategy (known as "M-step"); (4) update the likelihood function value based on the results obtained in E-step and M-step; (5) repeat steps (2) to (4) until convergence; (6) assign cluster based on the data point belonging to each Gaussian distribution.

GMM has shown its effectiveness in a wide-ranging of applications, including determining disease patterns [12], predicting customer purchasing behavior [13], and detecting running points in stock market [14], with its capability to handle clusters of different shapes, provide soft assignments of data points, and accommodate overlapping clusters, However, they are also very sensitive to initialization, and are computationally intensive for large datasets. In order to minimize the risk of converging to the local optimal solutions, it is often suggested to randomly initiate the GMM multiple times, and select the model achieving the best performance in practical applications.], with its capability to handle clusters of different shapes, provide soft assignments of data points, and accommodate overlapping clusters, However,

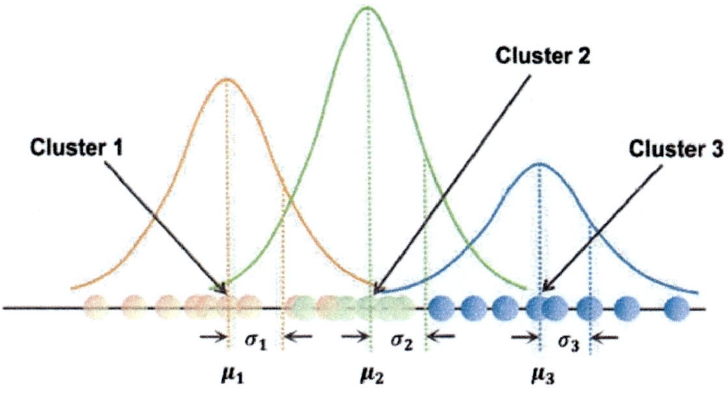

Fig. 4 An example of Gaussian Mixture Model

they are also very sensitive to initialization, and are computationally intensive for large datasets. In order to minimize the risk of converging to the local optimal solutions, it is often suggested to randomly initiate the GMM multiple times, and select the model achieving the best performance in practical applications.

2.3 Supervised Learning

In supervised learning, machine learning algorithms learn the intrinsic properties and patterns of the data, from labeled training samples, i.e., learning with prior knowledge. Classification is the most extensively studied supervised learning task. In classification, machine learning algorithms learn data patterns to model the boundaries between data from different categories, so that they can assign the unseen data samples to their respective categories correctly. The models require to be trained with data samples that are clearly annotated with their actual categories, i.e., learning with prior knowledge of what the categories should look like. This section particularly introduces three most commonly used classification algorithms, namely support vector machine (SVM), decision tree, and random forest.

Support vector machine (SVM) is one of the most widely used in the supervised learning. Differing from traditional methods that target on minimizing the empirical training error, SVM focuses on minimizing an upper bound of the generalization error to achieve better generalization and robustness. It aims at maximizing the margin between the separating hyperplane between the data points [15].

SVM particularly works in binary classification task. Figure 5 illustrates an example of binary classification conducted by SVM. It projects data points into a high-dimensional space and create two parallel hyperplanes (dashed green line in Fig. 5) for the two classes, which are lines or surfaces in the high-dimensional space. It then finds a separating hyperplane (solid red line in Fig. 2.5) between them that maintains a maximized gap between the two classes. This leads to an optimization problem that can be solved with Lagrangian dual function [15], defined as follows:

$$L(\mathbf{v}, b, \alpha) = \frac{1}{2}||\mathbf{w}||^2 - \sum_{i=1}^{l} \alpha_i (y_i[\mathbf{w} \bullet x_i + b] - 1) \tag{5}$$

$$\overline{w} = \sum_{i=1}^{l} \overline{\alpha}_i y_i x_i \tag{6}$$

$$\overline{b} = -\frac{1}{2}\overline{w}.(x_r + x_s) \tag{7}$$

where α denotes the Lagrange multiplier, x_r and x_s are the data points with α_r, α_s>0, y_r=1, y_s= −1.

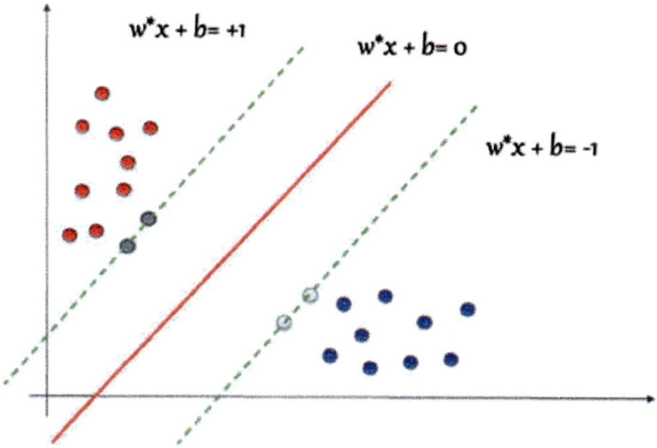

Fig. 5 An example of support vector machine

SVM utilizes kernel functions to transform data into a high-dimensional feature space and facilitate the seeking of separating hyperplane in that space [15]. The selection of an appropriate kernel function plays a significant role on the model performance when using SVM, and should be considered based on the nature of the data and task. The commonly adopted kernels in real applications include the linear kernel, polynomial kernel, and radial basis function (RBF) kernel [16]. Literally, the linear kernel computes and transforms the data linearly. Specifically, it compute the inner dot product of input feature vectors in the raw feature space, thus works for linearly separable data. Polynomial kernel can transform the feature vectors, and raise the inner dot computation to a higher-dimension space based on polynomial functions. Hence it is suitable for nonlinearly separable data. The RBF kernel is even able to map the data to an infinite-dimensional space, where the distance measurement between points is yielded based on Gaussian function. RBF kernel has strong flexibility and ability in handling different kinds of data, notably data with high-dimensional features, it thus is the most widely used kernel function in SVM. Its effectiveness has been demonstrated across many real-life applications [17–19].

Compared to traditional machine learning methods, SVM is more effective in modelling high-dimensional data and ensuring the generalization ability of the trained model, tackling the problem of over-fitting. The advantages of SVM also include the easy implementation. And more importantly, its computational complexity does not increase when the dimensionality of the input space increases. However, it is not suitable for large datasets, and performs poorly in cases where the number of features per data point exceeds the number of training data samples.

SVM has achieved outstanding performance with excellent generalization ability across diverse applications, including text detection and categorization, information retrieval, image processing, face detection and recognition, fault diagnosis, etc. [20–22]. In addition to being used for classification purposes, support vector machines can also be used for regression, for predicting discrete values. It has been successfully

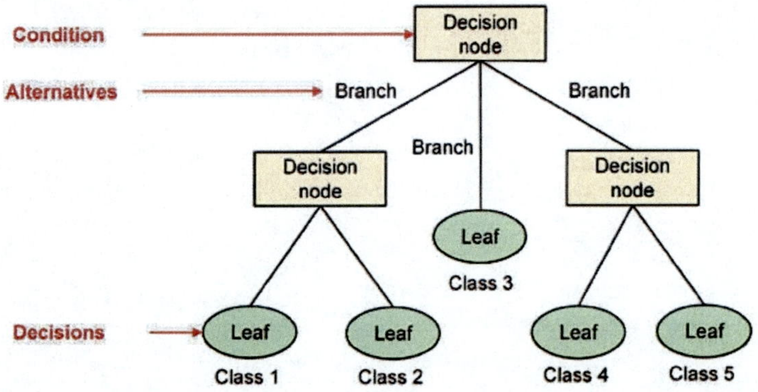

Fig. 6 An example of decision tree

applied in various applications such as stock trend prediction [23, 24], traffic accident prediction [25].

Decision tree is a non-parametric method that has been widely used in supervised learning [26]. Decision tree techniques are one of the most popular and effective machine learning methods, as such models closely resemble human reasoning, which is much easier to interpret and understand [27]. The well-known representatives of decision tree algorithms are ID3 [28], C4.5 [29] and CART [30]. Although the construction methods vary, the common goal of these algorithms is to create a tree-like structure to make predictions based on the input features. As shown in Fig. 6, a decision tree model is composed of a root node (the starting node), internal nodes and leaf nodes, where the root node starts the decision process, each internal node attributes a test based on a specific feature, branches say the test result and connect the nodes, and leaf nodes represent the end of the decision process, pointing out the prediction outcomes (e.g., class labels).

The basic process of creating a decision tree can be divided into two phases: the growth phase and the pruning phase. The growth phase is a top-down recursive partitioning of the training data [26]. The partitioning begins by assigning all training data to the root node. And then, on the basis of the properties test set, the training sample performed recursive partitioning and assigned to child nodes in order to improve the purity of child nodes. Within each node, attributes are selected based on their ability to divide the example into subsets with the purest possible class labels [31]. The selection methods for evaluating attribute partitioning ability include information gain, gain ratio, Gini coefficient, distance-based measurement, etc. [27]. The partitioning process is repeated on each node until the node meets certain stop criteria (such as the nodes of all examples from a single class), or further partitioning of leaf nodes will lead to children below a certain threshold, and the node is considered as a leaf node. The pruning phase generalizes the decision tree constructed during the growth phase in order to avoid overfitting the training data. This phase can be realized by post-pruning or pre-pruning during the growth phase[29]. Pre-pruning refers to estimating each node division during the decision tree generation process.

If the current node fails to improve the division of the decision tree for generalization ability, it stops and marks the node as a leaf node. Post-pruning involves a bottom-up examination of the internal nodes of a complete decision tree that has been generated from training data. If the internal node subtree is replaced by leaf nodes, which can improve the generalization ability of the decision tree, the subtree will be replaced by leaf nodes.

The advantages of decision trees are as follows: their tree structure provides easily understandable and interpretable information [27]. They can be used for both classification and regression [26]. They are non-parametric methods that do not need any distribution-based assumptions [32]. They can process both numerical data and classified data [33]. They are much less affected by outliers or missing values [29], and they are useful for revealing the most important variables and the relationships between two or more variables.

Decision trees also have some drawbacks. For example, it is easy to get overfitting, which often hampers the prediction accuracy on unseen data (generalization performance). They are seriously sensitive to the slight change in the training data set. Moreover, when working with continuous variables, they lose information during the categorization phase. It is also not guaranteed that the obtained model is the optimal model. Although there are some shortcomings, their advantages make them a good option for a variety of machine learning tasks.

Random forest is a collection of decision trees trained using the bagging method [26, 34]. Decision trees can potentially lead to local optima, resulting in certain parts of the tree having much lower quality than other parts of the tree. Even small changes in the training data can cause the initial node to select different attributes, which can lead to a completely different tree structure. Random Forest [34] was proposed to overcome these challenges, which combines a number of randomized decision trees during the training phase and aggregates their predictions through voting/averaging. Several studies have shown that using ensembles of trees, where each tree is grown based on a random parameter, can result in significant improvements in classification and regression accuracy [34, 35]. As a popular machine learning method, Random Forest has been widely applied in a variety of fields to develop prediction models [36–39].

Figure 7 illustrates a general architecture of random forest. Given a dataset that has N training samples, the specific steps of random forest algorithm are as follows: (1) bootstrap K subsets with M samples from the original dataset ($M < N$); (2) construct K decision trees, each of which is trained on one of the subsets; (3) when making predictions, use these K tress for prediction and obtain the final prediction result by taking the average (for regression tasks) or majority vote (for classification tasks) [34]. Specifically, when constructing each tree, random forest algorithms randomly select a subset of features for each node' split rather than using the whole features. The random sampling and feature selection helps in enhancing the model' generalization performance to avoid overfitting.

Random forests can produce better accuracy, handle datasets with a larger number of predictor variables, while have a lower risk of getting overfitting, compared to decision trees [34]. Random forests are generic and can be applied to a wide range

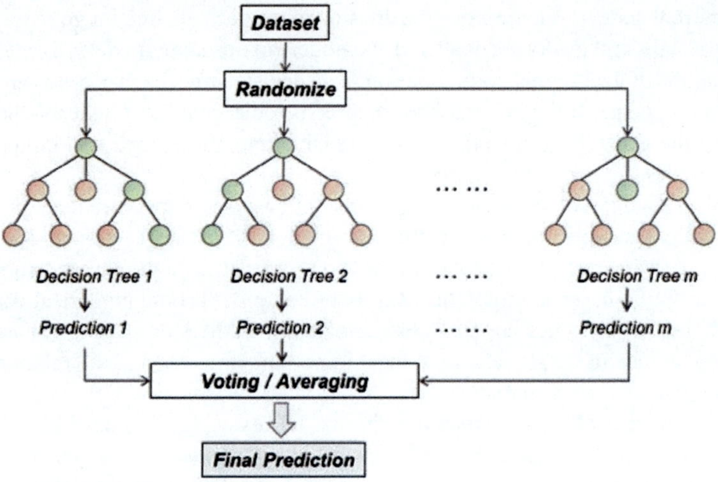

Fig. 7 General architecture of random forest

of prediction problems with only a few parameters to fine-tune [40]. Random forests can still get overfitting though, especially when they are trained with a noisy dataset. Utilizing multiple trees limits random forests' model interpretability, requires much more time for both model training and making predictions, and raises the risk of having bias in multiclass problems.

2.4 Feature Engineering and Selection

It is challenge to train machine learning models directly from raw data, which is often noisy and high-dimensional. The steps of data preprocessing, and feature engineering and selection are commonly applied to clean and prepare the data for more efficient machine learning training. Particularly, data preprocessing focuses on removing noise, handling outlier samples and missing data, transforming data format, and normalizing data, to ensure that the model has a reliable and well formatted dataset. While feature engineering and selection process aims at extracting the most relevant and informative features from the raw data.

Selecting more relevant and informative features also means guiding the model to ignore the interfered information achieving dimensionality reduction. This generally simplifies the learning process. For instance, in facial recognition the important features come from human facial landmarks and key points (e.g., the key points from eye, mouse, and nose contours), and any background related (e.g., texture of the dress, color of the background) should be disregarded. It is vital to identify the actually relevant features during this process. The absence of the relevant features can negatively affect the training process by omitting crucial information. Selecting too many features including the irrelevant ones, on the other hand, can result in less efficient training (e.g., high computational cost), and even miss-leading the training. Besides, there is no such a set as generally crucial features. Even for the same data, the relevant features can vary significantly from task to task. This further underlines the importance of engineering and selecting relevant features for each specific dataset and learning task for efficient model training. It often needs domain knowledge.

There are some commonly used feature selection methods, which can be broadly classified into three categories, namely: filter, wrapper and embedding feature selection methods. Filter feature selection methods score each feature's relevance individually, usually based on statistical analysis and heuristic knowledge, such as scoring based on their correlation with the prediction target. They rank the features based on the score and select the top-ranking features. Wrapper methods recursively search and include or exclude features from the initial feature set to form several subsets of features by a search strategy, such as forward and backward searching. They score each feature subset by training and evaluating a selected machine learning algorithm on them. The detection performances are used as the score for the feature subsets, and the subset with the best performance will be selected. Embedding selection methods incorporate the feature selection process into the machine learning algorithm and complete the feature selection during the model training process. In embedding-based methods, the machine learning algorithm can identify relevant features and adjust the parameters to extract the most informative representation on them in model training. Compare to filter feature selection methods, wrapper- and embedding methods are able to capture the global relations of different features. They thus often achieve better performance. Though they also require more extended computation and training time.

Sometimes, the dimensionality of the selected features is still excessively large even after feature selection, leading to expensive learning cost. Common dimensionality reduction methods, such as principal component analysis (PCA), are able to further encode the features into a sample space with lower dimensionality, thereby further enhance the training efficiency.

2.5 Training and Evaluation

Machine learning algorithms need to be trained with a set of profiled data with or without annotation to achieve the target function, such as modelling the patterns of different data categories for classification. The model performance (e.g., classification accuracy) is commonly evaluated by a different set of data, to test the model's ability in making prediction for unseen data. For model training and evaluation, the dataset is usually split into a training set, validation set and test set. Machine learning algorithms are trained on the data samples from the training set, and tested on the test set. The optimal machine learning algorithm along with the parameters can be determined based on their performance on the validation set. Specifically, machine learning algorithms with different parameters are trained on the training set and evaluated on the validation set respectively. The one achieves the best validation performance are selected (with the parameters) to be further evaluated on the test set and applied for real applications.

To achieve more trustworthy evaluation, cross validation technique is often adopted when training and evaluating machine learning models. In cross validation, the dataset is partitioned into several subsets. One of them is selected as the test set, and the model is then trained and validated on the others. This process is repeated until each of the subsets is selected and tested once. The overall model performance is computed as the average of the performance that the model yields on each of the subsets. The most used cross validation strategy is K-fold cross validation. It splits the dataset into K folds, usually by random, and trains and validates on $K - 1$ of them to test on the remaining one at each time.

It is common to refine and evaluate the model for several iteration to gradually improve the model by adjusting the hyperparameters, selected features, and even the model architecture and algorithm. The refining is usually conducted based on the performance that the model achieves in last round of iteration. This process helps to determine the optimal model settings and achieve satisfactory performance.

Overfitting and underfitting are common issues that negatively impact the generalization ability of machine learning models. In particular, overfitting occurs when the selected model is excessively complicated for the given task and dataset, such as containing an extreme number of parameters or handling with overabundant features. In such cases, the model may just simply memorize every single detail of the training samples, instead of learning their representative and informative patterns. This makes the model perform effectively on known data (i.e., data that has been seen by the model in the training phase) while predicting poorly for unknown data. Obtaining more training data is one of the typical solutions to tackle overfitting. Training with substantial data enables the model to learn from diverse kinds of scenarios, and eventually, enhance its generation ability.

However, it is challenging to acquire large-scale datasets in certain fields, including rare disease and extreme event (e.g., violence and fire events) detection [41, 42]. Data augmentation is an effective way to enlarge the originally collected dataset. For example, image dataset is commonly expanded by cropping, scaling, translating, flipping, and rotating the original images, to address overfitting issues in the field of computer vision. Another direction to attack overfitting is to simplify the model such as reducing the model parameters and/or used features. Moreover, techniques, such as regularization, could be applied to limit the memorizing capacity of the model or impose constraints on the memorized information.

On the other hand, when the model is too simple to capture adequate inherent patterns for the given task and dataset, the issues of underfitting arise. The most direct way to overcome underfitting is to boost the model complexity. This includes but not limited to using more sophisticated and nonlinear models or learning algorithms, increasing the model parameters (e.g., neural network layers), adding more features, and engineering more representative and informative features.

3 Deep Learning and Its Challenges

In recent years, deep learning has made significant development and shown its remarkable capability in achieving state-of-the-art performance in perceptual tasks, such as image and speech recognition. It thus keeps receiving considerable attention in both industry and research. By using neural network with multiple layers and a series of algorithms including activation functions and gradient backpropagation, deep learning is able to apply nonlinear transformations to extract and model high-level abstractions of the input data efficiently [43].

Deep learning exhibits its flexibility in employing different neural network architectures to handle diverse types of data, ranging from image and video to audio, text, time series sensory data, and even their combinations [44–46]. This empowers deep learning's effectiveness in different research areas, including the field of smart firefighting [42, 47]. It has become a valuable and essential tool for researchers and practitioners across various domains. To provide the brief introduction of deep learning, this section delineates the representative neural network architectures, and the general process of designing deep learning approaches with the related concepts and algorithms.

3.1 *Representative Neural Network Architectures*

Various neural network architectures are designed and adapted to effectively handle specific characteristics of different types of data. The selection of neural network architecture in deep learning depends highly on the nature of the input data. For example, convolutional neural networks (CNNs) are notably good at capturing representative spatial features from grid-like data, such as images. Recurrent neural networks (RNNs), on the other hand, are designed particularly for capturing temporal features from sequential data. They are the most commonly used neural network architectures in deep learning. This section will introduce them and their variants. them and their variants.

Convolutional Neural Networks (CNNs) have garnered widespread attention in recent years due to their success in computer vision tasks such as image recognition, object detection, and segmentation [48–50]. CNNs are composed of several crucial components that permit the processing and feature extraction of input data, particularly for grid-like data such as images. These components include convolutional layers, pooling layers, and fully connected layers, depicted in Fig. 8.

Convolutional layers are the fundamental building blocks of CNNs. They extract representative local features from the input data (e.g., images) based on convolution operations. Specifically, they apply and slide filters (also known as kernels) over the input image to perform element-wise multiplication and summation on the sliding local regions. This generates different feature maps that grab the relevant spatial

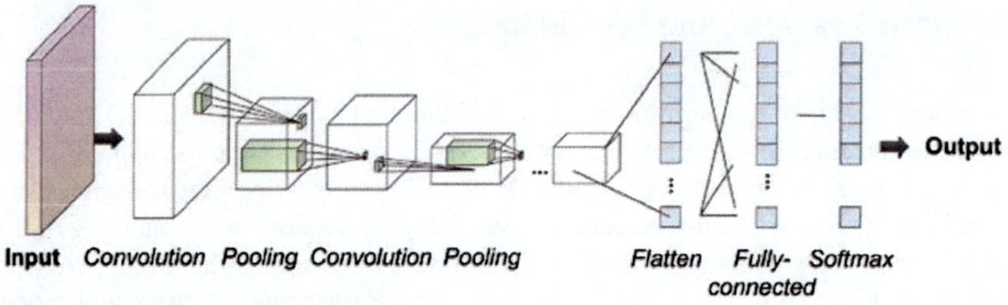

Figure 8 Example convolutional neural network architecture.

patterns in the input image and are utilized to build higher-level features in subsequent layers.

Pooling layers are utilized to diminish the spatial dimensionality of the data and enhance network computational efficiency. The most widely used pooling operation in CNNs is max pooling, which determines the maximum value within a small region of the input feature map and outputs the result in the output feature map. Though, pooling layers and convolutional layers are not always in strict pairs. The commonly used CNNs often place a pooling layer after a convolutional layer to achieve efficient modelling [51].

In fully connected layers, all neurons are linked to one another in the previous layer, allowing the network to grasp complex feature relationships. They are commonly utilized to complete the final classification or regression task of the features extracted based on the previous convolution layer and pooling layer in CNNs.

CNNs may also include other components, such as normalization layers, activation functions, and dropout layers. Normalization layers can improve training stability and speed by standardizing the previous layer's activations. Activation functions add nonlinearity within the network, permitting it to learn intricate functions. Dropout layers randomly eliminate some neurons during training, mitigating overfitting and enhancing generalization performance.

Given their capability in handling image and video data, CNNs widely are employed in computer vision applications, such as face recognition [52], pose estimation [53], eye gaze estimation [54], and group behavior analysis [44]. In addition, CNNs have also been applied to other domains, such as natural language processing, speech recognition, and audio analysis. CNNS are also used in conjunction with other technologies, such as transfer learning and data augmentation, to further improve their performance on challenging tasks.

However, well-performing CNNs often require to be trained with good quality and quantity training samples. Its performance may decrease when faced with insufficient or low-quality data. Overall, CNNs represent a powerful and versatile tool for a wide range of applications, and their development has led to significant advancements in the field of deep learning. Several state-of-the-art CNN architectures that have gained widespread popularity are ResNet (Residual Network), VGGNet, DenseNet, and MobileNet [55].

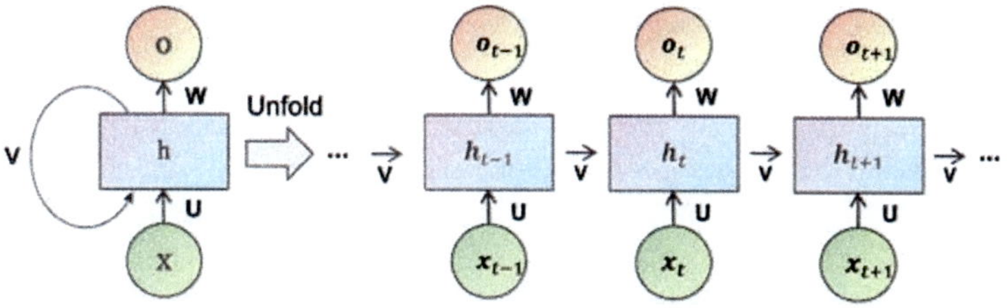

Fig. 9 Structure of Recurrent Neural Network

Recurrent Neural Networks (RNNs) are a neural network architecture specifically designed to tackle sequential data, including time-series and natural language data. In contrast to conventional feedforward neural networks, RNNs feature feedback connections that retain information across multiple time-steps, allowing them to model data sequences and capture temporal dependencies.

Figure 9 [56] illustrates the general structure of RNNs. Typically, an RNN is composed of three layers: the *input*, *hidden*, and *output* layer. Their functions are described as follows.

RNN input layers receive the input sequence (e.g., a sentence) and convert each of the sequence elements (e.g., a word) into an embedding vector respectively, which are the corresponding representations of the sequence elements that capture their semantic relationships. RNN's subsequent layers encode and model the sequence based on these embedding vectors.

RNN hidden layers use feedback connections to store the sequence information, enabling information to be transmitted between different time steps and capturing temporal dependencies in the sequence. The output of the hidden layer can be used as input for the next time step, and can also be passed to the output layer for prediction.

Finally, RNN output layers map the output of the hidden layer to the desired target space. Depending on the task, the output layer can use different activation functions and loss functions.

In the hidden layer of an RNN, each neuron receives an input and a hidden state from the previous time-step and produces an output and a hidden state that are passed on to the next time-step. The hidden state serves as a kind of memory that allows the network to maintain information about the previous inputs in the sequence, which can be used to predict the next output in the sequence. There are several variants of RNN, including the basic RNN, Long Short-Term Memory (LSTM) networks, and Gated Recurrent Units (GRUs). LSTMs and GRUs are popular variants of RNN that have been developed to address the problem of vanishing gradients during training, which can make it difficult for RNNs to learn long-term dependencies in the data. Due to the great advantage of RNN in processing sequential data, it has been applied to many fields, such as speech recognition [57], machine translation [58], text analysis [59], and mouse dynamics modelling [45].

3.2 Design of the Neural Network

When designing neural networks, it is suggested to carefully determine the components including network structure (backbone), activation function, initialization, optimization, and loss fuction, that is better suited for the given task, plays an essential role in achieving good performance. Different structures are suited for different types of data and tasks, and choosing the right architecture can improve accuracy and prevent overfitting or underfitting. Several factors should be considered when selecting a neural network structure. The type and complexity of the data is a critical consideration, as different structures are better suited for different data types. For example, convolutional neural networks (CNNs) are commonly used for image data, while recurrent neural networks (RNNs) are better suited for sequential data like text or speech.

The size of the dataset is also a crucial consideration. Complex structures have more trainable parameters, leading to higher capacity to learn and even memorize patterns from some specific samples that are not generalized enough (i.e., overfitting). Larger datasets provide more data samples with substantial diversity. This enables complex structures to learn intrinsic patterns that are generalized across a wide range of samples, avoiding overfitting. Smaller datasets have limited samples, and usually less data diversity. When training on these datasets, it is better to employ models with simpler structures (i.e., fewer parameters) to prevent overfitting.

The availability of pre-trained models is another important factor to take into account when determining neural network structures. It is believed that models pre-trained on public large-scale datasets learn the general and basic patterns related to the tasks and datasets. For example, CNN models pre-trained on ImageNet dataset [49], usually can capture the fundamental features of image data [44, 60]. They can be fine-tuned [61] or used as feature extractor [62] for similar tasks or datasets, to achieve efficient training with less data amount and computational resource requirement.

Finally, the availability of computational resources, such as graphics processing unit (GPU) or tensor processing unit (TPU) clusters, can impact the choice of architecture. Complex structures may be computationally expensive. When training these structures, it is suggested to consider utilizing serves equipped with high-performance GPUs.

In summary, selecting a suitable neural network structure requires careful consideration of the data type and complexity, the size of the dataset, and the availability of pre-trained models and computational resources.

Activation functions are commonly added after a neural network layer to transform its output before it is transmitted to the next layer. This enables neural networks to achieve nonlinear modeling. Without activation functions, neural networks can only express linear mapping. Even when multiple layers are employed, the input from the previous layer is passed to the next layer linearly. Such a network is equivalent to a single linearly connected layer. Therefore, it can also be considered that the deep neural network gains the ability to learn nonlinear mappings between different layers only after the activation function is added.

From the present point of view, most of the common activation functions express with exponential shapes, displaying the combination characteristics of segmented linear and nonlinear functions. Sigmoid function is one of the examples, which is inspired by and a simple simulation of biological neuron behaviors [63]. As its output values fall into a range of 0 to 1, sigmoid can also be applied to interpret the output as a probability or normalize the input. Mathematically, sigmoid function is expressed as the following equation:

$$f(x) = \frac{1}{1 + e^{-x}}$$

Tanh is another common activation function. It has an output range of -1 to 1, with a mean value of zero. This can lead to faster convergence and fewer training iterations in gradient-base optimization, compared to sigmoid function. Tanh function is expressed as:

$$f(x) = \frac{1 - e^{-2x}}{1 + e^{-2x}}$$

ReLU, expressed as follow, is another commonly used activation function. Compared to sigmoid and Tanh, it can better address the vanishing gradient problem and further accelerate the convergence speed.

$$f(x) = \max(0, x)$$

Initialization is a critical step when training neural networks, where initial values are assigned to the network parameters, such as the weight and bias of each neuron. It acts as a starting point of model training. Proper initialization can help speed up the convergence for the training process and significantly reduce the risk of getting stuck in sub-optimal solutions or local minima during optimization. There are several ways to do it. One naive method of initialization is to set all weights to the same value, for instance, zero or one. However, this is generally not recommended because it results in symmetry problems, where neurons in the same layer learn similar features during training. This limits models' capability to capture diverse features, making them suboptimal in handling intricate data. A more common approach is to initialize the weights randomly. This is known as random initialization and it helps in breaking the symmetry to ensure that features learned from different neurons have sufficient diversity enhancing models to capture complex data patterns and relationships [64]. The weights are typically initialized with small values, as large initial weights can lead to slower convergence [65].

Loss function measures the extent to which the predicted value of the model is different from the true value. Loss functions in neural networks should allow gradient computation for optimizing network parameters using gradient decent-based methods. Appropriate loss functions can help models to yield better performance efficiently. The selection of the loss functions largely depends on the nature of the data and task. Specifically, cross entropy loss and mean square error (MSE) are commonly used for classification and regression tasks respectively.

Cross entropy portrays the distance between two probability distributions and is more suitable for classification problems, because it expresses the probability that the predicted input sample belongs to a certain class.

Cross-entropy loss function for a multi-class task (C classes) is commonly expressed as follow:

$$L(Y, P) = -\frac{1}{N}\sum_i \sum_c y_i^c \log(p_i^c)$$

where N denotes the total number of samples, and the C classes are indexed as $\{1, \ldots, c, \ldots, C\}$. $Y = \{y_1, \ldots, y_N\}$ and $P = \{p_1, \ldots, p_N\}$ denote the actual label and the predicted output (probability distributions) for the samples respectively. Specifically, for sample i, the label is prepared in one-hot encoding $y_i = \{y_i^1, \ldots, y_i^C\}$, where $y_i^c = 1$ if sample i is belong to class c, and y_i^c is set to 0 if sample i is not belong to that class. Its probability distributions are represented as $p_i = \{p_i^1, \ldots, p_i^C\}$.

MSE measures the squared difference between the actual value and the predicted value. Training uses MSE aims at leading the model to make predictions that match the actual values as close as possible. It is better suited for regression tasks. Its formular is shown as follow:

$$L(Y, \widehat{Y}) = \frac{1}{N}\sum_{i=1}^{N}(y_i - \widehat{y}_i)^2$$

where $Y = \{y_1, \ldots, y_N\}$ and $\widehat{Y} = \{\widehat{y}_1, \ldots, \widehat{y}_N\}$ denote the actual label values and the predicted values respectively.

Optimization is an essential process in training neural networks, which involves tuning the parameters of the network in order to minimize the difference between the network's predictions and the actual outputs.

The most commonly used optimization algorithm in neural networks is known as gradient descent. It is an iterative optimization algorithm commonly used in neural networks to find the optimal set of parameters that minimize the loss function. It iteratively updates the network's parameters in the direction of steepest descent of the loss function by computing the gradients of the loss with respect to the parameters. Gradients from layer to layer in deep neural networks (multiple layers) are obtained based on backpropagation algorithm [66]. This process is repeated until convergence or until a predetermined number of iterations is reached. There are variations of gradient descent that differ in how the parameter updates are performed, such as Stochastic Gradient Descent (SGD) and Mini-Batch Gradient Descent [67].

These methods are enhanced with adaptive learning rate algorithms and regularization techniques to improve convergence speed and generalization performance to unseen data. Researchers continue to explore and develop new optimization algorithms to enhance the training process and achieve even better performance in neural networks.

Batch training in large dataset is used to speed up the convergence speed and enhance the generation performance. In batch training, the entire training set is divided into multiple batches with the same number of data samples. The model is trained, and the network parameters are updated on each batch respectively in every training iteration. Larger batch sizes face higher risk of using up the computational resource such as the GPU memory, and getting stuck at local optima. Smaller batch sizes, on the other hand, may result in slow convergence. The computer resources and convergence performance should be taken into account when determining the batch size.

These components and the hyperparameters play significant roles in the network performance. It is often suggested to carefully determine them with exploratory analysis on the specific dataset and task.

4 Applications and Data Analysis

Artificial intelligence (AI) and Machine learning (ML) methods have sparked a revolution across various industries, leveraging their ability to extract patterns and make predictions from vast amounts of data in real-life applications. This includes digital healthcare [62, 68, 69], finance technology, recommendation system [70], natural language processing [71], intelligent manufacturing, intelligent education, and smart firefighting [42, 72, 73]. Notably, in fire hazard prevention, AI is commonly applied to intelligently sense and model time-series (e.g., temperature signals) and image or video (e.g., satellite image) data to achieve real-time fire hazard prediction and/or off-line analysis. This section presents examples to illustrate the potential applications of AI and ML in these two domains. Video (e.g., satellite image) data to achieve real-time fire hazard prediction and/or off-line analysis. This section presents examples to illustrate the potential applications of AI and ML in these two domains.

4.1 Image/video Classification

Image/video classification is a fundamental computer vision problem, which aims at assigning images/videos to the respective categories. There are various applications in this field, including object recognition [74], face recognition [52, 75], scene segmentation and background subtraction [76, 77], motion tracking [78, 79], and action detection [41, 44], etc. Many factors should be considered when applying ML and DL in image and video classification tasks to yield effective models [80]. Convolutional neural networks (CNNs) are especially effective in modelling spatial patterns in grid-like data, they thus are commonly utilized in image and video classifications [48–50]. There are a wide variety of CNN structures, each designed for specific tasks and applications. Researchers must choose the appropriate network structures tailored to the specific image classification data and target, particularly considering factors such as data size, data domains, and expected outcomes for determining the optimal method for a specific task [56]. Moreover, it is also essential to take into account user requirements, compatibility with previous work, available image processing and classification algorithms, computational resources and time constraints, when devising the classification methods for practical applications [80].

Applying image classification models in fire fighting has been widely explored in previous studies. The models help in detecting, modelling, and predicting of fire, smoke, and human behaviors, aiming to prevent fire events and assist rescue activities. This includes the models that are constructed for thermal images [81, 82], satellite images [83], and standard color images and videos [84–86]. Their applications encompass kitchen fire prediction [86], forest fire detection [83–85], emergency person recognition [81], and intelligent firefighting robot [82], etc.

4.2 Time Series Data Mining

Time series data consists of a sequence of data points collected or recorded in a chronological order, which is commonly found in many various fields such as finance, medicine, environmental science, and social sciences [87]. The growing effectiveness and availability of machine learning techniques across diverse domains has led to their wide application in time series data [88, 89]. These techniques can provide valuable insights by identifying patterns, trends, and anomalies within time series data, facilitating decision-making processes [90]. For example, machine learning methods have been employed to identify patterns in temporal glucose monitoring data to help manage diabetes [91]. They also have also found applications in financial forecasting, such as predicting stock prices based on the historical data [92]. Features that capture the trend, energy, recurrent, and consistency patterns, and accommodate stretching or compressing of the time series data, are significantly productive for time series data mining tasks [93–95]. In deep learning, recurrent neural networks (RNNs) and the variants are commonly applied for time series data mining [56–59]. Similarly, factors, such as data type, data size, expected outcomes, user requirements, and time constraints, should be carefully considered when designing, developing and applying machine learning methods in time series data mining.

Machine learning-based time series data mining techniques have shown excellent performance in fire detection, particularly for modelling temporal sensor data such as temperature, smoke, and gas signals [72, 96]. These techniques extract and model the representative features of the temporal signals to detect and monitor the probability and severity of fire incidents. This results in a wide range of applications, such as cooktop ignition prediction [72, 96], and flashover forecast [97, 98].

5 Present Interests

The field of AI/ML develops rapidly. Particular focuses include exploring novel deep neural network architectures and techniques to boost prediction accuracy, generalization performance, and training efficiency. This section provides a glance at a novel architecture, named Transformer, and a recent training technique, named few-shot learning.

5.1 Transformer

Transformer is a sequence-to-sequence model that leverages an attention mechanism to capture interactions between elements within a sequence. This allows for effective modeling of long-term dependencies.

As shown in Fig. 10, a Transformer commonly consists of two main components: *Encoder* and *Decoder*, where *Encoder* converts the input sequence to a set of hidden representations and *Decoder* uses generates the output sequence based on those representations. In addition, Transformers use Positional Encoding to provide an encoding vectors which indicate the position information of each element (e.g., lexical unit) within a sequence. This allows the model to efficiently handle the sequence information.

1. ***Encoder*** is one of the core components in a Transformer, where each layer consists of multiple sub-layers, including a *Multi-Head Self-Attention* module and *Feedforward* layers. Specifically, *Multi-Head Self-Attention* module helps the model to capture the contextual information and inner dependencies between different elements within the input sequence. It generates hidden representations of the input sequence for further processing. *Feedforward* layers provide nonlinear mapping and dynamic processing for the hidden representations obtained from the attention layer, to yield the final representations of the input sequence. In addition, *Add & Norm* operations are also applied for input and output normalization in order to enhance the training efficiency, and improve the accuracy of the model.
2. ***Decoder*** is another crucial component in a Transformer, which works for generating output sequence based on the hidden representations of the input sequence. It again is composed of several sub-layers, including *Multi-Head Self-Attention* module,, *Feedforward* layers, and *Add & Norm* operations, that are mirroring structures found in the encoder. However, there is an additional *Encoder-Decoder cross-attention* module in the decoder, which leads the model to focus on the relevant parts of the input sequence when generating the outputs.

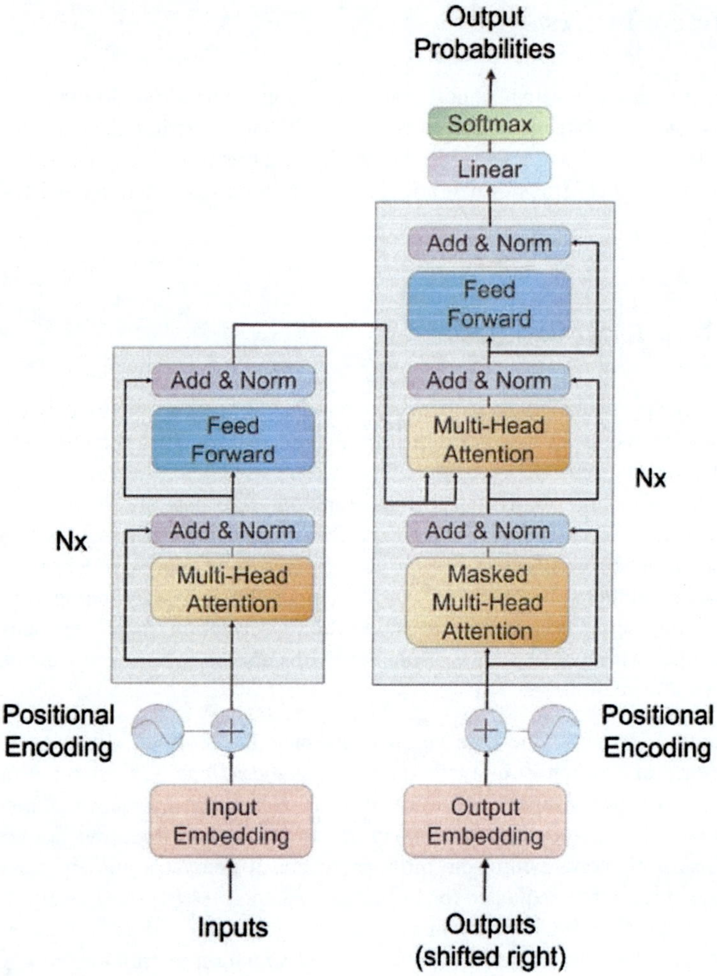

Fig. 10 Example transformer architecture

These structures and components allow Transformers to model the long-term dependencies more efficiently, resulting in impressive performance in diverse applications. Transformers thus gained popularity in various fields. For example, Transformers are often used for financial text classification and stock price prediction in finance domain [99–101]. In the realm of healthcare, they show effectiveness in lung CT scanning [102], ophthalmic medicine [103], and medical natural language processing tasks [104]. Transformers also have a wide range of applications in the field of education, including the analytics of students' classroom posture estimation [105], facial expressions and engagement [106], and learning performance [107].

Besides, Transformers have also been used for tasks such as image classification [108], cross-modal generation [103], target detection [109], pose estimation [110], and speech recognition [111]. The successful application of Transformers brings new ideas and methods to many domains, showcasing the potential capability of deep learning with significant theoretical and practical values.

5.2 *Data Insufficiency*

Deep neural networks (DNNs) have demonstrated remarkable results in various applications. However, the effectiveness of these algorithms depends heavily on large datasets, as the quality and quantity of the training data plays a critical role in the model performance. Unfortunately, obtaining a vast amount of data can be challenging, especially in fields such as fire fighting, human fights detection, and disease diagnosis [41, 42, 62, 69, 112].

Data augmentation is commonly applied to address the challenge, which involves a collection of techniques designed to enhance the volume and diversity of the data, thereby improving the model performance. Data warping involves applying various transformations to the original data is often used for data augmentation [113]. For example, in image classification tasks, geometric transformations, such as, rotating, flipping, and cropping, are often used to expand the image datasets. Figure 11 illustrates the example of rotating the image to achieve data augmentation. In addition, color perturbation techniques adjust the pixel values of image data to create more images with various colors and brightness. All of these techniques contribute to increasing the training data size and diversity to enhance the model generalization ability.

Few-shot learning is characterized by the ability to learn new knowledge for new tasks using with extremely small datasets. Compared to conventional machine learning, few-shot learning focuses more on the model's capability to generalize even when faced with a scarcity of training samples, effectively tackling the problem of

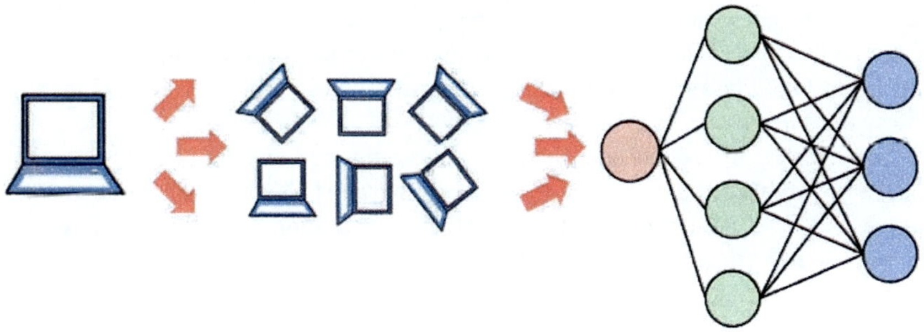

Fig. 11 Augmenting image data by rotating the image

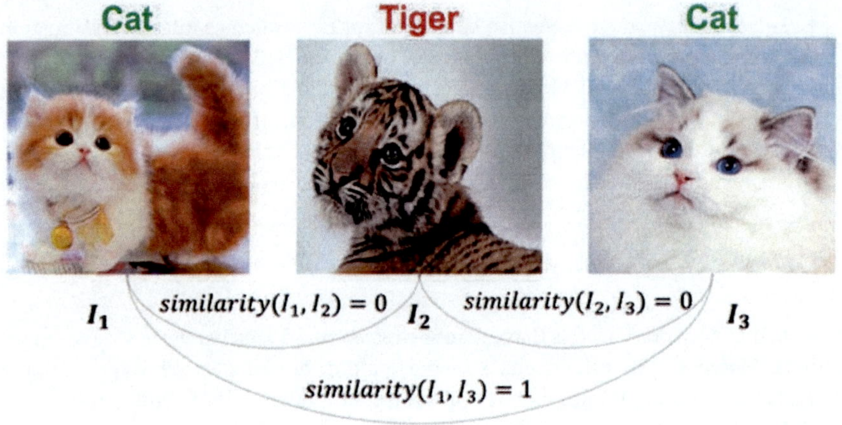

Fig. 12 An example of the similarity function in few-shot learning

data scarcity. It thus has become a research hotspot in machine learning and deep learning in recent years.

One the main goals in common few-shot learning is to learn a similarity function that maps the similarity between classes in the support (i.e., labeled training samples) and query sets (i.e., target samples for predicting). The similarity function typically outputs the probability values indicating the degree of the similarity. Figure 12 illustrates an example of similarity function in few-shot learning. Neural network, particularly with *Siamese Network* architecture, is often used to achieve similarity function in few-shot learning. *Siamese Network* works in a comparison-based learning strategy, where it processes two input samples to generate their feature vectors respectively, through twin neural networks with shared weights. These vectors are then passed through a metric space layer to calculate their similarity, effectively quantifying the likeness between the input samples.

The *Siamese Network*, with the parameters defining the similarity function, is trained on a large-scale labeled dataset. Subsequently, in the few-shot learning phase, the trained *Siamese Network* is applied to determine the similarity probabilities between the query and support samples, making classification for the query samples. Specifically, for each query sample, the few-shot model predicts its class label by identifying the class in the support set with the highest similarity to it. Figure 13 shows the overview of few-shot learning.

The capacity to address data scarcity issues enhance few-shot learning's performance in various tasks, including image classification [114, 115], image segmentation tasks [116], and target detection [117], etc. In addition to image data, few-shot learning is also effective in handling text data, and addressing label limitation issues in natural language processing tasks [118, 119].

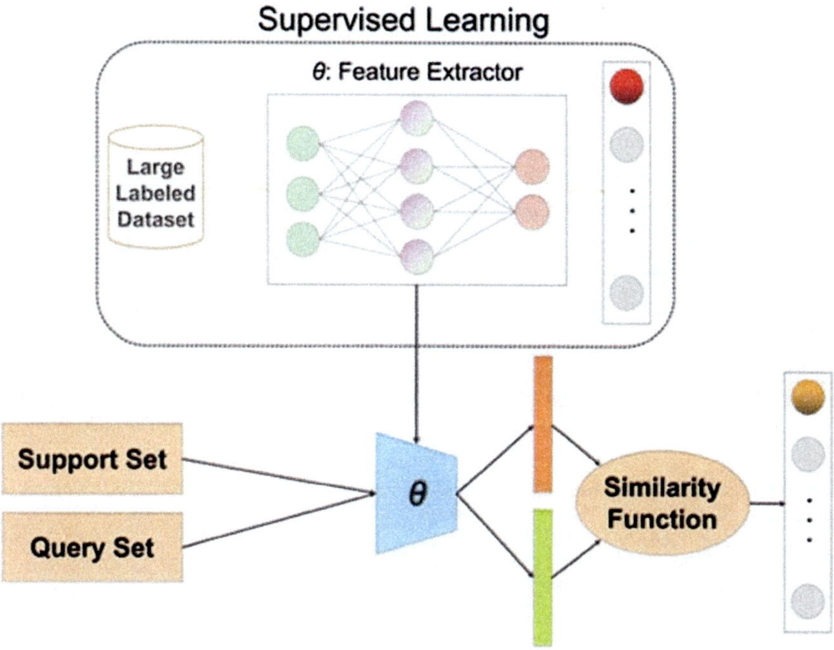

Fig. 13 Overview of few-shot learning, where similarity function achieved based on Siamese Network

6 Conclusions

Today, artificial intelligence (AI) keeps garnering attention and is being applied across a wide range of domains. It particularly helps to improve efficiency and minimize errors added by humans in tasks that involve recognition, analysis and decision making. This chapter provides a comprehensive overview of AI, with a special focus on machine learning (ML) and deep learning (DL). It starts by introducing the basic concepts and common algorithms (e.g., support vector machine) and model structures (e.g., convolutional neural network) of machine and deep learning. This is followed by the overview of their applications, particularly in image/video classification, and time series data mining. Additionally, a novel neural network, named Transformer, is also briefly discussed in this chapter, whose excellent performance has made it a popular model in recent years. Finally, the chapter presents some common solutions to the problem of data insufficiency, which is a trending topic in current research, since data plays an important role in AI.

The outstanding capabilities of AI and ML have been applied to make firefighting "smarter" in diverse kinds of applications. Particularly, in fire hazard prevention, AI and ML are commonly applied to intelligently sense and model time-series (e.g., temperature signals) and image/video (e.g., satellite image) data to achieve real-time fire hazard prediction and/or off-line analysis. This includes temperature monitoring, fire identification, fire risk assessment, smoke detection, and so on.

In anticipation of the future, there are still many research and applications in smart firefighting worth exploring. For instance, AI could be used in developing firefighting robots/drones to detect the presence of fire and extinguish it automatically, enhancing human safety. Besides, leveraging digital twins and semantic web technologies to achieve proactive control of smart firefighting systems in buildings during emergencies and disasters is highly desirable. AI can also work with other technologies such as internet of things (IoT), big data, and VR/AR, to further improve the performance and applicability of smart firefighting systems in practical applications. As an illustration, VR technology can create simulated extreme environments for training firefighters, while AI is able to monitor their training performance. Using the two technologies together can contribute to an enhanced virtual training system for firefighters.

Acknowledgements This work was in part supported by the Hong Kong Polytechnic University under Grant P0039489, and Shenzhen University.

References

1. A.L. Samuel, Some studies in machine learning using the game of checkers. IBM J. Res. Dev. **3**(3), 210–229 (1959)
2. Y. Yao, Y. Liu, Y. Yu, H. Xu, W. Lv, Z. Li, X. Chen, K-SVM: an effective SVM algorithm based on K-means clustering. J. Comput. **8**(10), 2632–2639 (2013)
3. R. Geng, J. Song, Z. Zhong, S. Ni, W. Liu, Z. He, J. Liu, Crosstalk of redox-related subtypes, establishment of a prognostic model and immune responses in endometrial carcinoma. Cancers **14**(14), 3383 (2022)
4. H. Tang, Y. Wang, S. Tang, D. Chu, C. Li, A randomized clustering forest approach for efficient prediction of protein functions. IEEE Access **7**, 12360–12372 (2019)
5. Y. Li, X. Chu, D. Tian, J. Feng, W. Mu, Customer segmentation using K-means clustering and the adaptive particle swarm optimization algorithm. Appl. Soft Comput. **113**, 107924 (2021)
6. J. Zhang, Y. Wang, B. Li, S. Hu, Privacy-Preserving accelerated clustering for data encrypted by different keys, in *Smart Computing and Communication: 5th International Conference, SmartCom 2020, Paris, France, December 29–31, 2020, Proceedings 5* (Springer International Publishing, 2021), pp. 156–162
7. A. Khatami, S. Mirghasemi, A. Khosravi, C.P. Lim, S. Nahavandi, A new PSO-based approach to fire flame detection using K-Medoids clustering. Expert Syst. Appl. **68**, 69–80 (2017)
8. M. G. Omran, A. Engelbrecht, A. Salman, An overview of clustering methods, Intell. Data Anal. (2007)
9. K. Khan, S. U. Rehman, K. Aziz, S. Fong, S. Sarasvady, DBSCAN: Past, present and future, in *The fifth international conference on the applications of digital information and web technologies (ICADIWT 2014)*, (IEEE, 2014), (pp. 232–238)
10. J. Tobin, M. Zhang, DCF: an efficient and robust density-based clustering method, in *2021 IEEE International Conference on Data Mining (ICDM)* (IEEE, 2021), pp. 629–638
11. H. C. Bao, Z. C. Juan, The research of speaker recognition based on GMM and SVM. in *2012 international conference on system science and engineering (ICSSE)* (IEEE, 2012), pp. 373–375

12. J.M. Górriz, F. Segovia, J. Ramírez, A. Lassl, D. Salas-Gonzalez, GMM based SPECT image classification for the diagnosis of Alzheimer's disease. Appl. Soft Comput. **11**(2), 2313–2325 (2011)
13. A. Salamzadeh, P. Ebrahimi, M. Soleimani, M. Fekete-Farkas, Grocery apps and consumer purchase behavior: application of Gaussian mixture model and multi-layer perceptron algorithm. J. Risk Financ. Manag. **15**(10), 424 (2022)
14. A. Aali-Bujari, F. Venegas-Martínez, G. Pérez-Lechuga, Impact of the stock market capitalization and the banking spread in growth and development in Latin American: a panel data estimation with System GMM. Contaduría y administración **62**(5), 1427–1441 (2017)
15. C. Cortes, V. Vapnik, Support-vector networks. Mach. Learn. **20**, 273–297 (1995)
16. B. Scholkopf, S. Mika, C.J. Burges, P. Knirsch, K.R. Muller, G. Ratsch, A.J. Smola, Input space versus feature space in kernel-based methods. IEEE Trans. Neural Netw. **10**(5), 1000–1017 (1999)
17. A. Razaque, B.H. Frej, M., Almi'ani, M., Alotaibi, M., & Alotaibi, B., Improved support vector machine enabled radial basis function and linear variants for remote sensing image classification. Sensors **21**(13), 4431 (2021)
18. M. Vatankhah, V. Asadpour, R. Fazel-Rezai, Perceptual pain classification using ANFIS adapted RBF kernel support vector machine for therapeutic usage. Appl. Soft Comput. **13**(5), 2537–2546 (2013)
19. K. Soman, A. Sathiya, N. Suganthi, Classification of stress of automobile drivers using radial basis function kernel support vector machine, in *International Conference on Information Communication and Embedded Systems (ICICES2014)* (IEEE, 2014), pp. 1–5
20. H. Byun, S. Lee, Applications of support vector machines for pattern recognition: a survey. in *International workshop on support vector machines* (Springer Berlin Heidelberg, Berlin, Heidelberg, 2002), pp. 213–236
21. P. Birzhandi, K.T. Kim, H.Y. Youn, Reduction of training data for support vector machine: a survey. Soft. Comput. **26**(8), 3729–3742 (2022)
22. L. Zeyang, Research on intelligent acceleration algorithm for big data mining in communication network based on support vector machine, in *2021 IEEE 4th International Conference on Automation, Electronics and Electrical Engineering (AUTEEE)*, (IEEE, 2021), pp. 479–483
23. Z. Tingyu, Short-term stock trend prediction based on improved SVR, in *2020 7th international conference on information science and control engineering (ICISCE)* (IEEE, 2020), pp. 1283–1286
24. G. Bathla, Stock Price prediction using LSTM and SVR. in *2020 Sixth International Conference on Parallel, Distributed and Grid Computing (PDGC)* (IEEE, 2020), pp. 211–214)
25. D. Wu, S. Wang, Comparison of road traffic accident prediction effects based on SVR and BP neural network, in *2020 IEEE International Conference on Information Technology, Big Data and Artificial Intelligence (ICIBA)*, Vol. 1. (IEEE, 2020), pp. 1150–1154
26. J. Han, M. Kamber, J. Pei, (2012). Data mining concepts and techniques third edition. In *University of Illinois at Urbana-Champaign Micheline Kamber Jian Pei Simon Fraser University*
27. R. C. Barros, M. P. Basgalupp, A. C De Carvalho, A. A. Freitas (2011) A survey of evolutionary algorithms for decision-tree induction. IEEE Trans. Syst. Man Cybernetics Part C (Appl. Rev.) 42(3), 291–312
28. J. R. Quinlan, Discovering rules by induction from large collections of examples. Expert Syst. Micro Electr. Age (1979)
29. J. R. Quinlan, *C4. 5: programs for machine learning*. Elsevier (2014)
30. L. Breiman, *Classification and regression trees* (Routledge, 2017)
31. S. Fletcher, M.Z. Islam, Decision tree classification with differential privacy: a survey. ACM Comput. Survey. (CSUR) **52**(4), 1–33 (2019)
32. K. P. Murphy, Machine learning: a probabilistic perspective. MIT press (2012)
33. J.R. Quinlan, Improved use of continuous attributes in C4. 5. J. Artif. Intell. Res. **4**, 77–90 (1996)
34. L. Breiman, Random forests. Machine Learn. **45**, 5–32 (2001)

35. L. Breiman, *Consistency for a simple model of random forests* (University of California at Berkeley, Technical Report, 2004), p.670
36. M. Dagdoug, C. Goga, D. Haziza, Model-assisted estimation through random forests in finite population sampling. J. Am. Stat. Assoc. **118**(542), 1234–1251 (2023)
37. D. Ghosh, J. Cabrera, Enriched random forest for high dimensional genomic data. IEEE/ACM Trans. Comput. Biol. Bioinf. **19**(5), 2817–2828 (2021)
38. A. Verikas, A. Gelzinis, M. Bacauskiene, Mining data with random forests: a survey and results of new tests. Pattern Recogn. **44**(2), 330–349 (2011)
39. S. Milanović, N. Marković, D. Pamučar, L. Gigović, P. Kostić, S.D. Milanović, Forest fire probability mapping in eastern Serbia: logistic regression versus random forest method. Forests **12**(1), 5 (2020)
40. E. Scornet, G. Biau, J. P. Vert, (2015) Consistency of random forests
41. E. Y. Fu, M. X. Huang, H. V. Leong, G. Ngai, Cross-species learning: a low-cost approach to learning human fight from animal fight. in *Proceedings of the 26th ACM international conference on Multimedia* (2018), pp. 320–327
42. W.C. Tam, E.Y. Fu, J. Li, R. Peacock, P. Reneke, G. Ngai, M.X. Huang, Real-time flashover prediction model for multi-compartment building structures using attention based recurrent neural networks. Expert Syst. Appl. **223**, 119899 (2023)
43. Y. LeCun, Y. Bengio, G. Hinton, Deep learning. Nature **521**(7553), 436–444 (2015)
44. E. Y. Fu, M. W. Ngai, Using motion histories for eye contact detection in multiperson group conversations, in *Proceedings of the 29th ACM International Conference on Multimedia* (2021), pp. 4873–4877
45. T. C. Kwok, E. Y. Fu, E. Y. Wu, M. X. Huang, G. Ngai, H. V. Leong, Every little movement has a meaning of its own: using past mouse movements to predict the next interaction. in *23rd International Conference on Intelligent User Interfaces* (2018), pp. 397–401
46. B. Guo, C. Zhang, J. Liu, X. Ma, Improving text classification with weighted word embeddings via a multi-channel TextCNN model. Neurocomputing **363**, 366–374 (2019)
47. L. Meng, Z. Peng, J. Zhou, J. Zhang, Z. Lu, A. Baumann, Y. Du, Real-time detection of ground objects based on unmanned aerial vehicle remote sensing with deep learning: Application in excavator detection for pipeline safety. Remote Sensing **12**(1), 182 (2020)
48. C. Li, Y. Wang, D. Li, D. Chu, M. Ma, An effective method of evaluating pension service quality using multi-dimension attention convolutional neural networks. Int. J. Software Eng. Knowl. Eng. **31**(04), 533–543 (2021)
49. A. Krizhevsky, I. Sutskever, G. E. Hinton, Imagenet classification with deep convolutional neural networks. Adv Neural Inform. Process. Syst. 25 (2012)
50. J. Ren, Z. Zhang, J. Fan, H. Zhang, M. Xu, M. Wang, Robust low-rank deep feature recovery in cnns: toward low information loss and fast convergence. in *2021 IEEE International Conference on Data Mining (ICDM)* (IEEE, 2021), pp. 529–538
51. K. O'Shea, R. Nash, An introduction to convolutional neural networks. *ArXiv preprint* arXiv: 1511.08458 (2015)
52. M. Coşkun, A. Uçar, Ö. Yildirim, Y. Demir, Face recognition based on convolutional neural network. in *2017 International Conference on Modern Electrical and Energy Systems (MEES)* (IEEE, 2017), pp. 376–379
53. D. Mehta, H. Rhodin, D. Casas, P. Fua, O. Sotnychenko, W. Xu, C. Theobalt, Monocular 3d human pose estimation in the wild using improved cnn supervision. in *2017 international conference on 3D vision (3DV)* (IEEE, 2017), pp. 506–516
54. K. Krafka, A. Khosla, P. Kellnhofer, H. Kannan, S. Bhandarkar, W. Matusik, A. Torralba, Eye tracking for everyone. in *Proceedings of the IEEE conference on computer vision and pattern recognition* (2016), pp. 2176–2184
55. A. Dhillon, G.K. Verma, Convolutional neural network: a review of models, methodologies and applications to object detection. Prog. Artificial Intell. **9**(2), 85–112 (2020)
56. Z. He, Deep learning in image classification: a survey report. in *2020 2nd International Conference on Information Technology and Computer Application (ITCA)* (IEEE, 2020), pp. 174–177

57. T. Mikolov, M. Karafiát, L. Burget, J. Cernocký, S. Khudanpur, Recurrent neural network based language model. In *Interspeech* , Vol. 2, No. 3 (2010), pp. 1045–1048
58. D. Bahdanau, K. Cho, Y. Bengio, Neural machine translation by jointly learning to align and translate. *ArXiv preprint* arXiv:1409.0473 (2014)
59. A. N. Jagannatha, H. Yu, Structured prediction models for RNN based sequence labeling in clinical text. in *Proceedings of the conference on empirical methods in natural language processing. Conference on empirical methods in natural language processing*, Vol. 2016 (NIH Public Access, 2016), p. 856
60. S.A. Siddiqui, A. Salman, M.I. Malik, F. Shafait, A. Mian, M.R. Shortis, E.S. Harvey, Automatic fish species classification in underwater videos: exploiting pre-trained deep neural network models to compensate for limited labelled data. ICES J. Mar. Sci. **75**(1), 374–389 (2018)
61. T. Alshalali, D. Josyula, Fine-tuning of pre-trained deep learning models with extreme learning machine. in *2018 International Conference on Computational Science and Computational Intelligence (CSCI)* (IEEE, 2018), pp. 469–473
62. E. Y. Fu, Z. Yang, H. V. Leong, G. Ngai, C. W. Do, L. Chan, Exploiting active learning in novel refractive error detection with smartphones. in *Proceedings of the 28th ACM international conference on multimedia* (2020) pp. 2775–2783
63. A.K. Jain, J. Mao, K.M. Mohiuddin, Artificial neural networks: a tutorial. Computer **29**(3), 31–44 (1996)
64. D. Ulyanov, A. Vedaldi, V. Lempitsky, Deep image prior. in *Proceedings of the IEEE conference on computer vision and pattern recognition* (2018), pp. 9446–9454
65. D. E. Rumelhart, G. E. Hinton, R. J. Williams. Learning internal representations by error propagation (1985)
66. L. Bottou, Stochastic gradient learning in neural networks. Proc. Neuro-Nımes **91**(8), 12 (1991)
67. S. Ruder, An overview of gradient descent optimization algorithms. ArXiv preprint arXiv:1609.04747 (2016)
68. J. Wang, E. Y. Fu, G. Ngai, H. V. Leong, H. V., Huang, M. X. (2019). Detecting stress from mouse-gaze attraction. in *Proceedings of the 34th ACM/SIGAPP symposium on applied computing*, pp. 692–700
69. J. Li, C. Brown, D.J. Dzikowicz, M.G. Carey, W.C. Tam, M.X. Huang, Towards real-time heart health monitoring in firefighting using convolutional neural networks. Fire Saf. J. **140**, 103852 (2023)
70. Y. Wang, S. C. F. Chan, G. Ngai, Applicability of demographic recommender system to tourist attractions: a case study on trip advisor. *in* 2012 IEEE/WIC/ACM international conferences on web intelligence and intelligent agent technology (IEEE, 2012), Vol. 3, pp. 97–101
71. Y. Wang, S.C.F. Chan, H.V. Leong, G. Ngai, N. Au, Multi-dimension reviewer credibility quantification across diverse travel communities. Knowl. Inf. Syst. **49**, 1071–1096 (2016)
72. W.C. Tam, E.Y. Fu, A. Mensch, A. Hamins, C. You, G. Ngai, va Leong, H., Prevention of cooktop ignition using detection and multi-step machine learning algorithms. Fire Saf. J. **120**, 103043 (2021)
73. W.C. Tam, E.Y. Fu, J. Li, X. Huang, J. Chen, M.X. Huang, A spatial temporal graph neural network model for predicting flashover in arbitrary building floorplans. Eng. Appl. Artif. Intell. **115**, 105258 (2022)
74. Y. Chen, J. Wang, B. Zhu, M. Tang, H. Lu, Pixelwise deep sequence learning for moving object detection. IEEE Trans. Circuits Syst. Video Technol. **29**(9), 2567–2579 (2017)
75. A.A. Ali, T.A. El-Hafeez, Y.K. Mohany, An accurate system for face detection and recognition. J. Adv. Math. Comput. Sci. **33**(3), 1–19 (2019)
76. S. Ren, K. He, R. Girshick, J. Sun, Faster r-cnn: towards real-time object detection with region proposal networks. in *Advances in neural information processing systems*, 28 (2015)
77. M. Babaee, D.T. Dinh, G. Rigoll, A deep convolutional neural network for video sequence background subtraction. Pattern Recogn. **76**, 635–649 (2018)

78. B. Heo, K. Yun, J. Y. Choi, Appearance and motion based deep learning architecture for moving object detection in moving camera. in *2017 IEEE international conference on image processing (ICIP)*, (IEEE, 2017), pp. 1827–1831
79. E. D. Tejada, P. A. Rodriguez, Moving object detection in videos using principal component pursuit and convolutional neural networks. in *2017 IEEE Global Conference on Signal and Information Processing (GlobalSIP)* (IEEE, 2017), pp. 793–797
80. D. Lu, Q. Weng, A survey of image classification methods and techniques for improving classification performance. Int. J. Remote Sens. **28**(5), 823–870 (2007)
81. M. Ulrich, T. Hess, S. Abdulatif, B. Yang, Person recognition based on micro-Doppler and thermal infrared camera fusion for firefighting. in *2018 21st international conference on information fusion (FUSION)* (IEEE, 2018, July), pp. 919–926
82. J.H. Kim, B.Y. Lattimer, Real-time probabilistic classification of fire and smoke using thermal imagery for intelligent firefighting robot. Fire Saf. J. **72**, 40–49 (2015)
83. S. Ganapathi Subramanian, M. Crowley, Using spatial reinforcement learning to build forest wildfire dynamics models from satellite images. Front ICT **5**, 6 (2018)
84. C. Yuan, Z. Liu, Y. Zhang, UAV-based forest fire detection and tracking using image processing techniques. In: *2015 International Conference on Unmanned Aircraft Systems (ICUAS)* (IEEE, 2015), pp. 639–643
85. S. Frizzi, R. Kaabi, M. Bouchouicha, J. M. Ginoux, E. Moreau, F. Fnaiech, Convolutional neural network for video fire and smoke detection. in *IECON 2016–42nd Annual Conference of the IEEE Industrial Electronics Society* (IEEE, 2016), pp. 877–882
86. M. Ngai, E. Fu, A. Tam, A. Yang, G. Ngai, Finding the signal from the smoke: a real-time, unattended fire prevention system using 3D CNNs. J. Student Res. 11(3) (2022)
87. D. E. Shasha, Y. Zhu, High performance discovery in time series: techniques and case studies. Springer Science & Business Media (2004)
88. R. Agrawal, C. Faloutsos, A. Swami, Efficient similarity search in sequence databases. in *Proceedings of the Fourth International Conference on Foundations of Data Organization and Algorithms* (Springer, Berlin, Heidelberg, 1993), pp. 69–84
89. T.C. Fu, A review on time series data mining. Eng. Appl. Artif. Intell. **24**(1), 164–181 (2011)
90. A. Toujani, H. Achour, A data mining framework for forest fire mapping. in *Handbook of Research on Geographic Information Systems Applications and Advancements* (IGI Global, 2017), pp. 214–237
91. L. Sacchi, C. Larizza, C. Combi, R. Bellazzi, Data mining with temporal abstractions: learning rules from time series. Data Min. Knowl. Disc. **15**, 217–247 (2007)
92. W. Bao, J. Yue, Y. Rao, A deep learning framework for financial time series using stacked autoencoders and long-short term memory. PLoS ONE **12**(7), e0180944 (2017)
93. P. Esling, C. Agon, Time-series data mining. ACM Comput Surv (CSUR) **45**(1), 1–34 (2012)
94. J. Lin, E. Keogh, S. Lonardi, B. Chiu, A symbolic representation of time series, with implications for streaming algorithms. in *Proceedings of the 8th ACM SIGMOD workshop on Research issues in data mining and knowledge discovery* (2003), pp. 2–11
95. D. J. Berndt, J. Clifford, Using dynamic time warping to find patterns in time series. in *KDD workshop*, Vol. 10, No. 16 (1994), pp. 359–370
96. A. E. Hamins, W. C. Tam, Z. J. Lu, K. Markell, C. You, M. Kupferschmid, Sensors and machine learning models to prevent cooktop ignition and ignore normal cooking. Fire Technol. 1–24 (2021)
97. J. Wang, W.C. Tam, Y. Jia, R. Peacock, P. Reneke, E.Y. Fu, T. Cleary, P-Flash–A machine learning-based model for flashover prediction using recovered temperature data. Fire Saf. J. **122**, 103341 (2021)
98. T. Zhang, Z. Wang, H.Y. Wong, W.C. Tam, X. Huang, F. Xiao, Real-time forecast of compartment fire and flashover based on deep learning. Fire Saf. J. **130**, 103579 (2022)
99. K. Mishev, A. Gjorgjevikj, I. Vodenska, L.T. Chitkushev, D. Trajanov, Evaluation of sentiment analysis in finance: from lexicons to transformers. IEEE access **8**, 131662–131682 (2020)
100. M. Maia, J. E. Sales, A. Freitas, S. Handschuh, M. Endres, A comparative study of deep neural network models on multi-label text classification in finance. in *2021 IEEE 15th International Conference on Semantic Computing (ICSC)* (IEEE, 2021), pp. 183–190

101. Hu, X. (2021, December). Stock price prediction based on temporal fusion transformer. In *2021 3rd International Conference on Machine Learning, Big Data and Business Intelligence (MLBDBI)* (pp. 60–66). IEEE.
102. D. P. Dao, H. J. Yang, N. H. Ho, S. Pant, S. H. Kim, G. S. Lee, Survival analysis based on lung tumor segmentation using global context-aware transformer in multimodality. in *2022 26th International Conference on Pattern Recognition (ICPR)* (IEEE, 2022), pp. 5162–5169
103. M. Li, W. Cai, K. Verspoor, S. Pan, X. Liang, X. Chang, Cross-modal clinical graph transformer for ophthalmic report generation. in *Proceedings of the IEEE/CVF Conference on Computer Vision and Pattern Recognition* (2022), pp. 20656–20665
104. L. Liu, X. Su, H. Guo, D. Zhu, A transformer-based medical visual question answering model. in *2022 26th International Conference on Pattern Recognition (ICPR)* (IEEE, 2022), pp. 1712–1718
105. Y. Fang, S. Shi, J. Fang, W. Yin, SPRNet: sitting posture recognition using improved vision transformer. in *2022 International Joint Conference on Neural Networks (IJCNN)* (IEEE, 2022), pp. 1–6
106. S. Mandia, K. Singh, R. Mitharwal, Vision transformer for automatic student engagement estimation. in *2022 IEEE 5th International Conference on Image Processing Applications and Systems (IPAS)* (IEEE, 2022), pp. 1–6
107. S.S. Kusumawardani, S.A.I. Alfarozi, Transformer encoder model for sequential prediction of student performance based on their log activities. IEEE Access **11**, 18960–18971 (2023)
108. J. Lanchantin, T. Wang, V. Ordonez, Y. Qi, General multi-label image classification with transformers. in *Proceedings of the IEEE/CVF Conference on Computer Vision and Pattern Recognition* (2021), pp. 16478–16488
109. D. Tu, X. Min, H. Duan, G. Guo, G. Zhai, W. Shen, End-to-end human-gaze-target detection with transformers. in *2022 IEEE/CVF Conference on Computer Vision and Pattern Recognition (CVPR)* (IEEE, 2022), pp. 2192–2200
110. B. Zhang, P. Cui, End-to-end 3D human pose estimation with transformer. in *2022 26th International Conference on Pattern Recognition (ICPR)* (IEEE, 2022), pp. 4529–4536
111. J. W. Hung, J. R. Lin, L. Y. Zhuang, The evaluation study of the deep learning model transformer in speech translation. in *2021 7th International Conference on Applied System Innovation (ICASI)* (IEEE, 2021), pp. 30–33
112. W. C. Tam, E. Y. Fu, R. Peacock, P. Reneke, J. Wang, J. Li, T. Cleary, Generating synthetic sensor data to facilitate machine learning paradigm for prediction of building fire hazard. Fire Technol. 1–22 (2020)
113. C. Khosla, B. S. Saini, Enhancing performance of deep learning models with different data augmentation techniques: A survey. in *2020 International Conference on Intelligent Engineering and Management (ICIEM)* (IEEE, 2020), pp. 79–85
114. R. Abdrakhmanov, D. Viderman, K. S. Wong, M. Lee, Few-shot learning based on residual neural networks for X-ray image classification. in *2022 IEEE International Conference on Systems, Man, and Cybernetics (SMC)* (IEEE, 2022), pp. 1817–1821
115. F. Zhou, L. Zhang, W. Wei, Z. Bai, Y. Zhang, Meta transfer learning for few-shot hyperspectral image classification. in *2021 IEEE International Geoscience and Remote Sensing Symposium IGARSS* (IEEE, 2021), pp. 3681–3684
116. Y. Kim, D. Kang, Y. Mok, S. Kwon, J. Paik, A review on few-shot learning for medical image segmentation. in *2023 International Conference on Electronics, Information, and Communication (ICEIC)* (IEEE, 2023), pp. 1–3
117. C. Wu, B. Wang, S. Liu, X. Liu, P. Wu, TD-sampler: learning a training difficulty based sampling strategy for few-shot object detection. in *2022 7th International Conference on Cloud Computing and Big Data Analytics (ICCCBDA)* (IEEE, 2022), pp. 275–279
118. M. Yu, X. Guo, J. Yi, S. Chang, S. Potdar, S., Y. Cheng, B. Zhou, Diverse few-shot text classification with multiple metrics. *ArXiv preprint* arXiv:1805.07513 (2018)
119. N. Muthukumar, Few-shot learning text classification in federated environments. In *2021 Smart Technologies, Communication and Robotics (STCR)* (IEEE, 2021), pp. 1–3

Intelligent Fire Safety Design

Artificial Intelligence Powered Building Fire Safety Design Analysis

Yanfu Zeng and Xinyan Huang

Abstract Emerging innovative architectural designs and new material inventions continue challenging building fire safety and calling for new design approaches. This chapter reviews the conventional approaches in building fire safety design and the latest smart design driven by Artificial Intelligence (AI) technologies. Today, fire engineers play a core while iterative role during the whole design process, including drawing, auditing, reviewing, and approving. Although computer-aided tools help conduct smoke management and evacuation analysis, the current design process is very time-consuming with many repetitive works and inevitable human errors. By training with massive data and past designs, AI can interpret the rules, code and patterns of fire safety design, so make the whole design more automatic and reliable with minimum human intervention. By lowering the overall cost of fire safety design and quickly identifying the design limit, the architectural and structural designs have more flexibility. The authority having jurisdiction can also use the AI tool to accelerate the review and approval process. AI will lead a revolution in building fire safety design and analysis achieving safer and more cost-effective solutions.

Keywords Building fire · Deep learning · AI software · Fire modelling · Smart firefighting

1 Introduction

Fire generates massive heat and toxic smoke and threatens human lives. Therefore, fire safety provisions are essential in the building design, including means of egress, fire compartmentation, fire protection systems, etc. To minimize fire-induced loss, many countries and associations have issued fire regulations that specify the fire safety requirements, e.g., the number of staircases, fire alarm zones, and fire-rated

Y. Zeng · X. Huang (✉)
Department of Building Environment and Energy Engineering, The Hong Kong Polytechnic University, Kowloon, Hong Kong
e-mail: xy.huang@polyu.edu.hk

X. Huang and W. C. Tam (eds.), *Intelligent Building Fire Safety and Smart Firefighting*, Digital Innovations in Architecture, Engineering and Construction,
https://doi.org/10.1007/978-3-031-48161-1_5

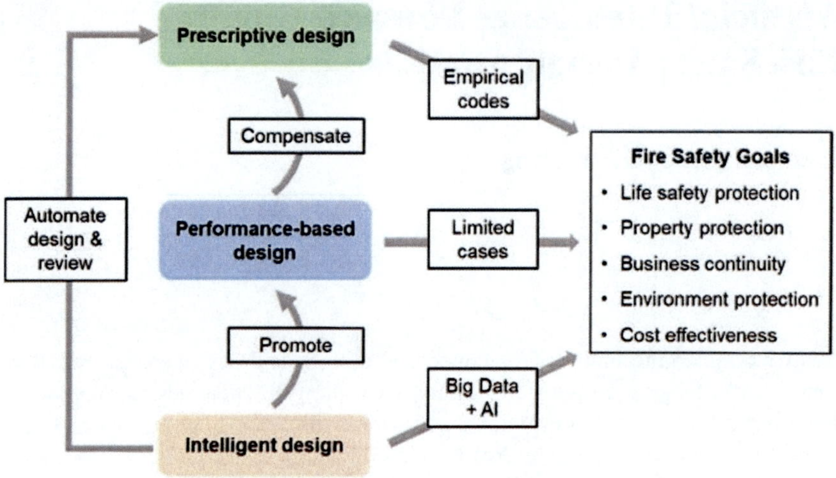

Fig. 1 Evolvement of design approaches for building fire safety

structure elements [1–4]. These codes provide common standards for the building designs and construction processes, and complying these standards and regulations is called the prescriptive design. However, these regulations are highly empirical and do not necessarily provide optimal solutions in terms of safety and cost [5].

With the emerging of many innovative designs and new construction materials, existing regulations and provisions may not meet the new needs of building stakeholders (e.g., owner, designer, and approval bodies). In order to address this issue, performance-based design (PBD) would be used for the non-compliant design issues, e.g. extended travel distance and reduced structural resistance rating, by performing fire engineering analysis, as shown in Fig. 1. PBD encourages innovation in building design without compromising safety and is now widely adopted globally [5].

The initial design process consists of iterations between architects and engineers. The architects provide the initial concept and space design of the building, while the engineers analyze the technical aspects, such as structural stability, fire safety provisions, and mechanical systems. Ideally, architects need to revise the architectural design based on engineering requirements or vice versa until both parties are satisfied with the final design. After that, the authority having jurisdiction (AHJ) will check the code compliance or review the PBD together with an expert panel. Often, one or more rounds of revisions are needed in order to pass the AHJ. While this iteration is critical to achieving the best possible outcome, there are some challenges:

1. **Time-consuming**: The design process can be time-consuming, especially if there are multiple revisions or changes requested by different parties. For complex architectural designs and large infrastructure projects, the design and approval processes often last for several years.
2. **Limited collaboration**: Architects and engineers are expected to work seamlessly to integrate all aspects of the design into a cohesive final product, but it can

be challenging in practice. A lack of effective communication and collaboration often results in errors, misunderstandings, and rework.

3. **Complex design standards**: Building codes and safety regulations can be complex, and designers may struggle to navigate the various requirements and guidelines. Hence, extra efforts for reviewing design are usually requested, further increasing the design expense.
4. **Limited creativity**: While collaborations are essential during the design iteration process, they can sometimes limit the creativity of either party. For example, engineer's technical requirements may constrain architect's artistic vision for the building.

Despite the availability of design-assist tools such as computer-aided design (CAD), building information modelling (BIM), and computational fluid dynamics (CFD) modelling, the current design process still heavily relies on human expertise and experience. Often, people's oversight and misunderstanding lead to safety risk, prolonged design cycles, increased costs, etc.

To facilitate the design process, intelligent and autonomous design process have proposed with rule-based systems and data-driven systems such as machine learning (ML) [6]. The rule-based system is an artificial intelligence (AI) that utilizes the expert knowledge or code to create a set of design rules and guidelines that the system follows. ML algorithms is not pre-programmed with given rules but learns hidden rules from the data. While rule-based systems can be effective in domains where the rules governing the system are well-understood, they can be limited by human knowledge and bias. In contrast, ML-based intelligent systems can adapt and evolve to new situations by analyzing past design experiences, making them capable of handling complex problems beyond human analysis.

Research on applying AI to fire safety issues has been greatly raised in recent years, such as predictive analytics [7–12], automated design assessment [13–15], and AI-driven fire simulation [16, 17]. Potentially, AI can conduct more accurate and efficient fire safety designs, reduce manpower, and lower costs associated with redesigns and errors. Also, AI can provide more insights from massive data, improve collaboration between architects and engineers, and accelerate the review process [18]. Ultimately, the AI-powered design can enhance the overall building fire safety and reduce costs. This chapter introduces the latest intelligent design with potential applications, and each section focuses on a specific task in building fire safety design.

2 Automated Design Drawing

Designers invest a significant amount of their time working with building drawings, i.e., the most essential design document. Architects take the lead in conceptualizing and proposing drawings, e.g., floor plans and sections, to illustrate the basic design ideas and spatial organization of the building. Fire engineers are responsible for adding fire service systems where appropriate to safeguard the building, such as sprinklers, fire detectors, emergency lighting, smoke ventilation, and other elements required by building fire codes (Fig. 2). These engineering drawings show the placement of fire protection equipment throughout the building. They may also indicate the location of fire emergency exits, stairwells, and other escape routes, as well as the fire compartmentation setup. The design drawings could be referred by the cost estimation and the construction guides.

Currently, the preparation of drawings still relies on the designers. In most of the practices, the design is proceeded in 2D based. Architects need to use CAD software to draft required documents including site plans, floor plans, elevations, sections, and details, to reproduce the spatial relationships of the building. Engineers need to add necessary engineering elements manually for each document provided by the architects. However, relying on manual inputs can be prone to human errors, especially when dealing with complex designs. More importantly, the process of generating fire safety drawings using manual efforts can be inefficient and time-consuming. Engineers need to spend a significant amount of time reviewing and updating the various design components by themselves, despite most of the work being repetitive throughout the entire design process.

The application of BIM allows a more advanced and automated design [19]. It creates a 3D building model that integrates all aspects of a project, including architectural, structural, mechanical, fire safety, and other systems. The 2D drawings can be automatically exported once the 3D model is well constructed. However, the construction of a 3D model requires a great effort and a good expertise, which are not equipped by most engineers [20]. Many architects and engineers still use 2D-based software due to its simplicity, familiarity, and cost-effectiveness. Additionally, some building projects may not require the complexity offered by BIM technology, making the 2D-based design more appropriate.

To address the abovementioned issues, AI-based design has been explored in recent years. Starting with the architectural design, researchers found that the AI models can generate façade design by inputting the building framework [21] (Fig. 3). It can also be trained to perform interior design based on the architectural plan [22]. This has paved the way for structural engineers to quickly adopt AI methods and achieve automated design for shear walls by feeding architectural layouts [23, 24]. The result shows that the proposed generative adversarial networks (GAN) model can provide comparable the shear wall design to the engineers, and more works are encouraged to achieve the smart design for other structural elements such as beams [25, 26]. Despite the great capability of AI algorithms, data collection becomes another challenge. The collected design drawings for the research team are limited since the industrial documents are not open access, which greatly restricts the performance of the data-driven AI model. To address this issue, researchers continue to upgrade the model with physical-enhanced [27, 28], knowledge-enhanced [29], and attention-enhanced [30] versions. The main objective is to incorporate AI training with expert domain knowledge in order to achieve better design performance with less training data.

Despite these advancements, it is noteworthy that there has been a lack of application of AI methods in fire engineering drawings (Fig. 2). This presents a significant

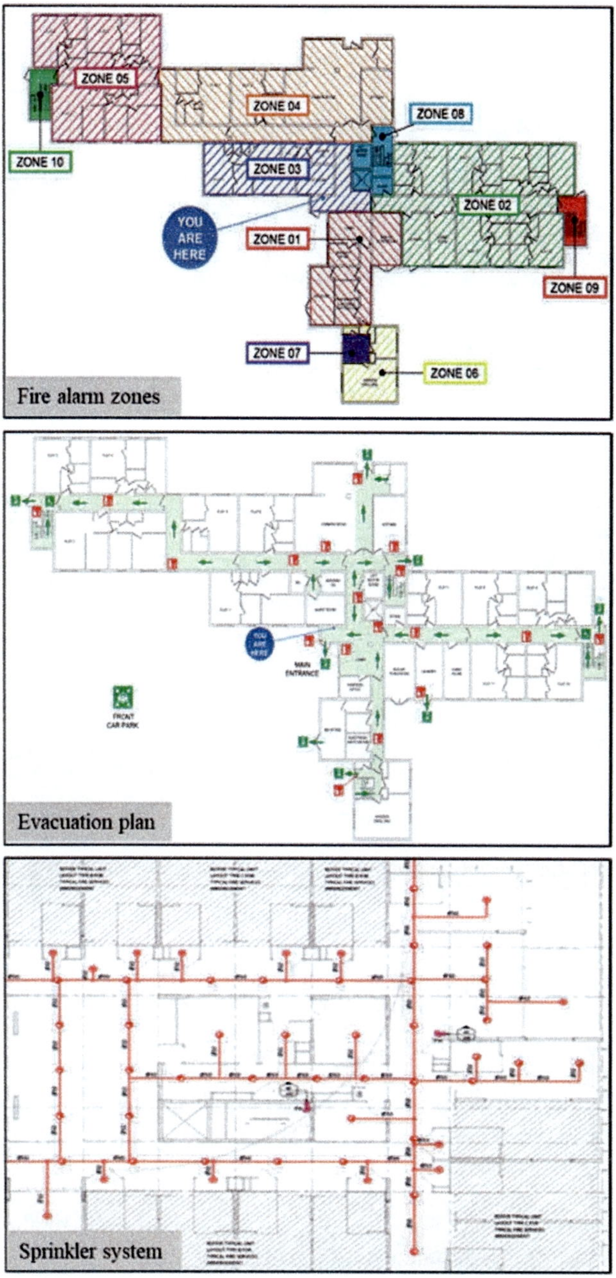

Fig. 2 Examples of fire safety designs

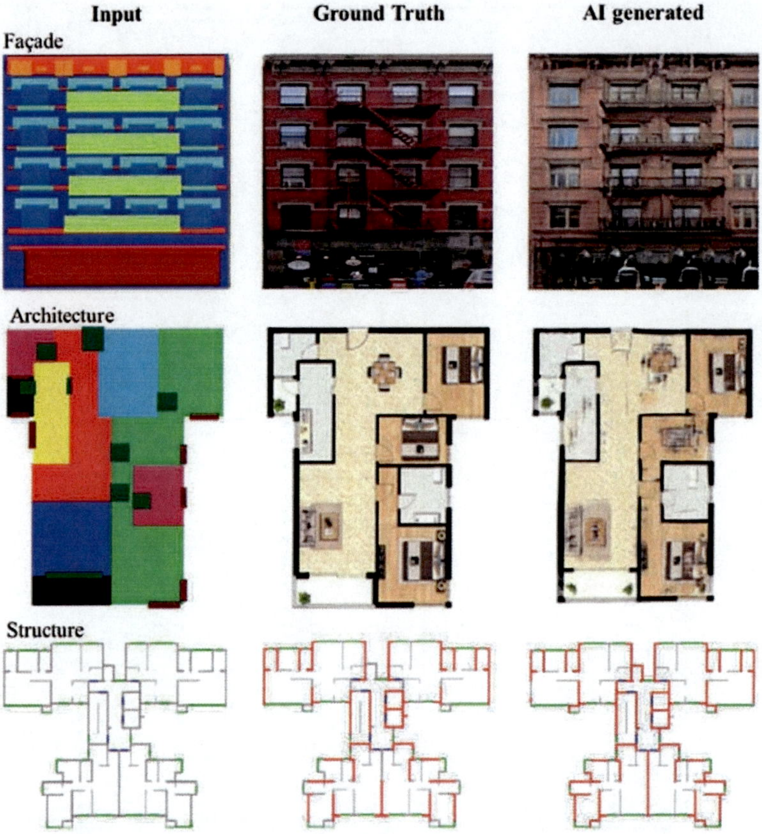

Fig. 3 Examples of AI generated building designs [21–23]

gap in the adoption of AI-based design solutions, which could potentially result in missed opportunities for optimizing fire safety measures in building designs. In fact, compared to the highly flexible architectural drawings, the design pattern behind the fire engineering is more standard. For example, there is a clear requirement for the maximum travel distance of escape route and spacing requirements for sprinklers/ detectors which are followed by the engineers. It means the pattern capturing and learning of the fire safety concerns by AI should be easier, and the design outcome can be more code-compliance.

Nevertheless, generating a code-compliance design drawing is just a baseline, and AI has a greater perspective on pursuing the optimal design. Despite the strict fire regulations imposed on building design, there is still a certain degree of flexibility in the design process. The cost-effectiveness of the design varies depending on the engineer's experience and their familiarity with the codes. In order to achieve an optimal design, engineers need to invest significant time and efforts. However, AI has the potential to identify the most effective design solutions and meanwhile maintaining an acceptable safety level. This can be achieved by enhancing the AI training process with optimal objectives. Here are some expectations for the AI-driven optimal solutions:

- Design an evacuation plan which allows a minimal required safe egress time (RSET) by optimizing the exit location/width, travel distance and occupant number distribution.
- Propose the detector/sprinkler layout with the minimum number of devices to reduce the construction cost.
- Divide the fire compartmentations by maximizing the area usage efficiency and compartment connection.
- Optimize smoke zones and ventilation system with less length of smoke curtain and duct, etc.

AI not only can generate the design drawings efficiently and accurately but also assist engineers in achieving the optimal scheme, making the whole design process more cost-effective. Therefore, the development of the AI-powered fire safety design is very promising and worth to be explored in the near future.

3 Smart Fire Detection Analysis for Complex Layout

Ceiling-mounted fire safety provisions, e.g., fire detectors and sprinklers, play a vital role in safeguarding the building and its occupants during a fire emergency, which are installed in almost every modern building. Their placement of bellowing the ceiling allows for an early response when the smoke ceiling jet arrives, as shown in Fig. 4. In many countries, buildings with sprinkler systems are given design relaxations like allowing for larger building volume/height and longer travel distance [1, 31, 32], which highlights the importance of sprinkler systems in building safety.

The activation analysis of sprinklers/detectors is valuable to assess the fire safety level of a building design. On one hand, sprinklers are expected to control the fire after activation, so the maximum fire size could be considered as the value at which the sprinkler activates. Delayed activation time leads to a more severe fire, resulting in less available time for safe evacuation and vice versa. On the other hand, the activation time of fire detectors is an important factor in assessing the time needed for people to evacuate safely. The shorter the detection time, the sooner the occupants can reach a safe place.

The activation time of the devices greatly depends on the fire-induced ceiling jet properties, i.e., gas temperature and flow velocity. Currently, engineers still rely on simplistic semi-empirical models, such as Alpert's semi-empirical correlations [33], to estimate the ceiling gas temperature and velocity. However, these models are primarily derived from the fire data collected at steady-burning stage with unconfined floor configuration, while most of the building fires vary with time and floorplan,

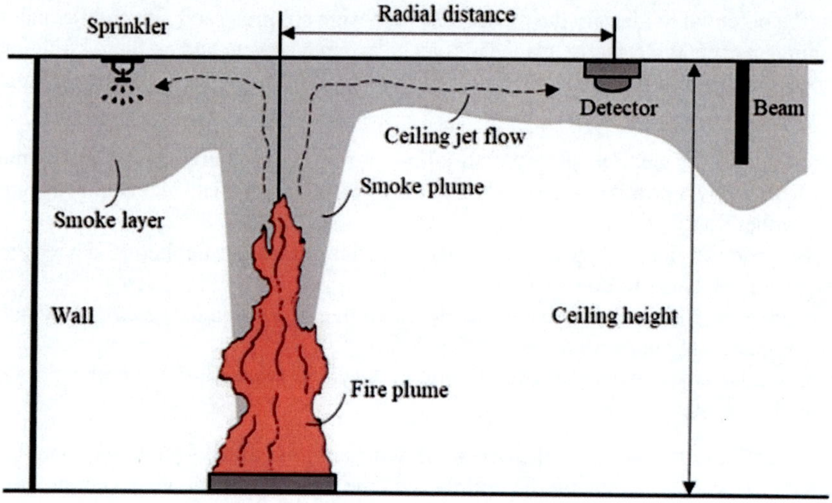

Fig. 4 Illustration of building fire smoke behaviour and detection

have great impacts on ceiling jet properties. The error of predicted detection time by semi-empirical model can be huge.

The development of computational fluid dynamics (CFD) modelling allows a more accurate prediction of the ceiling jet properties. Users can simulate the fire gas fields in a dynamic fire scenario with a specific building layout. However, it is not commonly utilized in design practices due to the significant amount of computing resources required. To model a 20 min fire development, it usually takes over 24 h on a 32-core server, which makes it not cost-effective enough from the design prescriptive.

The first attempt at AI-powered fire detection analysis was made by Zeng and Huang [15]. They proposed an AI model for the quick prediction of the unconfined ceiling jet properties under transient fire scenarios (Fig. 5a). A numerical database is firstly constructed with a total of 108 T-square fire scenarios to collect the fire data of unconfined ceiling jet flow, which is subsequently trained by a Long Short-Term Memory (LSTM) networks. By considering the key attributes of ceiling height, radial distance, radiative fraction, and depth below ceiling, the proposed AI model can provide prediction of the gas temperature and velocity based on a 20-s history of HRRs with an accuracy of 93%. It also has good performance when being applied to steady-burning and decay fires. To make the tool accessible to engineers, they have further developed an AI software, IFETool-Detection, which can provide the detection time result in 30 s. This work demonstrates that AI can provide comparable analysis results to CFD modelling but with less time. Nevertheless, it is still limited to describing the unconfined floor configuration.

To address the impacts of building layout, the authors further develop a GAN deep learning model which enables an accurate and real-time fire detection analysis for buildings with complex floorplans [34]. They first established a numerical fire

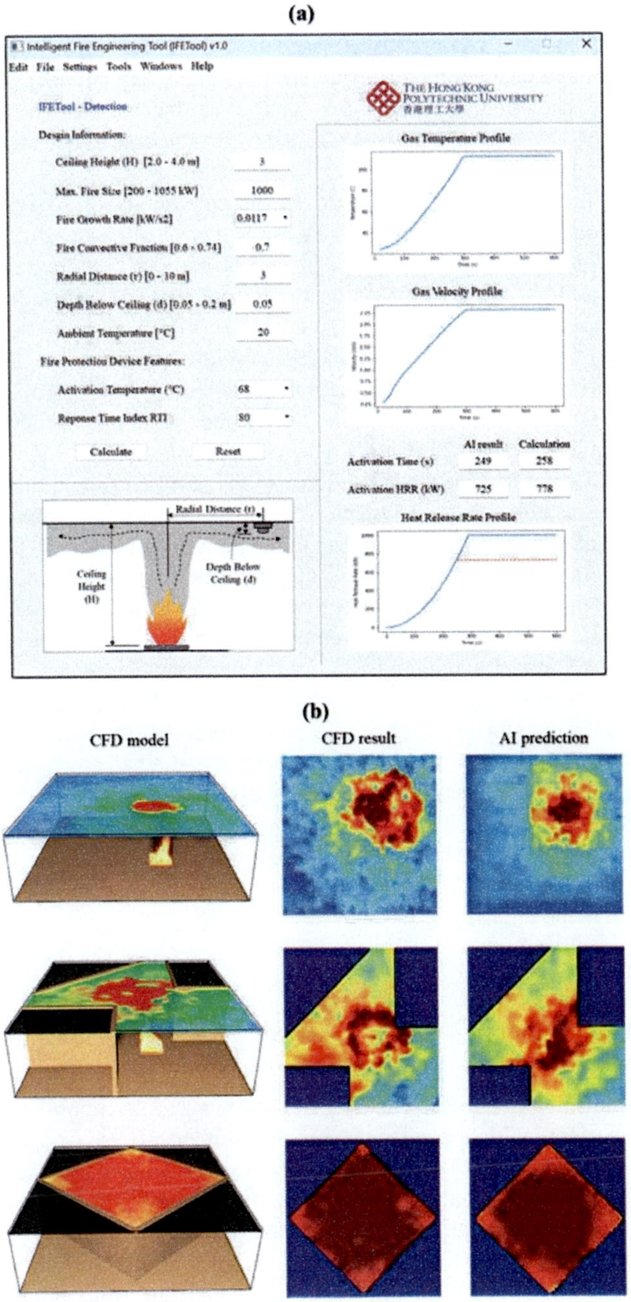

Fig. 5 **a** AI software for calculating sprinkler activation time in unconfined room [15]; **b** AI prediction of temperature field under different building configurations [34]

database comprising hundreds of floor plans, fire locations, and ceiling heights to train the GAN model. The pre-trained model is capable of identifying geometric characteristics and uncovering fire dynamics laws. With any building floorplan and fire detector distribution, the model can forecast ceiling temperature and velocity fields with 86% accuracy within a second, and detect the time of detection with 95% accuracy (Fig. 5b).

The abovementioned works have proven the potentiality of AI applications in promoting a smarter fire detection analysis. Nevertheless, there are still more efforts required before it can meet the practical demands. Firstly, the extension of the database is needed to include more scenario configurations, including both the current variables (e.g., ceiling heights, computational domain) and new attributes (e.g., different fire curves, boundary conditions, ventilation systems). This improvement is relatively easier to achieve because it only requires more simulation efforts while the methodology/algorithm remains the same. The main focus for database extension should be cost-effectiveness. The total case number will have exponential growth when more variables are considered, so the research should find out how to organize the data effectively and enable AI to learn the new pattern with minimal cases instead of blindly adding more data to the training database.

Secondly, the current GAN model only applies to buildings with complex 2D plans since both the input and output of the algorithm are 2D matrices. However, realistic buildings are 3D structures with more complicated spatial relationships. For example, the current GAN model is limited to describing buildings with flat ceilings but has no consideration of the structural components that affect the fire gas behaviour, such as beams, smoke curtains, and false ceilings with holes. If the building has a more complex vertical design, such as a declined ceiling and a balcony in the middle of the building, the fire gas behaviour will also be affected and subsequently change the temperature/velocity fields. To address this issue, not only the database needs to be extended, but also the AI architecture has to be upgraded in order to solve 3D assets. With these contributions, AI-powered smart fire detection analysis can be smoothly implemented in the industry.

4 Fire Evacuation Timeline Analysis

Emerging innovative architectural designs and new material inventions continue challenging the current building fire regulations. Performance-based design (PBD) is a design approach that focuses on achieving specific performance goals related to fire safety rather than simply complying with prescriptive codes and standards [5]. This approach exercises quantitative engineering analysis of the building performance under various fire scenarios and explores the optimal fire protection solutions to meet the desired performance objectives. For most of the buildings, the primary design goal is to protect the life safety of the occupants. It means the provided fire safety provisions, such as mechanical ventilation and exit, should allow all the occupants to escape to a safe place before the building reaches an untenable condition.

In this case, timeline analysis by comparing ASET (available safe egress time) and RSET (required safe egress time) is usually practiced (Fig. 6).

ASET refers to the time available for occupants to safely evacuate a building following the start of a fire, taking into account factors such as the rate of fire spread and the effectiveness of fire protection systems. In contrast, RSET refers to the amount of time required for occupants to safely evacuate a building, contributed by the availability of escape routes and occupant's conditions (e.g., occupant number, age distribution, and daily activities).

To complete a timeline analysis, fire engineers will use computational engineering softwares, e.g., Fire Dynamics Simulator (FDS) [35] and Pathfinder [36], to simulate the designed fire scenarios and estimate its associated ASET and RSET. However, there are several issues associated with this simulation-based approach. Firstly, FDS simulation is a time-consuming and resource-intensive process that can result in inconsistent results due to variations in computation settings and inappropriate model set-ups. It can take several days to simulate a 20-min fire scenario for an atrium [13] (Fig. 7), and the whole design process (analysis, audit, review, and approval) may take several years to complete. Secondly, the value of the simulation result is limited, as analysis is usually conducted at a later design stage with limited modellings, aiming for a design validation rather than optimization. Lastly, the workloads involved are often repetitive and tedious, requiring dedicated engineering analysis for each case, even for similar buildings, and these valuable engineering experience and knowledge are also difficult to be learned by junior engineers [15]. These issues limit the application of fire engineering PBD.

To address these challenges, we can establish a database of visibility profiles for atrium fires using CFD simulations [14]. The fire database should incorporate various scenarios such as different fire HRRs, ventilation conditions, and atrium sizes. The AI model, trained by a transposed convolutional neural network (TCNN), can accurately forecast the smoke visibility profile and associated ASET for different atrium designs (Fig. 8a). Compared to traditional CFD-based PBD carried out by professional fire engineers, the AI method provides more consistent and dependable results in a shorter time.

Then, we can code the AI algorithm into a more user-friendly software, Intelligent Fire Engineering Tool (IFETool) [13] for fire engineering design practice (Fig. 8b). The upgraded AI engine can consider more design variables, and the interested output was extended with gas temperature and CO concentration to meet

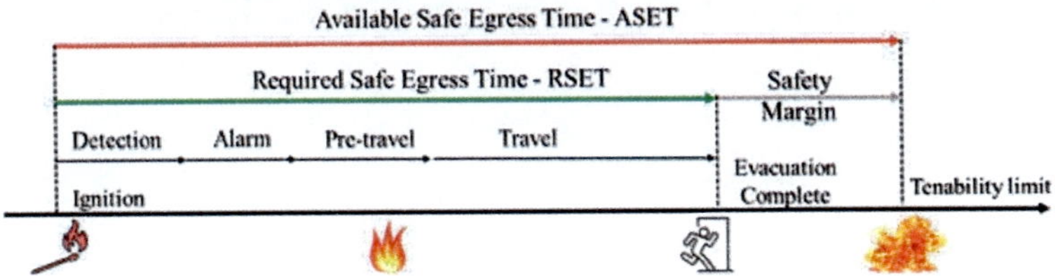

Fig. 6 Timeline analysis for life safety assessment in performance-based design

Fig. 7 Examples of large open space (atrium), typical atrium fire test and simulation

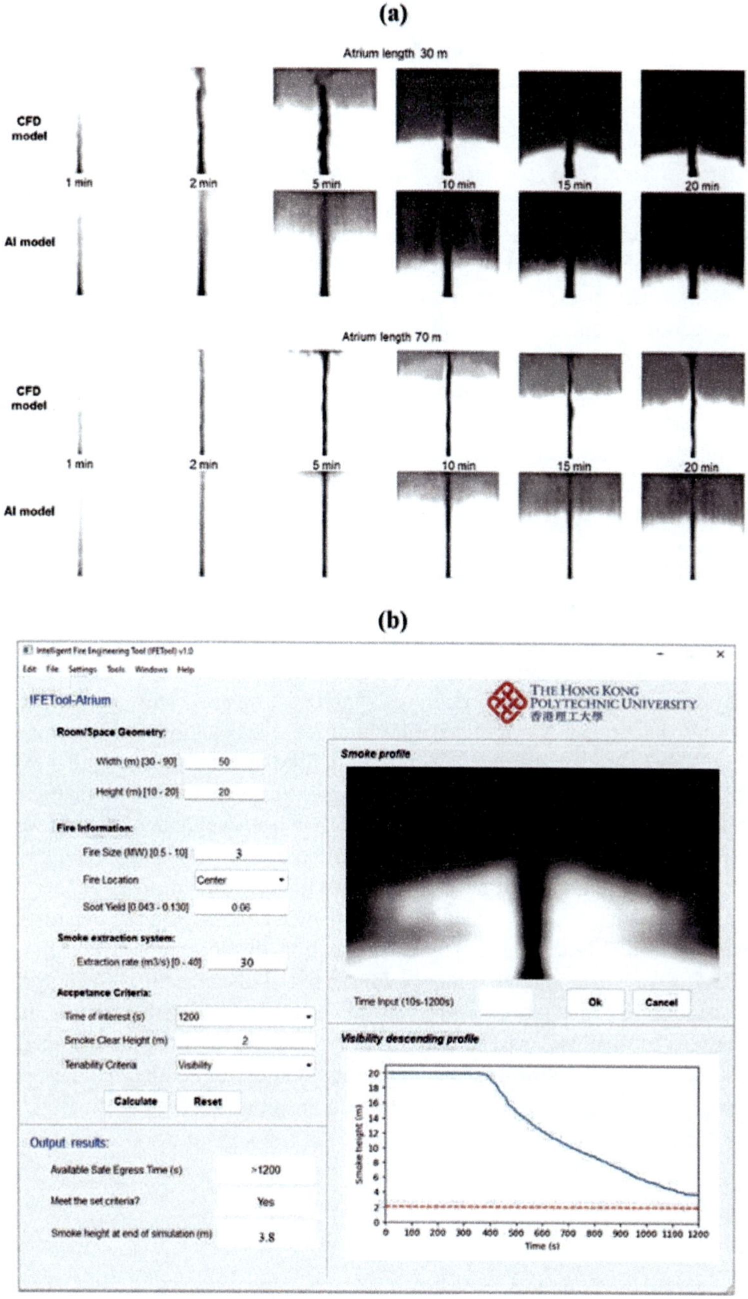

Fig. 8 **a** Comparison between CFD and AI predicted visibility profiles [14]; **b** First AI software (IFETool) incorporating pre-trained AI model to assist ASET assessment [13]

the design demands. The open-access online software (http://ifetool.firelabxy.com/) allows engineers to utilize advanced technology with good accessibility. It has demonstrated the ability to evaluate design safety, identify limiting conditions, and provide optimal solutions to speed up the architectural design review. Nevertheless, this work has almost the same limitations as the smart fire detection analysis mentioned in Sect. 3, i.e., the current model cannot apply to buildings with complex spatial designs, such as those with balconies that obstruct descending smoke layers or unique shapes like pyramids. As discussed before, overcoming these challenges will require more sophisticated data processing methods and deep learning architecture in the future. Moreover, the data processing algorithm can be updated to achieve a smarter identification of the optimal design solutions. In this perspective, the value of ASET can be set as a design objective instead of just an output, and AI is designed to find out the optimal solutions to meet the objective. For example, by inputting an ASET objective of 15 min, AI is expected to find out the minimum required smoke extraction rate for the ventilation system to achieve fire safety compliance under the targeted building.

Compared to the ASET assessment, the improvement of AI application on RSET estimation might be less significant. On one hand, the computational time of pedestrian modelling is much less than CFD modelling, normally ranging from 20 min to hours for single evacuation simulation, and the needed evacuation scenarios are limited. Regardless of where the fire occurs in the building, the evacuation strategy is that all personnel within the building are required to evacuate simultaneously, except for the extremely large-volume buildings which adopt a phased evacuation strategy. Sometimes the engineers might consider additional evacuation scenarios in which one exit is no longer available, considering the fire location is close to that exit, but the RSET of the design can be well determined through limited simulations. On the other hand, unlike CFD modelling which has been validated [37] and verified [38] against various experiments, the fidelity of pedestrian modelling results is often being questioned. It comes from the difficulties of data validation and uncertainties of human behaviors. For example, the occupants will normally find the nearest exit to evacuate in the simulation, but in a realistic situation, the occupants always run to the most familiar exit even though it takes a longer travel distance [39]. To compromise these uncertainties, a safety margin is usually required. Nevertheless, the benefits of AI engine to the RSET assessment are still considerable, imaging that engineers input some basic design information (e.g., total number of people, building layout, and design of escape routes) and can obtain the result in a few seconds.

5 Intelligent Design Review

Design review is a vital step in the current building design process, ensuring the proposed design provisions comply with the required fire regulations or the performance-based objectives. To avoid costly and time-consuming design amendments in later stages of design or even construction, engineers are supposed to enforce the suggested code compliance throughout the design process. Before granting approval for construction to commence, the AHJ may require dedicated code compliance checking by agencies or special design review process, e.g., peer review or expert panel meeting, if a PBD analysis is involved.

However, fire safety is a multidisciplinary field involving various disciplines, including architecture, mechanics, structures, transportation, etc. On one hand, there are numerous codes and regulations involved in the code compliance checking process of the building fire safety design, which makes the current manual process of compliance review to be time-consuming, inefficient, and prone to errors [40]. On the other hand, since the PBD has many design assumptions and quantitative analysis usually by simulation tools and experimental studies, it requires expert knowledge from the reviewers to check the rationality of these assumptions and analyze the results. The AHJ itself usually do not have such expertise, and, thus, they will assign the project to a third-party consultant company to conduct the peer review or organize an expert panel meeting by inviting many fire experts to discuss the feasibility of the design scheme. This process is even longer than the code-compliance checking process, which sometimes takes one to two years to complete.

To improve the code-compliance review process, many research efforts have been made to promote automated compliance checking (ACC) [40–45] (Fig. 9). The first challenge to achieve this goal is the requirement extraction from regulatory textual documents and formalizing them in a computer-processable rule representation, which is used as the baseline for the design review [46]. Developing a rule-based system is so far the most popular method for requirement extraction. For example, Dimyadi et al. [43] proposed a rule-based system to interpret performance-based codes in New Zealand. Delis [42] developed a system, Fire-Code Analyzer (FCA), to identify code violations with the same principle. However, these systems require manual effort to translate paper-based design requirements into computer-recognizable format, which still requires great effort. To address this challenge, the application of AI is of great interest. Specifically, Natural language processing (NLP) technology is proposed to allow computers to process human-like language, which has the potential to reduce the human effort required to develop systematic rules [47, 48].

On the other hand, the rule-based system relies on the system programmer to update built-in rules based on ongoing regulatory amendments, while the codes are revised periodically in response to the emergence of new construction approaches and technologies that come into the industry [43]. Thus, it takes considerable cost to maintain computational versions alongside paper-based versions using rule-based systems. The deep learning algorithms potentially have the ability to create systems which understand the intent of codes and standards as well as automate the translation into a computer-interpretable format [46].

Another restriction of the conventional method is that, after developing the rule-based ACC system, the targeted review document is usually the BIM model of the project instead of commonly used design drawings. The reason is that all the BIM models share the same open data schema, i.e. Industry Foundation Classes (IFC), to document their elements (e.g., wall, door, and floor) [49] (Fig. 10). With the consistent

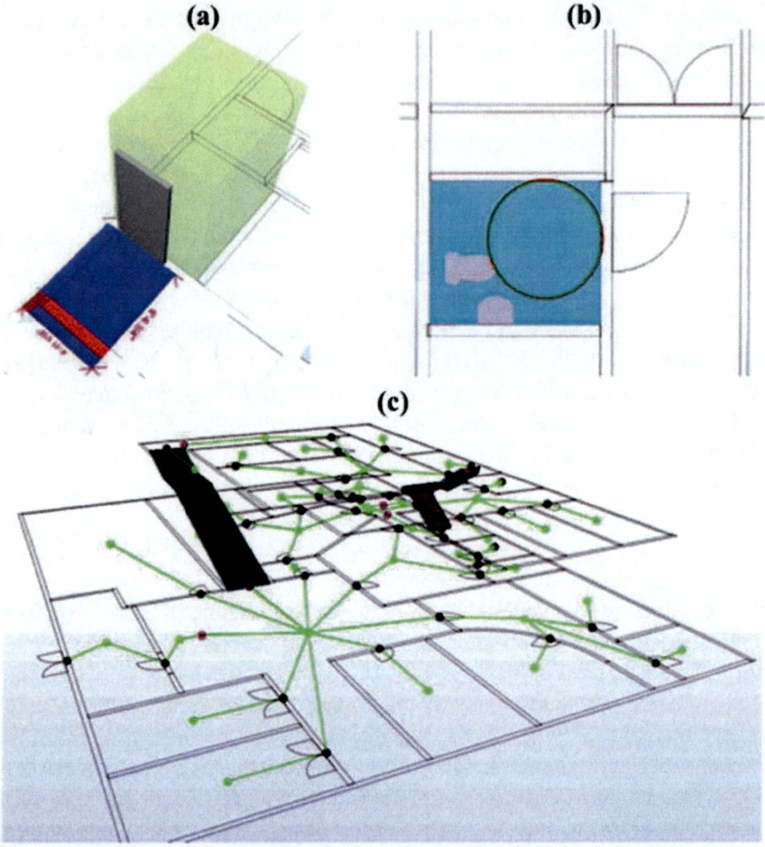

Fig. 9 Examples of code compliance checking: **a** overlap of door swing arcs; **b** wheelchair turning space checking in bathroom; and **c** spatial room connection [44]

standard of data source, the developed ACC system has good applicability for those projects with constructed BIM models. In contrast, the element representation in construction drawings has large variability.

For example, the legend of sprinklers can be different in different countries or even different design companies, as shown in Fig. 11. An experienced engineer can easily recognize if the element is a sprinkler without checking the design specifications. It is because the sprinklers are described in a similar pattern, although there are subtle differences. But for the rule-based system, the legend of sprinkler needs to be pre-programmed, and any difference can fail to recognize the sprinkler. Other fire engineering elements, such as fire doors, lifts, and smoke dampers, have the same recognition issue. Therefore, even if a rule-based system is developed to check the design drawings for one project, it cannot be directly used in another project

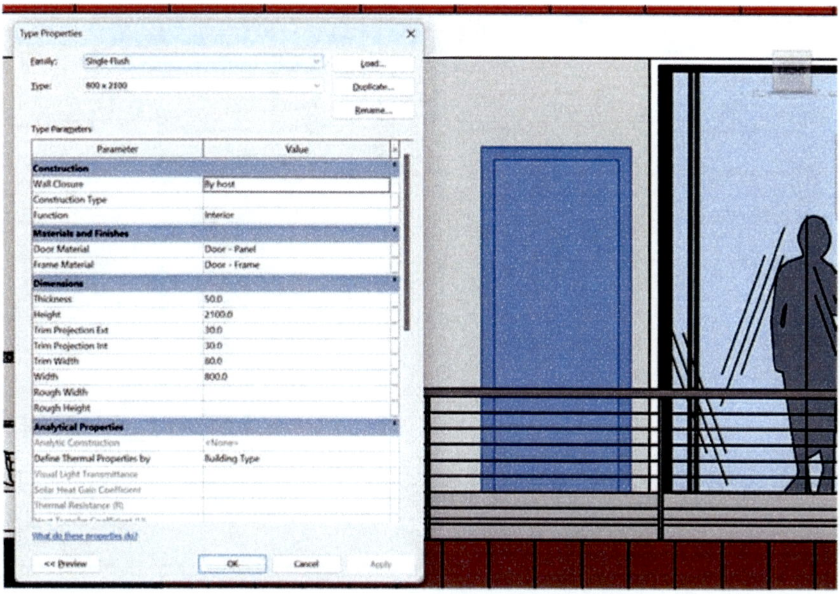

Fig. 10 Standard properties for a door in BIM model

unless all the element encodes are updated. This issue hinders the development of a generalized rule-based system for different projects.

As discussed before, however, BIM model is unavailable in many projects due to its high cost and competence requirements, especially for small-scale buildings with relatively simple designs. AI then becomes a promising tool to address the design review of 2D drawings. By training with different lists of legends, the AI is supposed to capture the overall features for each element, e.g., what should an exit or fire detector looks like, and recognize when it comes to a new one. The excellent identification capability of the AI models could contribute to the intelligent code compliance review for 2D drawings.

It should be noted that the necessity of all these efforts on code compliance checking is that there are unavoidable and considerable non-compliance issues in the error-prone manual design process. In the perspective of future smart design, however, the designs are generated by intelligent AI engines, as discussed in Sect. 2. AI-generated design is more comprehensive and expected with much fewer design issues or ideally no design issues because of its capability to solve complex problems (numerous design considerations from different codes in this case). Therefore, although the contributions to code-compliance checking can benefit the current manual design process, it is less significant to the future intelligent design process.

For the design review of the PBD, the availability of the smart analysis tools, as discussed in Sects. 3 and 4, allows an accurate and consistent engineering analysis result, and they can also be used to facilitate the PBD design review process. By saving

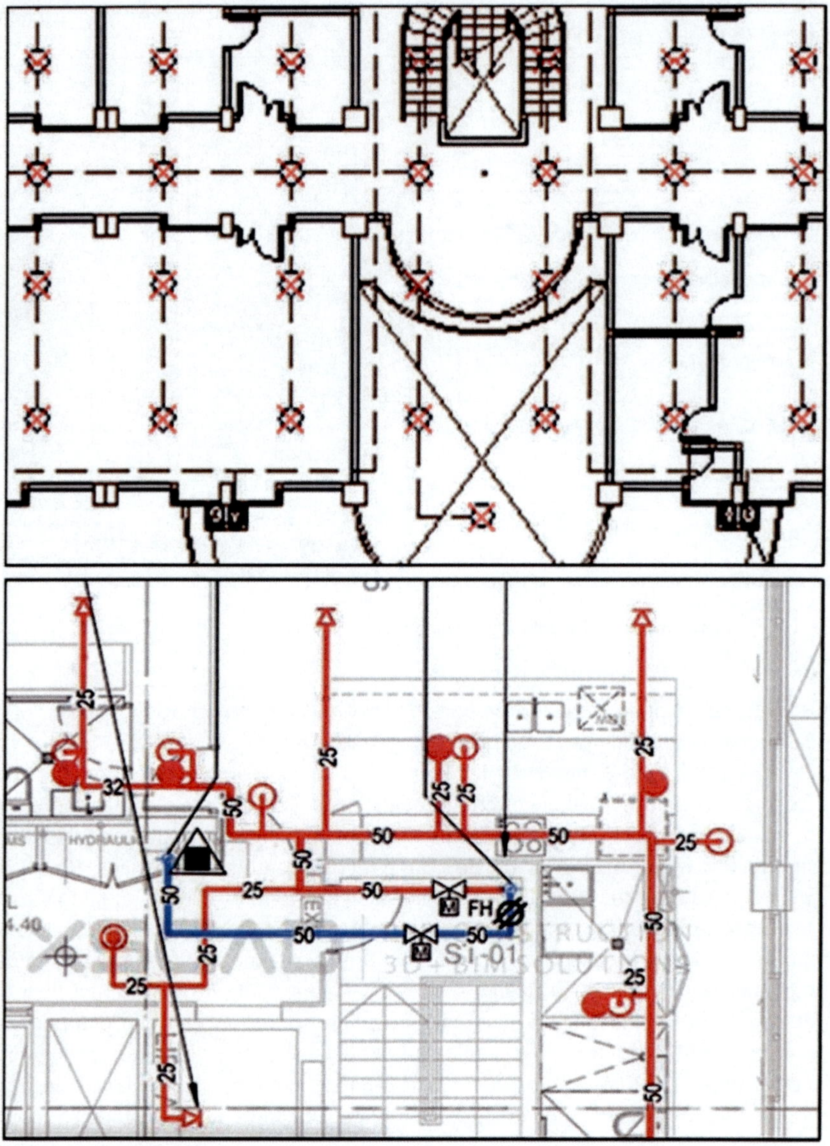

Fig. 11 Examples of sprinkler head legends in different drawings

great efforts on eliminating human errors and reproducing the analysis results, the reviewers or AHJ can focus more on the rationality of design assumptions made in the analysis scheme, such as the simulated soot yields based on the assumed fuel of the fire. This can significantly reduce the cost of the PBD design review and speed up the whole review and approval process.

6 Discussion and Perspectives

Today, fire safety engineers play a critical during the whole building design process, including drawing, auditing, reviewing, and approving. This process is very time-consuming, iterative, and repetitive and inevitably with many human errors. New methods with Artificial Intelligence (AI) have already achieved many advanced developments, but its application in the construction industry is slower. For using AI in fire safety design, it is still at an early stage.

Table 1 summarizes the key features of different fire safety design tasks. As the most time-consuming design task, automation in fire engineering drawing is urgent. The generative adversarial network (GAN) is a promising algorithm that, by training with massive data and past designs, can interpret the rules, code and patterns to accelerate the fire safety design. For the PBD or engineering analysis, the latest developments in the AI-driven prediction of fire detection time and ASET in relatively simple building layout have proven the capability of AI in facilitating the design. The next step is to address buildings with complex spatial designs and floorplans. The benefits of AI-powered design are not only improving the design efficiency and minimizing errors but also identifying the optimal solutions to maximize the building safety and lower the cost. Moreover, AI can support AHJ in the review process and avoid delay in construction projects.

Figure 12 proposes a flow diagram of the intelligent fire-safety design and review process powered by AI engines. By inputting the raw architectural designs, the automated design engine can complete the required fire safety designs, e.g., sprinkler layout, smoke detection distribution, evacuation plan, smoke zone division. This will free engineers from tedious and repetitive drawing tasks and conduct more challenging design tasks, such as estimating safety factor and cost reduction. Ideally, if the engine has well learned all the requirements from regulations, there is no need for AHJ to check the code compliance for the AI-generated design, which significantly shortens the design period.

The code-compliance design cannot meet the demands of all stakeholders. For example, architects want to remove one staircase around the central area, which can lead to an exceeded travel distance, so that fire engineers still need to use the PBD approach to ensure a sufficient safe egress time. With the AI analysis engine, fire engineers can assess the new architectural design quickly and find issues of non-compliance easily. At the same time, the review process by AHJ can also be accelerated with these AI tools, and it eventually becomes automatic and transparent without any human interference.

Table 1 Summarized key features of the different design tasks

Features	Design drawing	Fire detection analysis	Fire evacuation analysis	Design review
Conventional approach	Drew by people	Empirical correlations	CFD and pedestrian modelling	Peer or panel review
Main issues	• Error-prone • Time-consuming	• Over-simplified • large error for real fire	• High computational cost • Time-consuming	• Time-consuming • Lack of transparency
Intelligent approach	• GAN model for automated architectural and structural designs • No ML development in fire safety design	• LSTM model to predict ceiling jet profiles for open-floor conditions • GAN model to predict ceiling jet fields for buildings with complex floorplans	• TCNN model to predict tenability profiles in atrium • First AI software available	• NLP model for code requirement extraction • Rule-based systems for compliance review • No ML development in fire safety design
Main application	Design drawing generation	Activation time identification	ASET assessment	Code-compliance checking
Benefits	Improve design efficiency and accuracy	Accuracy improvement from 86 to 95%	• Real-time simulation • Design limit identification	Improve review efficiency and reduce cost
Limitations	No ML development in fire safety design	• Only applicable to temperature-sensed devices • Cannot predict buildings with complex 3D spatial design	• Limited to describing the cubic atrium • Limited available variables in database	• Rule-based systems are mainly applicable to BIM model • No ML development in fire safety design
Future perspectives	• Automated design for fire service systems • Providing optimal solutions	• Detection time prediction for smoke detectors • Achievable to buildings with complex 3D spatial design	• Achievable to buildings with complex 3D spatial design • Optimal design identification • RSET assessment	• Automated design review for fire service systems • Applicable to 2D design drawings

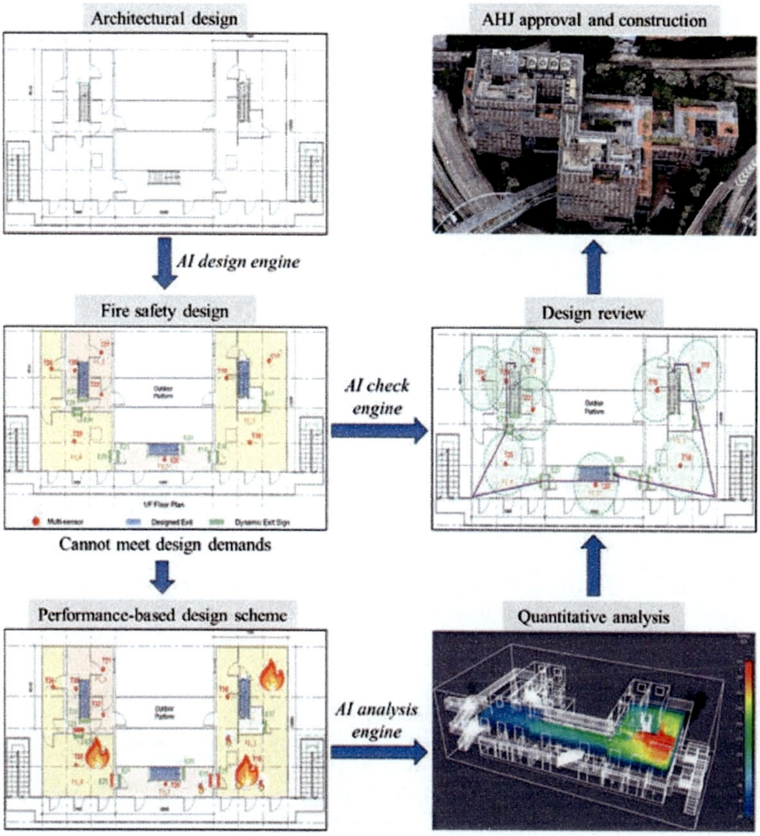

Fig. 12 Perspectives of the future intelligent fire safety design

In future, the workload of manual design can be significantly reduced by the use of AI design tools. This is similar to the transition in late 1990s, when the computer and CAD software completely replaced the human drawings with rules and pencils. Nevertheless, more architects and engineers are needed to design more advanced and complex buildings, as they are freed from low-efficient paper drawing. At the same time, the demand of creativity and problem-solving skill increases significantly, such as using CFD simulations. The use of AI can further free fire engineers from repetitive designs and allow them to conduct more accurate and customized design to further improve the building fire safety and reducing the construction codes, rather than blindly following the code and regulations, defined largely by intuition (if not randomly) a hundred years ago. New-generation of fire engineers will be used to work with AI tools during their university education, training and daily practices, just like current engineers who are all good at computer and CAD. In short, AI will lead a revolution in building fire safety design and analysis, achieving safer and more cost-effective solutions.

Acknowledgements This work is funded by HK Research Grants Council Theme-based Research Scheme (T22-505/19-N).

References

1. International Code Council (ICC). 2015 IBC International Building Code (International Code Council (ICC), 2014)
2. China M of H and U-RD of the PR. GB 50016–2014 (2018) Code for fire protection design for buildings (2018)
3. HM Government. Approved Document B – 2019 Edition including 2020 & 2022 Amendments 2022;6.
4. Building Department. Code of Practice for Fire Safety in Buildings (2011)
5. British Standard Institution, *BS 7974:2019 Application of fire safety engineering principles to the design of buildings - Code of practice* (BSI Standards Limited, London, 2019)
6. C. Grosan, A. Abraham. Intelligent Systems. Springer (2011)
7. W.C. Tam, E.Y. Fu, J. Li, X. Huang, J. Chen, M.X. Huang, A spatial temporal graph neural network model for predicting flashover in arbitrary building floorplans. Eng. Appl. Artif. Intell. **115**, 105258 (2022)
8. T. Zhang, Z. Wang, H.Y. Wong, W.C. Tam, X. Huang, F. Xiao, Real-time forecast of compartment fire and flashover based on deep learning. Fire Saf. J. **130**, 103579 (2022)
9. T. Zhang, Z. Wang, Y. Zeng, X. Wu, X. Huang, F. Xiao, Building Artificial-Intelligence Digital Fire (AID-Fire) system: a real-scale demonstration. J. Build. Eng. **62**, 105363 (2022)
10. Z. Wang, T. Zhang, X. Wu, X. Huang, Predicting transient building fire based on external smoke images and deep learning. J. Build. Eng. **47**, 103823 (2022)
11. Z. Wang, T. Zhang, X. Huang, Predicting real-time fire heat release rate by flame images and deep learning. Proc. Combust. Inst. **39**, 4115–4123 (2023)
12. X. Wu, Y. Park, A. Li, X. Huang, F. Xiao, A. Usmani, Smart Detection of Fire Source in Tunnel Based on the Numerical Database and Artificial Intelligence. Fire Technol. **57**, 657–682 (2021)
13. Y. Zeng, X. Zhang, L.C. Su, X. Wu, X. Huang, Artificial Intelligence tool for fire safety design (IFETool): Demonstration in large open spaces. Case Studies in Thermal Engineering **40**, 102483 (2022)
14. S. L. Su, W. Chu, X. Wu, X. Zhang, X. Huang, Smart performance-based design for building fire safety: Prediction of smoke motion via AI. J. Build. Eng. 43, 102529 (2021)
15. Y. Xeng, X. Huang, Smart building fire safety design driven by artificial intelligence. in M. Z. Naser, editor. Interpretable machine learning for the analysis, design, assessment, and informed decision making for civil infrastructure (Elsevier, New York, 2023)
16. X. Zhang, X. Wu, X. Huang, Smart real-time forecast of transient tunnel fires by a dual-agent deep learning model. Tunnell. Underground Space Technol. 129 (2022)
17. X. Wu, X. Zhang, X. Huang, F. Xiao, A. Usmani, A real-time forecast of tunnel fire based on numerical database and artificial intelligence. Build. Simul. **15**, 511–524 (2022)
18. X. Huang, X. Wu, A. Usmani, Perspectives of using artificial intelligence in building fire safety. In: M. Z. Naser (eds.) Handbook of cognitive and autonomous systems for fire resilient infrastructures (Springer, New York, 2022)
19. C. Eastman, K. Liston, R. Sacks, K. Liston, BIM handbook: a guide to building information modeling for owners, designers, engineers, contractors, and facility managers (2008)
20. A. Aibinu, S. Venkatesh, Status of BIM adoption and the BIM experience of cost consultants in Australia. J. Profess. Issues Eng. Educ. Practice **140**, 1–10 (2014)
21. P. Isola, J. Y. Zhu, T. Zhou, A. A. Efros, Image-to-image translation with conditional adversarial networks. in *Proceedings – 30th IEEE Conference on Computer Vision and Pattern Recognition, CVPR 2017* 2017 (2017), 5967–76

22. W. Huang, H. Zheng, Architectural drawings recognition and generation through machine learning (2018), 156–65
23. W. Liao, X. Lu, Y. Huang, Z. Zheng, Y. Lin, Automated structural design of shear wall residential buildings using generative adversarial networks. Autom. Constr. **132**, 103931 (2021)
24. W. Liao, Y. Huang, Z. Zheng, X. Lu, Intelligent generative structural design method for shear wall building based on "fused-text-image-to-image" generative adversarial networks. Expert Syst. Appl. **210**, 118530 (2022)
25. P. Zhao, W. Liao, Y. Huang, X. Lu, Intelligent beam layout design for frame structure based on graph neural networks. J. Build. Eng. **63**, 105499 (2023)
26. P. Zhao, W. Liao, H. Xue, X. Lu, Intelligent design method for beam and slab of shear wall structure based on deep learning. J. Build. Eng. **57**, 104838 (2022)
27. X. Lu, W. Liao, Y. Zhang, Y. Huang, Intelligent structural design of shear wall residence using physics-enhanced generative adversarial networks. Earthquake Eng. Struct. Dyn. 1–20 (2022)
28. W. Liao, X. Wang, Y. Fei, Y. Huang, L. Xie, X. Lu, Base-isolation design of shear wall structures using physics-rule-co-guided self-supervised generative adversarial networks. Earthquake Eng. Struct. Dyn. 1–23 (2023)
29. Y. Fei, W. Liao, Y. Huang, X. Lu, Knowledge-enhanced generative adversarial networks for schematic design of framed tube structures. Autom. Constr. **144**, 104619 (2022)
30. P. Zhao, W. Liao, Y. Huang, X. Lu, Intelligent design of shear wall layout based on attention-enhanced generative adversarial network. Eng. Struct. **274**, 115170 (2023)
31. Ministry of Housing and Urban-Rural Development of the People's Republic China. GB 51251–2017: Technical standard for smoke management systems in buildings (2017)
32. National Fire Protection Association. NFPA 5000: Building Construction and Safety Code. Boston (2018)
33. R.L. Alpert, Calculation of response time of ceiling-mounted fire detectors. Fire Technol. **8**, 181–195 (1972)
34. Y. Zeng, Y. Li, P. Du, X. Huang, Smart fire detection analysis in complex building floorplans powered by GAN. J. Build. Eng. **79**, 107858 (2023)
35. K. McGrattan, S. Hostikka, R. McDermott, J. Floyd, C. Weinschenk, K. Overhold, Fire dynamics simulator user 's guide (FDS). In *6th ed. National Institute of Standards and Technology (NIST)* (2021)
36. H. Wang, Q. Chen, J. Yan, Z. Yuan, D. Liang, Emergency guidance evacuation in fire scene based on pathfinder. In *2014 7th International Conference on Intelligent Computation Technology and Automation* (2014), 226–30
37. K. McGrattan, S. Hostikka, R. McDermott, J. Floyd, C. Weinschenk, K. Overholt, Fire dynamics simulator technical reference guide, vol. 1 (NIST, 2016)
38. K. McGrattan, S. Hostikka, R. McDermott, J. Floyd, C. Weinschenk, K. Overholt, in *Sixth Edition Fire Dynamics Simulator Technical Reference Guide Volume 1: Verification guide 1*, 1–147 (2015)
39. M. Kinateder, B. Comunale, W.H. Warren, Exit choice in an emergency evacuation scenario is influenced by exit familiarity and neighbor behavior. Saf. Sci. **106**, 170–175 (2018)
40. J. Zhang, N.M. El-gohary, A.M. Asce, Semantic NLP-Based Information Extraction from Construction Regulatory Documents for Automated Compliance Checking **30**, 1–14 (2016)
41. T.H. Beach, Y. Rezgui, H. Li, T. Kasim, A rule-based semantic approach for automated regulatory compliance in the construction sector. Expert Syst. Appl. **42**, 5219–5231 (2015)
42. A. Delis, Automatic fire-code checking using expert-system technology 3801 (2016)
43. J. Dimyadi, C. Clifton, M. Spearpoint, R. Amor, Computer-aided compliance audit to support performance-based fire engineering design. In *Proceedings of 10th International Conference on Performance-Based Codes and Fire Safety Design Methods* (2014)
44. C. Eastman, J. Lee, Y. Jeong, J. Lee, Automation in Construction Automatic rule-based checking of building designs. Autom. Constr. **18**, 1011–1033 (2009)
45. Z. Zheng, Y. Zhou, X. Lu, J. Lin, Automation in construction knowledge-informed semantic alignment and rule interpretation for automated compliance checking. Autom. Constr. **142**, 104524 (2022)

46. R. Amor, J. Dimyadi, The promise of automated compliance checking. Develop. Built Environ. **5**, 100039 (2021)
47. P. J. Tierney, International review of research in open and distributed learning processing and graph theory a qualitative analysis framework using natural language processing and graph theory (2022)
48. J. Zhang, N.M. El-Gohary, Integrating semantic NLP and logic reasoning into a unified system for fully-automated code checking. Autom. Constr. **73**, 45–57 (2017)
49. T. Liebich, Unveiling IFC2x4 – The next generation of OPENBIM. In *Proceedings of CIB W78 Conference* (2010), 124–31

Fire Safety Design of Confined Building Environment with Mechanical Ventilation

Xiao Chen

1 Introduction

Many built environments have confined spaces like nuclear power plants, ship cabins and underground spaces [1–4]. For these special confined spaces, it is difficult to exhaust smoke efficiently through the natural ventilation (see Fig. 1). Thus, the mechanical ventilation is required to control the spreading of smoke and remove the smoke out of the confined spaces. This chapter will provide some theoretical and experiment basis to understand the mechanical ventilation on fire development and smoke control in the confined spaces of buildings.

The smoke control of the special confined space mainly relies on the mechanical smoke exhaust system and the mechanical air supply system (see Fig. 2). When a fire occurs, the smoke control system is activated, and the original smoke flow of the confined spaces will change. Therefore, the law of smoke filling under forced ventilation conditions is of great significance for the study of the fire hazard of the confined spaces. Many scholars have carried out closed space fires under different forced ventilation conditions, and have studied the effect of forced ventilation on fire parameters such as fire source burning rate [5–8], temperature distribution [9–12], and cabin pressure [13–15]. The previous works mentioned above focused on the enclosure fire behaviors in naturally ventilated condition and mainly obtained by numerical simulation or scaled experiments. Large-scale experimental studies on the characteristics of enclosure fires under the mechanical ventilation are not enough and the effect of mechanical ventilation on the enclosure fire has not been fully revealed.

X. Chen (✉)
State Key Laboratory of Fire Science, University of Science and Technology of China, Hefei, China
e-mail: summercx@ustc.edu.cn

X. Huang and W. C. Tam (eds.), *Intelligent Building Fire Safety and Smart Firefighting*, Digital Innovations in Architecture, Engineering and Construction,
https://doi.org/10.1007/978-3-031-48161-1_6

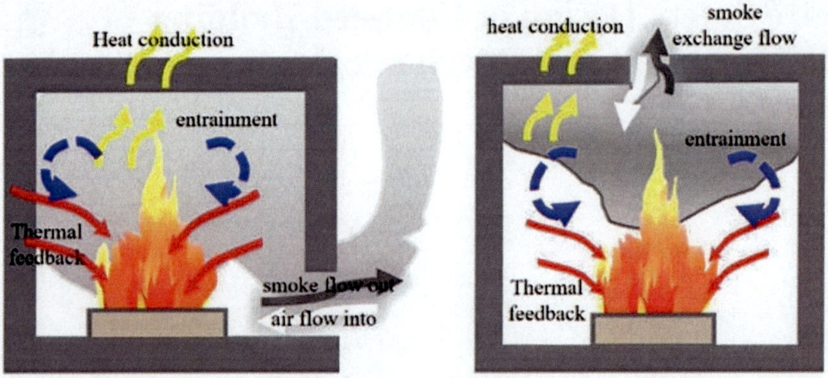

Fig.1 Exhaust smoke through the natural ventilation

Fig. 2 Exhaust smoke through the mechanical ventilation in the confined space

1.1 Fire Burning Under Mechanical Ventilation

The enclosure fire under mechanical ventilation condition was more complex than naturally ventilated condition. Generally, in a mechanically ventilated compartment, the mass loss rate of ignition source depended on two factors [8]: the oxygen concentration and the blowing effect on the oil pool, which was also controlled by the ventilation rate. A large number of mechanical ventilation tests showed that the burning rate increased with the increase in ventilation rate. The heat feedback of the fuel surface of a pool fire was considered to propose a theoretical formulation of the burning rate of fuels for pool fires in a confined compartment [16]. The burning characteristics of thin-layer pool fires with different initial fuel depths inside a confined compartment were investigated and the burning rate was presented the different trends in the free burning atmosphere and confined enclosure [17]. The analysis of scale effect [18] and low-frequency oscillation behavior [19, 20] on the burning rates were also conducted in the case of a fire scenario in a confined mechanically ventilated compartment.

1.2 Temperature Distribution in the Confined Spaces

The temperature distribution of pool fire in enclosures under different ventilated conditions has received much attention in the recent years. Ventilation volume and ventilation configuration could affect the compartment temperature [21], and the results showed that under the condition of low air inlet position, the ventilation rate was 2–3 times greater than the stoichiometric air volume required for fires (here called well-ventilated fires) produce a two-layer temperature distribution, whereas under-ventilated fires produce a single-layer distribution. The high-inlet location perturbs the two-layer temperature distribution of a well-ventilated fire. The mechanical effect of ventilation conditions hardly affected the temperature evolution with a low mechanical air renewal rate of 6.4 ~ 21.3 h^{-1} [22]. The M-Q-H method was also been adopted to predict the smoke layer temperature under mechanical ventilation condition [10]. However, these models assumed the heat loss coefficient for the compartment boundary as constant, which was actually affected by the heat release rate of the compartment fires [23]. The combined effect of ventilation rate and vent position has not been considered.

1.3 Smoke Height Under Mechanical Ventilation

Smoke layer height is one of the important parameters in compartment fire research. Existing calculation methods for fire smoke layer height include the classical N-percentage, integral ratio, intra-variance methods and so on. Some common models have been formed, such as the NFPA 92B model [24], the Milke Mowrer [25] model and the Tanaka [26] model. It has been more than 20 years since Zukoski [27] firstly proposed a simple analytical model of enclosed smoke filling based on energy and mass balance. Many researchers explored the smoke control conditions by experimental and numerical simulation. Increasing the height of the air inlet or reducing the height of the air inlet can destroy the formation of the smoke layer [4]. Based on the conservation equation of oxygen concentration, the oxygen exchange between the upper and lower layers due to convection and turbulent diffusion, especially under mechanical ventilation, the formation and the destruction of smoke layer was investigated by Hayashi [28] in CFD. However, the studies on the smoke layer height under mechanical ventilation conditions are still very limited.

2 Fire Phenomenon in the Confined Spaces with Mechanical Ventilation

2.1 Fires Development

In order to investigate the fire phenomenon in the confined spaces with mechanical ventilation, experiments of six ventilation rates (38 h^{-1}, 51 h^{-1}, 63 h^{-1}. 76 h^{-1}, 102 h^{-1}, 126 h^{-1}) and four air inlet heights (0.1 m, 0.4 m, 0.7 m, 1.0 m) were conducted to obtain the fire phenomenon. The test of no ventilation was also obtained for comparison. Detailed experimental setup could be found in [29].

Figure 3 presents the development of fire in confined spaces with no ventilation and a ventilation rate of 63 h^{-1} at the air inlet height of 0.7 m. The flame shape and combustion time of the test with ventilation was significantly different from the test without ventilation. The fire in the mechanical ventilation (Fig. 1b) ends a lot earlier due to adequate oxygen supply.

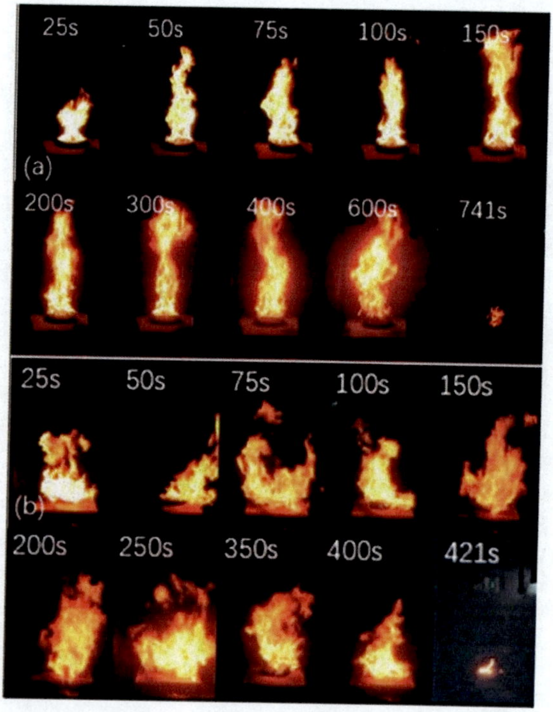

Fig. 3 The fire development in the confined space with **a** no ventilation and **b** a ventilation rate of 63 h^{-1} at the air inlet height of 0.7 m [29]

2.2 Fuel Burning Rate

Burning is a complex process accompanied by chemical reactions consuming both fuel and oxygen. In the special confined space fires under the mechanical ventilation, the heat transfer condition change, leading the complex change of the burning process and burning rate.

In order to quantitatively characterize the fire size of the confined space fire under different ventilation conditions, the mass burning rate in whole burning process was calculated by measuring the mass loss of liquid fuel over time. The mass burning rate varied with the time for different ventilation rates and different air inlet heights was shown in Fig. 4. It was obvious that the combustion process with no ventilation can be roughly divided into four stages: (1) growth stage; (2) steady stage; (3) peak stage; (4) extinguishing stage, which agreed with the steady burning and bulk boiling burning behavior in previous study [30]. However, there was no obvious steady stage in the combustion process with mechanical ventilation, and the combustion rate increased rapidly in the early stage of the fire.

The average values of mass burning rates varied with the ventilation rates for different air inlet heights were shown in Fig. 5. For the tests with ventilation rate ranging from 38 to 126 h^{-1}, the average mass burning rate first decreased and then increased with the ventilation rate. The mass burning rate of a 0.4 m square pool fire was mainly controlled by radiation feedback in still air. With a ventilation, the fuel vapor and air were well mixed as well as the radiation blockage effect would be reduced by the air flow, which led to the increase of mass burning rate. When the ventilation rate increased, the wind would cool the windward side of the pool, tilt the flame and lead to the decrease of radiation feedback to fuel surface, which would result in the decrease of mass burning rate. The dominant heat feedback mechanism is convection-controlled [31]. The second increase of mass burning rate resulted from the increasing ventilation rate was due to the enhanced convection, as explained by

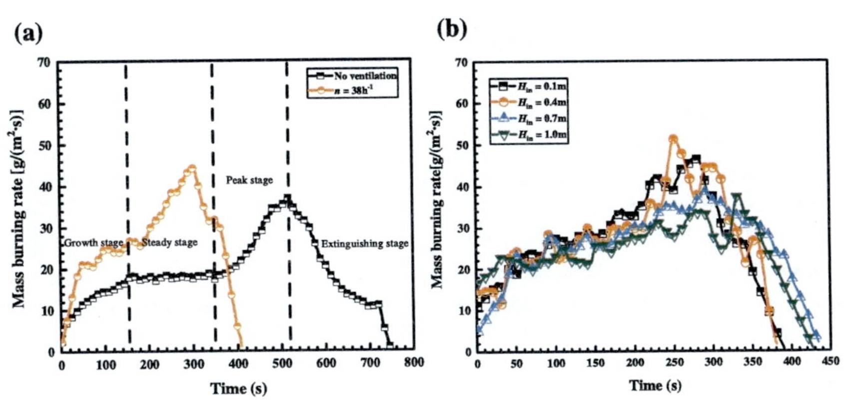

Fig. 4 Mass burning rates for **a** different ventilation rates (n) with $H_{in} = 0.7$ m and **b** different air inlet heights (H_{in}) with n = 63 h^{-1} [29]

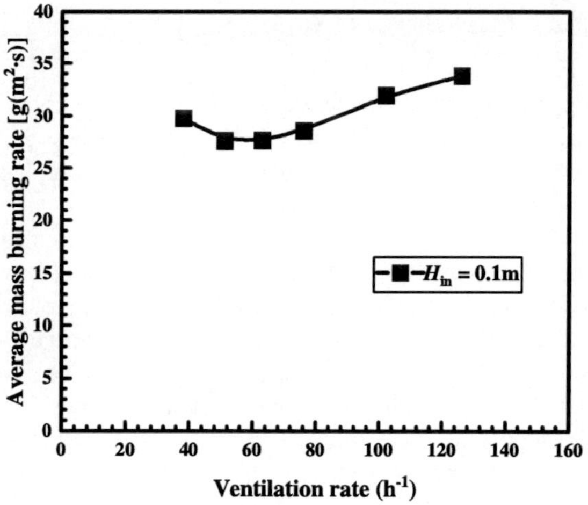

Fig. 5 Average mass burning rates varied with the ventilation rates (n)

the stagnant layer theory solution given by Hu et al. [32], as well as the enhanced conduction from the heating of the leeward side walls by the inclined flame [33].

Considering the linear relationship of average mass burning rate and ventilation rate, the enhancement rate of burning rate, R_n, was calculated by dividing the burning rate increment by ventilation rate, *n*, as,

$$R_n = \left(\dot{m}_n'' - \dot{m}_0''\right)/n \tag{1}$$

where $\dot{m}_0''$ was the average mass loss rate of no ventilation test and $\dot{m}_n''$ was the average mass burning rate of the test with the ventilation rate n.

The enhancement rates of burning rate varied with the ventilation rates were shown in Fig. 6. An exponential fitting of the enhancement rate of burning rate and ventilation rate was obtained as,

$$R_n = 428.99 + e^{\frac{(115-n)}{12.34}} \tag{2}$$

The dimensionless mass loss rate $\dot{m}^*$ can be defined as $\dot{m}^* = \dot{m}/\dot{m}_{th}$, where $\dot{m}$ is experimental mass loss rate, $\dot{m}_{th}$ is the mass loss rate of the oil pool of diameter D in the free atmosphere, $\dot{m}_{th} = \dot{m}_\infty''(1 - e^{-k\beta D})A_f$. The number of air changes N can be defined as $N = 3600V_s/V_c$, where V_s is the supply air volume, V_c is the cabin volume. By introducing a dimensionless length $f(D, V_c) = D \cdot {V_c}^{-1/3}$, the relationship between the dimensionless mass loss rate and the number of air changes and previous research results can be drawn on a graph, as shown in Fig. 7. It can be seen from the Fig. 7, under the conditions of the experimental cabin and various ventilation configurations, that the fire source mass loss rate and the supplementary air volume have the following relationship:

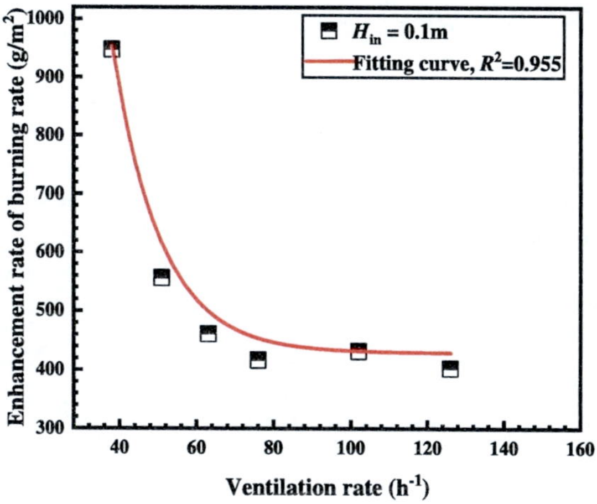

Fig. 6 Enhancement rates of burning rates varied with the ventilation rates (n)

$$\dot{m}^* f(D, V_c) = -0.21exp\left(-\frac{N}{36 \times 3.25}\right) + 0.24 \quad (3)$$

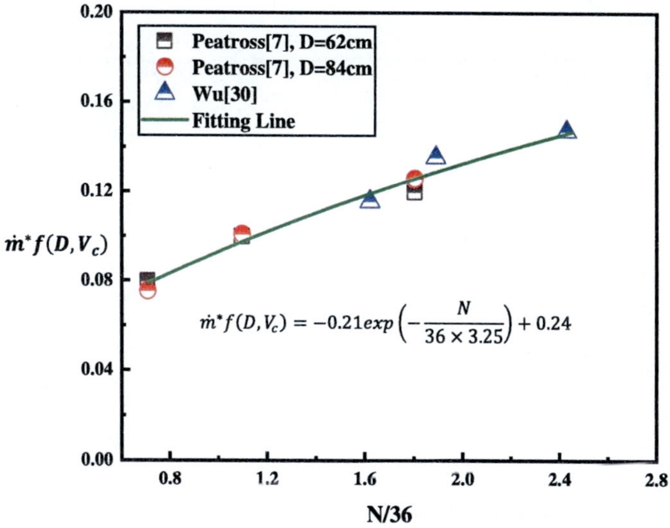

Fig.7 Relationship between dimensionless mass loss rate and dimensionless air volume [34]

2.3 Temperature Profiles

The temperature histories in the period from ignition to extinguishment for the confined space fires at different heights were recorded in Fig. 8. The temperature close to the pool fire could be used to characterize the most serious thermal damage of the confined space. The temperature histories of fires were mainly affected by the smoke produced from liquid pool fire.

In order to characterize the temperature distribution of the confined spaces under the mechanical ventilation, the time-averaged temperatures at different heights and positions were processed as,

$$\bar{T}_i(H_j) = \frac{1}{t_{ex}} \int_0^{t_{ex}} T_i(H_j, t)dt \tag{4}$$

where $\bar{T}_i(H_j)$ was the temperature measured by thermocouple at the position i and the height of H_j, t_{ex} was the extinguishing time.

The time-averaged vertical temperature distributions close to the pool fire for different ventilation rates and air inlet heights are illustrated in Fig. 9. It should be noted that the temperature at some positions was deviated from the consistent trend, which may be the result of the inclined flame directly contacting the thermocouple under the mechanical ventilation. Generally speaking, the vertical temperature distribution of confined space fires under the mechanical ventilation was presented to be

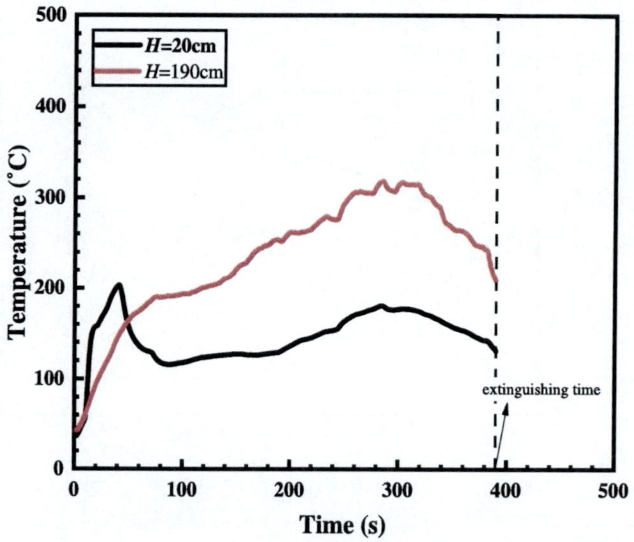

Fig. 8 Temperature histories at different vertical heights close to the pool fire for n = $63h^{-1}$

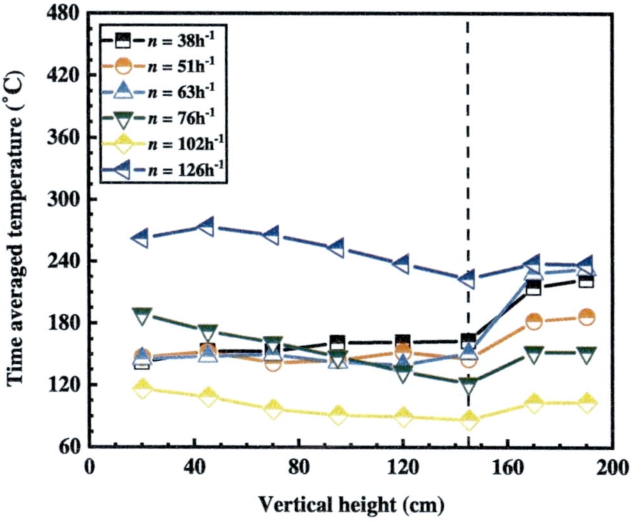

Fig. 9 Time averaged temperatures varied with vertical heights close to the pool fire for different ventilation rates (n) (H_{in} = 0.1 m) [29]

a single-zone distribution, which was consistent with the results obtained in the previous small-scale studies [18, 28].

In order to further investigate the effects of mechanical ventilation on the temperature characteristics of confined space fires, the time-averaged temperature at different positions was processed as,

$$\bar{T}_i = \frac{1}{n}\sum_{j=1}^{n} \bar{T}_i(H_j) \tag{5}$$

As shown in Fig. 10, the air inlet height had little effect on the average temperature of the entire confined space. With the increase of ventilation rate, the average temperature of the entire confined space obviously decreased. Therefore, the smoke control effect of confined space fire can be improved by adjusting the ventilation rate. For larger ventilation rate, the cooling effect of ventilation was greater than the combustion intensification effect, the average temperature of the entire confined space decreased with the increase of ventilation rate.

2.4 Smoke Height

In fire research, the common methods for solving smoke layer height mainly include: the extinction method, N percentage method, minimum integral ratio method and piecewise linear method. Among them, the minimum integral ratio method and the

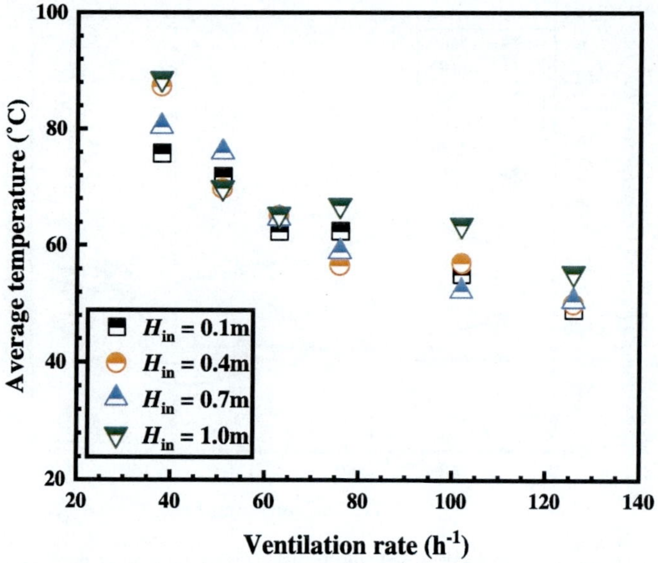

Fig. 10 Average temperature of the entire confined space varied with the ventilation rates (n) for different air inlet heights (H_{in}) [29]

piecewise linear method are used to process the temperature data. The minimum integral ratio method has a relatively clear definition of the neutral plane and a more rigorous calculation and derivation process, and the calculation results are more convincing. It defines the height of the smoke layer as the position where the sum of the minimum integral ratios of the upper and lower layers takes the minimum value. In the calculation, the integral ratios of the upper hot smoke layer and the lower air layer at time t are, respectively:

$$r_u = \frac{1}{(H-z)^2} \cdot \int_z^H T(z)dz \cdot \int_z^H \frac{1}{T(z)}dz \tag{6}$$

$$r_l = \frac{1}{z^2} \cdot \int_0^z T(z)dz \cdot \int_0^z \frac{1}{T(z)}dz \tag{7}$$

Then, the height z of the smoke layer at time t should satisfy:

$$r_t = \min(r_u + r_l) \tag{8}$$

It can be seen from the Fig. 11 that within the range of air volume and air supply port height, with the gradual increase in the forced air supply volume, the height of the smoke layer also gradually increases; with the increase in the air supply inlet height, the height of the smoke layer gradually decreases (Fig. 12).

Based on the above results, the empirical formula for the calculation of the dimensionless smoke layer height under the experimental conditions can be given:

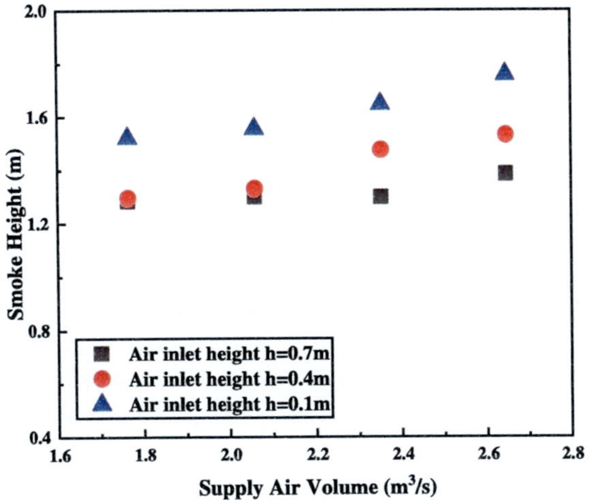

Fig.11 Comparison of cabin smoke layer heights under different supply air volume conditions [34]

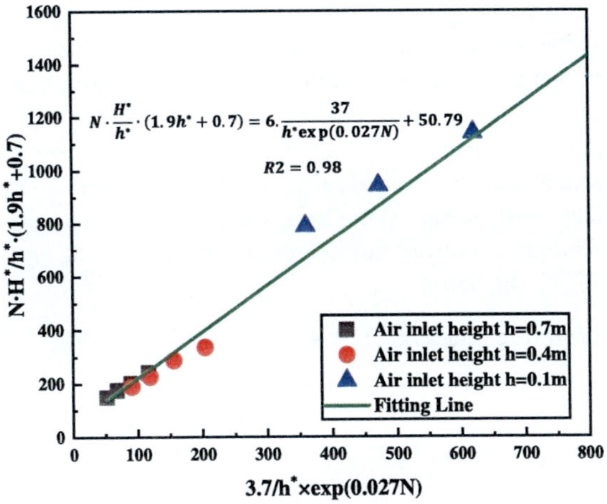

Fig.12 The relationship between the dimensionless smoke layer height and the dimensionless air supply volume [34]

$$H^* = \frac{6.37 \times exp(0.028N) + 50.79h^*}{(1.9h^* + 0.7) \times N} \tag{9}$$

3 Theoretical Analysis of Smoke in the Confined Spaces

3.1 Mass Balance Analysis

Previous work usually regarded the ignition source as a point source heat source in the study of confined space smoke filling. In actual confined enclosure fire, there is no clear boundary between the smoke layer and the air layer. Based on the theoretical analysis, the transition layer in the smoke filling process is usually ignored, and only the upper smoke layer and the lower air layer are considered. In addition, the effect of mechanical air supply on smoke deposition is mainly reflected in the following two aspects: the promotion of mechanical air supply to the fire source; the mechanical air supply promotes the mass mixing of the smoke layer and the lower air layer.

A physical model is assumed that the communication between the interior of the model and the outside world is only through the air supply vents and exhaust vents, and there are no other openings. If the smoke is used as the control body, as shown in Fig. 13, the control body always obeys the mass conservation equation during the smoke settling process. There are three main aspects in the process of quality change in the control body: (1) the smoke generated at the fire source enters the control body to increase the quality of the smoke in the control body; (2) the fresh gas provided by the mechanical air supply enters the smoke layer; (3) the upper mechanical exhaust port removes part of the smoke. The main energy transfer process in the confined spaces can be assumed to be that the heat generated by the fire source is divided into three parts: a part of the heat is used to heat the gas in the cabin; a part of the heat is dissipated outward through the wall; a part of the heat is dissipated outward through the smoke outlet.

Therefore, mass balance equation can be expressed as:

$$\dot{m}_p + \dot{m}_b - \dot{m}_e = \frac{d\rho_s V_u}{dt} \tag{10}$$

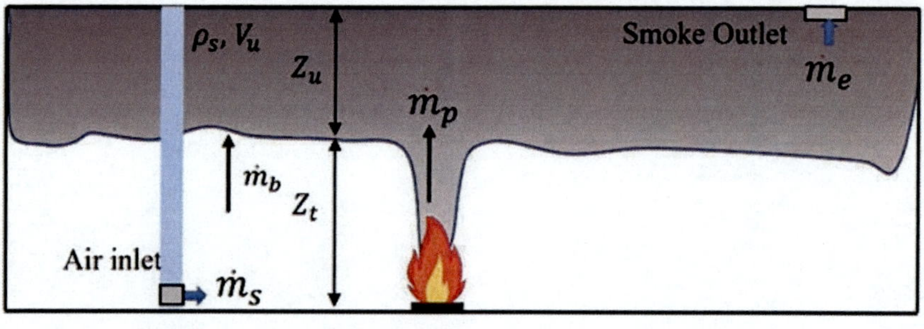

Fig.13 Schematic diagram of the smoke stratification in the confined space

where $\dot{m}_p$: the mass flow rate of the smoke entering the upper control body from the fire source area; $\dot{m}_b$: the mass flow rate of the air entering the smoke layer from other areas; $\dot{m}_e$: the mass flow rate of the smoke discharged by the mechanical exhaust; ρ_s: the smoke density, which is assumed to be constant and has the same value with air; V_u: the volume of the smoke layer. The volume change rate dV_u of the control body is the volume of the micro-element above the control body. Since the thickness dz of the micro-element is infinitely small, $dV_u = A_z dZ_u$, where Z_u is the smoke layer thickness, A_z is the cross-sectional area of the cabin at the lower boundary of the smoke layer.

During the combustion process, the hot smoke entrains the surrounding air during the upward movement to form a plume. In this model, Zukoski's axisymmetric plume model is mainly used, as shown in the following formula.

$$\dot{m}_p = 0.21\left(\frac{\rho_0 g}{C_P T_0}\right)^{1/3} \dot{Q}_C^{1/3} Z^{5/3} = C\dot{Q}_C^{1/3} Z^{5/3} \tag{11}$$

Based on previous studies, mechanical ventilation has a promoting effect on the power of the fire source. We can assume that the mass loss rate $\dot{m}$ of the fire source can be written as:

$$\dot{m} = f(V_s) \tag{12}$$

The fresh gas provided by the mechanical ventilation is not only entrained by the fire source, but the gas in other areas will be mixed with the upper layer smoke due to the gas flow. It is assumed that the proportion of the gas mixed with the upper layer smoke in the mechanical supply air volume is ψ. For different air supply heights, ψ is different; thus, it can be assumed that $\psi = g(h)$, then:

$$\dot{m}_b = \psi \dot{m}_s = g(h)\dot{m}_s = g(h)\rho_s V_s \tag{13}$$

During the process of smoke filling, the actual cross-sectional area of the cabin A_z is consistent. We introduce the height Z_t of the smoke layer, we obtain:

$$-\frac{dZ_t}{dt} = \frac{dZ_u}{dt} = \frac{C(0.7\chi \cdot f(V_s) \cdot \Delta H_c)^{1/3} {Z_t}^{5/3} + g(h)\rho_s V_s - \dot{m}_e}{A_z \rho_s} \tag{14}$$

Based on the results of the experiment and the above theoretical analysis, the results of the smoke layer height prediction are given below. Based on the results obtained from the dimensionless mass loss rate, namely, Eqs. (1) and (2), the fire source power can be written as:

$$\dot{Q}_c = 0.7\chi \cdot (-204.54 \cdot e^{-V_s/0.37} + 7.11) \cdot \Delta H_c \tag{15}$$

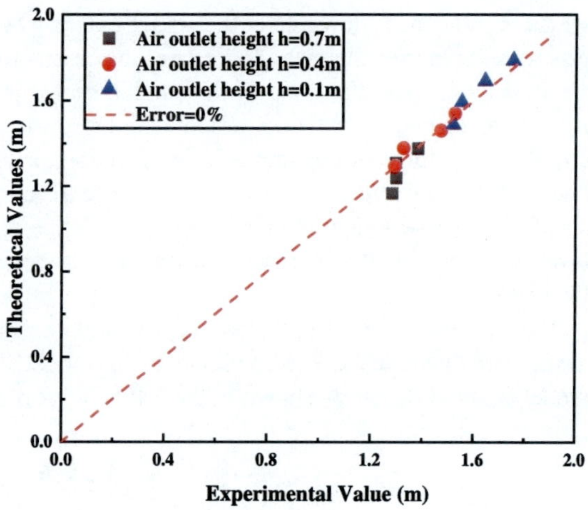

Fig.14 Comparison of predicted and experimental values of smoke layer height under different supply air volume conditions

The above formula can be applied to the fire source power under different air volume conditions to determine the height of the smoke layer. Formula (14) can be written as:

$$\frac{dZ_t}{dt} = -\frac{C[0.7\chi \cdot (-204.54 \cdot e^{-\frac{V_s}{0.37}} + 7.11) \cdot \Delta H_c]^{1/3} z^{\frac{5}{3}} + \psi V_s \rho_a - V_e \rho_s}{A_z \rho_s} \quad (16)$$

where C: according to Zukoski's model can be calculated, C = 0.076432; ψ: the mixing rate, a function of the air supply port height h, $g(h) = -0.25e^{-\frac{h}{0.64}} + 0.76$.

It is found that results in Fig. 14 within the range of the experimental air volume, the theoretical prediction results of smoke layer height are consistent with the experimental results. It can be considered that the theoretical prediction model given here has a good prediction effect.

3.2 *Thermal Balance Analysis*

Mechanical ventilation has two opposite heat transfer effects on confined space fires, combustion intensification effect and cooling effect. The combustion intensification effect will increase heat production by increasing the combustion rate or mass loss rate, while the cooling effect will increase the heat dispersion by sending in cooler fresh air and expelling hot smoke. In order to determine the effect of ventilation rate on the temperature characteristics of the confined space fires, we analyzed the heat production and dispersion of the confined space fires under mechanical ventilation.

In the confined space fires under mechanical ventilation, the heat of combustion is mainly used to: (a) heat the gas in the cabin; (b) dissipate heat through smoke exhaust; (c) dissipate heat through the bulkhead. Therefore, the conservation of heat of the confined space fires can be expressed as:

$$\dot{Q} = \dot{Q}_g + \dot{Q}_e + \dot{Q}_w \tag{17}$$

where $\dot{Q}$ is the heat release rate of pool fires; $\dot{Q}_g$ is the gas heating rate; $\dot{Q}_e$ and $\dot{Q}_w$ are the heat dispersion rates of smoke exhaust and bulkhead.

Previous studies show that the confined space fire temperature under mechanical ventilation can be characterized by a single zone model and the mass flow rate of smoke exhausted from the confined space can be assumed to be equal to the mass flow rate of fresh air. Thus,

$$T_g - T_{in} = T_{ex} - T_{in} \tag{18}$$

$$\dot{m}_{ex} = \dot{m}_{in} = \rho_{in} V \cdot n \tag{19}$$

where $\dot{m}_{ex}$ is the mass flow rate of smoke exhausted from the confined space; T_{ex} is the temperature of the smoke exhausted from the confined space; T_{in} is the temperature of the fresh air supplied the confined space.

The simplified conservation equation of heat under the quasi-steady state condition can be obtained by

$$\dot{Q} = \left(c_p \dot{m}_{ex} + h_T A_T\right)\left(T_g - T_{in}\right) \tag{20}$$

where h_T is the integrated heat transfer coefficient of bulkhead; A_T is the total superficial area of bulkhead.

For pool fires in a confined space under the mechanical ventilation, the oxygen level is sufficient during the whole burning process, and the combustion is a fuel-controlled burning. The heat release rate can be obtained by $\dot{Q} = \varphi A_f \dot{m}_n'' \Delta H_c$, where φ is the average combustion efficiency, φ = 1 for fuel-controlled burning; △Hc is the heat of combustion of fuel. The average mass burning rate with mechanical ventilation is significantly different from the mass burning rate with natural ventilation or no ventilation, and can be expressed as,

$$\dot{m}_n'' = \dot{m}_0'' + R_n n \tag{21}$$

Therefore, it is important to estimate the integrated heat transfer coefficient under mechanical ventilation before establishing the prediction model for average temperature of the confined space fires. Based on the Equations above, the integrated heat transfer coefficient can be calculated by

$$h_T = 1/A_T\left[\dot{Q}/\left(T_g - T_{in}\right) - c_p \rho_{in} V n\right] \tag{22}$$

Figure 15 presented the relation of integrated heat transfer coefficient of bulkhead and the ventilation rate. For most of ventilation condition, the integrated heat transfer

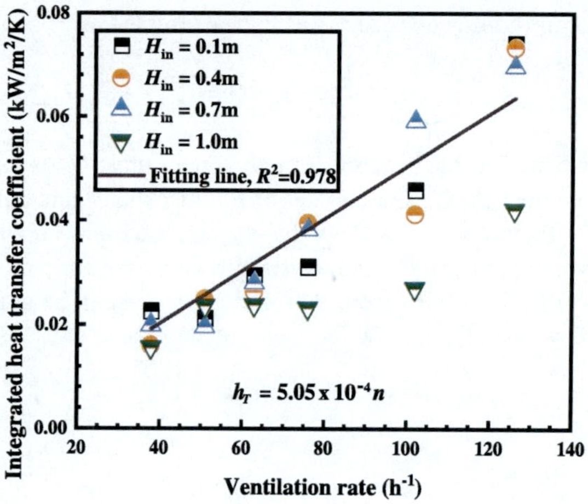

Fig. 15 Integrated heat transfer coefficients varied with the ventilation rates (n) for different air inlet heights (H_{in}) [29]

coefficient of bulkhead was linearly related to the ventilation rate. The actual integrated heat transfer coefficient of bulkhead deviated from Eq. (22) and decreased to a lower value, which supposed to be the calculated value of a lower ventilation rate. This indicated that $H_{in} = 1.0$ m was too high to obtain good cooling effect for the entire confined space (Fig. 15).

After estimating the integrated heat transfer coefficient, the average temperature of the confined space fires under the quasi-steady state condition can be calculated by

$$T_g = T_{in} + \frac{\chi A_f \Delta H_c}{\left(c_p \rho_{in} V + k_T A_T\right)} \left(\frac{\dot{m}_0''}{n} + R_n\right) \tag{23}$$

Substituting the above parameters into Eq. (23), the calculated average temperature of the entire confined space can be obtained by,

$$T_g = 35^\circ C + 58.41 \left[\frac{19.75}{n} + \frac{428.99 + e^{\frac{(115-n)}{12.34}}}{3600}\right] \tag{24}$$

It can be seen from Fig. 16a that the average temperature calculated by Eq. (24) in this work was closer to the actual measured temperature. The error of the measured average temperature of the entire confined space and the calculated average temperature by the prediction model was within ±20% as shown in Fig. 16b. It may be due to linear relation between integrated heat transfer coefficient and ventilation rate in the prediction model overestimated the dissipate heat through the bulkhead, while the actual heat transfer coefficient changes nonlinearly with ventilation rate under the complex effects of mechanical ventilation.

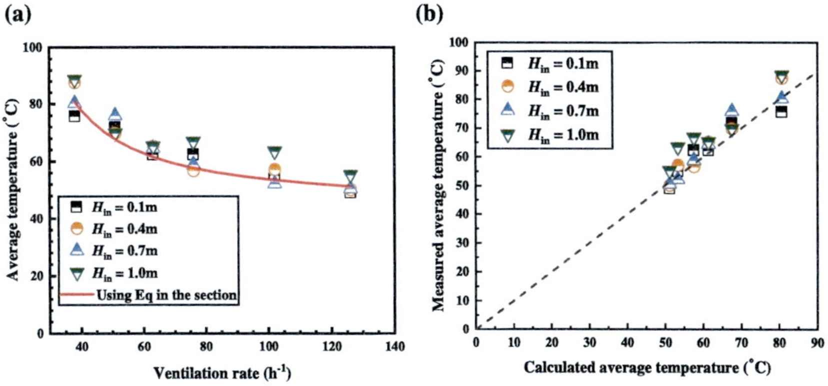

Fig. 16 Prediction model of average temperature of the confined space [29]

4 Fire Safety Design in the Special Confined Spaces

4.1 Mechanical Ventilation Design

The goal of smoke extraction is to replace 95% of the smoke-laden air produced by a fire with fresh air. According to different driving forces, smoke exhaust can be divided into natural smoke exhaust and mechanical smoke exhaust. Natural smoke exhaust refers to the use of the thermal buoyancy of the flame smoke itself to remove the smoke, no additional equipment is required but the smoke exhaust efficiency is low, and it is not suitable for tall spaces; mechanical smoke exhaust requires the use of fans to remove the smoke in the fire area, the smoke exhaust efficiency is high, but it is often necessary to set up supporting pipes, etc. Different smoke exhaust methods are often used for different places, and for some special confined spaces such as (nuclear power plants, cabins and underground spaces, etc.), it is difficult to naturally exhaust smoke through openings, so most of them use mechanical exhaust to remove the hot smoke generated by the fire.

For a fire compartment, turning on mechanical ventilation for smoke exhaust may increase the oxygen concentration in the compartment, intensify combustion and increase the fire risk in the compartment. According to the previous research, when the closed confined space is exhausted during the disaster, the smoke exhaust system has a great influence on the development of the internal fire. For ventilation-controlled fires, when the smoke exhaust system is turned on, the supplementary air will provide fresh air to the closed cabin to increase the heat release rate of the fire; while for fuel-controlled fires, turning on the smoke exhaust system can effectively reduce the temperature and visibility inside the cabin and concentrations of harmful gases. Aiming at the smoke control strategy in the confined chamber, the Stanford Research Institute further carried out experimental research on the impact of mechanical ventilation on the fire hazard of the confined space. The researchers defined a parameter that couples the volume of the cabin and the type of fuel:

$$E_v = \frac{37V}{E \cdot \Delta H} \tag{25}$$

Researchers believe that when the amount of fuel in the cabin is lower than $2E_v$, the aggravation effect of ventilation on combustion is negligible, and ventilation and smoke exhaust can be used at this time. The report also uses the experimental data to verify the exponential decay law of the smoke concentration with the ventilation times after the fire extinguished.

$$T = \frac{V}{\dot{Q}} \cdot \ln\frac{N_0}{N} \tag{26}$$

N_0 is the smoke concentration at the time of fire extinguishing, N is the smoke concentration expected to be achieved by ventilation, T is the time required to reach the expected smoke concentration, V is the cabin volume, and $\dot{Q}$ is the ventilation volume flow rate.

Beji and Meric [35] found that the burning rate of a pool fire in an enclosed cabin was affected by the dynamic interaction between the fuel mass loss rate and the ventilation rate, and the burning rate increases first and then decreases with the ventilation rate. Previous work analyzed three combinations of air supply and exhaust: (1) low air supply and high exhaust air (2) low exhaust air and high air supply (3) ceiling height air supply and exhaust air, the impact of mechanical ventilation.

The smoke exhaust fan shall meet the requirement of continuous operation for 30 min at 280 °C. The smoke exhaust fan shall be interlocked with the smoke fire damper at the entrance of the fan. When the valve is closed, the smoke exhaust fan shall be able to stop running. The smoke exhaust pipe and its connecting parts shall be able to ensure its structural integrity at 280 °C for 30 min continuously. The wind speed at the exhaust outlet should not be greater than 10 m/s.

4.2 Extinguishing

For special confined places such as confined spaces and underground spaces, due to the high risk of fire, it is difficult for personnel to enter and carry out fire extinguishing operations after a fire occurs. Therefore, physical barriers and automatic fire extinguishing systems are often used to control the spread of fire. There are many automatic fire extinguishing systems that can be used in such restricted places, the more common ones are as follows: water system fire extinguishing system, foam fire extinguishing system, gas fire extinguishing system, dry powder (ultrafine dry powder) fire extinguishing system, etc.

The water fire extinguishing system extinguishes the fire by spraying water directly to the fire point by absorbing heat and isolating oxygen. The widely used water sprinkler system and fine water mist system have a good fire extinguishing effect in the early stage of fire. However, this technology is difficult to be used in places with large number of electrical equipment.

Foam fire extinguishing system high-expansion foam is formed by uniformly mixing water and high-expansion foam fire extinguishing agent according to the specified volume ratio, and then using a generator to blow a large amount of air into the foam. When extinguishing a fire, large number of high-expansion foams will close or flood the fire area in a dense state, separating combustibles, flames and air, thereby stopping the combustion reaction. At the same time, the aqueous solution in the high-expansion foam not only has a cooling effect, but also after the water is evaporated into water vapor, the water vapor is mixed with the air to greatly reduce the oxygen content of the surrounding air, which has a certain suffocation effect. But the foam fire extinguishing agent is not suitable for the fire of live equipment, nor can it be used for the fire of water-soluble flammable liquids such as alcohols, and it will pollute the equipment.

Gas fire extinguishing system mainly includes CO_2 fire extinguishing system and halon fire extinguishing system. Among them, it uses the inert gas characteristics of CO_2. When extinguishing a fire, it absorbs heat and reduces the oxygen content around the combustion to cool down and suffocate the combustion to achieve the purpose of extinguishing the fire. At the same time, because the molecular weight of CO_2 is larger than that of air, it is easy to sink to isolate the air from the combustion to stop the combustion reaction. CO_2 is cheap and easy to obtain and prepare, but it is easy to cause suffocation to firefighters. After the halon fire extinguishing agent with extremely high fire extinguishing efficiency in the halon fire extinguishing system was banned, researchers developed alternative products such as heptafluoropropane. This kind of fire extinguishing agent can react with the active elements produced by combustion to block the chain reaction to achieve the effect of fire extinguishing, and has better fire extinguishing performance. However, the shortcoming is that the product after its fire extinguishing has certain harm to human body.

The ultra-fine dry powder fire extinguishing system plays a fire extinguishing role by contacting the dry powder sprayed after pressurization with the flame. The inorganic salt in the dry powder can react with the free radicals generated by combustion and inhibit the chain reaction. A covering layer is formed underneath to play the role of isolating the air. The ultra-fine dry powder fire extinguishing system has high fire extinguishing efficiency, and has no harm to the environment, and has good environmental protection benefits.

5 Summary

For the purpose of further clarifying the influence of mechanical ventilation conditions on the burning process, temperature distribution characteristics and smoke layer in confined space, the fire phenomenon in the special confined spaces has been introduced. The characteristics of confined fires under the mechanical ventilation have been introduced and the effect of mechanical ventilation on the confined fires have been revealed. This chapter presented the fire phenomenon in the special confined spaces under mechanical ventilation, and the latest advance in the fire safety research of special confined spaces. This chapter provided the theoretical and experimental analysis in the confined spaces for the evaluation of fire smoke control, as well as the fire safety design of the special confined spaces.

The relevant content of this chapter is more conducive to readers to understand the phenomenon of fire development and smoke flow in restricted cabins, have relevant sensory cognition, and is more conducive to the research and design of fire protection, which can provide basic theory and basic data for researchers and designers, and further lay a solid foundation for intelligent fire protection design.

References

1. E. Braun, D.L. Lowe, W.W. Jones et al., Comparison of full-scale fire tests and a computer fire model of several smoke ejection experiments. NASA STI/Recon Tech. Rep. N **93**, 23280 (1992)
2. F.W. Williams, E.W. Forssell, P.J. DiNenno et al., Shipboard smoke control tests using forced counterflow air supply. Naval Research Lab Washington (1994)
3. C. Andersson, D. Säterborn, Smoke control systems aboard-a risk analysis of smoke control systems in accommodation spaces on passenger ships. LUTVDG/TVBB–5093–SE (2002)
4. B. Zhang, X. Wang, S. Lu et al., Effects of air inlet configuration on forced-ventilation enclosure fires on a naval ship. Fire Technol. **52**(2), 547–562 (2016)
5. N.J. Alvares, K.L. Foote, P.J. Pagni, Forced ventilated enclosure fires. Combust. Sci. Technol. **39**(1–6), 55–81 (1984)
6. W.K. Chow, W.L. Chan, Experimental studies on forced-ventilated fires. Fire Sci. Technol. **13**(1+2), 1_71–1_87 (1993)
7. M.J. Peatross, C.L. Beyler, Ventilation effects on compartment fire characterization. Fire Saf. Sci. **5**, 403–414 (1997)
8. W. Le Saux, H. Pretrel, C. Lucchesi et al., Experimental study of the fire mass loss rate in confined and mechanically ventilated multi-room scenarios. Fire Saf. Sci. **9**, 943–954 (2008)
9. H.E. Mitler, Zone modeling of mechanical ventilation fires. Combust. Sci. Technol. **39**(1–6), 83–106 (1984)
10. K.L. Foot, P.J. Pagni, N.J. Alvares, Temperature correlations for forced ventilation compartment fires. In: *Fire safety science proceedings of the first international symposium* (1986), pp. 139–148
11. S. Deal, C. Beylert, Correlating pre-flashover room fire temperatures. J. Fire. Prot. Eng. **2**(2), 33–48 (1990)
12. W.K. Chow, Modelling of forced-ventilation fires. Math. Comput. Model. **18**(5), 63–66 (1993)
13. F. Bonte, N. Noterman, B. Merci, Computer simulations to study interaction between burning rates and pressure variations in confined enclosure fires. Fire Saf. J. **62**, 125–143 (2013)
14. H. Pretrel, P. Querre, M. Forestier, Experimental study of burning rate behaviour in confined and ventilated fire compartments. Fire Saf. Sci. **8**, 1217–1228 (2005)
15. H. Prétrel, W. Le Saux, L. Audouin, Pressure variations induced by a pool fire in a well-confined and force-ventilated compartment. Fire Saf. J. **52**, 11–24 (2012)
16. A. Nasr, S. Suard, H. El-Rabii, et al., Experimental study on pyrolysis of a heptane pool fire in a reduced scale compartment. In: *7th mediterranean combustion symposium* (2011)
17. X. Chen, S.X. Lu, Z.W. Ding, Initial fuel depth effect on the burning characteristics of thin-layer pool fire in a confined enclosure: Bench scale experiment and thermal equilibrium analysis. J. Therm. Anal. Calorim. **139**, 1409–1418 (2020)

18. H. Prétrel, L. Bouaza, S. Suard, Multi-scale analysis of the under-ventilated combustion regime for the case of a fire event in a confined and mechanically ventilated compartment. Fire Saf. J. **120**, 103069 (2021)
19. H. Pretrel, S. Suard, L. Audouin, Experimental and numerical study of low frequency oscillatory behaviour of a large-scale hydrocarbon pool fire in a mechanically ventilated compartment. Fire Saf. J. **83**, 38–53 (2016)
20. M. Mense, Y. Pizzo, H. Prétrel et al., Experimental and numerical study on low-frequency oscillating behaviour of liquid pool fires in a small-scale mechanically-ventilated compartment. Fire Saf. J. **108**, 102824 (2019)
21. J. Backovsky, K.L. Foote, N.J. Alvares, Temperature profiles in forced-ventilation enclosure fires, in *Lawrence livermore national laboratory* (1988)
22. J. Li, H. Prétrel, S. Suard et al., Experimental study on the effect of mechanical ventilation conditions and fire dynamics on the pressure evolution in an air-tight compartment. Fire Saf. J. **125**, 103426 (2021)
23. J.Q. Zhang, S.X. Lu, C.H. Li et al., Performance of overall heat transfer coefficient and exploring heat transfer through the ceiling vent of compartment fire in ship structures with A60 constructions. Ships Offshore Struct. **10**(3), 328–334 (2015)
24. NFPA92B, Guide for smoke management systems in malls, atria and large areas, in *National fire protection association-quincy, MA, U.S.A.* (1995)
25. F.W. Mowrer, R.B. Williamson, Estimating room temperatures from fires along walls and in corners. Fire Technol. **23**, 133–145 (1987)
26. T. Tanaka, T. Yamana, Smoke control in large scale spaces. Fire Sci. Technol. **5**(1), 31–40 (1985)
27. E.E. Zukoski, T. Kubota, B. Cetegen, Entrainment in fire plumes. Fire Saf. J. **3**(3), 107–121 (1981)
28. Y. Hayashi, Y. Hasemi, A. Ptchelintsev, Smoke layer formation by fires in mechanical ventilation enclosure. Fire Saf. Sci. **6**, 805–816 (2000)
29. X. Wu, et al., Large-scale experimental investigation of effect of mechanical ventilation on smoke temperature in ship engine room. Fire Technol. 1–25 (2022)
30. T. Ma, *Ignitability and explosibility of gases and vapors* (Springer, New York, 2015)
31. B. Karlsson, Modeling fire growth on combustible lining materials in enclosures (1992)
32. L. Hu, C. Kuang, X. Zhong et al., An experimental study on burning rate and flame tilt of optical-thin heptane pool fires in cross flows. Proc. Combust. Inst. **36**(2), 3089–3096 (2017)
33. L. Hu, J. Hu, S. Liu et al., Evolution of heat feedback in medium pool fires with cross air flow and scaling of mass burning flux by a stagnant layer theory solution. Proc. Combust. Inst. **35**(3), 2511–2518 (2015)
34. X. Wu, Y. Zhang, J. Jia, X. Chen, W. Yao, S. Lu, Experimental and theoretical analysis of the smoke layer height in the engine room under the forced air condition. Fire **6**(1), 16 (2023). https://doi.org/10.3390/fire6010016
35. T. Beji , B.Merci, Assessment of the burning rate of liquid fuels in confined and mechanically-ventilated compartments using a well-stirred reactor approach. Fire Technol. **52**, 469–488 (2016)

Automation of Structural Fire Resistance Design

Mhd Anwar Orabi, Zhuojun Nan, and Asif Usmani

1 Structural Fire Engineering

1.1 The Process of designing a structure

Designing and constructing a building is a complex process with many stakeholders, constraints, and requirements. Generally, a property development firm would hire an architect to produce a concept design that is meant to achieve the client objectives. After a few iterations, and if the development firm and client succeeded in obtaining initial funding for the project, they move on to hire design, construction, and possibly project management firms. The structural engineering designer is one of the various engineering teams involved in the project and is responsible for ensuring that the structure is designed with sufficient capacity to be safe under all expected loading conditions [1].

Typically, the structural engineer operates under the supervision of the architect and the project manager and liaises with the contractor and other engineering disciplines such as those dealing with fire, ventilation, plumbing, and electrical systems. Naturally, each of these engineering disciplines have their own objectives for their part of the project, and thus conflicts often arise and are resolved by interdisciplinary compromise and cooperation. Structural fire engineering is unique in its nature as it is, by definition, a cross-disciplinary enterprise that is concerned with both fire and structural behaviour. Unfortunately, current practice is that structural fire design is performed as a "check" rather than a part of a holistic design approach. This

M. A. Orabi · Z. Nan (✉) · A. Usmani
Department of Building Environment and Energy Engineering, The Hong Kong Polytechnic University, Hong Kong SAR, China
e-mail: zhuojun.nan@connect.polyu.hk

M. A. Orabi
School of Civil Engineering, The University of Queensland, Brisbane, Australia

X. Huang and W. C. Tam (eds.), *Intelligent Building Fire Safety and Smart Firefighting*, Digital Innovations in Architecture, Engineering and Construction,
https://doi.org/10.1007/978-3-031-48161-1_7

means that the structural fire engineer is often given the constraints of a mostly finished structural design that they need to ensure remains safe under any potential fire scenarios. The time given for such a critical undertaking is usually in the range of one to three weeks. This chapter will cover how automation of the most repetitive and time-consuming parts of structural fire engineering may enable the engineer to perform a thorough structural fire analysis within the tight limitations of realistic project timelines.

1.2 Fire-Induced Structural Loading

Fire is an important structural design consideration, without which a development would never get the necessary regulatory approvals for construction. Typical loading scenarios are those that most structures are expected to withstand, and they include both permanent and variable components as classified by the EN 1990 [1] and summarised in Fig. 1. Permanent actions, sometimes referred to as dead load, refers to the permanent weight of the structural members and other non-transient components such as the weight of false ceilings and flooring. Variable actions are concerned with "imposed load" that change over time due to the change in the weight and position of people and furniture. There are also other types of variable load including the load imposed by wind, rain, snow, and even earthquakes. Accidental action, such as blast and fire, is a special category of load that the structure is rarely, if ever, expected to experience. As an accidental load, structural fire design is often not regarded the same level of priority as the other load types. Regardless, consideration of fire loading is still required for regulatory approval before construction.

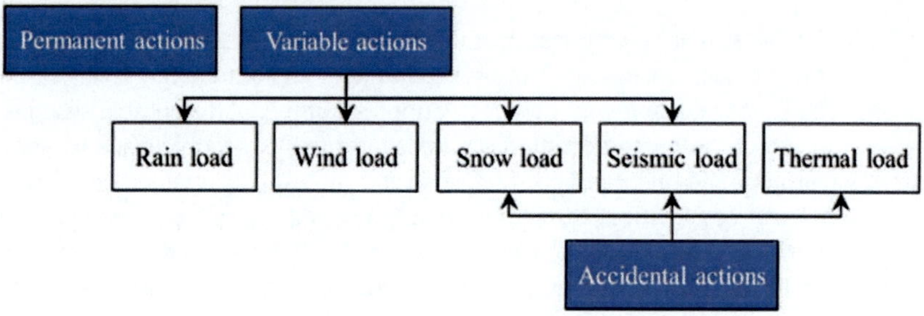

Fig. 1 General classification of structural loading according to EN1990

1.3 Prescriptive design solutions

The fire and structural fire regulatory environment and prescriptive guidelines have been long established in the industry and their origins can be traced as far as the great fire of London in 1666 [2]. After that catastrophic fire, regulations were put in place to limit the type of materials allowed for construction and the proximity of structures to one another to limit fire spread. Today, most fire resistance ratings for structural members are based on the "standard fire" temperature–time model. Based on experimental data obtained as early as 1896 [3], an early form of this model appeared in 1916. At the time, the prevailing idea was that if a structural member, in isolation, survives the standardised heating conditions in a furnace, then it would provide the required level of safety for the building in a real fire. As such, most prescriptive design codes consider the "fire resistance rating" of a building component as the duration this isolated member would survive the standard temperature–time curve in a furnace. This fire model is simply an exponentially increasing temperature that continues to grow with time as shown in Fig. 2.

This prescriptive approach has been the primary solution to ensure structural fire safety at a time where there was no alternative. Today, however, and after years of fundamental research on structural fire behaviour, it is clear that the response of a structure is exceedingly more complex than the temperature of its individual components in isolation. Furthermore, the standard temperature time curve is representative of data collected from over a century ago and does not represent the complex geometries and fuel and ventilation conditions of modern buildings. Because of these issues, and as architectural forms continue to evolve, an increasing number of buildings are falling outside the scope of traditional prescriptive design. Indiscriminate adherence to prescriptive approaches in modern structures results in over utilisation of fire protection with a false expectation that this would result in a fire safe structure. In reality, prescriptive approaches do not offer insight on how a structure, especially a

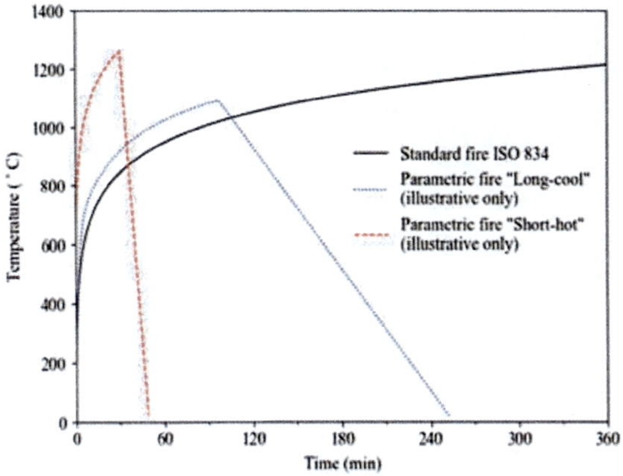

Fig. 2 Comparison between the standard temperature–time curve and the parametric temperature–time curve

modern one, would perform in fire. Designing a structure for a performance objective requires a performance-based structural fire engineering approach as will be discussed next.

2 Performance-Based Engineering in SFE

2.1 *Designing for Performance*

The design process in performance-based engineering (PBE) aims to meet various requirements related to the multimodal performance of a structure, which includes considerations such as life safety, usability, and downtime for repair. PBE approaches gained widespread acceptance in structural engineering first through their adoption for seismic design. The performance-based earthquake engineering methodology developed by the Pacific Earthquake Engineering Research Center (PEER PBEE) in 1997 is among the most popular and widely-accepted of the many PBE methodologies developed over the last three decades [4, 5]. PEER PBEE approach has four primary steps: (1) modelling the seismic load, (2) analysing the response of the structure, (3) assessing the post-hazard state of the structure, and (4) estimating the damage in terms of loss of life, dollar-amount, and repair duration. While different approaches may differ in application details, the steps of the PEER framework offer a good representation of the philosophy and tasks involved in most PBE methodologies.

Both earthquake and structural fire engineering aim to ensure the safety of the built environment in the face of a hazard that is difficult to quantify and predict. This similarity makes PBE an attractive methodology for structural fire engineering. There are currently many ongoing research projects around the world aiming to tailor PBE methodologies for structural fire engineering. These efforts include development of probabilistic material models [6], direct reliability-based damage assessment of structural members [7], and even adaption of the PEER fragility curves for structural fire engineering of bridges [8]. Underpinning all of these endeavours is the ability to (1) characterise the fire hazard, (2) assess the thermal load caused by the fire, and (3) analyse the thermomechanical response to the thermal load, which are covered in more detail in Sects. 2.2, 2.3, and 2.4, respectively.

2.2 *Modelling Fire*

An earthquake can generally be modelled by considering three primary variables: amplitude, frequency content, and duration [4]. Moreover, as a load, an earthquake is applied in a relatively uniform manner to the entire structure. A fire and the thermal load associated with it, on the other hand, are characterised by thousands of parameters and are never uniform throughout a building. An important distinction ought to be made here between the fire and the thermal load. The fire refers to the hot gases, flames and soot generated by the combustion of fuel within a structure. Conversely, the thermal load is the increase in temperature of the structural members that is caused by the fire. This section will discuss how fire is usually modelled in structural fire engineering, while Sect. 2.3 is concerned with how the thermal load on the structure is calculated for a given fire or fire model.

The parametric fire model accounts for various factors that can result in different thermal exposure compared to the standard fire as shown in Fig. 2. These include compartment geometries, ventilation conditions, fire loads, and compartment boundary materials [9]. In addition to these considerations, parametric fire curves also represent the pre-flashover fire, the fully developed fire, and the cooling phase. Depending on the compartment parameters, the length of the modelled fire phases can change resulting in variations of thermal exposures. For example, and as shown in Fig. 2, the parametric temperature time curve may model short and hot or long and relatively cool fires. However, the parametric temperature time curves are only applicable to a small proportion (8%) of buildings [10]. Most others are outside the limitations of this model as denoted by Eurocode [11]. To address this, localised fire models can be used to represent the fire impact prior to the fully-developed stage or for pre-flashover fires. Hasemi et al. [12] proposed a correlation to approximate heat fluxes over the structural boundary in localised fire scenarios based on pool fires, which was adopted by Eurocode 1 [9] and recently introduced to represent the fire impact of localised burning in large open plan compartments [13] for PBE.

"Travelling" fires have also been observed in many fire accidents, which can result in complex and interconnected structural fire behaviour. Various theoretical representations of travelling fire models have been developed, including Clifton's travelling fire model [14], Travelling Fires Methodology (TFM) [15] and its subsequent refined versions (i.e., Improved Travelling Fires Methodology (iTFM) [16] and Travelling Fires Methodology with flame extension (fTFM) [17]), and the Extended Travelling Fire Methodology (ETFM) framework [13, 10]. To enhance the flexibility of fire scenario design and enable the reproduction of time-variant fire development, the authors have developed a natural fire model in OpenSees for fire, which incorporates test-based travelling behaviour models and localised fire models of various modes [18]. While Computational Fluid Dynamics (CFD) is an alternative method to characterise fire severity for structural design, it is computationally expensive and requires professional fire science knowledge for interpretation, making it impractical for day-to-day routine design for structural engineers.

2.3 Structure-Fire Interface

While the fire is the source of thermomechanical loading, the load itself is the change in the temperature of the structural members. This requires calculations based on the exposure conditions, cross section parameters, and the material type. One way to do this, is to simply assume that the temperatures of the hot gases in the compartment are the same as those of the structure. This approximate approach works very well for small-sized, thin, and fully exposed steel and aluminium structural members. This is because the "thermal thickness" of these members is so small that there is barely any lag between the gas and structure temperatures. However, this is not the case for larger structural members, those covered by fire protection, or those composite with floor slabs. In these scenarios, a more sophisticated approach for estimating the structural temperature is needed, such as the tabular lumped temperature calculation method from Eurocode [19]. Yet, the assumption of a uniform temperature for a structural member is very rarely sufficient for thermomechanical analysis. For that, a more sophisticated heat transfer analysis is often required to estimate the detailed thermal profile across the cross section of the structural member. Currently, the most common way to do this is by manually building finite element models of each heated element. The results of these analyses are then exported and used as thermal loads for structural analysis.

2.4 *Structural Response*

Thermomechanical analysis investigates the response of the structural system of a building to thermal load, as manifested by thermally induced deflections, stresses, and forces. A proper thermomechanical analysis must also consider boundary conditions, thermal expansion, temperature profile, and material properties, as these factors govern the structural response to thermal load. One of the important shortcomings of prescriptive approaches is their over-emphasis on the degradation of material properties at elevated temperatures. This is often done with limited consideration of the effects of boundary conditions and thermal expansion. It has been shown in the literature that restrained expansion and boundary conditions are just as, or even more important than material degradation [20]. Under a pure uniform increase in temperature, as in a single temperature value ΔT, beams and slabs expand laterally and are restrained by the surrounding structure. This results in the development of very large internal compressive stresses due to the restraining forces as shown in Fig. 3a. These forces can result in the failure of the beam or the surrounding structure. Such a pure uniform increase in temperature, however, is not a realistic condition which most structural components would be subjected to, as the temperature across the depths of the members tend to vary and as such cannot be represented with a single ΔT value.

In the vast majority of cases, beams are composite with concrete slabs on profiled decking. In these situations, the different properties of the concrete slab and the beam will lead to a different rate of temperature increase resulting in the steel beam being significantly hotter than the slab. The difference in temperature profiles can be considered as a thermal gradient, $T_{,y}$, across the depth of the composite floor system. This gradient results in the development of tensile forces and large deflections, δ, in the structural member as shown in Fig. 3b. Under realistic exposure, all structural members experience a combination of the thermal gradient and the uniform increase

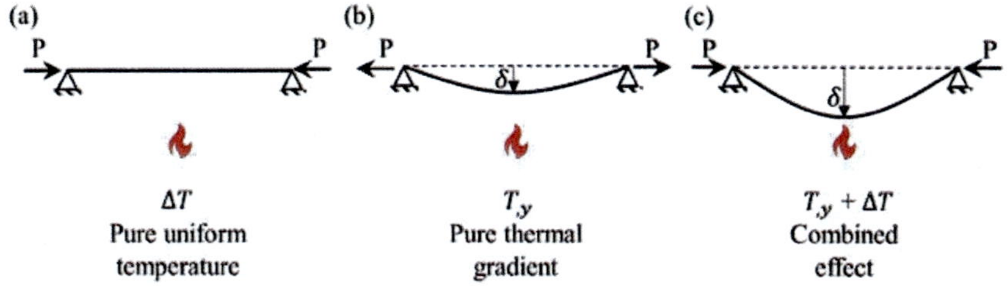

Fig. 3 Effect of different thermal loading paradigms on structural members

in temperature. This results in a reduction in the compressive forces induced into the surrounding structure and an increase in deflections as shown in Fig. 3c [20].

Boundary conditions also play a decisive role in determining the magnitude and type of the internal thermally induced forces and the corresponding deflections. As such, thermomechanical analysis must carefully consider both the boundary conditions and the temperature profile within a given structural member. Under uniform fire exposure, as expected in a post-flashover compartment, estimating the thermal load of the structural components is relatively straightforward as all members are exposed to the same, largely uniform, fire conditions such as in Fig. 4. In this kind of situation, the structural fire engineer only needs to perform a single heat transfer analysis for each different size of member, and then apply the resulting temperature profile to all similar members in the fire compartment.

For cases with non-uniform exposure, such as those modelled by travelling fire models, localised fire models or even CFD analyses, calculating the thermal load is exceedingly more difficult to do. In these scenarios, the temperature profile of each segment of each individual member is dependent on its local fire exposure conditions, as shown in Fig. 4b. The structural fire engineer will need to perform a heat transfer analysis for each segment, and then apply the resulting temperature profile to only that segment. This requires performing hundreds or even thousands of heat transfer analyses and then applying the temperature profiles to each segment. In practical

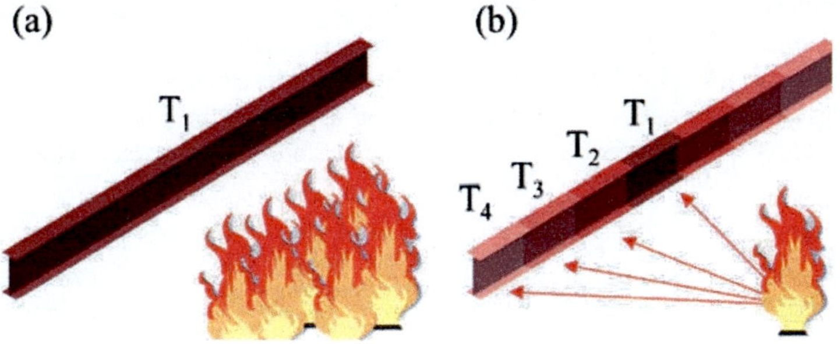

Fig. 4 Assessing the thermal exposure of structural members under different conditions, **a** uniform compartment temperatures, and **b** non-uniform fire exposure

engineering applications, this is rather unfeasible, and the engineers generally use simplified and uniform fire models to perform structural fire design. Section 3 of this chapter will cover how estimating the thermal load for structural members under nonuniform exposure can now be automated. This enables the structural fire engineer to consider realistic fire scenarios more accurately.

3 Automation of Analysis and Design

3.1 Numerical Simulation Tools

There are many advanced numerical simulation tools that can be used for structural fire engineering, most of which are general-purpose finite element programs such as Abaqus, Ansys and LS-DYNA [21–23]. There are also other in-house, specialised finite element programs such as FLEX which was developed by Weidlinger Associates [24], and some proprietary academic solutions including ADAPTIC developed at Imperial College London [25]. General-purpose tools are usually powerful, but their wide scope of application often introduces difficulties in the simulation workflow for structures in fire, which is shown in Fig. 5. Specialist engineers will often need to spend many hours building and debugging the many different kinds of models needed, and then spend days running the simulations before finally getting a chance to inspect the results. As discussed in the beginning of the chapter, this is not feasible in realistic structural fire engineering timelines. Bespoke solutions such as FLEX and ADAPTIC, on the other hand, are purpose-built, and may offer a more streamlined workflow. However, as they are proprietary by nature, they are not open for use by the wider industry and academic community.

OpenSees is an open-source finite element tool originally designed for earthquake engineering, and extended for structural fire engineering over the last decade [26–28]. It offers an alternative tool for academia and industry that allows source-code modifications and customisation. The structural fire capabilities of OpenSees include a heat transfer engine [29], fire library [30], and thermomechanical material library and elements [31, 32]. One of the primary problems with using OpenSees was the unavailability of a capable pre- and post- processor. This problem was solved for earthquake engineering by researchers at the Aristotle University of Thessaloniki who developed an open-source interface to be used with the general purpose pre and post processor GiD [33, 34]. Likewise, the authors of this chapter extended this

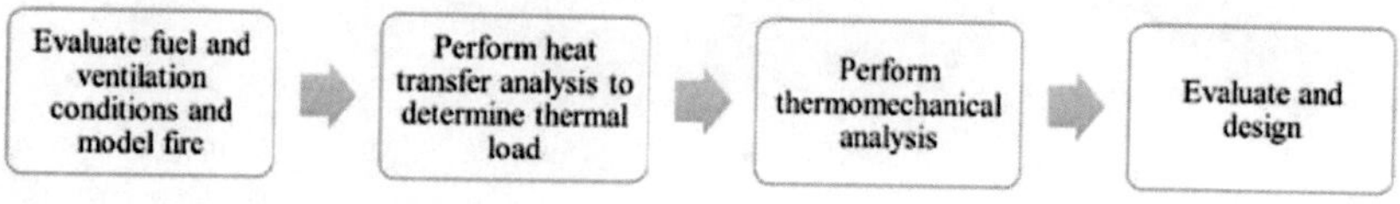

Fig. 5 Simulation workflow for structures in fire

open-source interface to cover structures in fire and used it as a powerful tool for automation as will be discussed next [35].

3.2 Integrated Simulation

One of the most time-consuming hurdles in structural fire engineering is that heat transfer needs to be performed for each member, and sometimes at different locations for the same member. To overcome this issue, the National Institute for Standards and Technology (NIST) developed the Fire Structure Interface (FSI) tool, which couples structural models built in ANSYS with results from fire dynamics simulator (FDS). This tool helps overcome the differences in spatial and temporal domains between FDS and ANSYS, and automates the heat transfer analysis process [36, 37]. SAFIR, a structural fire computational tool developed at the University of Liege, has also automated the process of performing heat transfer analyses for non-uniform fires [38]. This tool has been widely used both in academia and industry, and is a relatively mature although proprietary solution [39–41].

In the open source domain, there are currently two options: OpenFIRE [42–44], and the Integrated Simulation Environment (ISE) for OpenSees [35]. OpenFIRE is a stand-alone tool that performs one-way coupling between FDS, and both Abaqus and OpenSees. As a coupling tool, it provides the communication infrastructure to bridge the temporal and spatial discrepancies between FDS and FEM. It is also used by the ISE to communicate with FDS.

The general workflow of the ISE for automating heat transfer analysis associated with non-uniform fire is shown in Fig. 6. As the figure shows, the first step is to import the geometric properties of the structure into the extended GiD and OpenSees interface, and into FDS if it is to be used to simulate the fire. Structural modelling in the ISE consists of both mechanical and thermomechanical steps. The mechanical step includes creating and assigning OpenSees materials, sections, and mechanical loads to the structure. In the thermomechanical step, the structural fire engineer needs to determine the resolution for which they want to consider the thermal loading. For example, the temperature of profile of the composite beam shown in Fig. 7 can be calculated at one point, two points, or any number of points the engineer chooses by assigning "thermocouples". For uniform temperature exposure, one thermocouple per structural member is sufficient. For nonuniform exposure, such as that due to complex fires simulated with FDS, the engineer is able to choose a level of discretisation that they see fit for the problem. It was found in previous work that one to two thermocouples per meter are able to adequately represent the temperature fields [35, 45]. The ISE stores the location of each of the thermocouples and then exports this information as a text file that is read by OpenFIRE to retrieve formatted temperature output files from FDS. These temperature files are then used to perform a 2D heat transfer analysis at each thermocouple location and apply the resulting temperature profile to the corresponding elements.

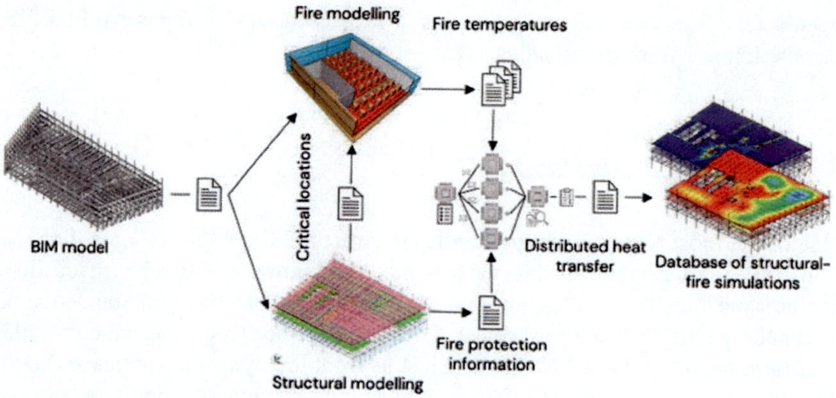

Fig. 6 General workflow of the ISE

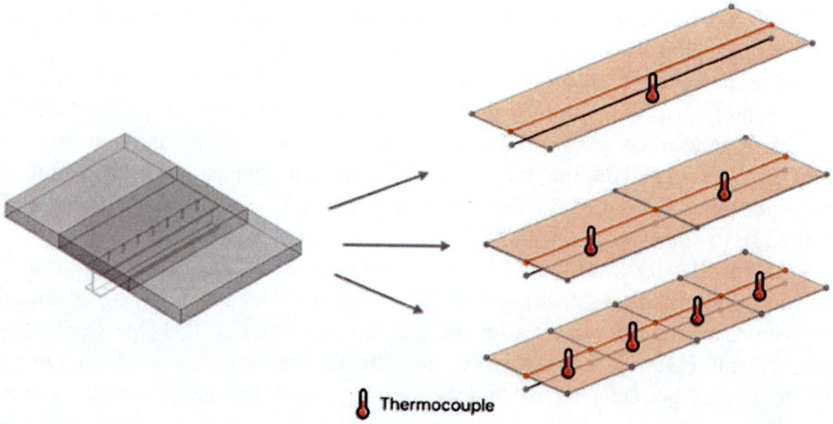

Fig. 7 Spatial resolution for heat transfer analysis

The automation of heat transfer is possible because, generally, there are only a limited number of section types used in most structures. For example, only nine different sections were needed to perform the large scale collapse analysis of WTC7 [45]. These section types are predefined in the ISE and shown in Fig. 8. There are many more types of sections in construction, but the heat transfer models share many aspects in common such as the geometric shapes, types of boundary conditions applied, and the basic material properties. This makes heat transfer analysis suitable for parametrisation as was done in the ISE. Each type of section can be defined in terms of simple parameters that can be used to automate generation of many different heat transfer models to include most sections engineers use in construction. After building a sufficiently large library of different section types in addition to

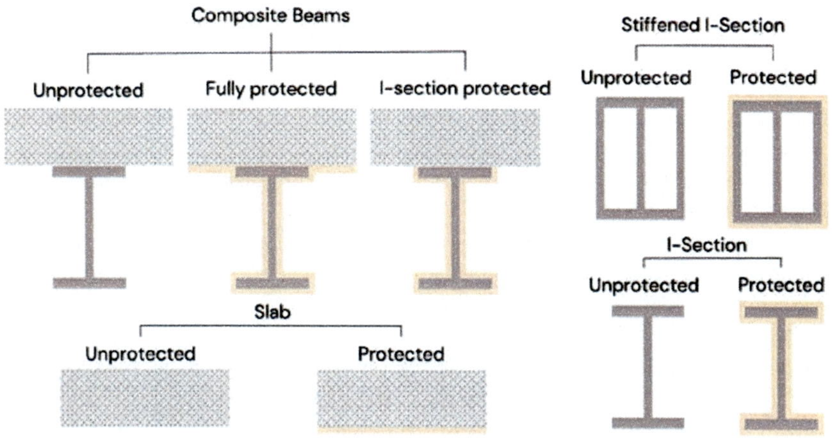

Fig. 8 Predefined section types for automated heat transfer in the ISE

those shown in Fig. 8, it is possible to fully automate most heat transfer analyses for non-uniform fire exposure for any type of structure.

The ISE uses gas temperatures generated from the fire models discussed in Sect. 2.2 or from FDS. These temperatures are converted to heat fluxes using Eqs. (1) and (2), which are then applied as boundary conditions to perform the heat transfer analysis.

$$\dot{q}''_{conv} = h\left(T_g - T_s\right) \tag{1}$$

$$\dot{q}''_{rad} = \varepsilon_s \sigma \left(T_g{}^4 - T_s{}^4\right) \tag{2}$$

In these equations, T_g and T_s are the gas temperature and surface temperature of the structural element, respectively. h is the convective heat transfer coefficient, and $\dot{q}''_{conv}$ and $\dot{q}''_{rad}$ are the convection and radiation heat fluxes respectively. ε_s is the surface emissivity, and σ the Stephen Boltzman constant. The convective heat transfer coefficient is defined along with the thermocouple, and so the engineer may choose to use different coefficients for different parts of the structure. In future, it is aimed to enrich the ISE with an interface to use a variable heat transfer coefficient. This is currently accepted as input by the heat transfer engine of OpenSees.

3.3 Multi-simulation Automation

As discussed in Sect. 2.2, fire is a unique type of accidental loading that may lead to vastly different consequences depending on where it occurs in the structure. That is, a fire with the same intensity in two different locations in the same building may result in completely different outcomes. Additionally, even for the same compartment and fuel load distribution, the fire may change if the ventilation conditions are changed by having the door or windows open or closed. As such, it is never sufficient to design a structure with thermal load by considering only one fire scenario. Likewise, if optimal utilisation of fire protection is one of the objectives for the structural fire simulation, then it is necessary to study the response of the structure with the consideration of different fire protection layouts.

The standard approach provided by most structural engineering tools requires the designer to perform each simulation separately where each fire scenario constitutes an independent model. Given the limited timeframe the engineer has for structural-fire analysis, which is discussed in Sect. 1.1, this results in simplifying the fire scenarios and only studying a handful of cases that the engineer deems "critical". What constitutes a "critical" fire scenario is completely left to the engineer's professional judgement or intuition. While informed by experience and in-depth domain knowledge, it does not constitute an omnipotent view of the potential structural fire performance. The ISE was thus extended with the ability to manage running different fire scenarios and fire protection layouts for the same model using built-in "parametric case" tools that automate this process. This would allow the engineer to cover a wide range of potential fires and thus better ensure that their structural fire design is indeed robust.

The ISE is able to automate analysing multiple cases by decoupling the fire scenarios and fire protection layouts from the thermomechanical model, as shown in the workflow depicted in Fig. 9. This workflow relies on the interface and data as shown in Fig. 10. Using the ISE, the engineer starts by applying the thermocouples from Sect. 3.2. After that, they enter the number of fire scenarios into the interface shown in Fig. 10 which then creates a directory for each case. Each case directory contains the definition for all heat transfer sections as well as a directory for the corresponding results. To perform the analysis, gas temperature data from FDS or a fire model needs to be placed into the relevant case directory. This is currently being done by using a Python script [46, 47], with a more streamlined processes being currently under development and to be added in the near future. All heat transfer simulations are run with a single command, and all thermomechanical analyses are performed with a single button.

The interface in Fig. 11 is used for applying different fire protection layouts. The different thicknesses for each layout are entered during the process of defining the "parametric SFRM thickness" material, and then the material properties are applied to the members. More than one definition of parametric thicknesses can be specified and different parametric thicknesses can be applied to different members at the same time. The ISE will create $N_{fires} \times N_{layouts}$ case directories and will perform combinations of analyses so that each fire protection layout is analysed under all fire scenarios.

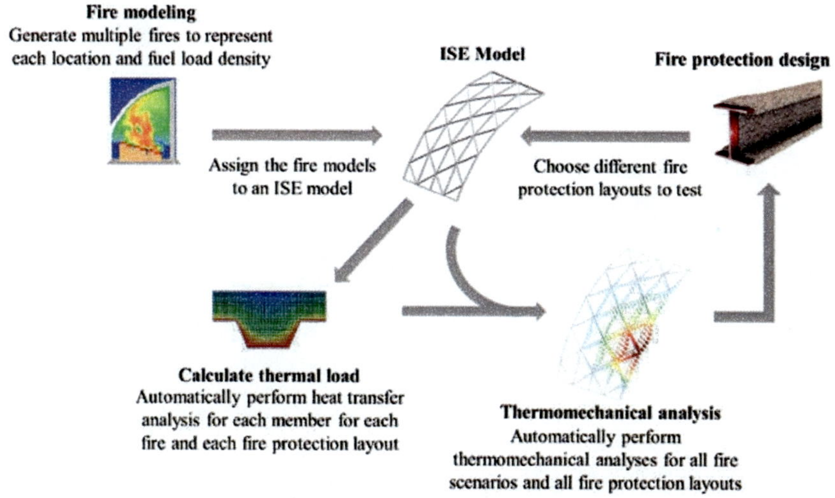

Fig. 9 The workflow needed for PBE and automated by the Integrated Simulation Environment

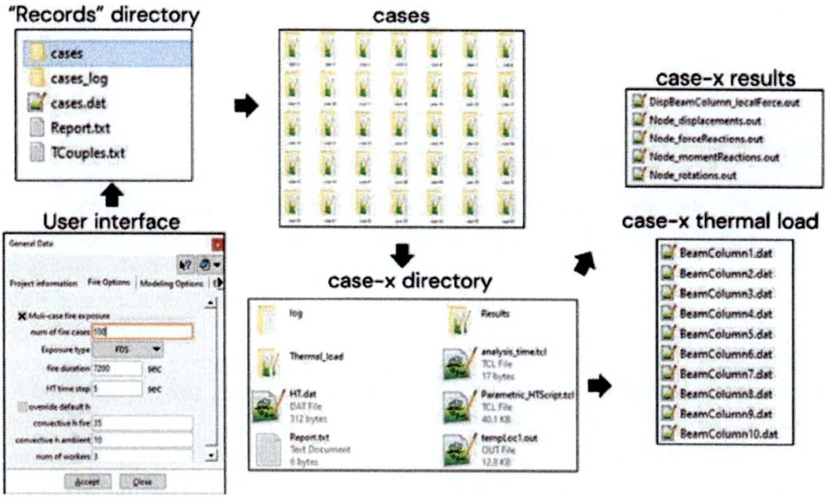

Fig. 10 Workflow for generating and analysing multiple fire scenarios

3.4 *Challenges*

Given the various computational tools such as ISE, we now have the capabilities to automatically perform a large number of concurrent simulations to analyse the response of a structure, with different fire protection layouts and materials, to many different fire scenarios. This is helpful as it eliminates the vast majority of repetitive

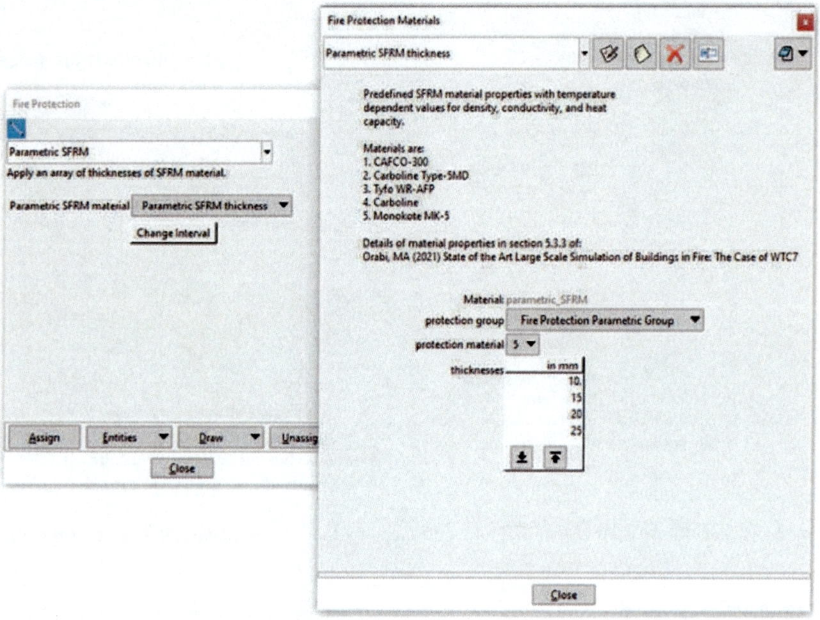

Fig. 11 Interface for applying different thicknesses of fire protection layouts

manual work the engineer needed to do in the past to directly assess structural fire performance. This, however, also introduces new complexities. Now, the engineer is inundated with the results of hundreds of numerical simulations which they need to first evaluate and then use for design. There is currently no simple way to deal with post processing hundreds of results. What is currently done by the authors is to select a few representative measurements, such as the deflection of middle of selected composite beams and compare these parameters for all performed analyses. Evaluation of the results is a particularly important step in the structural fire engineering process, and the authors hope to develop a data visualisation and analysis tool to streamline this part of the workflow in the future.

PEER PBEE offers valuable insight on how making sense of large structural analysis datasets could be tackled. The work by Lange [48] and that by Khorasani [8] are excellent examples how the PEER PBEE framework could offer a way to produce fragility curves for structural fire that allow for cost and performance-conscious decision making by the engineer. There remain many issues to be resolved, however. While the automation discussed earlier allows for analysing the response of a structure to a library of fires, there is currently no tool available to automatically decide and generate suitable fire scenarios: this largely remains at the discretion of the fire and structural fire engineers. The authors currently use Python scripts to generate these. Additionally, the criteria for assessing the damaged state of a structure in fire are still an active area of research that is need of further exploration. Establishing a unified and widely accepted measure for fire-induced structural damage and for evaluating structural fire performance are especially important issues. When it comes to earthquake engineering, specialised tools like ETABS and Perform3D have automated the majority of the PEER PBEE framework allowing for generation of the required curves with minimal effort from the designer [49, 50]. This is not the case for structural fire engineering, which is largely due to the aforementioned knowledge gaps.

That being said, perhaps the most important obstacle to automation in structural fire engineering is not technical, but rather regulatory. To this day, the biggest hurdle in application of performance based structural fire engineering is that prescriptive solutions offer a much faster path to obtaining the necessary regulatory approvals. At present, most overseeing authorities around the world have not had the necessary experience nor time to develop the required competencies to thoroughly evaluate performance-based solutions and spot when prescriptive approaches are not sufficient. As such, checking whether a design meets the prescriptive requirements documented in design codes remains the default path towards obtaining the much-needed clearances. This greatly incentivises property developers and pressures engineers to always fall back to prescriptive approaches, making it difficult to make the business case for the development of automated tools for performance based structural fire engineering. By developing the automated tools discussed in this chapter, the authors hope that academics and engineers are empowered to explore structural fire more thoroughly and help make performance based structural fire engineering ubiquitous around the world.

4 Summary

While structural fire engineering is essential for ensuring the safety of buildings, it is highly time consuming and difficult to perform well within the severely limited timelines allocated to it. Computational tools developed over the last decade can help alleviate this problem by allowing for automation of the most repetitive and time-intensive aspects of the design process. This chapter began by first introducing the environment in which structural fire engineers operate. Structural fire engineering is generally done at critical points within the design cycle and is subject to severely restrictive timelines of about one to three weeks. Because of this, prescriptive solutions are often preferred because they can be applied successfully within the time restrictions. Performance-based engineering was then introduced as well as how it is traditionally performed and why it needs to be automated. Performance-based engineering requires detailed thermomechanical analyses, which in turn necessitate performing heat transfer analyses for each part of the structure especially when considering non-uniform fire exposure. This process was automated by some tools, especially the Integrated Simulation Environment developed by the authors. The last section provided details of current tools that are used for automation in the field, with special focus on the integrated simulation environment that allows for automated multi-simulation. The integrated simulation environment allows the engineer to perform hundreds of simulations with most of the repetitive model-building tasks taken care of by automation. There remain challenges when it comes to dealing with the huge databases of results that are generated by these simulations, which the authors hope to develop a data analysis tool to deal with in the future. Even when that is done, however, there remains a critical knowledge gap in what constitutes "acceptable performance" for structures in fire. Current criteria, such as deflection-based limits, are woefully insufficient. A rigorous, PEER-like framework is thus imperative for the future of performance based structural fire engineering.

References

1. BSi, EN 1990 Eurocode 0: Basis for structural design (2002)
2. M. Luo, X. Sun, S. Li, A historical review of fire engineering practice and advances in China. Fire Technol. (2022). https://doi.org/10.1007/s10694-022-01300-8
3. J. Gales, B. Chorlton, C. Jeanneret, The historical narrative of the standard temperature-time heating curve for structures. Fire Technol. **57**, 529–558 (2021). https://doi.org/10.1007/s10694-020-01040-7
4. V.E. Saouma, M.A. Hariri-Ardebili, Aging, shaking, and cracking of infrastructures (2021). https://doi.org/10.1007/978-3-030-57434-5.
5. K.A. Porter, An overview of PEER's performance-based earthquake engineering methodology, in *9th International Conference on Application of Statistics and Probability in Civil Engineering* (2003), pp. 973–980
6. R. Qureshi, S. Ni, N. Elhami Khorasani, R. Van Coile, D. Hopkin, T. Gernay, Probabilistic models for temperature-dependent strength of steel and concrete. J. Struct. Eng. **146**, 1–18 (2020). https://doi.org/10.1061/(asce)st.1943-541x.0002621
7. T. Molkens, R. Van Coile, T. Gernay, Assessment of damage and residual load bearing capacity of a concrete slab after fire: Applied reliability-based methodology. Eng. Struct. **150**, 969–985 (2017). https://doi.org/10.1016/j.engstruct.2017.07.078
8. Z. Zhu, S.E. Quiel, N.E. Khorasani, Bivariate structural-fire fragility curves for simple-span overpass bridges with composite steel plate girders. Struct. Saf. **100**, 102294 (2023). https://doi.org/10.1016/j.strusafe.2022.102294
9. BSi, EN 1991–1–2:2002 Eurocode 1: Actions on Structures - Part 1–2: General Actions - Actions on Structures Exposed to Fire (2002)
10. X. Dai, S. Welch, A. Usmani, A critical review of "travelling fire" scenarios for performance-based structural engineering. Fire Saf. J. **91**, 568–578 (2017). https://doi.org/10.1016/j.firesaf.2017.04.001
11. A.M. Jonsdottir, G. Rein, Out of range. Fire Risk Manag. 14–17 (2009). https://doi.org/10.5749/minnesota/9780816676262.003.0005
12. Y. Hasemi, Y. Yokobayashi, T. Wakamatsu, A.V. Ptchelintsev, Modelling of heating mechanism and thermal response of structural components exposed to localised fires, in *Thirteenth Meeting of the UJNR Planel on Fire Research and Safety* (1996)
13. X. Dai, S. Welch, O. Vassart, K. Cábová, L. Jiang, J. Maclean, G.C. Clifton, A. Usmani, An extended travelling fire method framework for performance-based structural design. Fire Mater. **44**, 437–457 (2020). https://doi.org/10.1002/fam.2810
14. C. Clifton, Fire models for large firecells, HERA Report (1996)
15. G. Stern-Gottfried, J. Rein, Travelling fires for structural design-Part II: design methodology. Fire Saf. J. **54**, 96–112 (2012). https://doi.org/10.1016/j.firesaf.2012.06.011

16. E. Rackauskaite, C. Hamel, A. Law, G. Rein, Improved formulation of travelling fires and application to concrete and steel structures. Structures **3**, 250–260 (2015). https://doi.org/10.1016/j.istruc.2015.06.001
17. M. Heidari, P. Kotsovinos, G. Rein, Flame extension and the near field under the ceiling for travelling fires inside large compartments. Fire Mater. **44**, 423–436 (2019). https://doi.org/10.1002/fam.2773
18. Z. Nan, A.A. Khan, L. Jiang, S. Chen, A. Usmani, Application of travelling behaviour models for thermal responses in large compartment fires. Fire Saf. J. **134**, 103702 (2022). https://doi.org/10.1016/j.firesaf.2022.103702
19. BSi, EN 1993–1–2:2005 Eurocode 3: Design of steel structures – Part 1–2: General rules – Structural fire design (2005)
20. A.S. Usmani, J.M. Rotter, S. Lamont, A.M. Sanad, M. Gillie, Fundamental principles of structural behaviour under thermal effects. Fire Saf. J. **36**, 721–744 (2001). https://doi.org/10.1016/S0379-7112(01)00037-6
21. Dassault Systèmes, Abaqus Analysis User's Guide (2014). http://130.149.89.49:2080/v6.14/books/usb/default.htm
22. Livermore Software Technology Corporation (LSTC), *LS-DYNA: keyword user's manual*, vol. I (2018)
23. ANSYS (2007)
24. Weidlinger Associates Inc., WTC 7 Collapse Analysis and Assessment Report (2010)
25. L. Song, B.A. Izzuddin, A.S. Elnashai, P.J. Dowling, An integrated adaptive environment for fire and explosion analysis of steel frames - part I: analytical models. J. Constr. Steel Res. **53**, 63–85 (2000). https://doi.org/10.1016/S0143-974X(99)00040-1
26. F. McKenna, G. Fenves, M. Scott, *OpenSees: open system for earthquake engineering simulation* (2000). https://opensees.berkeley.edu/
27. M.A. Orabi, A.A. Khan, A. Usmani, An overview of OpenSEES for fire, in *Proceedings 1st Eurasian Conference. OpenSEES OpenSEES Days Eurasia, Hong Kong, China, 2019*
28. J. Jiang, L. Jiang, P. Kotsovinos, J. Zhang, A. Usmani, F. McKenna, G.-Q. Li, *OpenSees* software architecture for the analysis of structures in fire. J. Comput. Civ. Eng. **29**, 04014030 (2015). https://doi.org/10.1061/(ASCE)CP.1943-5487.0000305
29. Y. Jiang, Development and application of a thermal analysis framework in OpenSees for structures in fire, The University of Edinburgh, 2012
30. X. Dai, An extended travelling fire method framework with an OpenSees-based integrated tool SIFBuilder, The University of Edinburgh, 2017
31. J. Jiang, Nonlinear thermomechanical analysis of structures using OpenSees, The University of Edinburgh, 2012
32. L. Jiang, M.A. Orabi, J. Jiang, A. Usmani, Modelling concrete slabs subjected to fires using nonlinear layered shell elements and concrete damage-plasticity material. Eng. Struct. **234**, 111977 (2021). https://doi.org/10.1016/j.engstruct.2021.111977
33. T. Kartalis-Kaounis, V.K. Papanikolaou, GiD + OpenSees, (2017). https://github.com/rclab-auth/gidopensees
34. CIMNE, GiD (2020). www.gidhome.com
35. M.A. Orabi, A.A. Khan, L. Jiang, T. Yarlagadda, J. Torero, A. Usmani, Integrated nonlinear structural simulation of composite buildings in fire. Eng. Struct. **252** (2022). https://doi.org/10.1016/j.engstruct.2021.113593
36. T. McAllister, R. MacNeill, O. Erbay, A. Sarawit, M. Zarghamee, S. Kirkpatrick, J. Gross, Analysis of structural response of WTC 7 to fire and sequential failures leading to collapse. J. Struct. Eng. (United States) **138**, 109–117 (2012). https://doi.org/10.1061/(ASCE)ST.1943-541X.0000398
37. K. Prassad, H. Baum, Fire structure interface and thermal response of World Trade Center towers , National Institute of Standards and Technology, Gaithersburg, MD, 2005. https://doi.org/10.6028/NIST.NCSTAR.1-5g
38. J.-M. Franssen, T. Gernay, User's manual for SAFIR (version 2022) - Part 1: general considerations, Liege University, 2022. https://www.uee.uliege.be/upload/docs/application/pdf/2022-04/manual_of_safir_2022_-_1_general.pdf

39. R. Zaharia, C. Vulcu, O. Vassart, T. Gernay, J.M. Franssen, Numerical analysis of partially fire protected composite slabs, Steel Compos. Struct. **14**, 21–39 (2013). https://doi.org/10.12989/scs.2013.14.1.021
40. J.M. Franssen, T. Gernay, Modeling structures in fire with SAFIR®: theoretical background and capabilities. J. Struct. Fire Eng. **8**, 300–323 (2017). https://doi.org/10.1108/JSFE-07-2016-0010
41. L. Lelli, J. Loutan, Advanced analyses of the membrane action of composite slabs under natural fire scenarios a case study of the JTI headquarters. J. Struct. Fire Eng. **9**, 77–90 (2018). https://doi.org/10.1108/JSFE-12-2016-0020
42. A.A. Khan, M.A. Khan, C. Zhang, L. Jiang, A. Usmani, OpenFIRE: an open computational framework for structural response to real fires. Fire Technol. (2021). In press
43. A. Alikhan, OpenFIRE (2021). https://github.com/aatif85/OpenFIRE
44. A.A. Khan, M.A. Khan, K.A. Cashell, A. Usmani, An open-source software framework for the integrated simulation of structures in fire. Fire Saf. J. **140**, 103896 (2023). https://doi.org/10.1016/j.firesaf.2023.103896
45. M.A. Orabi, L. Jiang, A. Usmani, J.L. Torero, The collapse of world trade center 7: revisited. Fire Technol. (2020). https://doi.org/10.14264/9f81895
46. Z. Nan, M.A. Orabi, X. Zhang, A.A. Khan, X. Huang, Rapid forecasting of the structural failure of a full-scale aluminium alloy reticulated shell structure in fire, in *12th International Conference on Structures in Fire, Hong Kong, China, 2022*
47. Z. Nan, M.A. Orabi, X. Huang, Y. Jiang, A. Usmani, Structural-fire responses forecasting via modular AI. Fire Saf. J. *140*, 103863 (2023). https://doi.org/10.1016/j.firesaf.2023.103863
48. D. Lange, S. Devaney, A. Usmani, An application of the PEER performance based earthquake engineering framework to structures in fire. Eng. Struct. **66**, 100–115 (2014). https://doi.org/10.1016/j.engstruct.2014.01.052
49. Computers & Structures Inc., ETABS (2016)
50. Computers & Structures Inc., Perform3D (2021)

Improved Fire Safety in the Wildland-Urban Interface Through Smart Technologies

Jeanette Cobian-Iñiguez, Michael Gollner, Shusmita Saha, Joseph Avalos, and Ehsan Ameri

Abstract Wildfire activity across the globe has increased in frequency and intensity in recent years. Alongside this increased fire activity has been rapid population growth bringing human settlements closer to wildlands. This has resulted in the growth of the wildland urban interface (WUI) where an increased risk to people and their property from fires exists. To better plan for increased WUI fire activity, various initiatives have focused on building and planning practices that decrease the chance of ignition and increase building survivability in the case of a fire. Crucial to understanding how to best protect the built environment during a WUI fire is a thorough understanding of the causes of WUI fire ignition and spread and the factors that lead to building survivability. In this chapter, we describe key factors leading to building ignition, namely ember ignition and direct flame contact, building features shown to enhance building survivability, land management and fuel treatment practices. We use these basic principles of WUI fire behavior to propose the adoption of artificial intelligence and machine learning to tackle the WUI fire safety problem.

1 Introduction

The rapid growth of the wildland-urban-interface (WUI) has brought increasing numbers of people and their property to areas at risk of harm when fires occur in nearby wildlands. The growth of the WUI has occurred in parallel to increases in uncontrolled wildfire activity driven by climate change and modern era shifts in land management [1]. Although many countries have adopted fire safety regulations, codes and standards aimed at reducing losses, and have strategies for improved planning, protection, and prevention, many fire vulnerabilities have not been thoroughly mitigated using current methodologies. In this chapter we propose *smart WUI fire*

J. Cobian-Iñiguez (✉) · S. Saha · J. Avalos · E. Ameri
Department of Mechanical Engineering, University of California, Merced, USA
e-mail: jcobian3@ucmerced.edu

M. Gollner
Department of Mechanical Engineering, University of California, Berkeley, USA

X. Huang and W. C. Tam (eds.), *Intelligent Building Fire Safety and Smart Firefighting*,
Digital Innovations in Architecture, Engineering and Construction,
https://doi.org/10.1007/978-3-031-48161-1_8

safety measures as those strategies that address the WUI fire safety problem by integrating the state of the art in traditional approaches such as land management and home hardening and, technological advances, particularly in the form of applications of artificial intelligence, machine learning and the Internet-of-Things concept. In this way, we argue that *smart WUI fire safety* provides an opportunity to generate improved fire safety in the areas of (1) planning, (2) monitoring and, (3) forecasting. We frame the chapter around the assumption that there is not a one size fits all strategy to designing a fire safety system for the wildland-urban-interface. Instead, we propose that the characteristics of the building type, the regional wildland and natural landscape, building materials, available fuels, settlement type and local terrain and climate must all be taken into consideration when creating an approach for monitoring, planning, and forecasting for fire-safe wildland-urban-interface communities. Moreover, 'smart' approaches must consider not only the regional characteristics of the WUI but must also consider the resources available to deploy a advanced safety system. The latter may include access to power and high-speed internet, which may be more readily available in some regions than others.

To better contextualize smart WUI fire safety, this chapter begins by providing an overview of key terms and concepts related to the WUI. The following section describes home hardening strategies for WUI fire safety. Next, an overview of land management practices for wildland fuels management is provided. The following section reviews advanced technological methodologies in fire safety and their applications to the WUI. We conclude the chapter by providing recommendations for future work and implementation strategies.

2 WUI Fire Overview

2.1 Defining the WUI Fire Problem

Fires occurring within the WUI, defined as the region that exists at the intersection of the wildland and human settlements [2], are increasingly becoming a global phenomenon affecting millions of people yearly. In Canada, for instance, the last few decades of WUI fires have resulted in the loss of more than 2,400 structures representing losses of \$4.5 billion [3]. Moreover, the WUI in the conterminous US covers 719, 156 km^2 (9% of land area) and contains 44.8 million housing units (39% of all houses) with more than 35,000 structures lost to wildfire in the last decade [4]. A map of the WUI in the United States is presented in Fig. 1. In Australia an average of 118 houses have been burnt per year over the last 50 years [5]. According to the Commonwealth Scientific and Industrial Research (CSIRO), three reasons have been associated with increasing WUI fire risk in Australia [6]. The first are construction trends in the WUI, namely design and material choices which increase fire vulnerability in structures. Firebrand entry through the openings in structures also makes them more vulnerable to damage from bushfires. Moreover, the lack of minimum clearances adjacent to construction and the surrounding environment such as garden vegetation, sheds, and storerooms with flammable materials heighten risk. As will be observed in the following section, these WUI fire drivers have clear parallels in most WUI regions worldwide.

The WUI is typically classified according to the arrangement of structures in relationship to the wildland. An intermix WUI generally refers to an arrangement in which structures are embedded within the wildland. An interface of WUI, on the other hand, refers to arrangements in which structures are present at the border of wildlands [8]. The intermix and the interface of WUI may take various shapes across the globe or coexist for different communities. For instance, where in the outskirts of urban sectors an interface arrangement may be clearly identified, in more rural areas an intermix pattern may be dominant. Moreover, in places where agriculture prevails, vegetative fuels and agricultural-burn practices may influence WUI-fire patterns.

The WUI fire problem is one encompassing both structural and vegetative fuels. Both can act as ignition sources and as host fuels spreading a fire. As fires spread, burning vegetation or building fragments known as embers [9] often break away from their source fuels during combustion. When these particles become airborne and travel through the air, they are typically known as firebrands [10]. Fires in the WUI ignite and spread through firebrand transport, direct flame contact, and radiant exposure [11, 12]. Of these three, perhaps the leading source of WUI fire ignition and spread is firebrand driven hot spot ignition. Firebrands transport heat from vegetative or structural fuels to ignite homes and other structures within a community, eventually

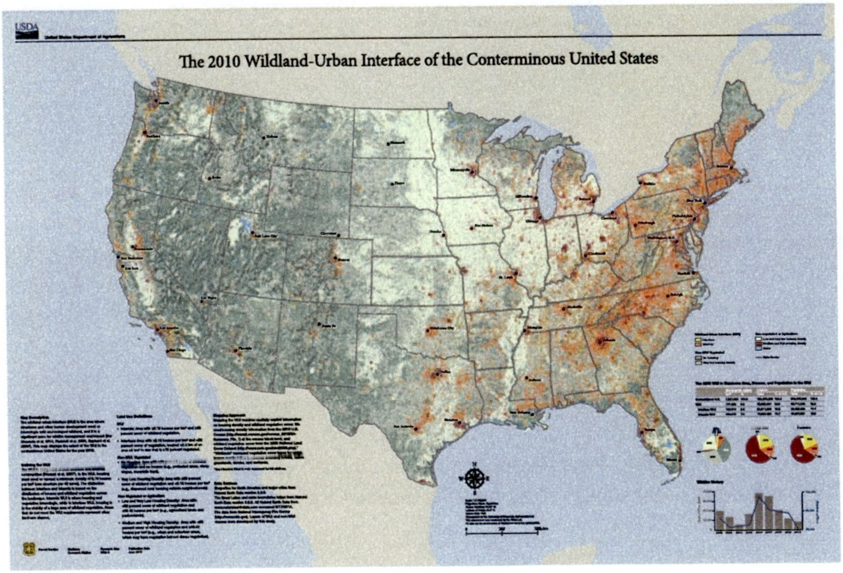

Fig. 1 Map of the wildland urban interface in the United States as mapped in 2017. *Source* Martinuzzi et al. [7]

igniting new 'spot' fires in vegetation or on structures. Whether a firebrand will ignite a structure will depend on the properties of the firebrand and the fuel that it has landed on. Thus, understanding how firebrands form and how they interact with WUI fuels is an essential element of the WUI fire problem. In the next section, we discuss how firebrands interact with various building elements to cause ignition and spread of fire in the WUI.

2.2 Fire Spread in the Wildland

Fire behavior in the wildland-urban-interface is a function of the dynamics of structural ignition and spread, community level spread and on fire dynamics within the wildland adjacent to human developments. For this reason, an understanding of wildfire dynamics is an important element of wildland-urban-interface fire safety. Wildfire behavior is shaped by the properties of fuels, topography and the weather. These drivers are represented and incorporated in classical wildfire spread models (cite Frandsen and Rothermel) and are integral parts of operational fire spread software include FARSITE and BehavePlus. Fuel models such as those developed by Anderson [13] and Scott et al. [14] group wildland fuel species into categories and through fire behavior estimates like expected flame length or fire spread, illustrate fuel-species effects on fire behavior.

Perhaps one of the leading weather drivers of fire behavior is wind which typically enhances the rate of spread [15]. The effect of wind on fire behavior has been explained in detail by canonical works in the field (see [15–19]). Works examining the effect of wind have shown that wind acts to promote fire spread to enhanced heat transfer to unburned fuels [19, 20]. Further, the effect of wind is often coupled and interacts with the effects of other key fire drivers such as fuel moisture [21, 22] and slope [23, 24].

In addition to fuels and weather drivers, topography is known to affect wildfire dynamics particularly through alterations to rate of spread and flame angle. In general, fire spread is expected to exhibit a faster rate of spread when moving upslope and a slower rate of spread when moving downslope [25]. Tihay et al. [26] for instance, found that in laboratory scale pine needle fuel beds, fire spread up a inclined surface was 2.7 times faster than fire spread through a flat surface. The effect of topography on fire dynamics is not only a function of whether a fire is moving uphill, downhill or a flat terrain but rather a function of the degree of slope inclination as shown by several works including those by Dysdale et al. [27] and Gollner et al [28].

The effect of terrain slope on fire dynamics can also extend to features like flame tilt angle and fire line shape as shown by Dupuy et al. [29]. Their work examined fire spread through a *Pinus halepensis* fuel beds up a spread surface slope of varying angles. They observed changes in fire line shape with increasing slope angle where in fire spreading through a flat surface, the fire line was a smooth curve and upon inclining the spreading surface to 10° the fire line shape changed from a smooth curve to a sharper shape with a developing angle. When the spreading surface angle was increased further to 20°, the shape of the fire line became a sharp V-shape and remained unchanged with slope thereafter. Moreover, their experiments pointed toward a correlation between spreading surface angle and flame tilt angle where for instance a 10° spreading surface inclination led to a 83° flame tilt angle, a 20° slope to a 62° flame tilt angle and a 30 slope to a 56° flame tilt angle.

2.3 *Mitigation Within the Home Ignition Zone*

Studies have found that the presence of flammable materials close to structures greatly enhance the probability of ignition during a WUI event [30]. Ensuring adequate separation between structures and possible ignition sources is often the first step towards reducing the ignition potential in WUI fires. This area around a home for mitigations to apply, including the home itself is often called the "home ignition zone" and represents the area surrounding a structure where embers, radiation or direct flame contact may play a role in the ignition of a structure and mitigation should occur. Defensible space is the term most often used to represent the area surrounding a structure where fuel modifications can be applied to prevent fires from coming closer to the structure and reducing the chance of ignition [31, 32].

Defensible space strategies do not necessarily involve complete clearing of fuels surrounding structures. They often are defined by multi-layered approaches, with the closest region near a home (~1.5 m) being completely clear of all flammable materials including mulch and wooden fences, out to ~10 m including more sparse fuels that are not susceptible to ignition and are adequately spaced so as to not spread fire to the structure, followed by wider regions where the goal would be to reduce any potential fire intensity. These approaches work well in more sparse regions; however, structure separation is also critically important as neighboring structures can produce huge fire radiation and ember load [33, 34]. Many current codes and standards in the United States [35, 36] cite 10 m as an acceptable spacing between structures to avoid ignition, however the actual spacing is not yet known and current studies [37] are devoted to better defining adequate spacing to prevent home-to-home spread in the WUI.

3 WUI Fire Safety Through Home Hardening

Developing 'smart' fire safety approaches for design, monitoring and planning, requires advanced knowledge of the performance of the various building features and materials when exposed to ignition sources. In this section, we provide an overview of WUI structure flammability as it relates to building features and construction material selection.

In building structures exposed to the elements, natural fuels such as leaves, twigs, litter, and other debris may accumulate and become a highly flammable fuel bed for ignition sources like firebrands. This may include fences, roofs, decks, windows, and various decorative architectural features. Figure 2 includes a listing of major building features involved with WUI fire ignition and spread. In fences for instance, studies have shown that clearing out accumulated debris and fuels along edges can reduce the potential of fence ignition and the consequential productions of firebrands that can occur if the fence successfully ignites [11, 38]. This is particularly important as fences can be the first point of contact between flames or firebrands produced by wildland fuels and potential structural fuels.

Although natural fuel accumulation in external portions of structures represent an important ignition concern, fuels located inside of a building may also represent a fire risk. For instance, insulation used within home walls and attic spaces such as polyurethane foam, extruded, or expanded polystyrene foam and loose fill cellulose may by easily ignited by firebrands or direct flame contact. This is especially true for insulation located in attics that are connected to the building exterior by vents that may facilitate the entry of firebrands during a WUI fire. Thus, one approach to reducing the potential of firebrand entry into attics and other building components is to cover vents with a mesh. Covering vents with fine meshes has been shown to decrease the number of locations where spot ignition occurs and the potential for smoldering to flaming transition [39, 40].

In addition to fuel and litter accumulation, the presence of overhanging leaves and branches above or next to a roof is a factor that leads to increase ignition potential. Some states implement the usage of closed soffits, which are typically metal sheathings that are installed directly underneath a roof's overhang/eave (see Fig. 3). This soffit protects the wooden overhang from the elements, and therefore flames/ firebrands, however, due to the gaps and vents built into the product, some firebrands can still enter and accumulate inside [11].

Another fire mitigation strategy in the WUI is selecting external materials that are fire retardant or fire-resistant. For this reason, many studies have focused on investigating the flammability of traditional and new building materials when exposed

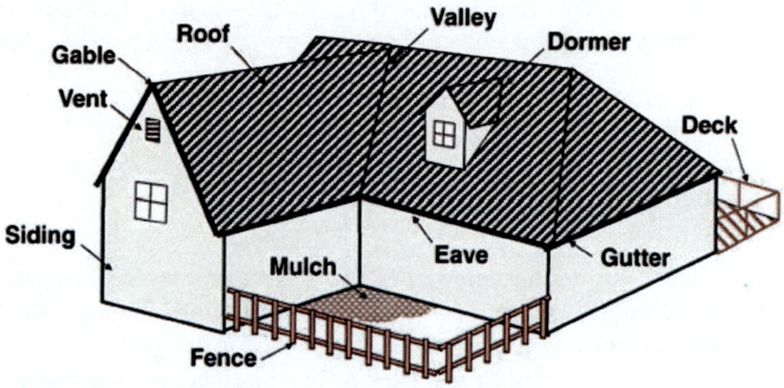

Fig. 2 Diagram of major building features involved in WUI ignition and spread (Diagram from Hakes et al. [39])

Fig. 3 Example of soffit eave (Top) (WildlifeIllinois.org) versus open eave (Bottom) [11]

to firebrands and direct flame contact. Roofing material flammability for instance, has been widely studied within the context of WUI fire safety due to their vulnerability to firebrand accumulation. There are three roofing categories, Class A, Class B, and Class C as defined by ASTM E108 [41] where Class A is typically considered the most fire resilient in the WUI [39]. Works including those by Syphard et al. [42], and Maranghides [30] have shown that replacing roof materials by non-wood based roofing materials such as synthetic roof shake, concrete based construction, synthetic/ceramic shingles or metal shingles may increase the building resiliency when impacted by firebrands. Both of these studies examined fire behavior in the WUI in San Diego, California where class A roofs are required by the Chap. 7A of the California Building Code [43]. Moving on from roofs to fences, some have suggested that replacing old wood-based fences with non-wood-based materials can help mitigate fence ignition. Butler et al. [44] found that fences constructed from combustible materials like wood exhibited greatest ignition potential than vinyl-based fences. The higher flammability of wood fences was increased if a layer of mulch was present. Figure 4 shows the differences in fire behavior for each of these fence materials.

Material selection is also critical in reducing the ignition potential of exterior building sidings. For instance, Manzello et al. [45] compared the ignition and smoldering-to-flaming deterrence of wood (cedar), vinyl and polypropylene siding boards and observed that vinyl and polypropylene sidings did not exhibit transition to

Fig. 4 Time sequence of wood-plastic composite without mulch at low windspeed (Top), Vinyl fence without mulch at high windspeed (Bottom) [44]

flaming and that only vinyl exhibited some firebrand penetration. The cedar materials on the other hand, underwent flaming ignition. It is worth noting that a key process in the firebrand ignition problem is that of smoldering-to-flaming transition which occurs when a fire spreading burning in the smoldering regime experiences a change in behavior leading up to a transition to flaming fire. In WUI fire, this is cause for concern because although a weak ignition source may produce a smoldering ignition, which burns at lower temperatures and slower increasing heat release rates, a sudden change in conditions such as increased wind speed may induce transition to flaming which will quickly accelerate and aggravate a fire. Thus, smoldering-to-flaming transition mitigation strategies are often desirable. One strategy investigated by Manzello [46] was to add bird-stops underneath clay roofing tiles which, according to their findings, can effectively reduce the potential of smoldering-to-flaming transition by preventing firebrand accumulation. While a certain fire or spotting damage is not preventable, by preventing the transition to flaming the extent of building damage and destruction may be decreased. An added benefit of the strategy proposed by Manzello [46] is the potential reduction of firebrand production due to the replacement of shingles with non-firebrand producing materials.

One of the most critical building elements contributing to increased fire ignition potential are windows. When assessing fire risk associated with windows it has been argued that the highest risk is associated with the glass and frame regardless of the materials chosen [11]. Because of this, in structures within WUI areas, double-paned panels are often required and typically at least one of the panes should be constructed from tempered glass. Double-paned windows allow the exterior tempered pane to provide shattering protection for the indoor pane and by doing so may prevent flames and embers from entering the building. As far as window frames, vinyl-framed windows deform at much lower temperatures than what it takes to break the glass, while wood window frames are a fuel source for accumulated firebrands and direct flame contact. Typically, either aluminum or wood is suggested as they perform similar to one another under standardized testing [11].

As mentioned previously the removal of fine fuels, debris, and all litter surrounding a home's edge and property (e.g., defensible space) plays a crucial role in preventing building features/components from ignition. While a certain damage on the structure or its surroundings is not preventable, the risk of a structure being ignited and destroyed from a distance via firebrand showers can be decreased substantially by ensuring that nearby fuels are managed to reduce fire intensities. Large fires can still contribute radiation over long distances (e.g., in timbered areas) and fuel treatments must be applied for a wider area around communities. This is because even sparse vegetation (e.g., chaparral) can contribute firebrands that reach multiple kilometers into communities and igniting the structures. Therefore, to reduce the potential ignition from fires in the WUI, proper land managements must be followed to avoid the spread of fires from the wildland to structures.

4 Fire Safety Through Land Management

A wildland fuel treatment for our purposes can be defined as a technique that can be used to change the wildland such that when a fire occurs it will be less intensive, easier to control, and less likely to spread into WUI regions. The main objectives of wildland fuel treatments are often to lessen the negative ecological impacts of a wildfire, reduce the danger of injury, and decrease the expense and effort of fighting wildfires (by decreasing the spread rate, intensity, or flame length of a fire) [46]. The spread and intensity of wildfire can be significantly decreased with a few treatments. By keeping the fire on the ground (i.e. preventing transition to crown fire) and reducing the density of fuel by methods such as thinning, cutting trees, or eliminating ladder fuels, firefighters have a better chance of suppressing the fire [47]. Further, fuel treatments may provide benefits to WUI communities by lowering the cost of wildland fire preparedness and suppression [46]. The primary types of fuel treatments currently used are listed below; brief descriptions are also provided.

Fig. 5 Ignition of a prescribed burn using a drip torch [48]

4.1 Prescribed Fire

A prescribed fire, as shown in Fig. 5, is a controlled fire designed to mitigate the intensity of wildfires by lowering the amount of fuel available and altering its structure [48]. In locations where smoke might pose risks to human health and transportation, controlled burning is avoided [49]. Prescribed burning is typically restricted by factors including air quality, wildlife, permitting, weather, and staff availability [50].

4.2 Mechanical Treatments

In addition to prescribed burning, one of the most well-known fuel treatment strategies is a mechanical treatment. Mechanical treatment is defined as removing vegetation that has grown to unsafe levels or rearranging the fuels in a wildland fire. Although mechanical treatments cannot completely substitute burning methods, their use in particular situations, including in the wildland-urban interface, can reduce costs and liability [50]. Mechanical fuel reduction treatments are an effective substitute for prescribed burning in certain situations, either to replace it or to set the stage for the reintroduction of controlled fire [49]. There are various types of mechanical treatment practices currently being used for land management including hand labor, thinning and chipping. These are now briefly described.

Hand labor methods are focused on manually removing, pruning or cutting trees, shrubs and grasses (see Fig. 6). The common objective of hand labor operations is to reduce the fuel load by physical removal of fuels. In the case of tall forests or shrubs, the crowns part of the fuel may become inaccessible by expanding the gap between the crown and the surface fuels and increasing the crown base height through hand labor methods like pruning or selective shrub removal [51]. Through hand labor, land can be cleared of vegetation with little disruption if just individual shrubs and

Fig. 6 Crew removing fuels using hand labor methods (CAL FIRE [53])

trees are manually cut or grubbed. On steep or rocky terrain, when clearing with bulldozers or other machinery is not practical, hand operations can be used. In areas close to habitations where other treatments, such as herbicides, are either inadvisable or illegal, brush regrowth can be easily managed by hand treatments. Furthermore, if numerous trees or shrubs are to be kept for landscaping or timber production, hand-cutting is especially suitable for thinning woody vegetation [52].

Thinning, refers to the removal of tree stems and other elements [54]. This technique can be effective for developing fire-resistant stand thinning and it can be typically divided into three categories: low, crown, and selection. While crown thinning concentrates on mid-canopy trees and selection thinning on the largest trees in the stand, low thinning eliminates trees from lesser diameter categories [55].

Chipping is a method through which machines known as chippers are used to change the size, shape, or distribution of fuels in the wildland [49]. Chipping has a number of benefits for getting rid of woody debris. It may be done at any time of the year as clearing advances; in fact, slash is frequently more readily chipped when it is still green. Chipping eliminates the issues associated with burning stacked brush in the winter and the potential for fire escape at other times of the year. Pollution issues are prevented when there is no smoke. Chipping can also improve esthetic values since slash disposal lessens the likelihood that fire would harm trees meant to be left standing and the chance that burned-out pines will prompt beetle infestation hotspots. Cost is a significant drawback to chipping, unless the chips can be used commercially [52].

4.3 *Fuel Breaks*

Moving on from mechanical treatments, a fuel break is a process through which a portion of a vegetated area is thinned or reduced. In the WUI, fuel breaks can be designed to prevent or slow down the encroachment of a fire towards homes and other structures which are considered to be critical in WUI fire safety planning. In addition to applying fuel breaks to the landscape as a fire management technique, landscape features may serve as natural fuel breaks, or buffers, shielding WUI structures from fire. These include agricultural areas, outdoor sports fields, managed open spaces, areas with targeted grazing, among others. Fuel breaks are particularly useful nearby WUI communities as they can be used as buffering patters to protect agricultural areas, outdoor sports fields, managed open spaces, managed open spaces with targeted grazing and other human settlements. These buffer areas can stop the fire from moving forward and reach to WUI structures [56].

4.4 Grazing

On the other hand, grazing, a technique in which grazing animals consume, break off, or trample vegetation to lessen the amount or density of fuels, is most effective in grasslands or shrublands [57]. Figure 7 shows a typical grazed landscape as presented by Moritz et al. [58]. Rangeland managers and planners can strategically combine fuel management techniques like cattle grazing with other natural resource objectives and management practices like controlled fire in order to lessen wildfire dangers [58].

Fig. 7 Examples of grazing animals from Moritz et al. [56]

4.5 Mastication

Mastication is a fuel treatment practice in which material is ground, shredded or chopped to reduce fuel particle size. Trees or bushes can both be masticated. However, each plant type's species, diameter, and height will dictate how difficult the particles will be to chop and how much power will be required to complete the procedure effectively. Masticated fuels can achieve various treatment goals in addition to eliminating canopy fuels and lowering the likelihood of canopy fire. To function as mulch and guard the soil surface from moisture loss or erosion, certain fuels are cut into tiny, regularly shaped pieces and spread up in very dense layers on the ground. For small animals to use as shelter or passageways, other masticated fuels including mixtures of minute shreds and huge logs can be used. Additionally, masticated fuels have been used to improve soil nutrients, enhance the diversity of species in the understory, reduce soil warmth and moisture, and decrease the stability of the soil aggregate [59].

4.6 Other Land Management Techniques

Chemical treatments as land management tools are mostly used to stop regrowth of plants, especially in areas that are inaccessible to heavy machinery and where soil disturbance is to be prevented. Some herbicide treatments for instance, are potentially successful for the long-term treatment of annual grass-dominated rangelands, but if they are not filled by desired perennials, treatments will leave a void and they will probably return to annual herbaceous communities [60].

Although each of the fuel management mentioned in this section may be used alone to control, reduce or reorganize fuels, combining two or more of these techniques may increase the effectiveness of land management campaigns. For example, combining mechanical and prescribed fire treatments can most effectively limit the possibility of active crown fires [61].

4.7 Ornamental Plant Maintenance and Selection

Fuel treatments, prescribed burning and other land management practices mentioned in this section, contribute to the reduction or management of fire risk in wildland fuels near WUI structures. In the WUI however, wildland fuels are only part of the natural fuels landscape as ornamental plants located near structures represent potential fuel sources for spreading WUI fires. Ornamentals are considered decorative plants and objects typically placed against a structure or in its surroundings. As grasslands and shrublands have some of the largest natural rates of spread, they are dangerous to nearby structures. As such, the surrounding fields should be maintained constantly.

The type of ornamental used can be the difference between them igniting and spreading flames onto a structure or not. Findings from Ganteaume et al. [62] for instance, showed differences in flammability between various species of ornamental plants used in southeastern France. In their study, plant species with a higher leaf area-to-volume ratio exhibited greater flammability than species with a lower area-to-volume ratio. This increase in flammability may pose a risk in the WUI as highly flammable ornamentals may facilitate ignition to nearby structures. Thus, such work highlights the importance of considering plant species when selecting ornamentals for WUI communities. Moreover, like in wildland fuel species, the moisture content in ornamental plants affect flammability as studies have shown that lower moisture content may lead to enhancements of burning through increases in heat release rate [63].

5 Fire Adapted Communities

In addition to building planning and land management, community level interventions that integrate social dimensions to fire-preparedness have become popular approaches to advanced fire safety. This has taken shape in what are known as Fire Adapted Communities (FAC), groups that bring together residents, land managers, fire professional and other stakeholders to develop strategies for fire planning, response and recovery [64].

In the United States, the Firewise program for instance, is a community-led fire mitigation program organized by the National Fire Protection Association and co-sponsored by the USDA Forest Service, the U.S. Department of the Interior and the National Association of State Foresters (https://www.nfpa.org/Public-Education/Fire-causes-and-risks/Wildfire/Firewise-USA, accessed 21 April 2023). In this program, communities self-organize and recruit members that then reach out to Firewise to coordinate a fire risk assessment conducted through a site visit. Communities then collaborate with Firewise to develop fire mitigation activities such as community-led fuel treatments, educational workshops and community cleanup days [65]. The impact and effectiveness of these programs is still an ongoing subject of study but recent works have shown encouraging preliminary findings including lower home destruction rates in communities near Firewise communities [66].

6 Advanced Technology Applications for WUI Fire Safety

There is a growing body of work leveraging artificial intelligence and machine learning for wildland-urban-interface applications. Zohdi [67] for instance, used machine learning to estimate firebrand transport dynamics. The method used physical models and the Digital Twin concept to develop an ember tracking methodology for rapid deployment of hand-held devices. Tohidi et al. [68] developed a computer vision-based methodology to detect and characterize firebrands. Similarly, Hedayati et al. [69] developed an automated edge detection algorithm for firebrand characterization.

Although the aforementioned and other works have used 'smart' techniques for WUI fire safety, the number of works in this field is still limited. As the field progresses to increasingly embrace 'smart' technologies, researchers and practitioners may find insightful guidance from the fields of structural and compartment fire safety where successful applications of machine learning and artificial intelligence are more numerous. Huang et al. [70] provides guidance on the application of 'smart' techniques for structural fire safety which we now summarize as these have useful parallels to WUI fire safety. In general, they propose three areas of potential for the use of artificial intelligence for building fire safety: (1) AI based fire engineering design, (2) Implementation of Digital Twin technologies for building fire monitoring and, (3) Super real-time forecasting of fire evolution. These areas highlight potential applications for use of 'smart' technologies at various stages of fire safety engineering. For instance, the development of data driven and intelligent design strategies that address fire vulnerabilities in buildings can be a critical element of proper fire safety planning (see Chap. 11). The Digital Twin approach on the other hand, represents an innovative means of fire monitoring and could be an important ally in firefighting and evacuation (see Chap. 13). In the Digital Twin method, a digital twin of a physical building integrates data from monitoring devices active in a physical building (the physical twin) to run fire scenario simulations [70]. These fire scenario simulations can be used to predict fire evolution in the physical building. Similar applications are proposed for wildland fire spread and suppression modeling and thus, along with the Digital Twin, represent innovative means of fire monitoring. Furthermore, within the realm of fire forecasting, numerous studies have aimed to develop real-time data-driven fire modeling capabilities, integrating observations into their output [71–74]

Although approaches addressing the WUI and wildland fire problem are in their infancy, they have already been applied in structural and compartment fire providing a preliminary framework, which once adapted to the unique characteristics of the WUI, could result in the development of effective strategies for smart fire safety. We argue that the three areas above can be categorized into methods of, (1) fire planning, by ways of fire-resilient design, (2) monitoring and, (3) forecasting. We therefore propose that these same areas are key areas of growth for 'smart' fire safety in the WUI, this section will touch upon approaches to fire safety in each of these three areas.

We conclude the introduction to this section by making a note on the Big Data Paradigm within the context of fire safety engineering. One of the most important requirements for the implementation of artificial intelligence and machine learning to any system is a very large dataset. This is why the *Big Data Paradigm*, and its three V's, volume, velocity, and variety, are closely tied to the successful integration of smart technology. Like in many other engineering applications, producing large volumes of real-world data can be resource intensive. A cost saving alternative that some researchers have advocated for is the generation of synthetic datasets through numerical modeling. This approach is not exclusive of fire safety engineering. Synthetic data alternatives provide the opportunity to implement methods like machine learning and artificial intelligence to solve challenging problems without being restricted by experimental or field data availability. A word of caution however is provided later in this chapter on the over reliance on synthetic data especially when it is completely devoid of validation with real data. There, we will present an argument for collaborative experimental research projects focused on generating such datasets.

We begin this section with an overview of sensing technology and current leverage of the internet of things philosophy to fire safety. We continue with a discussion of computer vision applications followed by a description of current ember production and firebrand dynamics works. The chapter is concluded with key takeaways and notes on future work.

6.1 Artificial Intelligence of Things (AIoT)

The concept of Internet of Things (IoT) refers to the integration of everyday physical objects into the virtual world [75]. The physical objects that perhaps are the most elemental for implementation of IoT frameworks are sensors which have the role of perceiving the physical world and translating it to the virtual world by ways of signals and data. In this way, IoT technology bridges the gap between the physical and virtual world and provide a means through which sensor data can be integrated into virtual networks. Because of this, IoT systems represent particularly useful technological advances for disaster situations, including fire, where situational awareness and monitoring are essential for development of adequate response and mitigation actions [76]. These sensors could provide information on fire location as well as local conditions such as temperature, humidity, smoke, etc. Several recent studies have leveraged this potential to develop systems for fire monitoring and prediction in various fire situations. In this section we examine how these strategies have been effectively implemented for various fire situations and comment on how future approaches may leverage IoT and sensors to plan for fire prevention and, monitor and forecast fire evolution in the WUI.

As with other field-deployed monitoring systems, the use of low-cost systems has many advantages ranging from reduction of overall system costs, greater data volume due to the capacity to deploy more sensors, and increased spatial and temporal resolutions. In addition, systems built from low-cost components can make fire safety advances more accessible to homeowners who although may appreciate innovative methodologies developed to keep their homes safe and protected may not be able to afford them. In this way, low-cost fire monitoring systems may extend ‘smart’ fire safety to the greater numbers of end-users worldwide.

Designing and optimizing low-cost fire safety systems is a balance between accuracy and cost and therefore requires careful selection of affordable sensors and data-acquisition devices. Thus, reliable systems must be tested against real or reliable modeled data. A low-cost fire detection system for compartment fires presented in Antunez et al. [77] for instance, integrated sensing, data acquisition, processing, and communications with other fire safety systems such as sprinkler networks. Central to their low-cost approach was the use of affordable off-the shelf far infrared sensors which can be orders of magnitude less expensive than infrared cameras. In addition, data acquisition and controls were conducted using a Raspberry Pi microprocessor, adding to the affordability of the instrumentation package. To better understand the effectiveness of the system to detect laboratory scale experimental fires, IR images and temperature profiles produced by the system were compared to visual images overlayed with IR temperature detections. The resulting analysis provided a qualitative assessment of system performance and showed that overall, the system was able to detect fire and identify the occurrence of critical temperatures indicative of fire conditions.

In a similar study, Rosas et al. [78] developed a fire detection system that used standard temperature and environmental sensing devices to monitor fire conditions inside buildings. Their system integrated communications infrastructure using a mobile application that provided push notifications to users when critical temperature thresholds were reached. Although system performance was not tested, their proof-of-concept design showed how relatively simple and affordable sensors could be used alongside user-friendly mobile applications to enable users and stakeholders to monitor building health. Several proposals have been made to implement wireless sensor networks to detect fires both in the backcountry and WUI communities [79]

Moving from active fire monitoring to prediction and post-fire analytics, a relatively new application of machine learning and artificial intelligence in fire focuses on predicting future fire scenarios and in conducting forensic investigations based on data on structural damage and destruction. In the area of prediction, several efforts in structural and compartment fire are worth mentioning as the methods and applications are potentially transferable to WUI fire safety. The first notable application is for the identification of fire location or fire characteristics given sensor data. Obtaining this information during an active fire can help firefighters better target areas with active fire and plan evacuations. A good example of such approach is that presented in Wu et al. [80] where an AI-based methodology to predict the location of a fire source within a compartment fire was developed. There, the focus was the interior of tunnels where vehicles may become trapped during a fire. The study investigated the use of deep learning methods through a type of recurrent neural networks (RNN) known as long short-term memory (LSTM). The goal was to simulate a great number of tunnel fire scenarios and form a large database; the database was used to train a LSTM model that was in turn used to characterize the location and size of a fire within the compartment. In addition to identifying fire location and size, their study provided insight into the use of LSTM to optimize sensor location. Because the optimum location of the sensor will vary greatly depending on the configuration and geometry of the actual compartment, their methodology provides a potential framework for designing of sensor networks for building monitoring not only for compartment fire applications but other applications such as WUI fire. Here in lies one of the greatest takeaways from this methodology for WUI fire applications, optimized design of sensor locations. As we project into the future of 'smart' WUI fire buildings and landscaping, monitoring, and designing smart buildings and landscaping will require greater and better instrumentation and sensing capabilities. To achieve this, we can leverage AI methodologies to simulate parametrized WUI buildings and landscaping fire scenarios where we vary sensor placement.

Lessons learned from Wu et al. [80] show that deep learning methods such as neural networks are uniquely equipped to support the large dataset required to thoroughly examine the wide range of location sensor location scenarios required for an optimized design. In the case of WUI fire, sensors could be placed outside of a structure to quickly identify the landing of firebrands and approaching flames. Moreover, by extending the parameter space, one may consider optimizing sensor location for key building elements known to accumulate firebrands thus prioritizing building elements with high fire risk.

Similar to the work in Wu et al. [80], Cabrera et al. [81] developed a methodology for post-fire investigation using AI. Their approach sought to determine fire scenarios in a compartment given a set of potential ignition sources. In their scenario, three packages were placed in a room and one of them was ignited while the rest were tested to check whether they reached ignition conditions. The task for the AI was to test various hypotheses as to what will happen once an initial ignition occurs thus providing insight into fire evolution in the compartment once an ignition has occurred. Their method of choice was a Bayesian framework which tested various ignition hypotheses and calculated the probability of likely scenarios by examining sensor data. The sensors used in the study were flame thermometers placed around the compartment to capture heat flux emanating from the potential ignition sources. The methodology was able to identify correct scenarios with a good degree of accuracy where part of the variation was due to the location of sensors. Worth noting from this study is that in contrast to common usages of AI in fire where data inputs are based solely on simulations including CFD, their work leveraged experimental data, thus creating an interesting framework for the successful integration of experimental data into AI-based fire behavior methodologies.

Whether it is using experimental or numerical data to monitor or forensically investigate fires, a common feature in 'smart' fire safety methodologies is the development of methods to back-calculate fire behavior features occurring during a fire. In addition to estimating fire location or ignition behavior, identifying potential fire spread behavior can be an important contribution of AI to the field of fire safety. In Buffington et al. [82] for instance, an artificial neural network framework was developed to formulate heat release rate scenarios given temperature profiles for compartment fires simulated using the Fire Dynamics Simulator (FDS), a CFD based fire behavior software. The opposite problem was also addressed, that is, formulating temperature profiles from heat release inputs. In this approach, the input data, whether it was HRR or temperature, was first simulated using a Gaussian process and FDS or CFAST. Once the temperature or HRR curves are generated, these are sent to an artificial neural network (ANN) framework for training. The ANN will map out an estimation for the correct HRR or temperature output as could be generated by FDS. Overall, their method was effective in producing temperature predictions with an error of less than 2 degrees Celsius. The estimation of the HRR curves had an error of 6% on average. Hence, their method had a greater degree of accuracy for the temperature estimation than the HRR estimation.

Physical data collected by sensors provide real-time data for system monitoring or fire behavior prediction in which facilitate decision-making based on actual fire conditions and characteristics. Some approaches have been proposed to fill in this data, such as using crowd sourcing [83]. However, relying on physical data may not be feasible or the most effective method, especially in situations where fire prediction may be needed for fire safety application. In this case, synthetic data may provide a useful avenue for generation of simulated scenarios which may lead to rapid response to future fire evolution or even for planning of future fire scenarios. In this way, some studies have explored the use of synthetic data for fire evolution analysis. Tam et al. [84] for instance, developed a method to generate synthetic data for use in compartment fire simulations. Their method focused on generating building fire input parameters related to room geometry and thermal properties. The input parameter space was generated using expected probability distribution functions, namely normal, lognormal, and uniform distributions, and fed into a simulation engine to produce fire behavior outputs. To test the potential use of their data generation system, they tested the inputs generation machine using three machine learning methods: support vector machine, decision tree, and random forest. Their machine learning models showed good agreement between ML predictions of location and the ground true location. Following required machine learning practice, the input data parameter space was divided into a test set and a validation set. Results showed agreement between predictions produced when applying machine learning methods with their test set and the metrics provided by the validation set. However, the method was not validated with real data.

Overall, the compartment fire studies mentioned, show how AI and ML have been used to characterize fire behavior given simulated or experimental inputs. It is now fairly common to find visible or infrared video used as a fire detection source, however there are many opportunities to further optimize this approach with improved detection algorithms, placement, etc. [85–87]. To apply more of these methodologies to WUI fire, additional approaches that integrate features particular to the WUI would be necessary. For instance, contrary to compartment or building fire where the focus is on a single structure, the focus of a WUI fire safety framework is on multiple structures as structure-to-structure ignition is common. Moreover, structures can also become ignition sources themselves. Thus, the concept of sensor location optimization for fire monitoring and firefighting could be an asset for WUI fire safety applications. In WUI fire, the sensor network would likely need to extend past the main structure and would encompass annexed structures, such as storage sheds, and the landscaping present between the structure and the wildland. Because the primary fire risk of structure ignition during a WUI fire is on the exterior of a structure, sensors would need to be placed in the exterior walls and elements of a structure. Artificial intelligence can therefore be used for data processing and feature characterization as well as optimization of sensor location and structural element design. Additionally, integrating video and infrared cameras to monitor approaching flames in the wildland could further improve WUI fire risk estimations. More information on computer vision and image processing is provided later in this section.

Moving on from fire behavior prediction to post-fire analysis, it is worth noting that a particularly valuable source of forensic information for WUI fire analysis is provided by datasets like those originating from damage inspection reports compiled in the aftermath of large fires. In California for instance, the Damage Inspection (DINS) dataset by the California Department of Forestry and Fire Protection (CAL FIRE) provides valuable information on structural loss during WUI fires. Analysis of these datasets through statistical or machine learning methodologies may maximize our understanding of the mechanisms leading to greater structure loss during WUI fires and thus lead to improved planning and data driven home hardening efforts. In work by Troy et al. [85], for instance, the DINS dataset for the 2018 Camp Fire was analyzed using logistic regression; their study found that structural survival was associated with factors including the type of roof, whether double-pane windows were used and the existence of vents. Similarly, in Syphard et al. [42], they analyzed damage inspection data compiled by the County of San Diego Department of Planning and Land Use on fires occurring in San Diego, California in October 2003 and 2007. They used linear multiple regression models and classification trees to better understand the factors that led to structure destruction and survival. Their findings pointed towards vinyl window framing, tile roofing material and double pane windows, among other factors, as being conducive to greater probability of structural survival. Works like those by Troy et al. [85], Syphard et al. [42, 86], and Knapp et al. [87] highlighted the potential of conducting analysis of damage inspection data to better understand the factors leading to home survival. Thus, planning for home hardening and WUI-safe building design could be increasingly driven by advanced analysis of structural survival data using statistical methods.

In the future, machine learning may be leveraged to develop or parameterize models to understand future structural properties and community arrangement scenarios or to produce estimates for WUI scenarios in communities where a damage inspection dataset may not be readily available. This is particularly important for greater integration of region-specific building approaches and community arrangements such as those mentioned in Sect. 1 where neighborhood arrangement, building material types, and types of building elements can vary greatly from one country to another and even within a country or region. Further, in terms of building monitoring for fire safety, current approaches focus almost solely on compartment and structural fire and although these have transferrable applications to WUI fire safety, they do not however, address the WUI fire problem holistically. This is particularly the case because the WUI fire problem includes more than structures in isolation but also clusters of structures and a landscape physical connection with the wildland. In this sense considering elements such as structure-to-structure ignition, vegetation-to-structure ignition and fire spread in nearby vegetation are important features often not being integrated in structural/building fire safety but critical to WUI fire safety. Planning, monitoring, and forecasting for firebrand generation, transport, and ignition, should be the critical components of any smart WUI fire safety scheme.

Artificial intelligence could be used to understand which locations and structure elements should be prioritized for fire monitoring. Whether it is on walls, roofs, decks, or high-risk structural elements like vents, data from CFD simulations or experiments focusing on WUI fire could be used as inputs to AI algorithms to determine sensor location. Approaches to fire safety in building fire in non-WUI scenarios provide important insight into smart fire detection, monitoring and planning. The fire safety community has taken advantage of advances on the IoT approach to building health monitoring by developing an array of systems for fire detection and the communication of fire detection with other building sensing systems.

For locations which have been identified with high fire vulnerabilities in structures, adding sensor networks could prove essential to providing monitoring of risk and early warning during a fire and valuable data for post-fire investigations. As mentioned in Sect. 2, building features known to affect building ignition risk and survivability during a WUI fire include roofs, windows, vents, fences, and exterior sidings. The mechanisms through which these facilitate firebrand penetration to buildings, and firebrand and combustible debris accumulation are explained in detail in Sect. 2. Given, as described in Sect. 2, the mechanisms via which these building elements facilitate building ignition and spread include facilitating firebrand penetration and the accumulation of firebrands and combustible debris, we propose that sensors can be deployed in key locations of vulnerability on structures to monitor building maintenance (i.e., to promptly identify risky situations like debris accumulation) and fire risk situations like the entry of firebrands into high-risk areas. Determination of the type of sensors and data acquisition systems would depend on factors like targeted data sources which could include simple building maintenance or active fire detection. In the latter case, temperature sensors, heat flux gauges, radiometers, and low-cost air quality sensors could provide essential information for an Internet-of-Things frameworks. Further, we propose that these systems would benefit from large sensor networks that would better maximize integration of data acquisition from multiple strategic points around buildings. For this reason, these systems would most likely have to composed of low-cost sensors and data acquisition systems that would maximize the volume of data acquired while maintaining a reasonable cost.

6.2 Computer Vision

Visual fire data provides important qualitative information such as that related to ignition detection and can be used to obtain quantitative data on flame parameters such flame height, tilt and spread. In WUI fire, image analysis can be used to monitor ignition and spread in vegetation and structures. Methodologies using image analysis for fire protection and monitoring are varied although most are based on visual imagery. Perhaps the most basic type of image processing techniques for fire is edge detection which as opposed to more advanced methods that will be described later, rely heavily on human interaction for processing steps including labeling, classification, and tuning. Edge detection consists of isolating a flame usually by way of conversion of a real-color image to a black and white image, also known as intensity image. The isolated flame in black and white format is then used to identify the contours, or edges of the flame. The resulting flame edge data can then be used to calculate flame features such as flame height, length, and tilt. Works applying edge detection for fire image analysis include those by Gupta et al. [88], Qiu et al. [89], and Cobian-Iñiguez et al. [90, 91]. In these types of approaches, the user may interact with the image processing code to determine critical thresholds for converting RGB images to binary and to crop non-fire elements as ways of manual labeling. An example of edge detection process can be found in Cobian-Iniguez et al. 2019; there edge detection was used to calculate flame height and tilt in experimental chaparral crown fires (see Fig. 9).

Although 'manual' approaches of image analysis may be useful for research teams in which a limited number of experiment images will be processed, it may be less useful for real-time fire monitoring in the WUI or wildland. This is because the data volume in real-world fire observations may be orders of magnitude greater than that of experimental case analysis. In this case, deep learning, which is a class of machine learning techniques, becomes a critical ally as it can allow for the reduction or replacement of human input from the image analysis process. There, the human can be replaced by systematic machine-driven image classification. Thus, in recent years, multiple approaches have emerged that use deep learning in flame image detection and classification. A common approach is the use of neural networks for image classification and labeling. For instance, convolutional neural networks have been used to classify images and detect fire [92–95] and smoke [96–98] from footage of outdoor fires. An example of a convolutional neural network approach as presented by Frizzi et al. [96] is reproduced in Fig. 8.

Examples of methods other than neural networks used for fire detection are presented by Toreyin et al. [99] et al. and Celik et al. [100] as well as the review by Bu et al. [101].

An interesting fire image processing approach is to use infrared imagery either as a standalone method or in conjunction with RGB imagery. In Martinez-de Dios et al. [102] for instance, infrared and visual imagery were used to detect the presence of fire at the landscape scale and to estimate flame features including flame height and

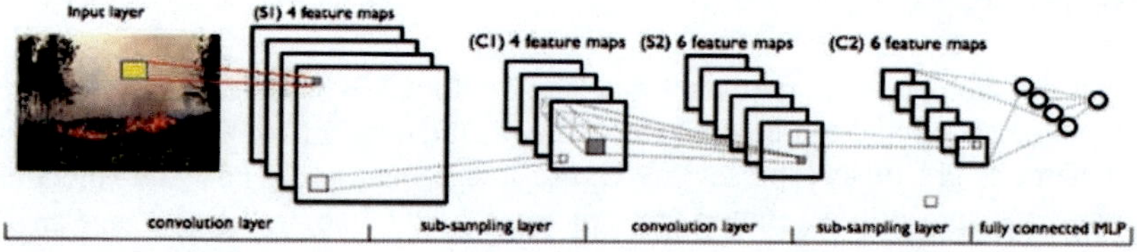

Fig. 8 Convolutional method developed in Frizzi et al. [96] (image as presented in original source)

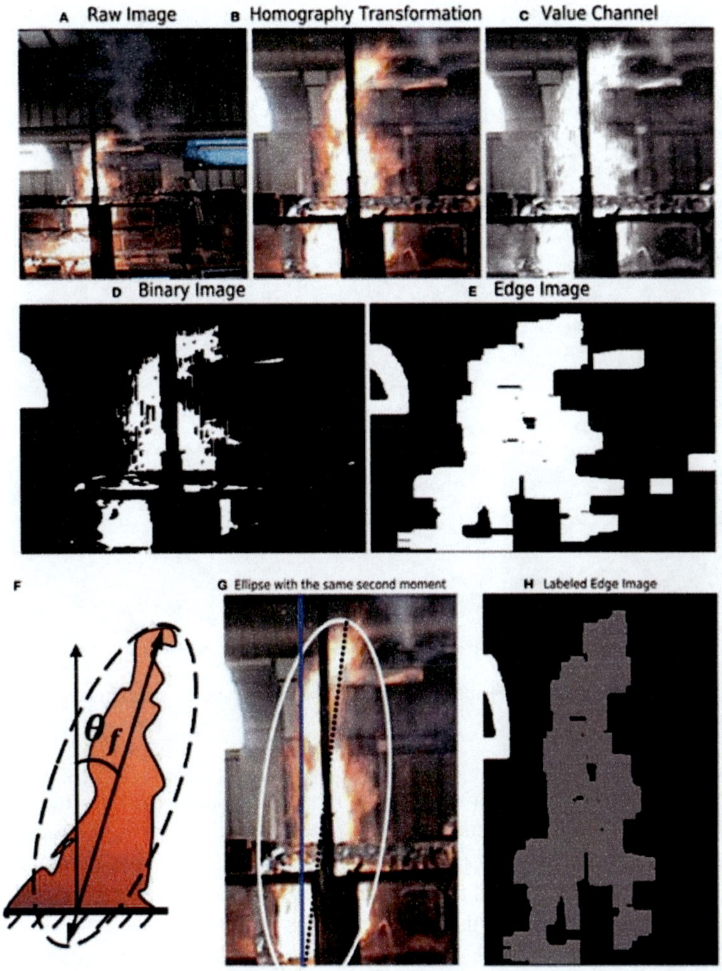

Fig. 9 Flame tilt estimation algorithm presented in Cobian-Iñiguez et al. [90] where the image processing steps are as follows (clockwise from the top left): **a** raw RGB image obtained from a single video frame, **b** homographic transformation, **c** Value extraction, **d** Binary black and white image, **e** Edge image, **f** Illustration showing the direction of the tilt angle with respect to the vertical axis, **g** labeled image showing the ellipsoidal envelope around the flame, **h** labeled edge image

tilt angle (see Fig. 9). Similarly, Töreyin et al. [103] used infrared imagery to detect fires in images from a diverse set of outdoor fire settings (Fig. 10).

Infrared technology provides many advantages in fire detection such as the enhanced ability to isolate flames from an image due to the added temperature detection feature of infrared imagery, one common drawback is the high cost of infrared instrumentation. Thus, a potential alternative is the development of lower costs

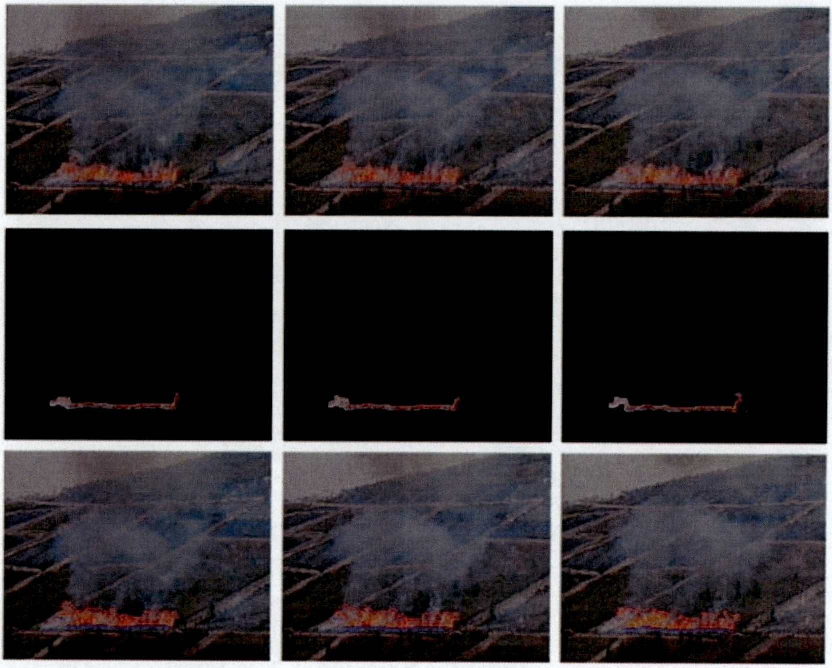

Fig. 10 Example of image processing approach for landscape scale fires presented in Martinez-de Dios [102]

systems based on off-the-shelf sensors as opposed to expensive infrared cameras. A good example of this is the study by Xavier et al. [104] where by leveraging pyroelectric infrared (PIR) sensors and deep neural networks, they developed a system that uses features detection and machine learning to perform classification and identification of fire scenarios. In this approach, PIR is preferred over classical computer vision approaches for image classification and fire detection because, as they argue, processing PIR data is less computational expensive than processing visual data using computer vision. To better understand the performance of this method various experimental scenarios were tested, two fire scenarios, two human motion scenarios and a scenario simulating background activity. Out of 322 fire test scenarios detected, their system was able to correctly identify 307 cases. For human motion, 777 out of 841 cases were correctly identified and, 348 out of 349 cases testing background activity detection were correctly identified. Accuracy in feature detection seemed to be affected by distance from sensor as prediction accuracy decreased as distance to sensor increased.

Although greater integration of computer vision and other image processing technologies for WUI fire planning, monitoring and response benefits from advances in machine learning and image processing for outdoor, compartment and wildfire applications, further research is needed to incorporate peculiarities of the WUI fire problem. For instance, critical to WUI fire safety is structure-to-structure and vegetation-to-structure spread which as has been mentioned before in this chapter is driven by direct flame contact and firebrand ignition. Monitoring fire activity in vegetation surrounding the WUI, in this sense is critical for early detection and warning systems. For instance, deployment of CCTV or similar equipment in wildland areas in the vicinity of structures could provide vital information for homeowners, authorities, first responders and other stakeholders. Advancement in research resulting in increased precision and reduction of false alarms may yield these systems more effective and impactful.

The use of computer vision for 'smart' fire protection, however, is still in its early stages and much work is required not only to optimize hardware and software for image detection but also to develop systems that protect the privacy and safety of homeowners. This is because as opposed to wildfire detection which is presumable conducted away from the presence of humans, WUI fire safety is inevitable intertwined with human settlements and thus much work is still needed to ensure that systems being deployed for fire detection complies with cyber-security protocols and practices. Further, although it will not be covered in this chapter, it is important to note that there are important advancements and applications of remote sensing for WUI fire safety. The use of satellite instruments such as VIIRS and MODIS [105, 106], aerial observations and other types of remote sensing allow for the safe and early retrieval of critical fire safety information including the occurrence of hot spots and approaching flames. For a sampling of remote sensing studies for WUI fire safety the reader is referred to Skowronski et al. [107], McNamara et al. [108] and Badia et al. [109].

6.3 *Firebrand Production and Transport*

Up to now we have discussed methods for monitoring fire activity in buildings and, forensic data analysis for planning and forecasting. Most of the methods examined and some parallel solutions proposed for the WUI have in common a focus on monitoring the status of an ignition source such as a firebrand or the analysis of structure damage or destruction occurring after a fire. Preceding the entrance of a firebrand and the spread of a fire through a structure and the surrounding vegetation, is the production and transport of firebrands that facilitate such ignition and spread. Therefore, understanding the mechanisms that lead to firebrand productions, the factors that determine firebrand characteristics, and the firebrand flight dynamics is therefore key concerns for fire safety that are unique to WUI and wildland fire. Important insights on this have been produced because of several experimental studies conducted at the laboratory and field scale addressing firebrand production and transport.

At the laboratory scale for instance, Manzello et al. [110, 111] conducted experimental burns investigating the size and mass distribution produced by burning Douglas fir and Korean pine trees respectively. At the building and community scale, Suzuki et al. [112] and Manzello et al. [113] analyzed firebrand characteristics from a controlled structural burn and from a real fire, the 2007 Angora fire occurring in California, respectively. Their results provided insight into structural and vegetative firebrand characteristics. Filkov et al. [114] investigated firebrand characteristics occurring during a prescribed fire therefore providing information on firebrand features produced by vegetation. An example of their thermal images is provided in Fig. 11. The wealth of data and information produced by these and other studies on firebrand mass and size distribution produced by structural and vegetative fuels, represent important opportunities to leverage machine learning techniques that can provide greater more efficient data feature and pattern extraction. A good example of this is the work by Jha et al. [115] which used machine learning to process firebrand production data obtained from a series of experiments conducted at the Insurance Institute for Business and Home Safety Research Center in Richburg, South Carolina (full report on data and methods available in Zhou [116]). Their machine learning-based approach focused on predicting the number and mass of firebrands landing on a target area. The independent variables considered in the study were firebrand travel distance, wind speed and fuel type (vegetative and structural fuels). Their analysis showed that given the non-linear nature of the firebrand data, the most suited machine learning method for the process was a non-linear, non-parametric method which in their case was the K-nearest neighbor method. This latter point is an example of the importance of properly selecting the best methods to match input data features and patterns.

Characterizing firebrand production characteristics enables forecasting on firebrand geometry, density and mass occurring as a result of features like fuel species and moisture content. In addition to firebrand feature characterization, understanding the flight path and dynamics of firebrands provides greater understanding on the potential of firebrands to land on high-risk areas such as vents, decks, or roofs and as a consequence is critical for a comprehensive characterization of firebrand ignition potential in the WUI. Further, much like in fire detection, computer vision and edge detection techniques have been used in firebrand detection. Similar to flame detection,

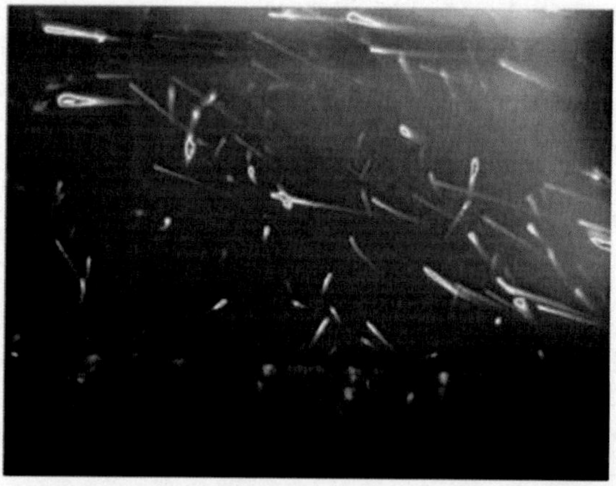

Fig. 11 Thermal image of firebrands used to study firebrand production in Filkov et al. [114]

most techniques use either visual imagery or infrared photography to characterize firebrands. Current research has enabled the characterization of various firebrand features including firebrand area [69, 117], trajectory [118–120] and flux [68, 69].

7 Conclusions and Future Work

We conclude with some takeaways and recommendations for future work. First, we observe methods that have been widely adopted for fire safety applications, notably, neural networks. Neural networks methods belong to the category of machine learning methods known as unsupervised methods. In contrast to supervised methods where the input and output features are given by the dataset, in unsupervised methods, only the inputs are given [121]. In unsupervised methods the algorithm identifies trends in the input data and can categorize or cluster the data based on identified trends [122]. This makes unsupervised methods attractive for fire safety applications such as those related to image labeling and processing, post-fire investigations and others. On the other hand, regression models which are supervised learning methods [122] have been successfully used to analyze fire damage inspection data [42, 85]. Because supervised learning may be less computationally expensive than unsupervised learning, regression approaches may be well suited for efficient implementation of machine learning to datasets where the inputs and outputs are well known such as those created from post-fire investigations.

Second, the use of artificial intelligence and machine learning has been primarily focused on compartment and structural fire problems, therefore adoption to the WUI requires examination of the particularities of the WUI fire problem, many of which were described here in Sects. 1 and 2. To better describe how to successfully transfer the knowledge and methodologies developed by compartment and structural fire safety works to WUI fire applications we refer to the set of recommendations provided in Naser et al. [121, 123] for applying machine learning and artificial intelligence in fire science and engineering. Key ideas from those works are now mentioned. To select an algorithm, Naser et al. [123] recommends first understanding the nature of the fire problem and understanding which category of machine learning algorithms best fits the problem (i.e., supervised or unsupervised). Moreover, they argue for a balance between data quantity, quality and range and suggest that simply having more data does not guarantee a quality dataset. They recommend prioritizing experimental data over synthetic data as inputs for machine learning. This is because experimental data may contain random effects that synthetic datasets such as those produced by methods like numerical modeling may not be capturing. Including these random effects is important as these may in fact be representative of physical system features.

In this chapter we have presented the 'smart' WUI safety issue as one that can benefit from lessons learned from the of application land management, home hardening practices, artificial intelligence and machine learning in compartment and structural fire science and engineering and, from current successful applications such as those related to ember production and computer vision. We conclude by recommending that future works and new directions in 'smart' fire safety in the WUI should integrate the particularities of the WUI including the properties of the data, and should leverage the most suited machine learning methods for the application. Designing 'smart' fire safety schemes that meet these challenges will require large high-quality datasets which may be achieved through coordination between engineers, scientists, and practitioners. Thus, we argue that as recommended by recent works [124] removing silos and enhancing data sharing architecture may facilitate the application of 'smart' methodologies in WUI fire safety.

Acknowledgements The authors would like to acknowledge support from the UC Merced School of Engineering for startup funding used to support work by Jeanette Cobian-Iñiguez and her research group.

References

1. J.T. Abatzoglou, A.P. Williams, Impact of anthropogenic climate change on wildfire across western US forests. Proc. Natl. Acad. Sci. **113**(42), 11770–11775 (2016). https://doi.org/10.1073/pnas.1607171113
2. L. Johnston, R. Blanchi, M. Jappiot, Wildland urban interface, in *Encyclopedia of Wildfires and Wildland-Urban Interface (WUI) Fires*, 2020th ed. (2019), pp. 1–13
3. K.M. Hoffman, A.C. Christianson, R.W. Gray, L. Daniels, Western Canada's new wildfire reality needs a new approach to fire management. Environ. Res. Lett. **17**(6) (2022). https://doi.org/10.1088/1748-9326/ac7345
4. M.M. Godoy, S. Martinuzzi, H.A. Kramer, G.E. Defossé, J. Argañaraz, V.C. Radeloff, Rapid WUI growth in a natural amenity-rich region in central-western Patagonia, Argentina. Int. J. Wildl. Fire **28**(7), 473–484 (2019). https://doi.org/10.1071/WF18097
5. J. McAneney, K. Chen, A. Pitman, 100-years of Australian bushfire property losses: Is the risk significant and is it increasing? J. Environ. Manage. **90**(8), 2819–2822 (2009). https://doi.org/10.1016/j.jenvman.2009.03.013
6. P. Deb et al., Causes of the widespread 2019–2020 Australian bushfire season. Earth's Futur. **8**(11) (2020). https://doi.org/10.1029/2020EF001671
7. S. Martinuzzi, S.I. Stewart, D.P. Helmers, M.H. Mockrin, R.B. Hammer, V.C. Radeloff, *The 2010 wildland-urban interface of the conterminous United States* . Research Map NRS-8. Newtown Square, PA: U.S. Department of Agriculture, Forest Service, Northern Research Station, 124 p. includes pull-out map
8. V.C. Radeloff, R.B. Hammer, S.I. Stewart, J.S. Fried, S.S. Holcomb, J.F. McKeefry, The Wildland – urban interface in the United States. Ecol. Appl. **15**(3), 799–805 (2005)
9. A.C. Fernandez-Pello et al., Spot fire ignition of natural fuel beds by hot metal particles, embers, and sparks. Combust. Sci. Technol. **187**(1–2), 269–295 (2015). https://doi.org/10.1080/00102202.2014.973953
10. V. Babrauskas, Firebrands and Embers. Encycl. Wildfires Wildland-Urban Interface Fires 431–444 (2020). https://doi.org/10.1007/978-3-319-52090-2_3
11. S.L. Quarles, Y. Valachovic, G.M. Nakamura, G.A. Nader, M.J. de Lasaux, *Home survival in wildfire-prone areas: building materials and design considerations* (University of California, Agriculture and Natural Resources, 2010)
12. S.E. Caton, R.S.P. Hakes, D.J. Gorham, A. Zhou, M.J. Gollner, Review of pathways for building fire spread in the Wildland urban interface part I: exposure conditions. Fire Technol. **53**(2), 429–473 (2017). https://doi.org/10.1007/s10694-016-0589-z

13. H.E. Anderson, Aids to determining fuel models for estimating fire behavior, in *USDA Forest Service General Technical Report INT-122*, pp. 1–22 (1982). http://digitalcommons.usu.edu/cgi/viewcontent.cgi?article=1147&context=barkbeetles
14. J.H. Scott, R.E. Burgan, Standard fire behavior fuel models: a comprehensive set for use with rothermel's surface fire spread model. *General Technical Report RMRS-GTR-153*, pp. 1–80 (2005); U.S Forest Service General Technical Report RMRS-GTR-153
15. T. Beer, The interaction of wind and fire. Boundary-Layer Meteorol. **54**(3), 287–308 (1991). https://doi.org/10.1007/BF00183958
16. W.L. Fons, Analysis of fire spread in light forest fuels. J. Agric. Res. 3 (1946)
17. R.C. Rothermel, A mathematical model for predicting fire spread in wildland fuels, in *USDA Forest Service Res. Pap. INT-115*, p. 40 (1972)
18. F.A. Albini, *Estimating windspeeds for predicting wildland fire behavior*, vol. 221. Intermountain Forest and Range Experiment Station, Forest Service, US (1979)
19. F.A. Albini, A model for fire spread in wildland fuels by radiation. Combust. Sci. Technol. **42**(5–6), 229–258 (1985). https://doi.org/10.1080/00102208508960381
20. R.J. Krupar, Wind, in *Encyclopedia of Wildfires and Wildland-Urban Interface (WUI) Fires*, pp. 1–4 (2019)
21. H.E. Anderson, R.C. Rothermel, Influence of moisture and wind upon the characteristics of free-burning fires. Symp. Combust. **10**(1), 1009–1019 (1965). https://doi.org/10.1016/S0082-0784(65)80243-0
22. N.P. Cheneyl, J.S. Gouldl, W.R. Catchpole, The influence of fuel, weather and fire shape variables on fire-spread in grasslands, vol. 3, pp. 31–44 (1993)
23. F.A. Albini, A model for fire spread in wildland fuels by- radiation † a model for fire spread in wildland fuels by radiation. Combust. Flame 42 (1985). https://doi.org/10.1080/00102208508960381
24. D.R. Weise, X. Zhou, L. Sun, S. Mahalingam, Fire spread in chaparral—'go or no-go?'. Int. J. Wildl. Fire **14**, 99–106 (2005). https://doi.org/10.1071/WF04049
25. D.X. Viegas, Slope and wind effects on fire propagation. Int. J. Wildl. Fire **13**(2), 143–156 (2004)
26. V. Tihay, F. Morandini, P.-A. Santoni, Y. Perez-Ramirez, T. Barboni, Study of the influence of fuel load and slope on a fire spreading across a bed of pine needles by using oxygen consumption calorimetry. J. Phys: Conf. Ser. **395**(1), 12075 (2012)
27. D.D. Drysdale, A.J.R. Macmillan, Flame spread on inclined surfaces. Fire Saf. J. **18**(3), 245–254 (1992)
28. M.J. Gollner, X. Huang, J. Cobian, A.S. Rangwala, F.A. Williams, Experimental study of upward flame spread of an inclined fuel surface. Proc. Combust. Inst. **34**(2), 2531–2538 (2013). https://doi.org/10.1016/j.proci.2012.06.063
29. J.-L. Dupuy, J. Maréchal, D. Portier, J.-C. Valette, The effects of slope and fuel bed width on laboratory fire behaviour. Int. J. Wildl. Fire **20**(2), 272–288 (2011)
30. A. Maranghides, W. Mell, A case study of a community affected by the Witch and Guejito wildland fires. Fire Technol. **47**(2), 379–420 (2011). https://doi.org/10.1007/s10694-010-0164-y
31. J.D. Cohen, Preventing disaster: home ignitabilitiy in the wildland-urban interface. J. For. 15–21 (2000)
32. J.D. Cohen, Relating flame radiation to home ignition using modeling and experimental crown fires. Can. J. For. Res. **34**(8), 1616–1626 (2004). https://doi.org/10.1139/X04-049
33. A. Maranghides, E.L. Johnsson, Residential structure separation fire experiments. NIST Technol. Note **1600**, 42p (2008)
34. S.L. Manzello, S. Suzuki, Y. Hayashi, Exposing siding treatments, walls fitted with eaves, and glazing assemblies to firebrand showers. Fire Saf. J. **50**, 25–34 (2012). https://doi.org/10.1016/j.firesaf.2012.01.006
35. National Fire Protection Agency, Standard for reducing structure ignition hazards from wildland fire (2018)
36. International Code Council, International wildland-urban interface code (2021)

37. A. Maranghides, E.D. Link, S.L. Quarles, D.J. Gorham, NIST Technical Note 2205 WUI Structure/Parcel/Community Fire Hazard Mitigation Methodology (2022)
38. E.L. Johnsson, A. Maranghides, Effects of wind speed and angle on fire spread along privacy fences, Gaithersburg, MD, 2016. https://doi.org/10.6028/NIST.TN.1894
39. R.S.P. Hakes, S.E. Caton, D.J. Gorham, M.J. Gollner, A review of pathways for building fire spread in the wildland urban interface part II: response of components and systems and mitigation strategies in the United States. Fire Technol. **53**(2), 475–515 (2017). https://doi.org/10.1007/s10694-016-0601-7
40. S.S. Wessies, M.K. Chang, K.C. Marr, O.A. Ezekoye, Experimental and analytical characterization of firebrand ignition of home insulation materials. Fire Technol. **55**(3), 1027–1056 (2019). https://doi.org/10.1007/s10694-019-00818-8
41. C. Schevin, ASTM E108 standard test methods for fire test of roof coverings. Retrieved from palmex international (2013. http://www.palmex.fr/wp-content
42. A.D. Syphard, T.J. Brennan, J.E. Keeley, The importance of building construction materials relative to other factors affecting structure survival during wildfire. Int. J. Disaster Risk Reduct. **21**, 140–147 (2017). https://doi.org/10.1016/j.ijdrr.2016.11.011
43. State of California, California building code: materials and construction methods for exterior wildfire exposure (2012)
44. K. Butler et al., Wind-driven fire spread to a structure from Fences and Mulch (2022). https://doi.org/10.6028/NIST.TN.2228-upd1
45. S.L. Manzello, S. Suzuki, Y. Hayashi, Enabling the study of structure vulnerabilities to ignition from wind driven firebrand showers: a summary of experimental results. Fire Saf. J. **54**, 181–196 (2012). https://doi.org/10.1016/j.firesaf.2012.06.012
46. W.E. Mell, S.L. Manzello, A. Maranghides, D. Butry, R.G. Rehm, The wildland-urban interface fire problem - current approaches and research needs. Int. J. Wildl. Fire **19**(2), 238–251 (2010). https://doi.org/10.1071/WF07131
47. Texas Forest Service Mitigation and Prevention Department Managing Vegetation in the WUI, "Vegetation Managementin the Wildland Urban Interface," Texas A&M (2022)
48. T.J. Duff, J.G. Cawson, T.D. Penman, *"Prescribed Burning", in encyclopedia of wildfires and wildland-urban interface (WUI) fires* (Springer International Publishing, Cham, 2018), pp. 1–11
49. D. Mitchell, M. Smidt, Costs of mechanical fuel reduction treatments, in *Encyclopedia of Wildfires and Wildland-Urban Interface (WUI) Fires* (Springer International Publishing, 2019), pp. 1–7
50. S.L. Stephens et al., The effects of forest fuel-reduction treatments in the United States. Bioscience **62**(6), 549–560 (2012). https://doi.org/10.1525/bio.2012.62.6.6
51. G. Xanthopoulos, M. Athanasiou, V. Varela, K. Kaoukis, Deliverable 3.3.–fuel management scenarios (2021)
52. L.R. Green, *Fuelbreaks and other fuel modification for wildland fire control*, vol. 499. US Department of Agriculture, Forest Service (1977)
53. CALFIRE, Fuels Reduction Guide (2021)
54. A.M. Van Der Bles et al., Communicating uncertainty about facts, numbers and science. R. Soc. Open Sci. **6**(5), 181870 (2019)
55. J.K. Agee, C.N. Skinner, Basic principles of forest fuel reduction treatments. For. Ecol. Manage. **211**, 83–96 (2005). https://doi.org/10.1016/j.foreco.2005.01.034
56. M.A. Moritz et al., Beyond a focus on fuel reduction in the WUI: the need for regional wildfire mitigation to address multiple risks. Front. For. Glob. Chang. **5**, 71 (2022)
57. I. Wildland Resource Management and Baseline Environmental Consulting, "Wildfire Hazard Reduction and Resource Management Plan (Plan), FUEL TREATMENT METHODS, Part 5," East Bay Hills (2009)
58. F. Ratcliff et al., Cattle grazing reduces fuel and leads to more manageable fire behavior. Calif. Agric. **76**(2), 60–69 (2022)
59. P.G. Sikkink, Masticated fuels, in *Encyclopedia of Wildfires and Wildland-Urban Interface (WUI) Fires* (Springer International Publishing, 2018), pp. 1–7

60. D.J. Shinneman et al., Future direction of fuels management in sagebrush rangelands. Rangel. Ecol. Manag. **86**, 50–63 (2023)
61. S.L. Stephens et al., The effects of forest fuel-reduction treatments in the United States. Bioscience **62**(6), 549–560 (2012)
62. A. Ganteaume, M. Jappiot, C. Lampin, M. Guijarro, C. Hernando, Flammability of some ornamental species in wildland-urban interfaces in southeastern France: laboratory assessment at particle level. Environ. Manage. **52**(2), 467–480 (2013). https://doi.org/10.1007/s00267-013-0067-z
63. D.R. Weise, R.H. White, F.C. Beall, M. Etlinger, Use of the cone calorimeter to detect seasonal differences in selected combustion characteristics of ornamental vegetation. Int. J. Wildl. Fire **14**(3), 321–338 (2005). https://doi.org/10.1071/WF04035
64. T.B. Paveglio, C.M. Edgeley, Fire adapted community. Encycl. Wildfires Wildland-Urban Interface Fires 1–9 (2020). https://doi.org/10.1007/978-3-319-51727-8_114-1
65. NFPA, Firewise toolkit, in *Wildfire Prep.* (2016)
66. H.A. Kramer, M.H. Mockrin, P.M. Alexandre, S.I. Stewart, V.C. Radeloff, Where wildfires destroy buildings in the US relative to the wildland-urban interface and national fire outreach programs. Int. J. Wildl. Fire **27**(5), 329–341 (2018). https://doi.org/10.1071/WF17135
67. T.I. Zohdi, A machine-learning framework for rapid adaptive digital-twin based fire-propagation simulation in complex environments. Comput. Methods Appl. Mech. Eng. **363**, 112907 (2020). https://doi.org/10.1016/j.cma.2020.112907
68. A. Tohidi, M. Gollner, C. Standohar- S. Quarles, Computer vision techniques for firebrand detection and characterization (2017)
69. F. Hedayati, B. Bahrani, A. Zhou, S.L. Quarles, D.J. Gorham, A framework to facilitate firebrand characterization. Front. Mech. Eng. **5**, 1–14 (2019). https://doi.org/10.3389/fmech.2019.00043
70. X. Huang, X. Wu, A. Usmani, Perspectives of using artificial intelligence in building fire safety, in *Handbook of Cognitive and Autonomous Systems for Fire Resilient Infrastructures* (2022)
71. O. Rios, M.M. Valero, E. Pastor, E. Planas, *A data-driven fire spread simulator: validation in vall-llobrega's fire*, vol. 5 (2019), pp. 1–11. https://doi.org/10.3389/fmech.2019.00008
72. J. Mandel et al., A dynamic data driven wildland fire model. Comput. Sci. **4487**(1), 1042–1049 (2007). https://doi.org/10.1007/978-3-540-72584-8_137
73. M.C. Rochoux, B. Delmotte, B. Cuenot, S. Ricci, A. Trouvé, Regional-scale simulations of wildland fire spread informed by real-time flame front observations. Proc. Combust. Inst. **34**(2), 2641–2647 (2013)
74. M. Gollner et al., Towards data-driven operational wildfire spread modeling: a report of the NSF-funded WIFIRE workshop (2015)
75. F. Mattern, C. Floerkemeier, From the internet of computers to the internet of things, in *Active Data Management to Event-Based Systems and More. Lecture Notes in Computer Science*, ed. by K. Sachs, I. Petrov, P. Guerrero. (Springer, Berlin, 2010), pp. 242–259
76. L. Yang, S.H. Yang, L. Plotnick, Technological forecasting & social change how the internet of things technology enhances emergency response operations. Technol. Forecast. Soc. Chang. **80**(9), 1854–1867 (2013). https://doi.org/10.1016/j.techfore.2012.07.011
77. M. Antunes, L.M. Ferreira, C. Viegas, A.P. Coimbra, A.T. De Almeida, Low-cost system for early detection and deployment of countermeasures against wild fires, in *IEEE 5th World Forum Internet Things, WF-IoT 2019 - Conference Proceedings* (2019), pp. 418–423. https://doi.org/10.1109/WF-IoT.2019.8767331
78. J.C. Rosas, J.A. Aguilar, C. Trip-Barba, R. Espinosa, P. Aguilar, A Mobile-Sensor Fire Prevention System Based on the Internet of Things. Comput. Sci. Its Appl.-ICCSA **2017**, 274–283 (2017)
79. M. Hefeeda, M. Bagheri, Wireless sensor networks for early detection of forest fires, in *IEEE International Conference on Mobile Adhoc and Sensor Systems* (2007), pp. 1–6
80. X. Wu, Y. Park, A. Li, X. Huang, F. Xiao, A. Usmani, Smart detection of fire source in tunnel based on the numerical database and artificial intelligence. Fire Technol. **57**(2), 657–682 (2021)

81. J.-M. Cabrera, O.A. Ezekoye, R.D. Moser, Bayesian inference of fire evolution within a compartment using heat flux measurements. Fire Technol. **57**(6), 2887–2903 (2021)
82. T. Buffington, J.-M. Cabrera, A. Kurzawski, O.A. Ezekoye, Deep-learning emulators of transient compartment fire simulations for inverse problems and room-scale calorimetry. Fire Technol. **57**(6), 2859–2885 (2021)
83. X. Zhong, M. Duckham, D. Chong, K. Tolhurst, Real-time estimation of wildfire perimeters from curated crowdsourcing. Sci. Rep. **6**(1), 1–10 (2016)
84. W.C. Tam et al., Generating synthetic sensor data to facilitate machine learning paradigm for prediction of building fire hazard. Fire Technol. (2020). https://doi.org/10.1007/s10694-020-01022-9
85. A. Troy et al., An analysis of factors influencing structure loss resulting from the 2018 Camp Fire. Int. J. Wildl. Fire **31**(6), 586–598 (2022). https://doi.org/10.1071/WF21176
86. A.D. Syphard, J.E. Keeley, M. Gough, M. Lazarz, J. Rogan, What makes wildfires destructive in California? Fire **5**, 133 (2022)
87. E. Knapp, Y. Valachovic, S.L. Quarles, N.G. Johnson, Factors associated with single-family home survival in the 2018 Camp Fire , California. Fire Ecol. 1–27 (2021)
88. P. Gupta, V. Gaidhane, A new approach for flame image edges detection, in *International Conference on Recent Advances and Innovations in Engineering (ICRAIE-2014)* (2014), pp. 1–6. https://doi.org/10.1109/ICRAIE.2014.6909178
89. T. Qiu, Y. Yan, G. Lu, An autoadaptive edge-detection algorithm for flame and fire image processing. IEEE Trans. Instrum. Meas. **61**(5), 1486–1493 (2012). https://doi.org/10.1109/TIM.2011.2175833
90. J. Cobian-Iñiguez, A. Aminfar, D.R. Weise, M. Princevac, On the use of semi-empirical flame models for spreading chaparral crown fire. Front. Mech. Eng. **5**(August), 1–13 (2019). https://doi.org/10.3389/fmech.2019.00050
91. J. Cobian-Iñiguez, A. Aminfar, S. Saha, K. Awayan, D.R. Weise, M. Princevac, The transition and spread of a chaparral crown fire: insights from laboratory scale wind tunnel experiments. J. Combust. (2022)
92. Y. Zhang, S. Lim, J.J. Sharples, Modelling spatial patterns of wildfire occurrence in South-Eastern Australia, *Geomatics*. Nat. Hazards Risk **7**(6), 1800–1815 (2016). https://doi.org/10.1080/19475705.2016.1155501
93. J. Sharma, O.C. Granmo, M. Goodwin, J.T. Fidje, Deep convolutional neural networks for fire detection in images. Commun. Comput. Inf. Sci. **744**, 183–193 (2017). https://doi.org/10.1007/978-3-319-65172-9_16
94. K. Muhammad, J. Ahmad, S.W. Baik, Early fire detection using convolutional neural networks during surveillance for effective disaster management. Neurocomputing **288**, 30–42 (2018). https://doi.org/10.1016/j.neucom.2017.04.083
95. Y. Wang, L. Dang, J. Ren, Forest fire image recognition based on convolutional neural network. J. Algorithms Comput. Technol. **13** (2019). https://doi.org/10.1177/1748302619887689
96. S. Frizzi, R. Kaabi, M. Bouchouicha, J.M. Ginoux, E. Moreau, F. Fnaiech, Convolutional neural network for video fire and smoke detection, in *IECON Proceedings (Industrial Electronics Conference* (2016), pp. 877–882. https://doi.org/10.1109/IECON.2016.7793196
97. J. Ryu, D. Kwak, A study on a complex flame and smoke detection method using computer vision detection and convolutional neural network. Fire **5**(4) (2022). https://doi.org/10.3390/fire5040108
98. P. Li, W. Zhao, Image fire detection algorithms based on convolutional neural networks, in *Case Studies in Thermal Engineering*, vol. 19 (2020). https://doi.org/10.1016/j.csite.2020.100625
99. B.U. Töreyin, Y. Dedeoğlu, U. Güdükbay, A.E. Çetin, Computer vision based method for real-time fire and flame detection. Pattern Recognit. Lett. **27**(1), 49–58 (2006). https://doi.org/10.1016/j.patrec.2005.06.015
100. T. Celik, H. Demirel, H. Ozkaramanli, M. Uyguroglu, Fire detection using statistical color model in video sequences. J. Vis. Commun. Image Represent. **18**(2), 176–185 (2007). https://doi.org/10.1016/j.jvcir.2006.12.003

101. F. Bu, M.S. Gharajeh, Intelligent and vision-based fire detection systems: a survey. Image Vis. Comput. **91**, 103803 (2019). https://doi.org/10.1016/j.imavis.2019.08.007
102. J.R. Martinez-de Dios, B.C. Arrue, A. Ollero, L. Merino, F. Gómez-Rodríguez, Computer vision techniques for forest fire perception. Image Vis. Comput. **26**(4), 550–562 (2008). https://doi.org/10.1016/j.imavis.2007.07.002
103. B.U. Töreyin, R.G. Cinbis, Y. Dedeoglu, A. enis Cetin, Fire detection in infrared video using wavelet analysis. Opt. Eng. **46**(6), 067204 (2007). https://doi.org/10.1117/1.2748752
104. K.L.B. Xavier, V.K. Nanayakkara, Development of an early fire detection and deep neural networks. Fire Technol. **58**(6), 3529–3552 (2022). https://doi.org/10.1007/s10694-022-01319-x
105. L. Giglio, L. Boschetti, D. Roy, A.A. Hoffmann, M. Humber, Collection 6 MODIS burned area product user's guide, version 1.0 (2016), p. 26. https://modis-land.gsfc.nasa.gov/pdf/MODIS_C6_BA_User_Guide_1.0.pdf
106. W. Schroeder, L. Giglio, NASA VIIRS land science investigator processing system (SIPS) visible infrared imaging radiometer suite (VIIRS) 375 m & 750 m active fire products: product user's guide version 1.4 (2018)
107. N.S. Skowronski, S. Haag, J. Trimble, K.L. Clark, M.R. Gallagher, R.G. Lathrop, Structure-level fuel load assessment in the wildland-urban interface: a fusion of airborne laser scanning and spectral remote-sensing methodologies. Int. J. Wildl. Fire **25**(5), 547–557 (2016). https://doi.org/10.1071/WF14078
108. D. McNamara, W. Mell, Towards the use of remote sensing for identification of building damage, destruction, and defensive actions at wildland-urban interface fires. Fire Technol. **58**(1), 641–672 (2022). https://doi.org/10.1007/s10694-021-01170-6
109. A. Badia, P. Serra, S. Modugno, Identifying dynamics of fire ignition probabilities in two representative Mediterranean wildland-urban interface areas. Appl. Geogr. **31**(3), 930–940 (2011). https://doi.org/10.1016/j.apgeog.2011.01.016
110. S.L. Manzello, A. Maranghides, W.E. Mell, Firebrand generation from burning vegetation. Int. J. Wildl. Fire **16**(4), 458–462 (2007). https://doi.org/10.1071/WF06079
111. S.L. Manzello, A. Maranghides, J.R. Shields, W.E. Mell, Y. Hayashi, D. Nii, Mass and size distribution of firebrands generated from burning Korean pine (Pinus koraiensis) trees. Fire Mater 21–31 (2009). https://doi.org/10.1002/fam.977
112. S. Suzuki, S.L. Manzello, M. Lage, G. Laing, Firebrand generation data obtained from a full-scale structure burn. Int. J. Wildl. Fire **21**(8), 961–968 (2012). https://doi.org/10.1071/WF11133
113. S.L. Manzello, E.I.D. Foote, Characterizing firebrand exposure from wildland-urban interface (WUI) fires: results from the 2007 Angora Fire. Fire Technol. **50**(1), 105–124 (2014). https://doi.org/10.1007/s10694-012-0295-4
114. A. Filkov et al., Investigation of firebrand production during prescribed fires conducted in a pine forest. Proc. Combust. Inst. **36**(2), 3263–3270 (2017). https://doi.org/10.1016/j.proci.2016.06.125
115. A. Jha, A. Zhou, Applying machine learning for firebrand production prediction. Fire Technol. **58**(5), 3261–3290 (2022). https://doi.org/10.1007/s10694-022-01309-z
116. A. Zhou, S. Quarles, D.R. Weise, Fire ember production from wildland and structural fuels (2019)
117. S. Prohanov, A. Filkov, D. Kasymov, M. Agafontsev, V. Reyno, Determination of firebrand characteristics using thermal videos. Fire **3**(4), 1–17 (2020). https://doi.org/10.3390/fire3040068
118. N. Bouvet, E.D. Link, S.A. Fink, A new approach to characterize firebrand showers using advanced 3D imaging techniques. Exp. Fluids **62**(9), 1–14 (2021). https://doi.org/10.1007/s00348-021-03277-6
119. D.K. Thompson et al., Quantifying firebrand production and transport using the acoustic analysis of in-fire cameras. Fire Technol. **58**(3), 1617–1638 (2022). https://doi.org/10.1007/s10694-021-01194-y

120. A. Filkov, S. Prohanov, Particle tracking and detection software for firebrands characterization in wildland fires. Fire Technol. **55**(3), 817–836 (2019). https://doi.org/10.1007/s10694-018-0805-0
121. M.Z. Naser, Mechanistically informed machine learning and artificial intelligence in fire engineering and sciences **57**(6); Springer US (2021)
122. G. Rebala, A. Ravi, S. Churiwala, in *An Introduction to Machine Learning* (Springer, 2019)
123. M.Z. Naser, Demystifying ten big ideas and rules every fire scientist & engineer should know about blackbox, whitebox and causal artificial intelligence. Fire Technol. **58**(3), 1075–1085 (2022). https://doi.org/10.1007/s10694-021-01210-1
124. J.K. Shuman et al., Reimagine fire science for the anthropocene. PNAS Nexus **1**(3), 1–14 (2022). https://doi.org/10.1093/pnasnexus/pgac115

New Fire Protection Materials

Andre L. Thompson

1 Fire Barriers

1.1 Background

Fires involving residential upholstered continue to be the leading cause of civilian fire deaths in the United States [1]. Fire statistics reported by the National Fire Protection Agency (NFPA) between 2013 and 2017 indicates that residential fires in which a RUF item was first ignited caused an estimated annual average of $243 million in direct sources in deadly fires, where RUF is the first item ignited; they account for 55% of home fire deaths [2, 3]. According to the NFPA [2], the vast majority of fire deaths (about 95%) occur when fire spreads beyond the first item ignited (i.e., the upholstery furniture item) and about two-thirds occur when fire spreads beyond the room of origin; which can only be explained by the presence of flaming [4].

There are two alternative ways to decrease RUF flammability risk: (a) by flame prevention (i.e., reducing the probability that flaming will occur by increasing the flaming or smoldering ignition resistance of the item) or (b) by flame mitigation (i.e., limiting the consequences of a burning item, such as flashover, by reducing fire growth) [3]. To enhance ignition resistance in the presence of smoldering ignition sources, tests like Upholstered Furniture Action Council (UFAC) methodology, ASTM E1353, NFPA 260 and California Technical Bulletin 117 2013 (TB 117–2013) [5–8], have been adopted. TB 117–2013 has been adopted as a federal regulation in 2021 [9], and compliance can be met without using chemical fire retardants (FRs) [10].

A. L. Thompson (✉)
Fire Research Division, Engineering Laboratory, National Institute of Standards and Technology, Gaithersburg 20899, USA
e-mail: andre.thompson@nist.gov

X. Huang and W. C. Tam (eds.), *Intelligent Building Fire Safety and Smart Firefighting*, Digital Innovations in Architecture, Engineering and Construction,
https://doi.org/10.1007/978-3-031-48161-1_9

In the past few decades, FRs chemicals have been largely used in attempt to reduce RUF flammability; however, concerns about possible harmful effects on human health and the environment, posed by specific FRs, led to public backlash and a reassessment of costs/benefits associated with the use of such chemicals [11–15]. In 2017, the state of Maine (LD 182) banned any chemical compound for which "fire-retardant" appears on the material safety data sheet [16]. In 2018, California passed a law (AB2998) that banned halogenated, organo-phosphorous, organo-nitrogen chemicals, and nanofillers [17]. At the federal level, the Consumer Product Safety Commission recommended refraining from intentionally adding hazardous additives such as nonpolymeric organo-halogen FRs and has an ongoing project to evaluate additive organo-halogenated flame retardants [15, 18].

Due to the extreme limitations in the use of FRs, fire safety researchers have been looking for alternative approaches to improve fire safety [19]. In this regard, fire barriers, i.e., protective layers designed to separate the flame from a combustible substrate to limit its flame involvement, offer a potential solution. In the context of RUF, fire barriers are usually fabrics placed between the padding materials, like flexible polyurethane foam (FPUF) or polyester fiber fill, and the cover fabric. Barriers have been shown to mitigate the RUF fire hazard by reducing the fire growth and limiting the fire size [20–32].

Passive barriers operate by physical mechanisms of action including heat/mass transfer reduction, endothermic decomposition and dilution effects associated with the release of non-flammable gases (e.g., water vapor) during the decomposition of the barrier [31, 33]. Active barriers operate by both physical and chemical mechanisms. Chemical mechanisms include flame quenching in the gas phase and charring in the condensed phase. Passive barriers adopt fire resistant fibers (e.g., glass, carbon, polysilicic acid/rayon fibers); active barriers use combinations of fibers and/or coatings with gas-phase-active flame retardants for flame inhibition [34]. Because active barriers might be subject to restrictions in RUF depending on the type of fire retardant they adopt, research at NIST is mainly focused on passive barriers.

This chapter has several aims: (a) review the performance of fire barriers in reduced scale "cube test" design and the importance of the wetting point for the prediction of full-scale performance [31] (b) review the effect of mass transfer of liquid products through the bottom of a burning chair to better understand the key parameters affecting the performance and failure of barriers at a full scale [29] (c) review to what extent passive barriers (expected to comply with the aforementioned US State regulations because they do not contain FRs) can delay the fire growth of RUF when exposed to a strong (18 kW) ignition source (equivalent to five sheets of crumpled newsprint) [29] (d) review how the failure mechanism of barrier fabrics in chair mock-up experiments also applies to commercial RUF in furnished compartments, so the use of the cube test can be possibly extended to predict the performance of barrier fabric in realistic fire scenarios [30, 32] (e) demonstrate how new fire protection materials delay fire growth allowing more time for safe egress and allows more time for first responders to intervene and prevent flashover.

1.2 Materials, Density, and Air Permeability

The fire barriers used throughout this chapter will be a combination of one cover fabric (C0) and six barriers (B1 to B6) used in two previous studies [29, 31]. Barriers were selected from commercial fabrics known to be inherently flame resistant [36] and did not contain any halogenated, organo-phosphorous, organo-nitrogen, or nanoparticle-based flame retardants according to the manufacturers', thus, were expected to be compliant with California Assembly Bill 2998 [17]. The selected fabrics were intended to represent—as much as possible and within the number of barriers tested—a variety of commercially available products (nonwoven/woven, thermally thin/thick, high/low air permeability) and a broad range of materials including cellulosic fibers, E-glass fibers [37], para-aramid fibers [38], polysilicic acid fibers [34, 39], oxidized polyacrylonitrile fibers [40], and combinations thereof. Figure 1 shows photos of representative 2 cm wide strips of each fabric. The air permeabilities of each fabric was measured in accordance to the ASTM D737-18 [41]. A breakdown of the fabrics' composition, areal density, and air permeability are summarized in Thompson et al. [29].

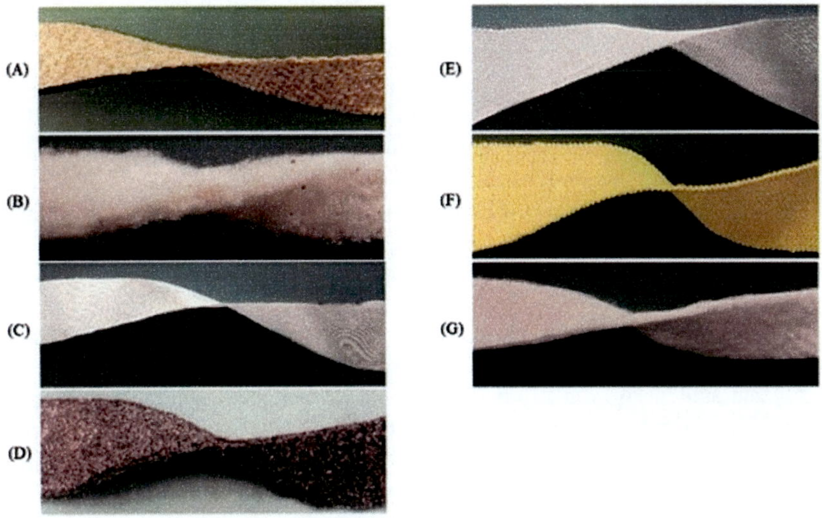

Fig. 1 Pictures showing a 2 cm wide strip sample of: **a**, cover fabric C0; **b** barrier B1, **c** barrier B2; **d** barrier B3; **e** barrier B4; **f** barrier B5; and **g** barrier B6 [29]

1.3 Chemical Analysis of Barriers and Cover Fabric

An elemental analysis of all fabrics was performed by independent laboratories according to the following techniques: ion chromatography (IC) for chlorine (Cl) [42]; the Kjeldahl method [43] for nitrogen (N); inductively coupled plasma atomic emission spectroscopy (ICP-AES) for boron (B), bromine (Br), phosphorus (P), sulfur (S), antimony (Sb), and silicon (Si) according to ASTM E1479-16 [44]; and X-ray fluorescence (XRF) spectroscopy [45] for semi-quantitative screening of antimony, bromine, chlorine, and phosphorus.

The presence of commonly used chlorinated, brominated, and organo-phosphorus FRs was investigated in all fabrics by organic solvent extraction and gas chromatography-mass spectrometry (GC–MS) analysis [46] of the extracted solution. FRs tested for by GC–MS analysis were tributyl phosphate (TBP), tri(2-chloroethyl) phosphate (TCEP), tris(2-chloroisopropyl) phosphate (TCPP), tris(1,3-dichloroisopropyl) phosphate (TDCPP), triphenyl phosphate (TPP), and 1, 2, 5, 6, 9, and 10-hexabromocyclododecane (HBCD). Additionally, B1 was independently tested by aqueous extraction followed by inductively coupled plasma optical emission spectroscopy (ICP-OES) [47] to rule out the presence of boron-based FRs in the inner layer (IL). Details about sample preparation are reported in Data S1 of Thompson et al. [29].

According to manufacturers' specifications, the barriers used in this study did not contain FRs based on halogenated, organo-phosphorous, organo-nitrogen chemical, or nanoparticles, thus they were expected to be compliant with California Assembly Bill 2998 [17]. The barriers were screened for: (a) the presence of elements, which are typically used in FRs, by XRF, IC, Kjeldahl method, ICP-OES, and ICP-AES, and (b) the presence of a number of FRs by GC–MS [48]. This screening methodology for FRs is summarized in Fig. 2 and the results of the screening methodology are summarized in Thompson et al. [29]. For barriers with a bilayer structure like B1 and B6, certain analyses were run on the outer layer (OL) and IL separately.

A full detailed breakdown on the chemical analysis results of each fire barrier and cover fabric is explained in Thompson et al. [29]. To summarize, XRF detected significant amount of P on B (OL and IL) and B6 (OL), but according to Petreas et al. [48], XRF is not reliable screening method for P, and, even for concentrations above 50 (μg/g), caution is recommended when interpreting XRF measurements for P. XRF analysis for Sb, which is considered reliable for concentrations above 40 μg/g [48], revealed Sb concentrations below 40 μg/g for all fabrics. Eventhough XRF detected Cl concentrations, XRF may not be a reliable indicator of chlorinated FRs as both false positive and false negative have been reported [48, 49] and other elemental analysis for Cl are preferable [50, 51].

The Kjeldahl method revealed N with B3 and B5 having the two highest concentrations of ≈180 000 and ≈70 000 μg/g, respectively. Since the Kjeldahl method [43] is applicable to amines, amides, amino acids, and their derivatives (all organonitrogen compounds), organo-nitrogen molecules are present in B3 and B5.

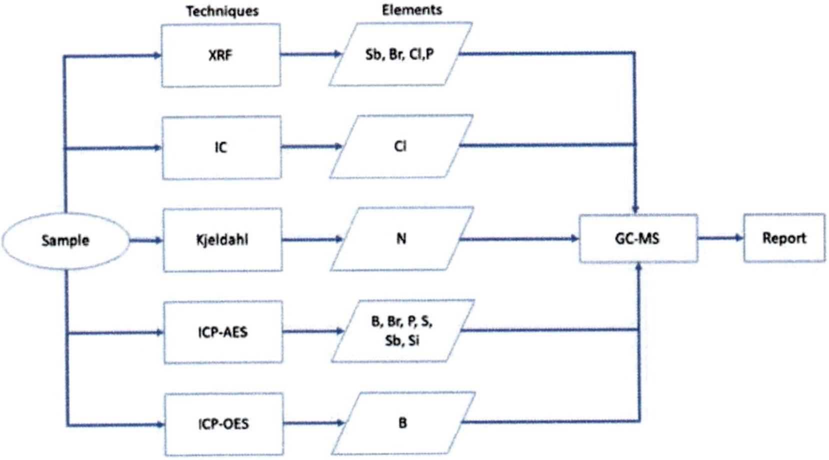

Fig. 2 Diagram illustrating the screening methodology for fire retardants [29]

ICP-AES indicated a significant content of B in B1, B2, B5, and B6 with concentrations ranging from ≈3700 to ≈21 000 μg/g. B2, B5, and B6 contain glass fibers, so B detection in these barriers was likely due to the presence of borosilicate glass [52–54]. B1 contains polysilicic acid and cotton but no glass according to the manufacturer specifications, and B detection (≈3700 μg/g) in this barrier could be associated with the use of additives such as B-based FRs, which are commonly used in combination with cotton [55]. ICP-AES is a more reliable technique to detect P elements than XRF as ICP-AES is very suitable in screening and quantification of P compounds in low concentrations [56]. ICP-OES did not reveal a significant boron content in B1IL (<1 μg/g), thus, the use of B-based FRs in B1IL was ruled out.

GC–MS was used to exclude the presence of some of the most used halogenated and phosphorus-based FRs. In particular, GC–MS analysis did not detect the presence of any of the target FRs. These data support the manufacturer's statement that no FRs were used in the barriers used in this study; however, a more comprehensive investigation based on nontargeted GC–MS [57] or liquid chromatography/time-of-flight mass spectroscopy operated with electrospray ionization [48] would be required to exclude the presence of other FRs that are not included in our list of targeted chemicals.

2 Fire Barriers in Reduced Scale Testing

2.1 Test Instrument/Cube Test Set-Up

This section will investigate the effect of mass transfer of liquid products through burning fire barriers to better understand the key parameters affecting the performance and failure of barriers on a reduced scale "cube test" design and investigate to what extent barriers can delay fire growth of RUF when exposed to a strong ignition source. The instrumentation (cone calorimeter) described in ASTM E1354, was used in all tests. Experiments were run in the cube test configuration [31].

A simplified schematic drawing showing the cube test setup, and an exploded view of the specimen holder and specimen for the Cube test configuration can be viewed in Zammarano et al. [31]. This specimen holder is designed to: (a) create an idealized representative cross-section of the composite product of interest; (b) allow liquid products, generated during pyrolysis, to percolate through the bottom fabrics of the specimen, wet the fire barrier, and form a drip pool that could ignite, and; (c) capture the time at which each of these events occurs. Views with dimensions and materials adopted for the top retainer frame, bottom retainer frame, and liner are provided in Appendix A of Supplemental Information of Zammarano et al. [31].

2.2 Cube Test Specimen Preparation and Procedure

For a detailed description of the cube test specimen preparation and procedure, refer to Zammarano et al. [31]. In summary, one specimen was a control specimen containing the FPUF core, the cover fabric (C0) as top fabric, and no fire barrier. Other specimens using barriers B1 to B6 contained one top fire barrier and one bottom fire barrier (selected among B1 to B6), the cover fabric C0 (above the top fire barrier), and the FPUF core. For a detailed description of the assembly sequence of the specimen refer to Appendix B in Supplemental Information in Zammarano et al. [31].

2.3 Effect of Heat Flux on Burning Behavior

In this test, a 75 kW/m^2 heat flux was adopted because, in full-scale tests on RUF, measured values of flame to surface heat flux exceeding 50 kW/m^2 and reaching up to 100 kW/m^2 have been observed [21, 58]. In addition, in the Cube test, there is no heat source underneath the specimen and the incident heat flux at the bottom of the specimen might not be high enough to support full pyrolysis of the FPUF (low heating rate in FPUF shifts its decomposition toward an oxidative rather than pyrolytic pathway) [59].

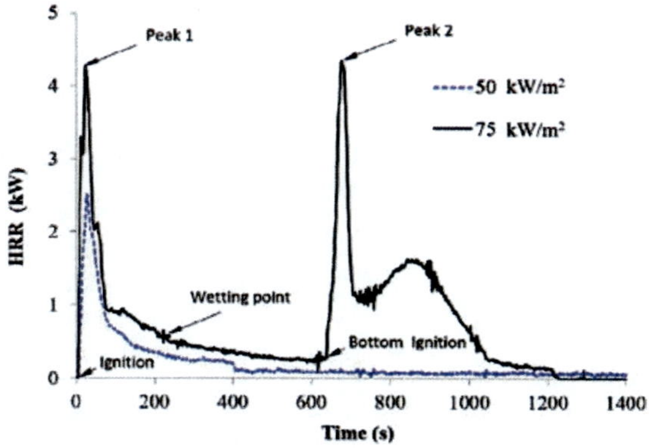

Fig. 3 Heat release rate (HRR) measured for one representative sample with fire barrier B6 at an external heat flux of 50 and 75 kW/m^2 [31]

Figure 3 shows representative HRR curves measured for specimens with barrier B6. In both curves of Fig. 3, the first peak of heat release rate (PHRR) was associated with the combustion of the cover fabric on top of the specimens. The cover fabric (C0) was rapidly consumed during the first peak and, after that, the HRR was mainly due to the combustion of gaseous products produced by foam pyrolysis. At this stage, the FPUF kept collapsing [60] and, thus, the incident heat flux on the FPUF surface and FPUF pyrolysis rate decreased. At 50 kW/m^2, the incident heat flux was not high enough to support flaming. At 75 kW/m^2, flaming was supported throughout the test, and liquid products percolating through the bottom FB became visible within 4 min. A second peak of HRR was observed when the bottom FB (now saturated with liquid products) ignited and, shortly after, flaming drips accumulated in the catch pan and generated a pool fire under the specimen.

Ultimately, at 50 kW/m^2, a significant fraction of the FPUF was converted into char (as typically observed in smoldering combustion) [61] and no liquid products were released under the specimen even after 30 min from test start, whereas at 75 kW/m^2 the FPUF produced visible liquid products underneath the specimen without any significant char production (better mimicking the decomposition pathway observed during the combustion behavior of RUF at full-scale where flaming drips are generated by FPUF) [21] within 4 min from test start.

2.4 Determination of Wetting Point

The presence of liquid products under the specimen at the wetting point is revealed by "wetting" of the bottom FB due to liquid products percolation, or, if no bottom FB is used, by the appearance of liquid products underneath the specimen. The time to "wetting point" is assumed to be a good indicator of the "time to failure" of the fire barrier in the Cube test since, in presence of an ignition source underneath the specimen, the liquid products would readily ignite and likely cause a large increase in HRR.

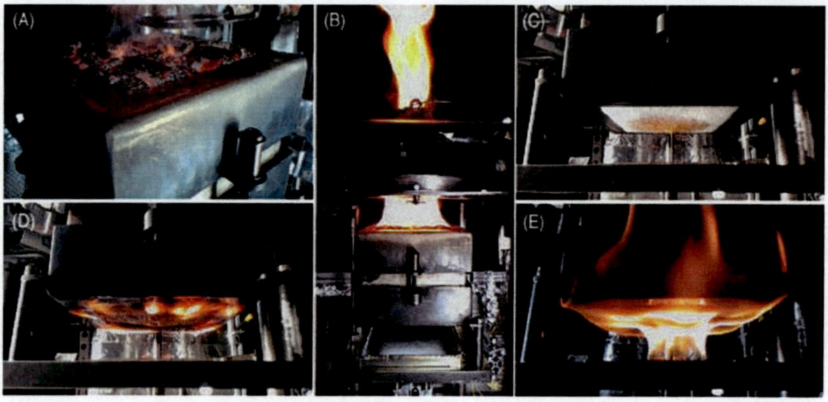

Fig. 4 Snapshots of a representative sample (FB B6 at 75 kW/m^2) displaying the characteristic stages of combustion in the Cube test: **a** ignition; **b** first peak of heat release rate; **c** about 50 s after wetting point; **d** bottom ignition, and **e** second peak of heat release rate [31]

In Fig. 9.2.1, the arrows on the 75 kW/m^2 heat release curve indicate the occurrence of ignition, first peak of heat release rate (peak 1), wetting point, bottom ignition, and second peak of heat release rate (peak 2). Snapshots displaying these characteristic stages of combustion for a representative sample with FB B6 in the Cube test are shown at time of ignition, time of first peak of heat release rate, about 50 s after time of wetting point, time of bottom ignition, and time of second peak of heat release rate (Fig. 4).

Heat release rate curses measured in the cube test for one representative sample are reported in Fig. 5. For each specimen, A reduced cone-calorimetry data are reported in Table 1. For the full table showing all conventional cone-calorimetry data, refer to Zammarano et al. [31]. Compared to C0, specimens containing FBs showed average increases of ≈1.6-fold in time of ignition, up to ≈1.9-fold in time to peak 1, up to ≈1.5-fold in $PHRR_1$, and an average decrease of ≈ threefold in $PHRR_2$, ≈1.3-fold in EHC and ≈1.2-fold in THR.

The significant reduction in EHC, induced by FBs, was not due to a free-radical trapping mechanism (FBs did not contain significant amounts of gas-phase FRs) but likely to the release of gaseous pyrolyzates through the bottom FB, where they did not burn for $t_w < t < t_{BI}$ (video recordings show that visible smoke—and presumably gas products—were released underneath the specimen starting at $t \approx t_w$ and that the smoke release rate visibly increased until $t \approx t_{BI}$), and possible flame quenching over the FB [62]. This fire scenario is not unrealistic and has been observed in full-scale RUF tests before the bottom of the item ignites [29]. The time required for the appearance of flaming on the bottom of the specimen, t_{BI}, ranged from an average value of 106 s to 658 s (for C0s and B6, respectively).

The PHRR is highly dependent on a multitude of factors related to liquid products' flaming (such as flow, release timing and release location of liquid products, type of substrate under the item, substrate-to-item distance, extent of feedback between

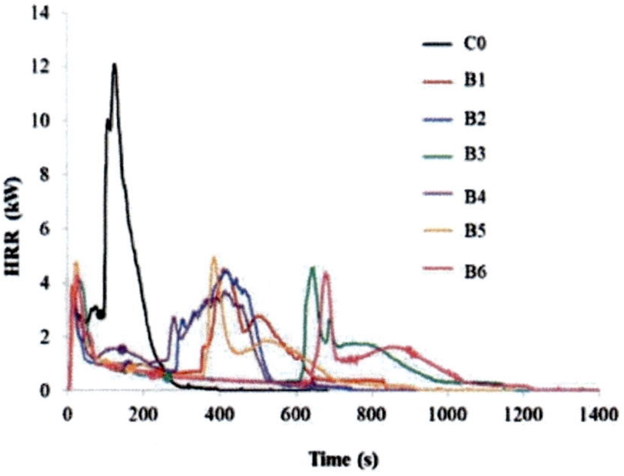

Fig. 5 Heat release rate (HRR) curves measured in the Cube test for one representative sample, cover fabric only sample, C0, and samples with barriers B1 to B6 at an external heat flux of 75 kW/m^2. For each curve, the circled data point indicates the wetting point [31]

pool-fire, residual item, etc.) [63] that even achieving a good experimental reproducibility for PHRR at full-scale may be challenging, and achieving a robust full-scale PHRR prediction by the only means of a reduced-scale test may be unfeasible. However, a conceptual analysis of the burning behavior of RUF (or cored composites in general) can highlight key stages of full-scale fire development and, as presented in Zammarano et al. [31, 32], it is apparent that the wetting point measured in the Cube test may be a valuable metric to predict the performance of fire barriers in RUF and how they can impact the transition between these stages.

3 Fire Barriers in Full-Scale Testing: Chair Mock-Ups

3.1 Chair Mock-Ups

This section has two aims: (a) to further investigate the effect of mass transfer of liquid products through the bottom of a burning chair to better understand the key parameters affecting the performance and failure of barriers at a full scale and (b) to investigate to what extent passive barriers (expected to comply with the aforementioned US State regulations because they do not contain FRs) can delay the fire growth of RUF when exposed to a strong (18 kW) ignition source (equivalent to five sheets of crumpled newsprint) [58].

A chair mock-up was composed of four cushions: a seat cushion, a back cushion, and two armrest cushions. Details about the dimensions of the cover fabric and chair construction can be found in Thompson et al. [29]. The control chair mock-ups (chair C0) included only the cover fabric C0, whereas all other chairs (chair B1 to B6) added a barrier (B1 to B6) as an interliner between the cover fabric and the padding material. Metal staples were used to close the barrier seam because seam failure was previously observed with polyamide threads [23], aluminum zippers held

Table 1 Reduced Cube test data at an external heat flux of 75 kW/m^2. For the full table is Zammarano et al. [31]

	t_{p1} (s)	$PHRR_1$(kW)	t_w (s)	Q_{Cube}(kW)	t_{BI} (s)	t_{P2} (s)	$PHRR_2$(kW)	THR (MJ)	EHC (MJ/Kg)
C0	14 ± 1	3.20 ± 0.10	90 ± 3	2.80 ± 0.10	106 ± 2	118 ± 12	11.5 ± 1.5	1.13 ± 0.02	25.2 ± 0.2
B1	15 ± 3	4.15 ± 0.27	235 ± 17	0.62 ± 0.03	381 ± 49	440 ± 48	4.3 ± 0.3	1.02 ± 0.05	22.1 ± 0.5
B2	16 ± 1	3.32 ± 0.12	153 ± 2	0.96 ± 0.02	278 ± 19	410 ± 24	4.1 ± 0.4	1.05 ± 0.03	23.8 ± 0.6
B3	27 ± 5	4.42 ± 0.25	258 ± 7	0.48 ± 0.02	591 ± 19	620 ± 22	4.7 ± 0.4	0.98 ± 0.02	19.0 ± 6.4
B4	16 ± 1	3.58 ± 0.14	143 ± 2	1.51 ± 0.08	260 ± 8	404 ± 11	3.8 ± 0.2	1.06 ± 0.01	23.9 ± 0.1
B5	21 ± 1	4.68 ± 0.47	164 ± 4	0.90 ± 0.09	354 ± 12	376 ± 12	4.9 ± 0.1	0.99 ± 0.06	23.3 ± 0.2
B6	26 ± 1	4.27 ± 0.13	222 ± 2	0.56 ± 0.06	658 ± 28	694 ± 25	4.4 ± 0.2	1.00 ± 0.05	22.2 ± 1.3

in place by polyamide webbing [22], thermoplastic thread and nylon zippers [21], and polyaramid thread [64].

3.2 Flammability Testing

Tests were conducted at NIST's National Fire Research Laboratory under a 6.1 m by 6.1 m exhaust hood [65]. The average expanded uncertainty (calculated with a coverage factor of 2 corresponding to an approximately 95% confidence interval) [66] for generic combustible fuel was 6.8% in the range from 100 to 3000 kW. The range of expanded uncertainty for the natural gas (fuel consumption) verification burners is 1.4% to 1.8%. Detailed information on the NIST calorimetry measurement system is provided by Bryant and Bundy [65].

Seven chair mock-up combinations (one control chair, chair C0, and six chairs with barriers, chairs B1 to B6) were tested in triplicates for a total of 21 tests; however, due to improper assembly of a B2 chair, only two replicate tests were accepted for B2 chairs. Chairs were assembled on a metal test frame shown in Thompson et al. [29]. Figure 6 shows the test setup used in the flaming tests. A detailed description of the test setup is explained in Thompson et al. [29].

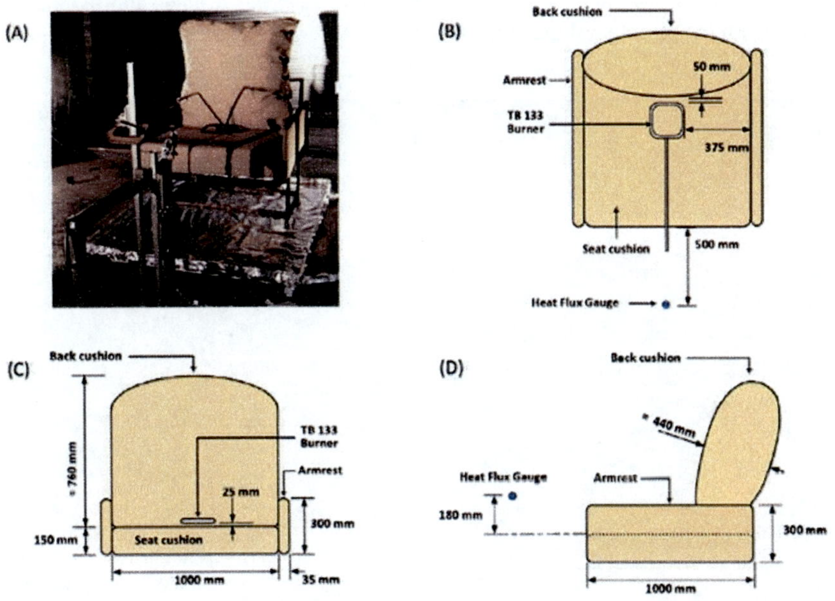

Fig. 6 Test setup used in flaming tests: **a** photo showing a chair mock-up assembled on the metal frame with a burner, heat flux gauge, and catch pan; **b** top view schematic; **c** front view schematic; and **d** side view schematic (nominal dimensions, back cushion not to scale) [29]

3.3 Effects of Barriers on Fire Growth

In order to illustrate the effects of barriers on fire growth in the given test scenario, Fig. 7 shows snapshots for representative chair mock-ups without a barrier and with barrier B6 at t = 30 s, t = 80 s (when the burner was turned off and removed), at the time of pool fire formation, and at the time of peak of heat release rate (PHRR). B6 was the most effective barrier in this study, delaying the intense burning of the chair by at least 20 min.

In a typical test without a barrier, flame jets from the burner simultaneously ignited the top surface of the seat cushion and the front bottom of the back cushion within a few seconds. The polypropylene cover fabric quickly melted and peeled away from the spreading flame front, exposing the padding material which quickly ignited. Shortly after the removal of the burner, the back cushion completely collapsed. Intense flaming over the seat cushion eventually generated a hole in the center of the cushion. Flaming liquid products (mainly, regenerated polyols, produced by the pyrolyzing foam, and possibly liquefied polyester) [35] poured through the hole onto the catch pan and quickly formed a pool fire. Heat flux feedback from the pool fire to the seat cushion increased the generation and release rate of flaming liquid, which in turn produced a larger pool fire and stronger heat flux feedback over the seat cushion and pool fire. This was a self-accelerating process that led to an abrupt increase in heat release rate and continued until the FPUF and regenerated polyol in the seat cushion and the liquid pool were depleted. The peak heat release was reached approximately 1 min from the initial pool fire formation.

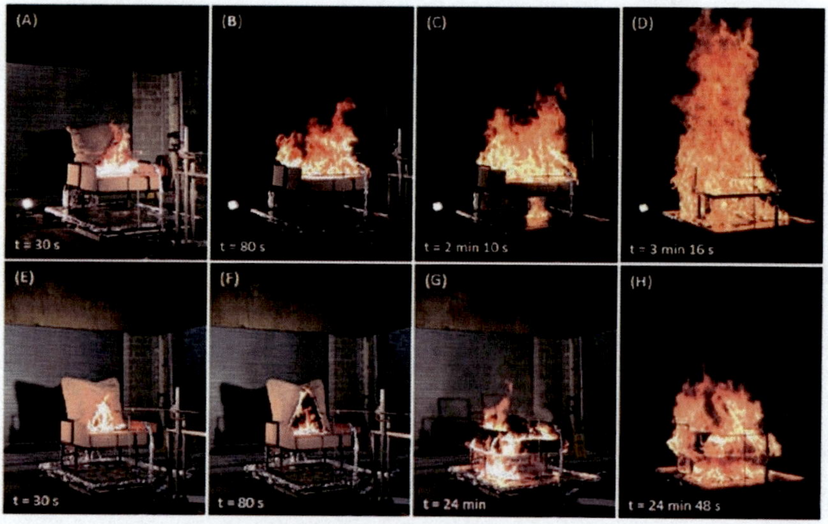

Fig. 7 Snapshots for representative chair mock-ups without a barrier (**a** at t = 30 s; **b** at t = 80 s; **c** at t = 2 min 10 s; and **d** at t = 3 min 16 s) and with barrier B6 (**e** at t = 30 s; **f** at t = 80 s; **g** at t = 24 min; and **h** at t = 24 min 48 s) [29]

In a typical test with a barrier, the burner ignited the cover fabric over the seat cushion and along the lower front bottom of the back cushion within a few seconds. The polypropylene cover fabric quickly melted/pyrolyzed along the flame front exposing the barrier, which remained intact and protected the padding material by decreasing the heat transfer from the flames. As a result, the pyrolysis rate of the padding material (hence, the heat release rate and flame size) decreased in the chair with barriers compared to the control chair. While the barrier was still intact, the thermoplastic polyester fiber fill inside the back cushion kept melting and/or pyrolyzing until the back cushion collapsed and tipped forward past the nichrome wire. Afterward, the rapid ignition of the cover fabric on the rear side of the back cushion and the increase in heat flux on the residual polyester fiber fill caused a temporary rise in heat release rate (HRR), which generated a first peak in the heat release rate ($PHRR_1$). Burning continued until an intense flaming appeared underneath the seat cushions, and, shortly after, flaming drips accumulated under the chair, generating a pool fire. The flame size increased until the second, most intense peak of heat release rate ($PHRR_2$) was reached about 1 min after initial pool formation.

Figure 8 shows representative HRR curves measured for the chair mock-ups with barrier (B1-B6) and the control chair mock-up with no barrier (C0). HRR plots for all repeat tests are also shown in Thompson et al. [29]. The control chair reached HRR values above 2500 kW within 3 min. Such rapid fire growth would most likely lead to flashover conditions in most residential fire scenarios [67]. A HRR of 1000 kW generally leads to flashover in a small bedroom and in dwellings where flaming can spread from RUF to other combustible items in the compartment, independent of the room size [68].

In comparison, all barriers significantly decreased the fire growth by generating a HRR plateau between the appearance of a first peak of heat release rate ($PHRR_1$)

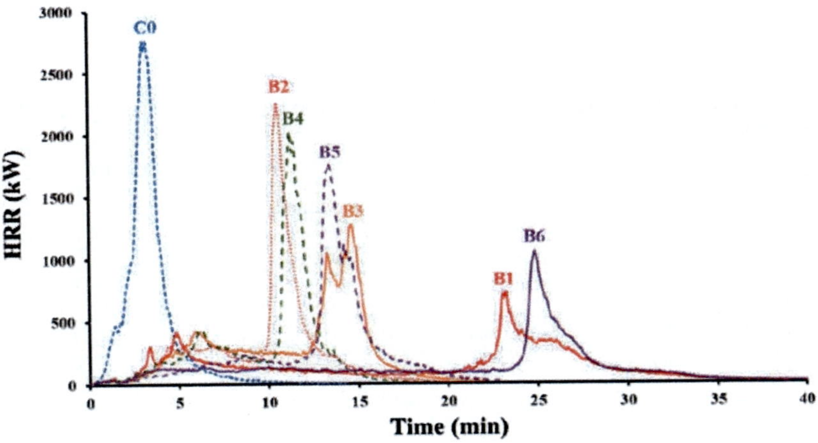

Fig. 8 Representative heat release rate curves for the mock-up chair with no barrier (C0) and the mock-up chairs containing barriers (B1-B6) [29]

and a second, higher peak ($PHRR_2$), which shortly followed the formation of a pool fire. This two-peak burning behavior in RUF and the role of pool fire on burning rate are consistent with previously reported studies [21, 22, 69, 70]. The intensity and timing of $PHRR_2$ are highly dependent on factors related to the flaming of liquid products (such as flow, release timing, and release location of liquid products, type of substrate under the item, substrate-to-item distance, extent of feed-back between pool fire, residual item, etc.) [71].

During the HRR plateaus, the HRR stabilized at roughly constant values that ranged between about 100 and 300 kW and lasted between 9 and 22 min, depending on the barrier. The intensity of $PHRR_1$ was presumably affected by the barrier effectiveness (i.e., extent of heat and mass transfer reduction) and the amount of combustible material (i.e., cover fabric and fiber fill) left in the back cushion when it fell forward past the nichrome wire. B6 chairs took the longest to collapse and no obvious $PHRR_1$ was detected (see Fig. 8). Table 2 shows the values of the flaming characteristics.

Depending on the barrier, time to peak 1 (t_{P1}) varied between 5 and 7 min and $PHRR_1$ between 300 and 400 kW; time to peak 2 (t_{P2}) increased from about 3 min for the control mock-up to about 22 min (≈sevenfold increase) for the mock-ups containing B1 and B6. In all chair mock-ups, $PHRR_2$ was by far the stronger peak, and its average value was reduced in the presence of barriers from about 2600 kW for the control chair to about 900 kW for barriers B1 and B6 (about a threefold decrease). The best barriers also induced a marginal reduction (under 15%) in THR, EHC, and ML, which was presumably due to an increase in the amount of FPUF charring; in fact, the lower the FPUF pyrolysis rate (due to the presence of the barrier), the higher the FPUF char yield [72].

Figure 9 shows heat flux curves for each chair type (the same representative tests are shown in Fig. 8). The flux gauge was positioned about 50 cm from the front of the chair and about 74 cm over the catch pan (see Fig. 6). The time at which heat fluxes of 20 kW/m^2 (t_{HF20}) and 2.5 kW/m^2 ($t_{HF2.5}$) were reached on the flux gauge is summarized in Table 3.

Table 2 Values of t_{p1}, $PHRR_1$, t_{P2}, $PHRR_2$, THR, EHC, and ML and barrier used in each chair mock-up [29]

	t_{P1} (min)	$PHRR_1$ (kW)	t_{p2} (min)	$PHRR_2$ (kW)	THR (MJ)	EHC (MJ/kg)	ML (%)
C0	–	–	3.3 ± 0.2	2610 ± 154	311 ± 5	24.3 ± 0.4	93 ± 3
B1	4.9 ± 0.9	294 ± 78	22.2 ± 2.2	888 ± 157	307 ± 12	22.4 ± 0.3	87 ± 1
B2	4.9 ± 0.6	288 ± 6	11.5 ± 1.4	2025 ± 329	288 ± 8	23.9 ± 0.1	82 ± 2
B3	5.0 ± 0.8	296 ± 46	14.1 ± 0.6	1030 ± 341	328 ± 3	23.7 ± 0.2	87 ± 2
B4	6.6 ± 0.4	392 ± 40	10.9 ± 1.0	1817 ± 222	304 ± 2	24.0 ± 0.2	89 ± 1
B5	7.0 ± 0.5	328 ± 59	12.6 ± 0.6	1648 ± 133	321 ± 3	23.8 ± 0.2	83 ± 1
B6	–	–	22.7 ± 3.4	991 ± 319	297 ± 6	22.2 ± 0.3	83 ± 1

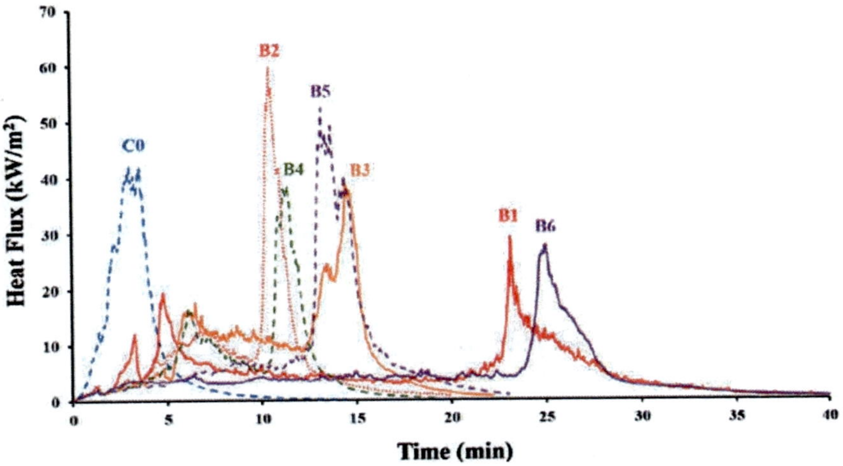

Fig. 9 Representative heat flux curves for each chair type (the same representative tests of Fig. 8 were used) [29]

Table 3 Time to reach heat flux of 2.5 kW/m^2 ($t_{HF2.5}$) or 20 kW/m^2 (t_{HF20}), and peak of heat flux (PHF) with time to PHF (t_{PHF}) [29]

Barrier	$t_{HF2.5}$ (min)	t_{HF20} (min)	PHF (kW/m^2)	t_{PHF} (min)
C0	0.6 ± 0.1	1.8 ± 0.1	53 ± 9	2.9 ± 0.1
B1	2.8 ± 0.7	21.7 ± 2.1	24 ± 5	22.0 ± 2.3
B2	2.7 ± 0.3	10.9 ± 1.3	52 ± 11	11.5 ± 1.5
B3	2.3 ± 0.3	13.3 ± 0.6	34 ± 6	14.1 ± 0.5
B4	1.5 ± 0.6	9.2 ± 2.9	42 ± 5	10.9 ± 0.9
B5	4.0 ± 0.9	12.1 ± 0.6	47 ± 5	12.9 ± 0.3
B6	2.8 ± 0.6	22.6 ± 3.1	29 ± 5	22.8 ± 3.5

Note Uncertainties are reported as ± one experimental SD calculated over three independent observations for each chair type, except chair B2 for which only two independent observations are available

At $t = t_{HF2.5}$, the conditions in proximity of the flux gauge are untenable. Occupants can tolerate a heat flux of 2.5 kW/m^2 for a short amount of time, but such conditions will quickly lead to incapacitation [73]. The value of $t_{HF2.5}$ ranged between a minimum of 0.6 (±0.1) min for chair C0 to a maximum of 4.0 (±0.9) min (almost a sevenfold increase) for B5. Except for incapacitated or immobile individuals that are unlikely to survive unless rescued, people can usually move away from a fire to areas that are tenable and possibly escape. However, untenable conditions will be eventually reached within the entire compartment. In ISO 9705 rooms, this happens when the HRR reaches roughly 400 kW and the hot smoke gas layer descends to about 1.2 m from the floor level [68]. This 400 kW criterion would suggest that tenable conditions would be maintained in the room until the end of the HRR plateau phase with most barriers (see Fig. 8).

The highest reductions in PHF were achieved with B1 and B6 (roughly twofold reductions compared to the C0 chairs). The reduction in heat flux during the plateau phase is expected to limit the fire growth in a compartment by preventing flame spread to surrounding items. A heat flux of about 20 kW/m^2, which is required for non-piloted flaming ignition, [74–76] was reached in less than 2 min in chair C0 but required over 20 min in chairs with barriers B1 and B6 (see t_{HF20} in Table 3).

3.4 Mechanism of Barrier Failure

All barriers significantly mitigated the fire hazard of the chair mockups by reducing the burning rate as compared to unprotected mockups until the end of the HRR plateau phase, when an abrupt increase in HRR occurred. The phenomena that led to this rapid increase in HRR (referred to as "barrier failure") at the end of the plateau phase is illustrated and discussed below.

Figure 10 shows video snapshots for a representative B6 chair, illustrating the sequence of events that occurred underneath the chair and led to the failure of the barrier. The flaming intensity varied depending on the barrier and was visibly enhanced in proximity of the armrests by heat flux feedback from the vertical surfaces. At this stage, flames were mainly sustained by the gaseous pyrolyzates produced by the padding material in the seat cushion, while the liquid products (mainly regenerated polyols [35]) accumulated inside the barrier envelope on top of the receding FPUF [77]. Eventually, the liquid products, now in contact with the barrier, started percolating through the bottom of the seat cushion.

This process is shown in Fig. 10. In particular, the inset in Fig. 10 highlights an area where the cover fabric has been removed due to melting, flaming, and dripping, and the barrier (normally white) shows brown areas that expanded over time due to liquid products percolating through and wetting the outside of the barrier. When wetting occurred, burning of the cover fabric sustained flaming underneath the chair and generated flaming drops which accumulated under the chair mock-up; meanwhile, smoke began to be released from the bottom of the chair and increased over time (see Fig. 10). The arrows in the Fig. 10 inset indicate regions where wetting occurred.

Eventually, flaming of the cover fabric led to the ignition of liquid products over the wetted areas. Bottom ignition was revealed by sustained flaming on the bottom of the seat cushion in regions where the cover fabric was no longer present (see Fig. 10). The inset in Fig. 10 zooms in on the location where bottom ignition occurred. Bottom ignition gave rise to more intense and prolonged flaming compared to the cover fabric flaming (where the flame front kept moving due to fuel depletion). Bottom ignition was identified by analyzing the videos recorded by the bottom camera for each test.

Bottom ignition led to an increase in incident heat flux over the bottom of the cushion, which in turn led to an increase in barrier degradation and in the pyrolysis rate of residual foam and liquid products; as a result, the percolation rate of liquid products

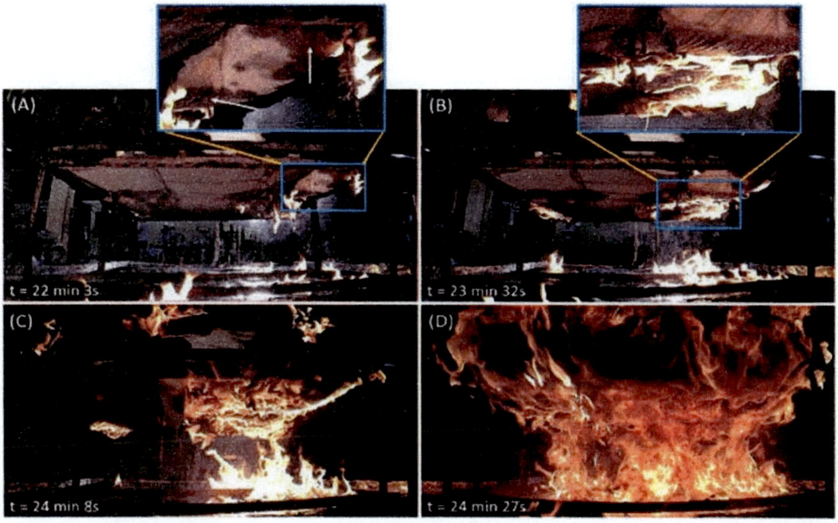

Fig. 10 Video frame grabs for a representative B6 chair mock-up illustrating the sequence of events that led to the failure of barriers; **a** "wetting" of the barrier due to the percolation of liquid products through the barrier; **b** ignition of liquid products ("bottom ignition"); **c** initial pool fire formation; and **d** feedback between the flame on the pool fire and the remaining fuel in the chair mock-up at the time of $PHRR_2$ [29]

increased and more fuel was available to feed flaming underneath the cushion and increase the heat release rate. Hereafter, the amount of liquid products percolating through the barrier exceeded the amount burned at the bottom of the cushion and a liquid pool began to form underneath the mock-up on the support stand. A pool fire was eventually generated (Fig. 10). Noticeably, at this stage, the barrier was still intact, yet the heat release rate kept rapidly increasing—with the self-accelerating feedback mechanism which was previously described until the peak value occurred. For each chair mockup, Table 4 reports bottom ignition characteristics.

For the control chairs C0, bottom ignition coincided with the appearance of flaming underneath the mock-up when liquid products immediately ignited, and flaming liquids started feeding the pool fire; t_{BI} occurred at (2.2 ± 0.1) min when the HRR was about 820 kW; it only took (1.1 ± 0.1) min to reach a $PHRR_2$ of (2610 ± 150) kW from t_{BI}; $AHRR_{BI}$ was (309 ± 35) kW and only (12.9 ± 1.7) % of THR was released by the time bottom ignition occurred (see Table 4). The addition of a barrier substantially delayed bottom ignition. t_{BI} increased from about 2 min for the control chair, to about 20 min for barriers (B1 and B6).

The difference in time, t_{P2}—t_{BI}, was about 2 min for all chairs except chair B4 (which had a t_{P2}—t_{BI} of about 4 min). Hence, barriers were able to moderate the HRR until bottom ignition. After that the HRR increased abruptly and reached its peak value within a couple of minutes. This is a strong indication that the barrier failure was triggered by bottom ignition.

Table 4 Time to bottom ignition (t_{BI}), elapsed time between bottom ignition and $PHRR_2$ (t_{P2}—t_{BI}), heat release rate at bottom ignition (HRR_{BI}), average heat release rate calculated between ignition and bottom ignition ($AHRR_{BI}$), and percentage of THR released between ignition and bottom ignition ($\%THR_{BI}$), as a function of the barrier used in the chair mock-up [29]

Barrier	t_{BI} (min)	t_{P2}—t_{BI} (min)	HRR_{BI} (kW)	$AHRR_{BI}$ (kW)	$\%THR_{BI}$ (%)
C0	2.2 ± 0.1	1.1 ± 0.1	817 ± 64	309 ± 35	12.9 ± 1.7
B1	20.0 ± 2.3	2.2 ± 0.9	118 ± 38	118 ± 9	46.1 ± 0.8
B2	9.9 ± 1.9	1.6 ± 0.5	185 ± 19	154 ± 17	31.5 ± 3.6
B3	9.7 ± 2.0	4.5 ± 1.4	290 ± 86	183 ± 28	32.2 ± 7.8
B4	9.2 ± 0.3	1.7 ± 0.8	221 ± 13	179 ± 10	32.2 ± 1.0
B5	10.5 ± 0.7	2.1 ± 0.1	187 ± 25	135 ± 17	26.3 ± 2.0
B6	20.2 ± 3.3	2.6 ± 0.9	101 ± 21	108 ± 22	43.2 ± 2.7

Note Uncertainties are reported as ± one experimental SD calculated over three independent observations for each chair type, except chair B2 for which only two independent observations are available

Further evidence is provided by Fig. 11, which clearly shows the effect of bottom ignition on HRR for a representative B6 mockup. The heat release rate stabilized to a value of HRR $\approx$ HRR_{BI} throughout the HRR plateau phase; the end of the HRR plateau phase coincided with the time of bottom ignition of the seat cushion. A similar trend was observed also with the other barriers, even though a moderate temporary increase in heat release rate ($PHRR_1$) was observed during the plateau phase. Ultimately, a large increase in HRR was observed for $t > t_{BI}$ in all tests, thus, it is reasonable to assume that: (a) t_{BI} is a good indicator of the "time to failure" for the barrier tested and (b) bottom ignition is the triggering mechanism that leads to barrier failure. Figure 11 also shows that HRR_{BI} can be used as a good approximation for the HRR at the plateau.

Compared to FPUF pillows, polyester fiber–filled pillows were shown to generate about half the total heat release and produce about five times more char [78]. Regenerated polyols account for about 70% of the total heat released by the FPUF, hence, understanding how and when regenerated polyols become involved in the burning process is extremely important [77].

In Fig. 11, the burning process of the chair mock-up is divided into stage I and stage II, where t_{BI} is the time at which stage I transitions to stage II. During stage I, HRR is mainly supported by gaseous pyrolyzates released through the barrier envelope, which remains intact. At the beginning of stage II ($t = t_{BI}$), liquid products percolating through the barrier envelope (mainly regenerated polyols) ignited and led to a rapid-fire growth. Most of the heat of combustion was released in this phase due to the involvement of regenerated polyols. The heat released during stage II ranged from a minimum average value of 54% THR for chair B1 to a maximum of 87% THR for chair C0.

The barrier failure mechanism described here largely agrees with the observation of so-called "basal melt fire," the failure mechanism previously described in the pioneering work of Ohlemiller and Shields [21]. In the basal melt fire mechanism,

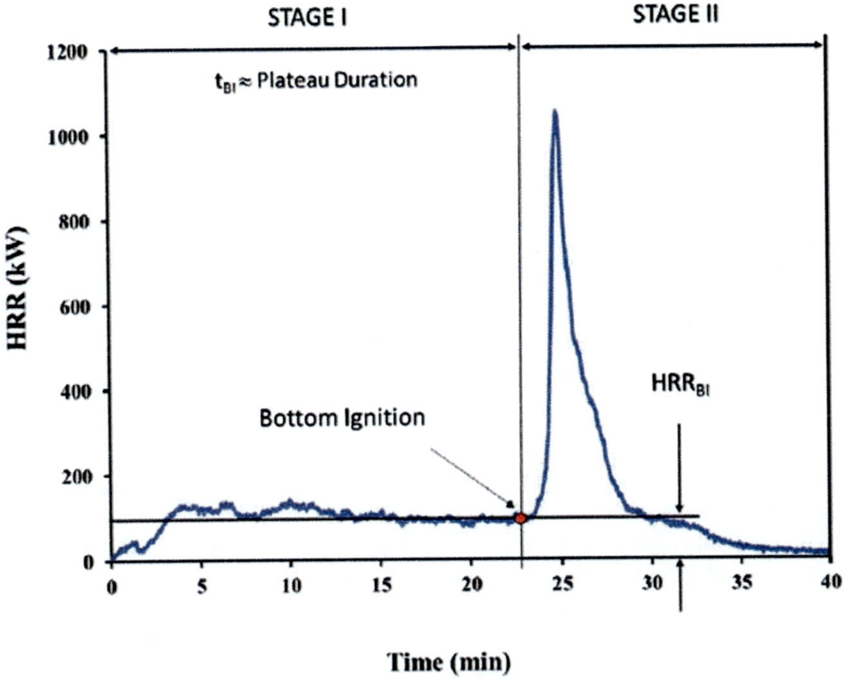

Fig. 11 Effect of bottom ignition on the heat release rate for a representative B6 chair [29]

flames become established on the bottom of a vertically oriented cushion (e.g., seat-back or arm cushion) whose barrier is completely intact. The fuel for the flames was initially the cover fabric but increasingly became the FPUF regenerated polyols as these liquid products were flowing down through the pores in the barrier at the base of the cushion. Based on these considerations, NIST recently developed a bench-scale test (the "Cube Test" mentioned in Sect. 9.2 of this chapter) aiming to predict time to barrier failure at full scale [31, 32].

3.5 *Impact of Barriers on Fire Risk*

Barriers were found to possibly fully extinguish flaming and prevent a complete combustion of the RUF item when small candle-like ignition sources were used [28]. However, flame extinguishment is not likely with larger ignition sources, like an 18 kW burner; in this scenario, barriers have been found to decrease the burning rate but not necessarily reduce the total heat release [21]. About 10% of RUF fire deaths are due to small flames and about 21% are due to open flames spreading from other items, where the flaming insult to the RUF is likely to be more severe than a candle-like flame [3]. RUF ignition over the seat cushion and alongside the back cushion led to a shorter time to peak of HRR compared to ignition alongside the arm or under the seat cushion [79].

Based on these considerations, in this study, we selected an 18 kW flame (mimicking the combustion of five sheets of crumpled newsprint) [58] as the ignition source, and the center–top of the seat cushion in proximity of the back cushion as the ignition location. A small bedroom typically requires HRR of about 1000 kW to reach flashover [68]. Flame spread to the surrounding combustibles is unlikely to occur for $t < t_{HF20}$ (i.e., when the heat flux at a distance of 50 cm from the RUF item is below 20 kW/m^2). In the presence of barriers, $t_{HF20} \geq t_{BI}$, and the HRR of the RUF item at $t = t_{BI}$ (HRR_{BI}) is always well below 1000 kW (HRR_{BI} ranges between about 100 and 300 kW). Based on these considerations, it can be concluded that flashover is not likely to occur for $t < t_{BI}$. Similarly, adopting the aforementioned 400 kW criterion for tenability, the conditions in the room will likely remain tenable for $t < t_{BI}$.

In a living room layout where multiple upholstered chairs are placed in close proximity, flame spread to the surrounding combustibles cannot be excluded for $t < t_{HF20}$ because the heat flux on the combustible can reach the critical value of 20 kW/m^2 for nonpiloted ignition. Also, piloted ignition (which requires a critical heat flux as low as 10 kW/m^2) [74] becomes likely for distances below 50 cm due to flickering flames or hot objects (projected by the RUF fire), which can act as an ignition source. Nevertheless, the use of barriers is expected to significantly reduce the fire hazard also in this scenario, especially if the second item ignited is also an RUF item protected by a barrier; in fact, in this case the second item ignited will also show a HRR plateau and consequent delay in peak of HRR. A 15-min delay time to flashover and 12 min increase in estimated available safe egress time (ASET) were achieved by using barrier B6 to protect a sofa in a fully furnished living room selected as case study; this reduction in fire growth would likely allow first responders to intervene and prevent flashover [77]. Due to the severity of the test conditions in the case study, it is not unrealistic to expect that first responders would be able to intervene before flashover in a large fraction of room fire scenarios where an effective barrier like B6 is adopted.

The dominant mechanisms by which the barriers mitigated the fire hazard of the chair mock-ups were: (a) delay in fire growth (the longer the time to bottom ignition, the longer the time available for egress before untenable conditions are reached and for first responders to arrive and attack the fire), and (b) a reduction in $PHRR_2$ (the lower the PHRR, the lower the likelihood of flashover and flame spread occurrence) [67, 80]. B1 and B6 were the best performing barriers in terms of t_{BI} and $PHRR_2$ (B6 being the best due to the absence of $PHRR_1$ and lower initial HRR). Interestingly, both B1 and B6 contain hybrid yarns made of regenerated cellulose and polysilicic acid, which are inherently fire resistant [36]. Polysilicic acid is an inorganic material that decomposes endothermically to release water and dilute combustible products of decomposition [36, 39]. In addition, the incorporation of molecular chains of silicic acids into the hybrid yarn is expected to improve barrier performance by increasing the cellulose char yield [39].

A delay of t_{BI} from about 2 min (control chair) to about 20 min (B1 and B6 chairs) can be contextualized by analyzing the objective (target) fire service response times. A properly placed smoke alarm can be activated within about 40 s even with RUF items using the best performing barrier B6 [30]. Based on NFPA 1710 [81] and NFPA 1720 [82], first responders are expected to intervene within about 12 min in urban areas, 13 min in suburban areas, and 17 min in rural areas [30]. An RUF with similar fire growth as for the chair mock-ups without barrier (i.e., $PHRR_2$ above 2500 kW and $t_{P2} \approx 3$ min) would most likely lead to flashover before responders intervention even in urban areas [67], whereas, an RUF with a reduced fire growth as for the chair mock-ups with B6 ($t_{BI} \approx 20$ min) would most likely allow responders to intervene before flashover even in rural areas because, as previously discussed, flashover is not likely to occur for $t < t_{BI}$.

4 Fire Barriers in Furnished Compartments

4.1 Test Setup and Procedure

The goal of Sect. 9.4 is to determine if the failure mechanism of barrier fabrics previously observed in the chair mock-up experiment (Sect. 9.3) also applies to commercial RUF in furnished compartments. B6 has been used in a commercial couch for full-scale compartment tests in a realistic fire scenario. The objectives of this section include: (1) quantifying the effect of barrier fabrics on the fire hazard (e.g., tenability conditions, heat release rate, time to flashover and smoke generation) of RUF ignited by a flaming-ignition source in a realistic living room scenario, (2) show that barrier fabrics, which can remarkably reduce the fire hazard without using chemically active fire retardants, not only exist but are commercially available, and (3) d etermine if the failure mechanism of barrier fabrics previously observed in the chair mock-up experiment also applies to commercial RUF in a realistic scenario.

RUF compositions used in this study can be found in Zammarano et al. [30]. Three fire tests, one per couch type, were conducted. In Test 1, the couch with the polyester covering was used. In Test 2, the cotton blend covering was used in combination with the barrier fabric. In Test 3, the same cotton blend covering of Test 2 was used without barrier fabric. The schematic drawing of the couch and additional items with dimensions, composition, mass, and quantity can be found in Zammarano et al. [30]. A picture of the test setup in shown in Fig. 12.

The instrumentation used in these experiments to measure HRR, incident heat flux, gas temperature, and description of air-cooled cameras under the loveseat and chaise sections of the couch are summarized in detail in Zammarano et al. [30].

Fig. 12 Picture of furnished compartment test setup under the 10 MW hood at the National Fire Research Laboratory [30]

4.2 *Observations and Heat Release Rate*

For all tests, the smoke alarm triggered within the first minute after fire ignition (42 s in Test 1, 47 s in Test 2 and 45 s in Test 3) presumably due to the rapid burning of the throw pillow. The burning behavior varied significantly between samples without a barrier fabric (Test 1 and Test 3) and the sample with a barrier fabric (Test 2). Figure 13 compares snapshots of the three tests at 2 min, 4 min, 6 min and 8 min from ignition. At 2 min, a similar smoke layer, flame spread and flame size were observed in all tests. In Test 1 and Test 3, after 2 min the flame spread and fire intensity increased continuously until flashover conditions were reached. In Test 2, after 2 min the flame spread decreased; 4 min after ignition, the smoke layer dissipated and did not reappear until approximately 10 min later in the test. The photographs in Fig. 13 qualitatively demonstrates the efficacy of the barrier fabric to reduce flame spread over the couch.

Figure 14 shows the heat release rates for Test 1, Test 2, Test 3, and Throw Pillow 1 during the first 500 s after ignition. A lower heat release rate was expected for Test 2 as compared to Test 3 due to the inclusion of the barrier fabric. The lower heat release rate for Test 1 as compared to Test 3, however, is surprising because cotton blend cover fabrics are typically expected to be less flammable than polyester cover fabrics even though the behavior of fabric blends are often unpredictable [21].

The heat release rate curves from ignition to extinguishment for Test 1, Test 2 and Test 3 are shown in Fig. 15. The values of peak heat release rate (PHRR), time

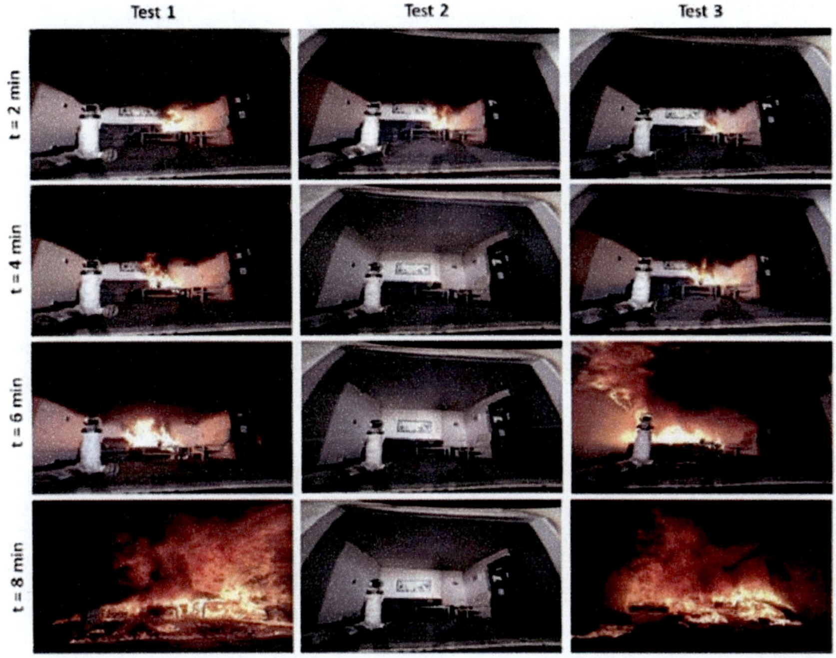

Fig. 13 Photographs of the compartments at 2 min, 4 min, 6 min and 8 min after ignition. These photographs qualitatively demonstrate the efficacy of the barrier fabric. At t = 2 min, a similar smoke layer, flame spread and flame size were observed in all tests. For t > 2 min, the flame spread and fire intensity increased continuously in Test 1 and Test 3, until flashover conditions were reached. In Test 2, the flame spread decreased; at t = 4, the smoke layer dissipated and did not reappear until approximately 10 min later [30]

Fig. 14 Heat release rate curves measured for Test 1 to Test 3 and for the Throw pillow 1 during the first 500 s after ignition [30]

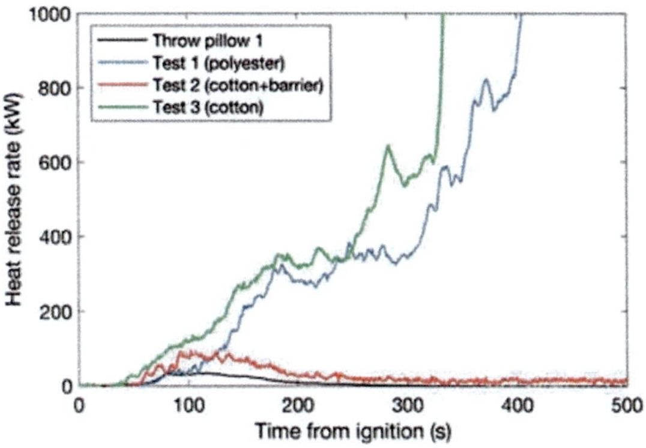

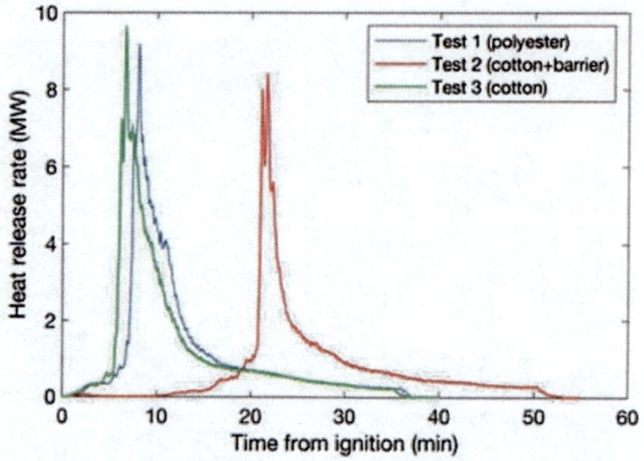

Fig. 15 Heat release rate curves for Test 1, Test 2, and Test 3 [30]

Table 5 Values of time to peak heat release rate t_{PHRR}, peak heat release rate (PHRR) and total heat release rate (THR) measured in Test 1, Test 2, and Test 3 [30]

	t_{PHRR} (min)	PHRR (kW)	THR (MJ)
Test 1	9.1 ± 0.2	9180 ± 550	2550 ± 160
Test 2	21.7 ± 0.2	8420 ± 500	2250 ± 140
Test 3	6.7 ± 0.2	9640 ± 580	2640 ± 160

to peak heat release rate (t_{PHRR}), and total heat release (THR) are summarized in Table 5. All tests were extinguished with water about (25 to 30) min after PHRR.

t_{PHRR} showed substantial variations. The t_{PHRR} was (6.72 ± 0.17) min for Test 3, (9.12 ± 0.17) min for Test 1, and (21.68 ± 0.17) min for Test 2. Compared to Test 3, the use of the barrier fabric with the same cover fabric in Test 2 offered a 300% increase in t_{PHRR} (i.e., it took approximately 3 times longer for the PHRR to be observed when a barrier fabric was used). This increase in t_{PHRR} was associated with the formation of a quasi-steady state burning with a minimum heat release rate plateau value of about 15 kW at $t \approx 8$ min in the early burning stage of Test 2 when the combustion of the throw pillow was complete. The THR in Test 2 decreased by about 15% compared to Test 3 due to the addition of the barrier fabric to the cotton blend cover fabric.

4.3 *Flashover*

Flashover is the last phase in fire growth that leads to a fully-developed compartment fire. Flashover occurs when the room is hot enough (or heat transfer from the upper smoke layer is large enough) that all combustible solids in the room begin pyrolyzing and producing gaseous volatiles, despite the presence or absence of direct flame impingement. Ignition of these gaseous pyrolyzates yields an abrupt increase in heat release rate [83]. In this study, it is assumed that flashover conditions in the compartment were reached when the magazines and the single upholstered chair ignited almost simultaneously. The time to flashover and the time to smoke alarm triggering for Test 1 to Test 3 are reported in Table 6.

The smoke alarm triggering times were similar for all tests. The time to flashover increased by less than 1 min when the cotton blend cover fabric (Test 3) was replaced by the polyester cover fabric (Test 1). Thus, the effect of the cover fabric was statistically significant, but small. In comparison, the influence of the barrier fabric on the time to flashover was large. The presence of the barrier fabric delayed the time to flashover by about 15 min in Test 2 (cotton blend cover fabric and barrier fabric) compared to Test 3 (cotton blend cover fabric only).

The rapid fire growth observed in Test 1 and Test 3 may present a significant hazard to life and property before first responders can arrive. Thus, a 15 min delay of flashover observed in Test 2 can drastically reduce this hazard. This can be contextualized by analyzing the objective (target) fire service response times in NFPA 1720 [82]. The target response time is 9 min in urban areas (for at least 90% of the responses), 10 min in suburban areas (for at least 80% of the responses), and 14 min in rural areas (for at least 80% of the responses). Response time begins upon completion of the dispatch notification to the fire department and ends at the time a prescribed number of staff members start mitigation measures at the fire scene. After ignition, assuming an average smoke alarm triggering time of 45 s, an alarm answer time of 15 s to 40 s and an alarm processing time of 64 s to 106 s (see NFPA 1710 for response objectives [82]), the response time will start about 2 min to 3 min after ignition. Thus, considering the target response time, the first responders are expected to intervene within 12 min in urban areas, 13 min in suburban areas and 17 min in rural areas. A graphic representation of time to flashover for Test 1, Test 2, and Test 3 versus response time is shown in detail in Zammarano et al. [30].

If these experiments are representative of available couches, the data suggests that first responders may not be able to intervene before flashover when a couch without barrier fabric is involved in the types of compartment fires represented by Test 1 and Test 3, even in urban areas. However, under the same conditions, first responders would likely be able to intervene and prevent flashover when a barrier fabric is adopted in the couch (Test 2) in urban, suburban, and rural areas. Flashover prevention is critical because it can prevent fire spread to adjacent rooms and buildings, which is where the majority of fire deaths occur in RUF fires [2].

Figure 16 shows video frames at flashover for Test 1, Test 2, and Test 3 captured by the water-cooled camera, the 360° camera, and the air-cooled camera under the chaise section of the couch. The water-cooled camera (top row in Fig. 16 shows that

Table 6 Time to smoke alarm triggering and flashover in Test 1, Test 2, and Test 3 [30]

	Time to smoke alarm triggering (s)	Time to flashover (min)
Test 1	42 ± 2	7.0 ± 0.2
Test 2	47 ± 2	21.0 ± 0.2
Test 3	45 ± 2	6.0 ± 0.2

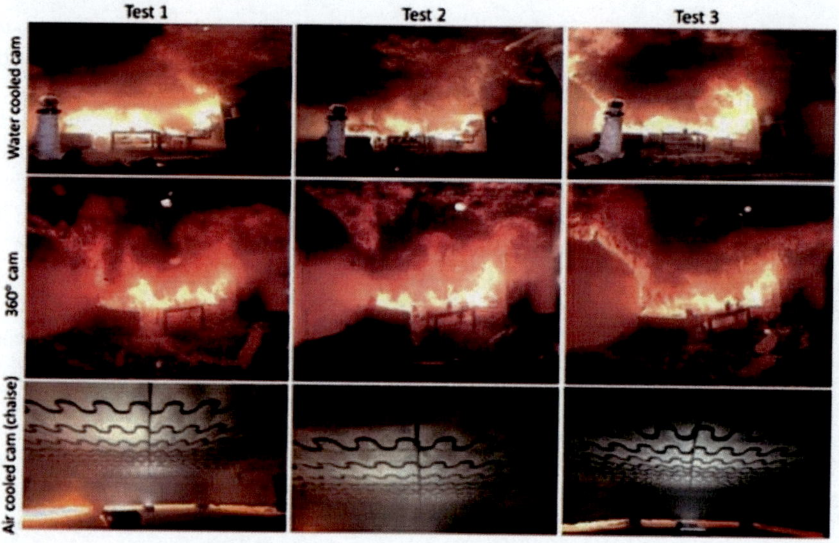

Fig. 16 Video frames at the time of flashover for Test 1 (left column), Test 2 (center column), and Test 3 (right column) captured by the water-cooled camera (top row), 360° camera (center row), and air-cooled camera under the chaise section of the couch (bottom row) [30]

by the time flashover occurred, flames appeared to cover the entire couch except the front vertical surface of the chaise section. The 360° camera (middle row in Fig. 16) confirms this for Test 1 and Test 3, however, for Test 2, it also shows that flame spread over the top surface of the chaise section was not complete, i.e., flaming was not present over the extremity of the top seat cushion which was opposite to the back cushion. The air-cooled cameras under the chaise section (bottom row in Fig. 16) show that there was no flaming underneath the chaise section at flashover for all tests.

The presence of flaming underneath the RUF item is important because it increases: (1) the incident heat flux on the couch padding material where most of the fuel, i.e., FPUF, is located and thus, (2) the rate of FPUF pyrolysis, (3) the rate at which combustible volatiles are released, and (4) the rate at which liquid product of FPUF decomposition (i.e., regenerated polyols [35]) are generated. All these phenomena promote an increase in heat release rate. Flaming under the loveseat section is initially due to the ignition of the polyester wadding and cover fabric on the bottom of the seat cushion and, afterwards, the ignition of regenerated polyols. The flaming ignition of regenerated polyol underneath the RUF item, referred to as "bottom ignition" as previously discussed throughout this chapter, is particularly important because about 70% of the FPUF heat released is due to regenerated polyol combustion [77]. Figure 17 shows video frames captured by the air-cooled cameras under the loveseat section at various times during Test 1 (top row), Test 2 (center row), and Test 3 (bottom row). In all cases, no obvious degradation was observed at $t = 6$ min (compare column 1 to column 2).

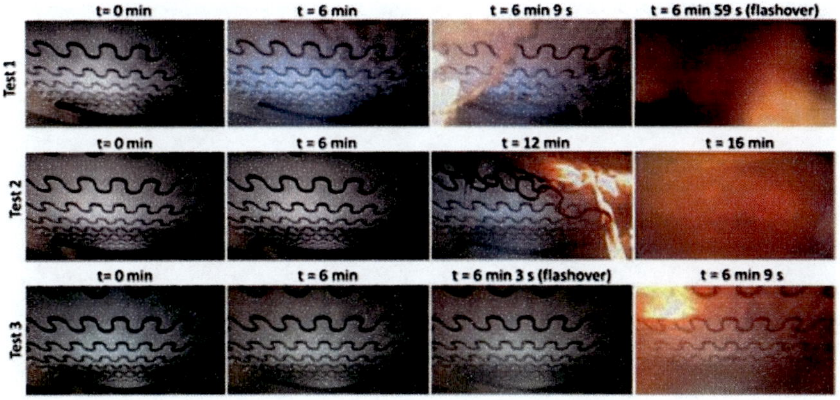

Fig. 17 Video frames captured by the air-cooled cameras under the loveseat section at various times during Test 1 (top row), Test 2 (center row), and Test 3 (bottom row) [30]

In Test 1, flaming became visible at about 6 min 9 s and was caused by the formation of a pool fire underneath the couch. The pool fire was generated by flaming liquids produced on the top of the seat cushion that flowed from the front of the loveseat section to the floor. In this test, flashover occurred about 50 s after flaming appeared under the loveseat section. At this time a significant flame spread underneath the loveseat section was observed in Test 1 (see video frame in the top row on the far right). Thus, bottom ignition briefly contributed to flashover conditions in Test 1, but the pool fire was mainly fed by the top of the seat cushion. In Test 3, flames appeared in the same location and by the same mechanism observed for Test 1 but a few seconds after flashover. Thus, flashover conditions could be reached in Test 3 without any contribution from bottom ignition. In Test 2, flames appeared at about 12 min in the back- right corner of the sofa, i.e., northwest corner of the couch. At this early stage, flaming under the loveseat section was likely supported by burning of the polyester wadding and the cover fabric but not by FPUF, which was protected by the barrier fabric. Flames continued to spread under the loveseat and by $t = 16$ min the bottom right of the loveseat was engulfed in flames.

To better describe the phenomena that led to flashover in Test 2, Fig. 18 shows four video frames captured by the 360° camera 6 min, 8 min, 10.2 min, and 18 min after ignition. At $t = 6$ min, the combustion of the throw pillow was complete and the heat release rate was below 15 kW. Flames were localized to the right side of the couch on the top of the seat cushion and front of the back cushion. At $t = 8$ min the heat release rate was still below 15 kW but the flames propagated to the back of the couch where the polyester wadding fabric, and wood frame were not protected by the barrier fabric and an increase in flame size was observed near the back right corner. Noticeably, only the padding material, i.e., FPUF and polyester fiberfill, were protected and covered by the barrier fabric. At $t = 10.2$ min, the flame spread further over the back of the couch and the HRR increased to about 30 kW. It was evident that the combustion of the back of the couch sustained flaming and promoted an

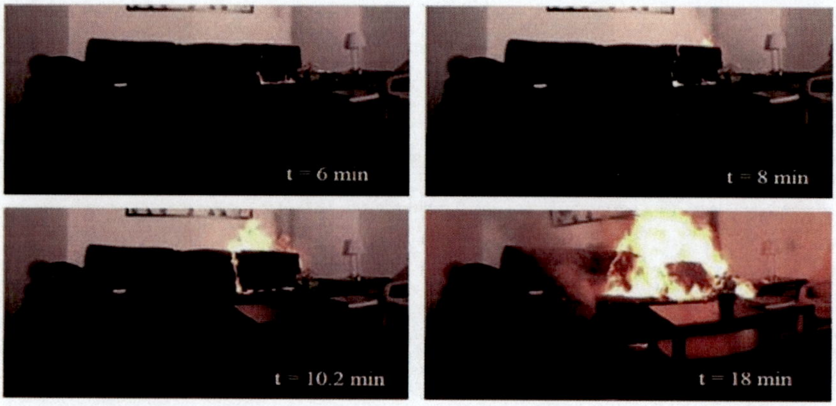

Fig. 18 Video frames captured during Test 2 at t = 6 min, 8 min, 10.2 min, and 18 min [30]

increase in HRR. These observations indicate that the addition of barrier fabric over the back of the couch could have further improved the fire performance of the couch. Shortly after at $t = 12$ min, flames had spread to the bottom of the loveseat. At $t =$ 18 min, an obvious pool fire formed underneath the loveseat section. This is a clear indication that, by this time in the test, bottom ignition occurred, and the regenerated polyols were involved in the burning process. An abrupt increase in HRR observed at approximately $t = 17$ min could be attributed to bottom ignition. Bottom ignition preceded flashover in Test 2, which occurred at about $t = 20.9$ min, by roughly 3 min. This finding agrees with previous work where the PHRR occurred about 2 min to 4 min after bottom ignition [29] discussed in Sect. 9.3.

4.4 Heat Flux and Tenability Conditions

Figure 19 shows the heat fluxes measured on the couch near the ignition location (PTC4). In Test 1 and Test 3, the heat flux reached about 75 kW/m^2 after about 4 min. These data suggest that values of incident heat flux over 50 kW/m^2 are advisable for a robust evaluation of the fire performance of barrier fabric at bench scale. Values of incident heat flux of 75 kW/m^2 were adopted in the Cube Test for ranking the performance of barrier fabrics [31] described in Sect. 9.2. In Test 3, the heat flux peaks at about 10 kW/m^2 at about 2 min when the throw pillow is still burning, and then drops to below 5 kW/m^2 until bottom ignition. After flashover the heat flux reached peak values above 150 kW/m^2 in all three tests. Noticeably, the highest peak value was achieved in Test 2 (about 200 kW/m^2) despite the presence of the barrier fabric, possibly due to the strong and prolonged flame heat feedback between the pool fire and the unburned portions of the couch. The heat fluxes along the wall

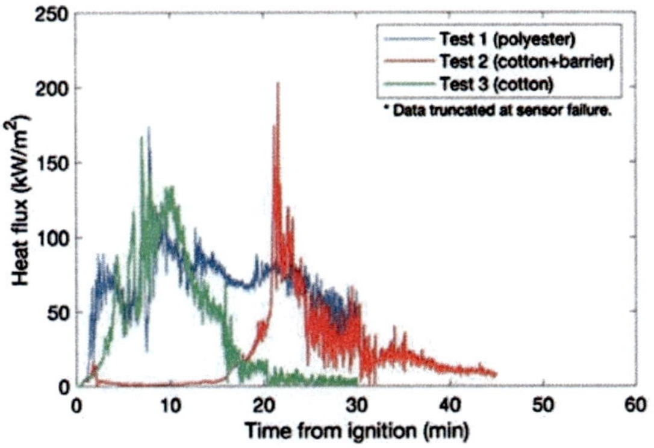

Fig. 19 Heat flux measured on top of the couch near the ignition location (PTC4) for Test 1, Test 2, and Test 3 [30]

reached the highest values at PTC3 close to the ceiling; the peak value of heat flux was about 175 kW/m^2 in Test 1 and Test 3 and about 150 kW/m^2 for Test 2.

During and immediately after a fire, conditions may become untenable such that the occupants are incapacitated and cannot escape by themselves, in which case they are likely to die unless rescued. The leading cause of incapacitation and death during and immediately after fires is exposure to asphyxiant gases: carbon monoxide (CO), hydrogen cyanide (HCN), and carbon dioxide (CO_2), with some additional effects from low-oxygen hypoxia and inhaled irritants [73]. The effect of fire effluents on incapacitation can be estimated using the Fractional Effective Dose [84]. In upholstery furniture flammability without chemical fire retardants, carbon monoxide is the main hazard, although the contribution of hydrogen cyanide is also significant [85, 86]. The CO concentration at which there is danger of incapacitation after approximately 5 min exposure in a person engaged in light activity is about 6000 ppm to 8000 ppm [73]. These values are never reached in these tests (see Fig. 20) due to the high ventilation factor of 0.14 $m^{1/2}$ but could be easily achieved in more realistic room scenarios [85].

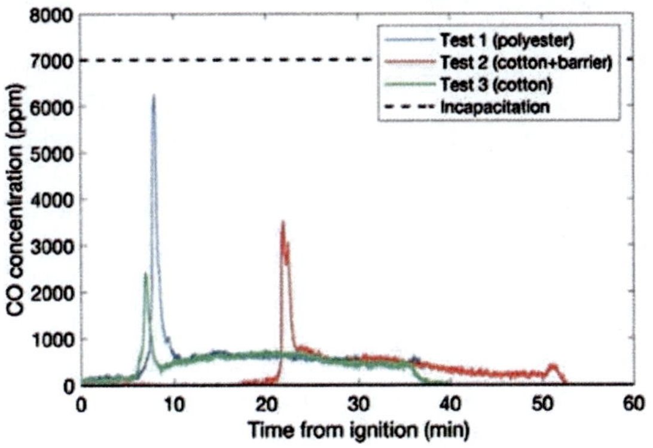

Fig. 20 CO concentration measured for Test 1, Test 2, and Test 3 and CO incapacitation level after approximately 5 min exposure [30]

Table 7 Time at which an upper layer temperature of 200 °C at $T_{ceiling}$ ($t_{200\,°C}$) and flux of 2.5 kW/m^2 at PTC2 ($t_{2.5\,kW/m^2}$) were met and calculated Available Safe Egress Time (ASET) [30]

	$t_{200\,°C}$ (min)	$t_{2.5\,kW/m^2}$ (min)	ASET (min)
Test 1	3.92	3.33	2.63
Test 2	16.03	16.56	15.25
Test 3	3.05	3.05	2.30

Alternate tenability criteria include: a maximum upper layer gas temperature of 200 °C and an exposure to radiant heat fluxes above 2.5 kW/m^2 [73]. For each test, Table 7 shows the times at which: (1) a temperature of 200 °C at $T_{ceiling}$ was reached ($t_{200\,°C}$), and; (2) a heat flux of 2.5 kW/m^2 was reached at PTC2 ($t_{2.5\,kW/m^2}$). The smaller between $t_{200\,°C}$ and $t_{2.5\,kW/m^2}$ is used to estimate the time at which untenable conditions were reached. These data indicate that untenable conditions were reached before flashover and, thus, heat exposure rather than gas toxicity seems to be the main hazard in this scenario.

Table 7 also shows the Available Safe Egress Time (ASET) in the enclosure fire calculated as the time between smoke alarm triggering (see Table 7) and the onset of untenable conditions [87]. Available Safe Egress Time values below 3 min for both Test 1 and Test 3 indicated an extremely high fire hazard. The adoption of a barrier fabric in Test 2 allowed a significant increase in the ASET value from about 3 min to about 15 min which would drastically increase the likelihood of survival for an occupant.

5 Conclusions

In the fire barriers in reduced scale testing section, a cone-calorimetry-based bench-scale test (the "Cube" test) was developed to capture mass and heat transfer phenomena at the top and bottom of a specimen. The Cube test was designed to mimic the burning process of cored composites where the flammable core material acts as the main fuel load and superficial layers may act as fire barriers. Results showed the mass transfer of flammable liquid products underneath the specimen (due to percolation of liquid products through the bottom fire barrier, which can lead to dripping and pool fire formation in upholstered furniture) captured in the Cube test.

Appropriate test conditions were determined in which the specimen decomposition pathway in the Cube test was representative of full-scale upholstered furniture fires. Specifically, an external heat flux of 75 kW/m^2 was required to "liquify" the foam to produce liquid pyrolyzates without significant char formation, thus allowing for measurement of the time required for the appearance of liquid pyrolyzates under the specimen as wetting of the fire barrier (time to wetting). Based on the values of time to wetting (t_w), the best performing fire barrier was B3 ($t_w = 258$ s $\pm$ 7 s), followed by B1($t_w = 235$ s $\pm$ 17 s), B6 ($t_w = 222$ s $\pm$ 2 s), B5 ($t_w = 164$ s $\pm$ 4 s),

B2 ($t_w = 153$ s $\pm$ 2 s), and B4 ($t_w = 143$ s $\pm$ 2 s). As expected, the control sample without fire barrier had the worst performance ($t_w = 90$ s $\pm$ 3 s).

In a fire scenario it was proposed that: (a) the time to wetting may be used to predict the time to "fire barrier failure" in full-scale tests (time at which a rapid increase in heat release rate is expected due to the percolation and ignition of flammable liquids beneath the RUF item), and; (b) the average heat release rate in the Cube test prior to t_w may be used to predict the average heat release rate before fire barrier failure in upholstered furniture items. An analytical approach to connect these bench-scale measurements to quantitative measurements of burning behavior in full-scale tests has been introduced in Zammarano et al. [31] and was assessed in another paper by Zammarano [32]. Even when the assumptions on the fire scenario are not perfectly fulfilled, the Cube test might still be adopted to rank the relative performance of fire barriers when used in full-scale experiments as it captures the mass and heat transfer effects of the fire barrier in a well-defined, realistic fire scenario.

In the fire barriers in full-scale testing chair mock-ups section, a twofold objective were explored: (a) reveal to what extent barriers, which do not contain chemical FR additives (passive barriers), can decrease the fire risk of upholstered chair mock-ups when exposed to a strong (18 kW) ignition source, and (b) investigate the mechanism leading to the failure of barriers. Six commercially available barriers (B1 to B6) contained inherently FR fibers, but no additional chemical FR according to the manufacturers. This was confirmed by independent chemical analyses, whose findings showed that there was an insufficient presence of common FRs to significantly affect the performance of the barriers.

This series of fire tests demonstrated that none of the barriers were able to prevent ignition of the padding material, and the cushions in all chairs eventually burned to completion with marginal differences in total heat release. However, all barriers significantly decreased the fire growth of the chair mock-ups by generating an initial plateau with reduced heat release rate. The second, dominant peak heat release rate was remarkably delayed from about 3 min for the control chair to as much as 22 min for the best performing barriers; similarly, the peak heat release rate was reduced from about 2.6 MW to approximately 1 MW. The heat release rate of a mock-up without a barrier was enough to lead to flashover in most rooms in less than 3 min, whereas, the use of an effective barrier sufficiently delayed the fire growth to likely allow occupants to escape and enable first responders to intervene before flashover and, thus, prevent many fire deaths. Additionally, barriers mitigated the fire hazard by increasing the time available for egress and by decreasing the likelihood of flame spread. Barrier B6 appeared to be the best performing barrier based on its low heat release at its HRR plateau ($<$200 kW), its low peak heat release rate ($\approx$1 MW), and its prolonged time to peak ($\approx$23 min).

A common failure mechanism was identified for all barriers. This was not a mechanical or thermal failure of the barrier; instead, it was associated with the so-called "bottom ignition," that is, the ignition of liquid pyrolizates which percolated through the barrier underneath the seat cushion. Based on these findings, a bench-scale test ("Cube test") aiming to predict time-to-barrier failure at full scale was being developed. The understanding of the barrier failure mechanism is also expected to assist manufacturers and scientists toward the development of barriers and upholstery products with improved fire performance.

In the fire barriers in full scale testing: furnished compartments section, a series of three full-scale compartment fire experiments helped characterize the potential fire hazard of residential upholstered furniture (RUF) and the potential ability of a barrier fabric to decrease such hazard. A representative living room was constructed and furnished identically except for three different couches that were the same except for the following: Test 1—couch with polyester cover fabric in Test 1; Test 2—couch with a cotton blend cover fabric and a barrier fabric (made of intrinsically fire-resistant fibers without any additional fire retardant; and Test 3—couch with only a cotton blend cover fabric (same cover fabric as Test 2).

A throw pillow on the right arm of the couch ignited using a 3 cm propane diffusion flame acted as an ignition source for each of the couches. From this set of experiments, the results demonstrated the use of a fire barrier without chemically active flame retardants can remarkably delay the growth of RUF fires. The two couches without a barrier fabric, which were compliant with TB-117–2013, posed a severe fire hazard with rapid fire growth leading to flashover within approximately 6 min of ignition. The addition of the barrier fabric delayed flashover by about 15 min from about 6 min to 21 min, and increased the estimated Available Safe Egress Time (ASET) from less than 3 min to over 15 min.

Considering target fire service response times, this delay is important because it allows first responders additional time to intervene before flashover (alarming time plus response time is typically 12 min to 17 min depending on whether the location is urban or rural) and potentially prevent flashover occurrence. Flashover prevention is critical because it can stop fire spread to adjacent rooms and building; where most fire losses occur. The barrier fabric in these experiments covered all padding material but not the back of the couch, where flaming rapidly spread over the wadding and cover fabric once ignited. Thus, a further improvement in fire performance could be expected if the back of the couch was also protected by additional barrier fabric.

The presence of the barrier fabric also affected the mechanism by which flashover conditions were reached and the role of pool fire formation. Without barrier fabric, flashover conditions could occur before a regenerated polyol pool formed on the floor with little to no flaming on the bottom of the couch. With the barrier fabric, flashover occurred several minutes after the appearance of flaming underneath the couch and the formation of a pool fire; pool fire increased the pyrolysis rate of the residual padding material and, in turn, caused an additional increase in heat release rate that led to flashover.

In summary, the set of data suggest that: (a) despite compliance with cigarette ignition tests, residential upholstered furniture may still pose a considerable fire hazard due to rapid fire growth and flame spread in presence of a flaming ignition source; (b) technologies promoting fire safety and health/environment safety for residential upholstered furniture are not mutually exclusive as these tests demonstrate that non-toxic technologies, which can meet both requirements, not only exist but are commercially available; (c) the presence of a barrier fabric resulted in a 15 min delay in time to flashover and 12 min increase in estimated ASET drastically decreased the fire hazard; (d) the mechanisms that lead to failure of the barrier fabric and consequent flashover in the compartment appear to be triggered by the ignition of liquid pyrolyzates percolating through the bottom barrier fabric in the residential upholstered furniture item.

In this chapter of new fire protection materials, an overall review of fire barriers was studied as a potential solution to reduce residential upholstered furniture (RUF) flammability. The fire barriers studied in this chapter did not contain chemical fire-retardant additives (passive barriers) and were shown to decrease the fire risk of RUF when exposed to a strong ignition source. Fire barriers were used in a reduced scale testing method (cube test), large scale testing method (chair mock-ups), and real-life scenario testing method (RUF in furnished compartments). In each testing method, the fire barriers were placed between padding materials and was shown to reduce fire growth rate after flaming ignition by limiting heat transfer to a flammable substrate and controlling the rate and location at which volatiles are released and able to burn. Additionally, barriers mitigated the fire hazard by increasing the time available for egress and by decreasing the likelihood of flame spread which provides additional time for firefighters to prevent flashover. The heat and mass transfer effects were physically based thus no chemical fire retardants were needed. A common failure mechanism was identified for all barriers. It was not a mechanical or thermal failure of the barrier; instead, it was associated with the ignition of liquid products which percolated through the barriers. The use of barriers is expected to significantly reduce the fire hazard, especially if the second item ignited is also an RUF item protected by a barrier. The data from the HRR curves in this chapter are useful for modeling second item ignitions to create more realistic fire cases to generate training data to develop more robust machine learning models. The understanding of the barrier failure mechanism is expected to assist manufacturers and scientists toward smart firefighting and the development of barriers and upholstery products with improved fire performance.

Acknowledgements The author gratefully acknowledges Mauro Zammarano, Leader, "Fire Barriers for Low Heat Release Rate Products" Project at the National Institute of Standards and Technology (NIST) for his mentorship and guidance throughout the various projects cited in this chapter. Also, the team at the NIST National Fire Research Laboratory—Matthew Bundy, Matthew Hoehler, Anthony Hamins, Marco Fernandez, Laurean DeLauter, Philip Deardorff, Anthony Chakalis, Michael Selepak, Artur Chernovsky, and Andrew Mundy—for their assistance in test setup, execution, and expertise made various experiments cited in this chapter possible. Members of the NIST Flammability Reduction Group—Ickchan Kim, John R. Shields, Isaac Leventon, Shonali Nazare, Rick Davis, William Pitts, and Richard Gann for their valuable contributions. Special thanks are given to the NIST Fire Department for their support for each burn conducted in NFRL.

References

1. M. Ahrens, *Home Structure Fires* (National Fire Protection Association, 2019)
2. M. Ahrens, *Home Fires that Began with Upholstered Furniture* (National Fire Protection Association Research, 2017).
3. J.R. Hall, Estimating fires when a product is the primary fuel but not the first fuel, with an application to upholstered furniture. Fire Technol. **51**(2), 381–391 (2015)
4. R.G. Gann, The challenge of realizing low flammability home furnishings. Interflam, in *Interscience Communications* (CD Version, London, England, 2013)
5. Anonymous: Requirements, Test procedure and apparatus for testing the smolder resistance of materials used in upholstered furniture, in *Bureau of Electronic & Appliance Repair Home Furnishing & Thermal Insulation, Technical Bulletin*, vol. 117 (Sacramento, CA, 2013)
6. Anonymous: filling/padding component test method—barrier test method—fabric classification test method. Upholstered Furniture Action Council (UFAC) (1990)
7. A. International, *E1353–16 Standard Test Methods for Cigarette Ignition Resistance of Components of Upholstered Furniture* (ASTM International, West Conshohocken, PA, 2016)
8. Agency NFPA: NFPA 260—Standard Methods of Tests and Classification System for Cigarette Ignition Resistance of Components of Upholstered Furniture (2013)
9. Anonymous: PART 1640—Standard for the Flammability of Upholstered Furniture, Federal Register/Vol. 86, No. 67/in 16 CFR Part 1640 ed Commission CPS (2021)
10. D. Clark, *What is TB117–2013?* (Home Furnishings Association, 2015)
11. I. Watanabe, S.I. Sakai, Environmental release and behavior of brominated flame retardants. Environ. Int. **29**, 665–682 (2003)
12. S.D. Shaw, M.L. Berger, D. Brenner, K. Kannan, N. Lohmann, O. Papke, Bioaccumulation of polybrominated diphenyl ethers and hexabromocyclododecane in the northwest Atlantic marine food web. Sci. Total Environ. **407**, 3323–3329 (2009)
13. S. Mizouchi, M. Ichiba, H. Takigami et al., Exposure assessment of organophosphorus and organobromine flame retardants via indoor dust from elementary schools and domestic houses. Chemosphere **123**, 17–25 (2015)
14. V. Babrauskas, A. Blum, R. Daley, L. Birnbaum, Flame retardants in furniture foam: benefits and risks, in *Paper presented at: Fire Safety Science-Proceedings of the Tenth International Symposium* (2011), pp. 265–287
15. M. Babich, C. Bevington, X. Chen, et al., Project plan: organohalogen flame retardant chemicals assessment. Consum. Prod. Saf. Comm. (2020)
16. An Act to Protect Firefighters by Establishing a Prohibition on the Sale and Distribution of New Upholstered Furniture Containing Certain Flame-Retardant Chemicals. 128th Maine Legislature House of Representatives 182 (2017)
17. Consumer Products: Flame Retardant Materials. State of California Office of Legislative Counsel Bloom 2998 (2018)
18. CPSC-2015–0022 CDN: Guidance document on hazardous additive, non-polymeric organohalogen flame retardants in certain consumer products, in *Commission CPS*, ed., vol. 82: Federal Register (2017)
19. C. Papaspyrides, P. Kiliaris, *Polymer Green Flame Retardants* (Elsevier, Amsterdam, Netherlands, 2014)
20. S. Nazare, W. Pitts, S. Flynn, J.R. Shields, R.D. Davis, Evaluating fire blocking performance of barrier fabrics. Fire Mater. **38**, 695–716 (2014)
21. T.J. Ohlemiller, J.R. Shields, Behavior of mock-ups in the California technical bulletin 133 test protocol: fabric and barrier effects. NIST Interagency/Internal Report (NISTIR) 5653 (1995)
22. W. Pitts, M. Werrel, M. Fernandez, A.M. Long, Assessing the Predictive Capability for Real-Scale Residential Upholstered Furniture Mock-Up Fires using Cone Calorimeter Measurements. Part 1: Real-Scale Experiments (Eisenberg EA, Runyon CD, 2020)
23. T. Fabian, Upholstered furniture flammability, in *Underwriter Laboratories* (2013)
24. W.D. Woolley, S.A. Ames, A.I. Pitt, K. Buckland, The ignition and burning characteristics of fabric covered foams. Build. Res. Establ. (1978)

25. K. Storesund, A. Steen-Hansen, A. Bergstrand, Fire safe upholstered furniture: alternative strategies to the use of chemical flame retardants. SP Fire Res. (2015)
26. A. Lock, *Memoranda on Full-Scale Upholstered Furniture Testing, 2014–2015. U.S. Consumer Product Safety Commission* (2016)
27. A. Lock, *Memoranda on Full-Scale Upholstered Furniture Testing, 2016. U.S. Consumer Product Safety Commission* (2017)
28. M. Black, A. Davis, D. Harris, P. Barry, R. Cohen, et al., *A Study of Chemical Exposure Risk and Flammability of Upholstered Furniture and Consumer Electronics. Underwriter Laboratories* (2019)
29. A.L. Thompson, I. Kim, A. Hamins, M. Bundy, M. Zammarano, Performance and failure mechanism of fire barriers in full-scale chair mock-ups. Fire Mater. **46**(1), 329–346 (2021)
30. M. Zammarano, M.S. Hoehler, J.R. Shields, A.L. Thompson, I. Kim, I. Leventon, M. Bundy, *NIST Technical Note 2129. Full-Scale Experiments to Demonstrate Flammability Risk of Residential Upholstered Furniture and Mitigation Using Barrier Fabric. NIST Technical Note* (2021)
31. M. Zammarano, J.R. Shields, I. Leventon, I. Kim, S. Nazare, A. Thompson, R.D. Davis, A. Chernovsky, M. Bundy, Reduced-scale test to assess the effect of fire barriers on the flaming combustion of cored composites: an upholstery-material case study. Fire Mater. **45**(1), 114–126 (2021)
32. M. Zammarano, *Fire Performance of Upholstery Materials: Correlation between Cube Test and Full-Scale Chair Mock-Ups. NIST Technical Note*, vol. 219 (2021)
33. S. Nurbakhsh, J. Mccormack, A review of the technical bulletin 129 full scale test method for flammability of mattresses for public occupancies. J. Fire Sci. **16**(2), 105–124 (1998)
34. S. Nazaré, R. Davis, A review of fire blocking technologies for soft furnishings. Fire Sci. Rev. **1**(1), 1 (2012)
35. M. Ravey, E.M. Pearce, Flexible polyurethane foam I. thermal decomposition of a polyether-based, water-blown commercial type of flexible polyurethane foam. J. Appl. Polym. Sci. **63**(1), 47–74 (1997)
36. E. Weil, S. Levchik, Flame retardants in commercial use or development for textiles. J. Fire Sci. **26**, 243–281 (2008)
37. Q. Qiu, M. Kumosa, Corrosion of E-glass fibers in acidic environments. Compos. Sci. Technol. **57**, 497–507 (1997)
38. X. Chen, W. Wang, S. Li, C. Jiao, Fire safety improvement of paraaramid fiber in thermoplastic polyurethane elastomer. J. Hazard. Mater. **324**, 789–796 (2017)
39. S. Heidari, A. Paren, P. Nousiainen, The mechanism of fire resistance in viscose/silicic acid hybrid fibres. J. Soc. Dyers Colour. **109**(7–8), 261–263 (1993)
40. T.-H. Ko, Characterization of PAN-based nonburning (nonflammable) fibers. J. Appl. Polym. Sci. **47**(4), 707–715 (1993)
41. ASTM D737–18 Standard Test Method for Air Permeability of Textile Fabrics (2018)
42. J.S. Fritz, Ion chromatography. Anal. Chem. **59**(4), 335–344 (1987)
43. R.B. Bradstrut, Kjeldahl method for organic nitrogen. Anal. Chem. **26**(1), 185–187 (1954)
44. ASTM. E1479–16 Standard Practice for Describing and Specifying Inductively Coupled Plasma Atomic Emission Spectrometers (2016)
45. B. Beckhoff, B. Kanngieber, N. Langhoff, R. Wedell, H. Wolff, *Handbook of Practical X-Ray Fluorescence Analysis* (Springer, Berlin, Germany, 2006)
46. O.D. Sparkman, Z. Penton, F. Kitson, *Gas Chromatography and Mass Spectrometry: A Practical Guide*, 2nd edn. (Academic Press, Cambridge, MA, 2011)
47. ASTM. C1875–18 Standard Practice for Determination of Major and Minor Elements in Aqueous Pore Solutions of Cementitious Pastes by Inductively Coupled Plasma Optical Emission Spectroscopy (ICP-OES) (2018)
48. M. Petreas, R. Gill, S. Takaku-Pugh et al., Rapid methodology to screen flame retardants in upholstered furniture for compliance with new California labeling law (SB 1019). Chemosphere **152**, 353–359 (2016)

49. H.M. Stapleton, S. Klosterhaus, A. Keller et al., Identification of flame retardants in polyurethane foam collected from baby products. Environ. Sci. Technol. **45**, 5323–5331 (2011)
50. C.-Y. Wang, S.D. Bunday, J.G. Tartar, Ion chromatographic determination of fluorine, chlorine, bromine, and iodine with sequential electrochemical and conductometric detection. Anal. Chem. **55**(9), 1617–1619 (1983)
51. J.F. Colaruotolo, R.S. Eddy, Determination of chlorine, bromine, phosphorus, and sulfur in organic molecules by ion chromatography. Anal. Chem. **49**(6), 884–885 (1997)
52. W.P. Shorrock, O.B. Yale, Inventors; Pilkington PLC, Merseyside, England, assignee. Borosilicate Glass Composition 55 (1993)
53. T. Furukawa, W.B. White, Raman spectroscopic investigation of sodium borosilicate glass structure. J. Mater. Sci. **16**, 2689–2700 (1981)
54. H.P. Hood, M.E. Nordberg, Inventors. Treated Borosilicate Glass (1938)
55. K.K. Shen, R. O'Connor, Flame retardants: borates, in *Plastics Additives: An A-Z Reference*, ed. by G. Pritchard (Springer, Dordrecht, The Netherlands, 1998), pp. 268–276
56. H.D. Brabandere, N. Forsgard, L. Israelsson et al., Screening for organic phosphorus compounds in aquatic sediments by liquid chromatography coupled to ICP-AES and ESI-MS/MS. Anal. Chem. **80**, 6689–6697 (2008)
57. E. Hoh, N.G. Dodder, S.J. Lehotay, K.C. Pangallo, C.M. Reddy, K.A. Maruya, Nontargeted comprehensive two-dimensional gas chromatography/time-of-flight mass spectrometry method and software for inventorying persistent and bioaccumulative contaminants in marine environments. Environ. Sci. Technol. **46**, 8001–8008 (2012)
58. T. Ohlemiller, K.M. Villa, NISTIR 4348 - furniture flammability: Sn investigation of the California bulletin 133 Test. Part 2. Characterization of the ignition source and a comparable gas burner. NIST (1990)
59. F. Rogers, T. Ohlemiller, Smolder characteristics of flexible polyurethane foams. J. Fire Flammability. **11**(1), 32–44 (1980)
60. R.H. Krämer, M. Zammarano, G.T. Linteris, U.W. Gedde, J.W. Gilman, Heat release and structural collapse of flexible polyurethane foam. Polym. Degrad. Stab. **95**(6), 1115–1122 (2010)
61. G. Rein, *Smoldering combustion. SFPE Handbook of Fire Protection Engineering* (Springer, New York, NY, 2016), pp. 581–603
62. X. Zhao, Y.-T.T. Liao, M.C. Johnston, T'J.S. Ien, P.V. Ferkul, S.L. Olson, Concurrent flame growth, spread, and quenching over composite fabric samples in low speed purely forced flow in microgravity. Proc. Combust. Inst. **36**(2), 2971–2978 (2017)
63. J. Sherratt, D.D. Drysdale, The effect of the melt-flow process on the fire behaviour of thermoplastics. Paper presented at: Interflam 2001. Edinburgh (2001)
64. I. Kim, A.L. Thompson, S.C. Kim et al., Demonstration of an all-in-one solution for fire safe upholstery furniture: a Benign backcoating for smoldering and flame-resistant cover fabrics. Fire Mater. **46**(4), 677–693 (2022)
65. R.A. Bryant, M.F. Bundy, The NIST 20 MW calorimetry measurement system for large-fire research. NIST Technical Note 2077 (2019)
66. B.N. Taylor, C.E. Kuyatt, *Guidelines for Evaluating and Expressing the Uncertainty of NIST measurement results. NIST Technical Note*, vol. 1297 (1994)
67. M.C. Bruns, Estimating the flashover probability of residential fires using Monte Carlo simulations of the MQH correlation. Fire Technol. **54**(1), 187–210 (2018)
68. B. Sundstrom, Combustion behavior of upholstered furniture. Important findings, practical use, and implications. Fire Mater. **45**, 97–113 (2020)
69. H. Denize, *The Combustion Behaviour of Upholstered Furniture Materials in New Zealand.* Civil Engineering, University of Canterbury ISSN 1173–5996 (2000)
70. S. Mehta, *Upholstered Furniture Full Scale Chair Tests-Open Flame Ignition Results and Analysis* (Consumer Product Safety Commission, 2012)
71. J. Sherratt, D. Drysdale, The effect of the melt-flow process on the fire behavior of thermoplastics, in *INTERFLAM* (2001)

72. M. Zammarano, S. Matko, W.M. Pitts, D.M. Fox, R.D. Davis, Towards a reference polyurethane foam and bench scale test for assessing smoldering in upholstered furniture. Polym. Degrad. Stab. **106**, 97–107 (2014)
73. D.A. Purser, J.D. McAllister, Assessment of hazards to occupants from smoke, toxic gases and heat. *SFPE Handbook of Fire Protection Engineering*, 5th edn. (Springer, New York, 2016), pp. 2308–2428
74. F.F. Chen, C. Fleischmann, *Radiant Ignition of New Zealand Upholstered Furniture Composites* (University of Canterbury, Christchurch, New Zealand, 2001)
75. SFPE, *Engineering Guide: Piloted Ignition of Solid Materials Under Radiant Exposure* (Society of Fire Protection Engineers, Gaithersburg, MD, 2002)
76. T.J. Ohlemiller, R.G. Gann, Estimating reduced fire risk resulting from an improved mattress flammability standard. NIST Tech. Note **1446**, 80 (2002)
77. R.H. Kramer, M. Zammarano, G.T. Linteris, U.W. Gedde, J.W. Gilman, Heat release and structural collapse of flexible polyurethane foam. Polym. Degrad. Stab. **95**, 1115–1122 (2010)
78. V. Babrauskas, Pillow burning rates. Fire Saf. J. **8**(3), 199–200 (1985)
79. H.E. Mitler, K-M. Tu, Effect of ignition location on heat release rate of burning upholstered furniture. Paper presented at: Annual Conference on Fire Research: Book of Abstracts (1994)
80. M.C. Bruns, Predicting the effects of barrier fabrics on residential upholstered furniture fire hazard. NIST Tech. Note 1920 (2016)
81. NFPA: NFPA 1710, *Standard for the Organization and Deployment of fire Suppression Operations, Emergency Medical Operations, and Special Operations to the Public by Career Fire Departments* (National Fire Protection Association, 2010)
82. NFPA, *NFPA 1720, Standard for the Organization and Deployment of Fire Suppression Operations, Emergency Medical Operations, and Special Operations to the Public by Volunteer Fire Departments* (2010)
83. T. Graham, G.M. Makhviladze, J.P. Roberts, On the theory of flashover development. Fire Saf. J. **25**(3), 229–259 (1995)
84. ISO 13571, Life-threatening components of fire—guidelines for the estimation of time to compromised tenability in fires (2012)
85. S.T. McKenna, R. Birtles, K. Dickens, R.G. Walker, M.J. Spearpoint, A.A. Stec, T.R. Hull, Flame retardants in UK furniture increase smoke toxicity more than they reduce fire growth rate. Chemosphere **196**, 429–439 (2018)
86. V. Babrauskas, B.C. Levin, R.G. Gann, M. Paabo, R.H. Harris, R.D. Peacock, S. Yusa, Toxic potency measurement for fire hazard analysis. Fire Technol. **28**(2), 163–167 (1992)
87. L.Y. Cooper, A concept for estimating available safe egress time in fires. Fire Saf. J. **5**(2), 135–144 (1983)

Smart Safety Design for Firefighting, Evacuation, and Rescue

Yuxin Zhang and Xinyan Huang

Abstract Fire is a vital threat to both occupants inside the building and firemen during their firefighting and rescue operation. Once a fire occurs, building environment changes rapidly, showing high temperature, toxic gas composition, low luminance and visibility. The occupants in a confined fire environment need to evacuate before reaching untenable conditions. Specific fire safety design of buildings has been required for fire evacuation over the last decades. To design a safe fire evacuation, conventional approaches rely on prescriptive codes or performance-based design. On top of that, with the booming of emerging technologies and thorough understanding of human behavior, smart design is increasingly welcome and applied to evacuation such as artificial intelligence evacuation modelling and real-time guidance systems However, few design considerations are given to firefighters who enter fire scenes and are exposed to more dangerous environment. Considering the firefighting and rescuing of trapped occupants, building fire safety design should include firefighting facilities and exclusive paths following specific codes, and corresponding safe firefighting principles for firefighters' operation according to principles for evacuation. Similarly, smart design should be applied in firefighting and rescue including automatic firefighting facilities, intelligent early warning systems. Thus, this chapter provides an overview of the fire safety design progress for evacuation, firefighting and rescue using both conventional and smart approaches. Specifically, it introduces smart fire safety design development and discusses their perspectives and challenges.

Y. Zhang (✉)
Department of Geotechnical Engineering, Tongji University, Shanghai, China
e-mail: yuxinzhang@tongji.edu.cn

X. Huang
Department of Building Environment and Energy Engineering, The Hong Kong Polytechnic University, Hong Kong, China

X. Huang and W. C. Tam (eds.), *Intelligent Building Fire Safety and Smart Firefighting*, Digital Innovations in Architecture, Engineering and Construction,
https://doi.org/10.1007/978-3-031-48161-1_10

1 Introduction

Today, fire is still the most common calamity in building. In case of a fire, the built environment could quickly alter dramatically, in terms of temperature, gas composition, brightness, and visibility. The occupants in confined fire environment need to evacuate before reaching untenable conditions, which is a key component of building fire safety design. Many fire disasters, such as the Grenfell Tower Fire in 2017 with 72 deaths and the Maolilong Road Tunnel Fire in 2019 with 36 deaths, have taught us lessons [1–3]. In addition, some firefighters suffered casualties such as two firefighters were dead, and 11 others were injured in the 2016 mini-storage fire in Hong Kong, while there was no occupant injury, see Fig. 1. Aiming to this issue, since the 1950s, several strategies have been developed to lower the danger of fire, such as active and passive fire protection systems, prescriptive rules, and performance-based fire safety designs [4, 5].

Two popular design approaches for safe evacuation techniques to protect people in a fire are prescriptive rules and performance-based fire safety solutions. The former regulates the detailed design parameters for various structure forms, which engineers must adhere to when designing buildings. For instance, the code establishes the size and separation between each exit [6]. Contrarily, the latter frequently apply to innovative structure designs and settings that are not covered by the current fire rules and codes. It is a goal-based approach where engineers, using numerical simulations and theoretical analysis, choose the specifics of the evacuation settings based on the

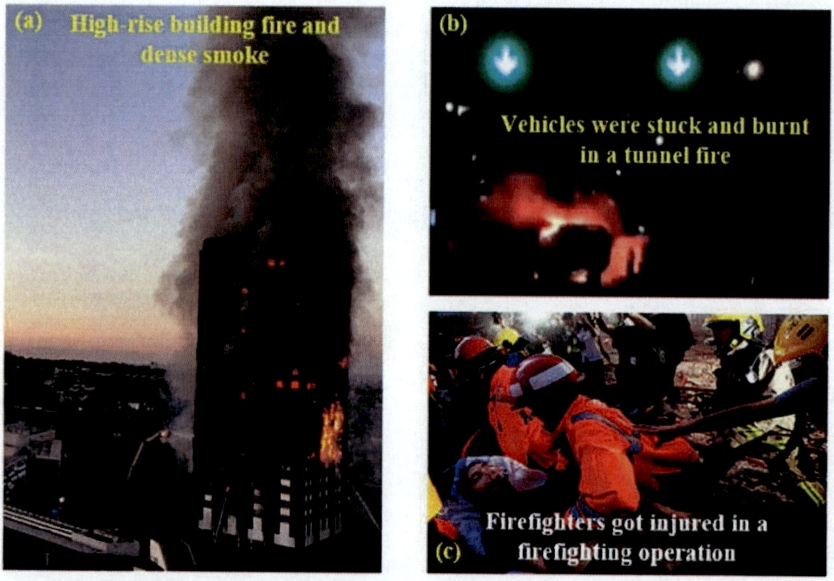

Fig. 1 Casualties of occupants and firefighters in fire. **a** Grenfell tower fire, **b** Maoliling tunnel fire and **c** Mini-storage fire

building's capacity and probable evacuation performance [7–9]. It requires clearance from the authority having jurisdiction (AHJ) even though there are no codes to follow. While it offers more flexibility, its design, documentation, and review processes are more expensive, time-consuming, and potentially riskier in the event of unavoidable fire events.

More recently, new smart technologies are used to improve the safety of fire evacuation by incorporating artificial intelligence (AI) into detection systems and real-time guidance systems such as the dynamic signages [10–13]. Engineers and scientists, for instance, have created a variety of active fire service systems, including sprinkler systems and automatic fire alarm systems. These systems increase the amount of time available for a safe evacuation, which further lower casualties and fire damage. The dynamic exit sign, unlike the static nature of the normal exit sign, uses many sensors in the buildings and smart algorithms to choose the best exit. In a complex constructed context, the dynamic sign aims to be more noticeable and adaptable to the rapidly changing fire scene [14, 15].

In addition to residents' safety, firefighters' safety is equally a crucial concern when they are battling unidentified fires and looking for trapped personals. Despite of the technology advances, inhabitants could scarcely flee from the fire scene on their own, [16] because of their insufficient knowledge on the building layout and exit routes, as well as the fire impact on their physical and mental stability [16]. Additionally, the intense fire smoke decreases visibility and makes evacuation more difficult [17]. As a result, firefighters' help in fire evacuation is necessary and crucial in reducing casualties. On the one hand, their intervention might stop the spreading of fire by setting up isolation zones or spraying extinguishers. On the other hand, they could locate and free trapped individuals. Carrying the injured greatly increases the challenge of their navigation inside the fire scene, and firefighters may run into severe scenarios like flashovers.

While performing their duties, firemen may experience injuries like comas, burns, and falling down [18, 19]. More research is needed to improve firefighter's safety, in terms of the codes of practices, the performance-based design of building, and the use of smart technologies [20]. Through numerical simulation and theoretical analysis, plus the real-time sensor data, we can evaluate the dynamic risk and body status of firefighters, based on the evolution history of fire scenarios and their in-site firefighting and rescue operations. One critical aspect is improving firefighters' safety by collecting, transferring, and processing real-time information through cloud services and releasing optimal operations via the IoT network [21–23]. It is worth noting that firefighters' operations and tactics can significantly affect the danger and casualty of fire. Therefore, defining firefighters' safe operation time and evaluating its safety level are of utmost importance.

Driven by decades of fire research and emerging technologies, fire safety design in evacuation, firefighting and rescue has developed largely. In this chapter, we summarize the fire safety design progress for evacuation, firefighting and rescue using both conventional and smart approaches, and introduce their current applications and future perspectives.

2 Fire Safety Design for Evacuation

When a building fire accident occurs, the most important thing is to ensure that all occupants inside the building can reach the safe area within a limited time, before engineers consider the safety of the building itself. On the one hand, some fire protection facilities could help control the fire development. On the other hand, occupants need to evacuate by themselves before they are stuck in the fire.

To reduce fire hazards and protect occupants and properties, building design must comply with the fire protection regulations based on the experience of current fire protection practice, latest research and test results, so that it could meet the minimum fire safety design requirements. However, with the rapid urbanization in developing countries, buildings with large spaces (e.g. terminal buildings), multi-functional (e.g. TOD transit-oriented development projects), and new forms (e.g. semi-open-air platforms) continue to emerge. New architecture design and new construction and building materials (e.g. Ethylene tetrafluoroethylene, photovoltaic panels, and electric vehicles) are also widely used in new buildings [24]. Thus, it is increasingly complex to meet fire safety requirements for evacuation. Fortunately, understanding fire and human behavior gradually, and multiple advanced technologies make it possible to achieve integrated fire safety design for evacuation.

2.1 Prescriptive-Based Codes for Evacuation Design in Fire

The majority of contemporary building codes and fire safety standards are prescriptive-based. Prescriptive-based codes trace their origins back to the nineteenth century, when major fires triggered the need for building provisions [6, 9]. These codes have not been effectively evaluated for their adequacy, excessiveness, or potential conflicts with other requirements at the beginning, and they have been updated based on empirical data, practical experience, and scientific understanding of fire behavior. Despite significant advances in fire safety research, many of these improvements have yet to be fully integrated into day-to-day fire safety practice.

The traditional basis of prescriptive-based life safety design centers around physical provisions for means of escape. Prescriptive-based methods for evaluating evacuation safety typically consider structural factors such as the number of exits, the maximum width and length of escape routes, the maximum time required for evacuation, and managerial strategies for maintaining safe and accessible escape routes. These methods generally assume that occupants will move at a certain speed (0.5–1.5 m/s), and that the time required for escape will range from 3 to 15 min depending on building types and fire scenarios [25, 26]. The prescriptive-based codes share lots of similarities in empirical data, scientific findings of fire and human behavior, etc. However, each country sets and follows its own codes. For example, GB50016 for China, BS 9999 for the United Kingdom and NFPA 1616 is for the United States, and some specific buildings such as tunnels have their own codes e.g. NFPA 502,

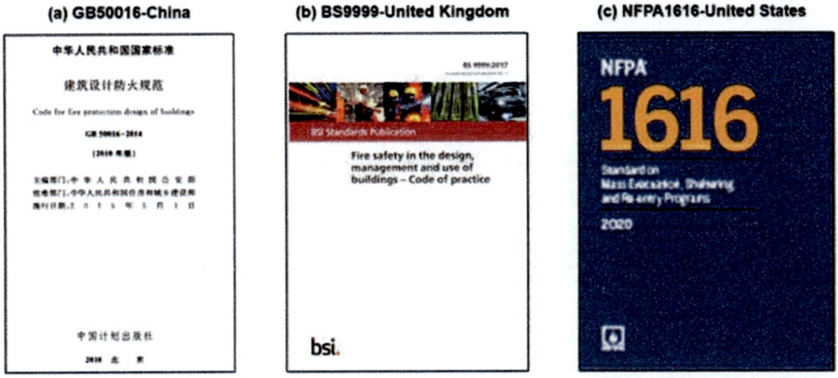

Fig. 2 Prescriptive-based codes for evacuation design in fire

the United States (Fig. 2). Apart from structure requirements, some fire protection facilities are also put forward, such as fire alarm, exit signs, fire detectors, fireproof doors etc., and the requirements are varied according to building types and regions.

2.2 *Performance-Based Design for Evacuation in Fire*

While regulations and codes instruct engineers how to design buildings to ensure occupants' safety during a fire evacuation, they are getting hard to satisfy the evacuation needs for complex buildings with a unique design. Thus, performance-based design could be an optimal choice compared to codes. Engineers design buildings as long as they satisfy the ultimate goal, that is occupants could evacuate safely. To evaluate the evacuation safety, researchers have proposed two key factors for occupants, namely "Available Safe Egress Time" (ASET) and "Required Safe Egress Time" (RSET) [27], and a number of researchers have further developed this concept in the calculation and numerical modelling in various fire safety design such as high-rise buildings and underground space [28, 29].

(1) Available safe evacuation time ASET (Available Safe Egress Time): Before the fire breaks out and develops to the stage of threatening personal safety, there is still a part of the time for people in the building to escape safely.
(2) Required safe evacuation time RSET (Required Safe Egress Time): the time required from the fire outbreak to the arrival of all personnel in the building to a safe place.

As illustrated in Fig. 3, the occupant safety could be guaranteed when the fire safety design satisfies Eq. (1) or increases the evacuation safety margin when maintaining the economic costs, see Eq. (2)

$$\mathrm{ASET} > \mathrm{RSET} \tag{1}$$

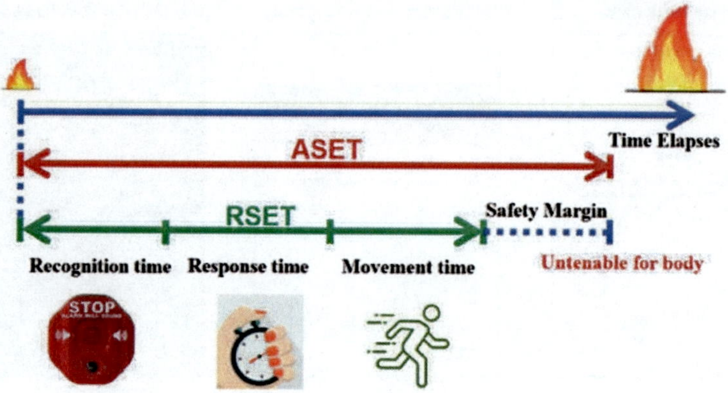

Fig. 3 The illustration of ASET and RSET

Table 1 Tenable criteria for ASET calculation at 2.0 m high [31]

Evaluation indicator	Threshold
Temperature (°C)	<60
Visibility (m)	>10
CO concentration (ppm)	<1000
Thermal radiation (kW/m^2)	<2.5

$$\text{Evacuation safety margin} = \text{ASET} - \text{REST}(> 0) \tag{2}$$

ASET is controlled by the fire scenarios and building design, and CFD fire modelling is widely applied for the computing [30]. Temperature, visibility, carbon monoxide (CO) concentration, and thermal radiation are important elements that affect occupant's safety, and their tenable criteria are listed in Table 1 [31]. RSET could be regarded as the issue of pedestrian evacuation consisting of three components, namely recognition time, response time and movement time.

These processes can be calculated through evacuation modeling using scientific programs like Pathfinder, FDS + Evac, etc., or academic programs containing social-force or cellular-automation models [42, 43].

3 Smart Design for Fire Evacuation

The performance-based design method is a more scientific and flexible approach to assess building evacuation safety in fire quantitatively, especially for those which own large volumes or special functional requirements. However, it has several problems:

- **High cost:** The design, research, archiving, review, approval and other processes of the evacuation timeline analysis need substantial amount of time and human resources.
- **Large uncertainty:** The analysis process adopts a large number of assumptions, such as fire loads, personnel tolerance limit thresholds, use of evacuation exits, etc., and the results from the analysis have certain uncertainties.
- **Wide limitations:** Due to high computational costs, most projects only simulate limited fire and evacuation scenarios to verify that current options meet design requirements, rather than to optimize designs.
- **Repetitiveness:** Most calculations are tedious and repetitive work, and due to commercial considerations, the calculation results of various projects cannot be effectively accumulated and inherited. Even if engineers encounter a case with similar design content, they need to repeat its calculation process.
- **Difficult to review:** The validity of fire and evacuation models and the reliability of simulation results are often questioned by reviewers from the authorities. Even with the source code provided, it is difficult for reviewers to point to specific design issues.

These problems greatly limit the application of performance-based design in practical evacuation application. Improving the design efficiency, ensuring the reliability of the scheme, and expanding the scope of the applications are the difficulties and challenges faced by the entire fire design industry. With the advent of the era of big data, artificial intelligence technology is favored in every aspect of life due to its advantages such as relatively low computational resources, being numerically efficient, and having promising modeling performance. Time-consuming and complicated design work is one of the best application scenarios for artificial intelligence, and smart design for evacuation has become a popular trend. On the one hand, smart technology could solve the issues of performance-based design during the design process by establishing intelligent evacuation modelling. Intelligent evacuation modelling using artificial intelligence is expected predict the evacuation process accurately and promptly by learning human behavior with deep learning algorithms [32]. For example, AI models can analyze past evacuation scenarios and learn patterns of crowd movement and behavior, and predict how people are likely to evacuate in a given situation. Real-time data such as fire location, building layout, and occupancy information could be analyzed to identify the most efficient evacuation routes [33, 34].

On the other hand, smart design could benefit to the real evacuation process such as smart guiding systems with dynamic signages. Those smart guiding systems take advantage of multiple sensors and Internet of Things [35], and choose optimal evacuation routes for occupants. Occupants could follow the guiding systems and reach the safety zones in a short period. Smart design for evacuation also could benefit from smart fire technology such as smart fire detection and prediction. For example, researchers have trained the latest deep learning models to quickly identify and predict the development of fire and smoke inside the building, which can be further used to estimate the safe evacuation time of personnel. AI models like convolutional neural networks (CNNs), long short-term memory (LSTM), and Generative adversarial networks (GAN) have been applied and demonstrated in flashover forecast [36], fire source detection [37], and estimation of fire power [38].

Apart from smart detection and recognition in fire source and development, new technologies have also applied in human detection and tracking in evacuation process based on computer vision and deep learning. For example, Chen et al. (2020) applied YOLO-based recursive neural network to the evacuation model in the design of public construction. Zhang et al. (2021) used YOLO to extract the individual's position and velocity data in the crowd for the development of an evacuation simulation system. Li et al. (2022) applied an object detecting algorithm to extract and model human movement characteristic parameters during evacuation.

With the advanced development in fire safety and human behavior, it is feasible to apply smart design for evacuation using AI engine with the support of real-time data collection (see Fig. 4). The whole process could be divided into four parts as below:

(1). Data collection: The data include fire behavior, demographics of occupants, their moving abilities and real-time evacuation process. To point out, the information of the building should be stored in advance. The data can be collected using sensors, cameras, and other IoT devices.
(2). Cloud server: Once the data has been collected, it can be transferred to and stored on a cloud server. This allows for easy access to the data from anywhere and further calculation, and ensures that the data is secure and backed up.
(3). AI Engine: The AI engine is the heart of the smart evacuation system. It uses the data collected to analyze the building in fire and occupants' evacuation, and to develop evacuation strategies that are tailored to the specific situation. Some widely applied and accepted AI approaches include statistical analysis, random forest, Region Convolutional Neural Networks (RCNN), Deep Neural Networks (DNN) etc.

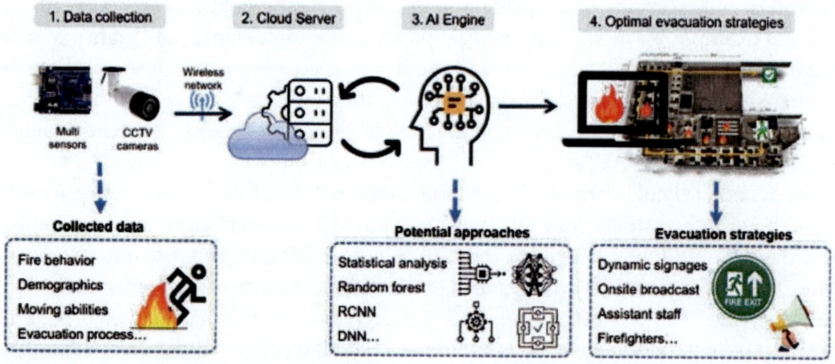

Fig. 4 AI-based smart fire evacuation design

(4). Optimal evacuation strategies: The AI engine outputs the optimal evacuation strategies based on the fire conditions and human behavior with AI and deep learning algorithms. For example, the strategies will control dynamic signages to show the safest route, guide occupants' movement via real-time broadcast. In addition, the strategies will help assistant staff and firefighters to lead the evacuation and rescue trapped people.

It is believed that in the next 10–20 years, scientific research will guide practical applications, continuously improve evacuation design efficiency, save design costs, and realize full intelligent evacuation design in fire.

4 Fire Safety Design for Firefighting and Rescue

Scientific research and engineering applications with CFD fire modelling and evacuation modelling promote the evacuation design in fire safety, and have solved a great number of evacuation safety problems in practice.

However, the majority of modern procedures ignore the necessity of firefighting, which cannot be regulated by codes or adequately explained by ASET and RSET. The complexity of buildings is detrimental to occupants' self-evacuation, and dynamic fires will cause additional difficulties by emitting irritating chemicals and reducing visibility [41, 42]. As a result, the fire department's help is advantageous and necessary. Thus, current applications of ASET and RSET are not sufficient, the safe firefighting time should be presented individually and considered in fire safety PBD.

4.1 Conventional Design Methods for Firefighting and Rescue

Firefighters are essential in fighting fires, conducting searches, and performing rescues. On the one hand, they can put out the fire with fire extinguishers or by creating firebreaks, which lessens the risk of a fire and gives people more time to flee. On the other hand, they may locate and free any trapped occupants. However, these firefighting and rescue operations significantly raise the danger to firefighters who are fighting close-range fires. Therefore, it is crucial to consider the complicated role of firefighting. It could give a scientific guideline in fire safety design which help firefighters' operation in fire. In addition, sensor data and AI algorithms could give instant and actionable warning to the firefighters of potential danger. The primary goal is to ensure that firefighters may leave the fire scene safely before it gets unmanageable, despite the fact that different facilities and procedures are available to them [43, 44].

Though it is well accepted that firefighters' safety is the priority during the process of firefighting and rescue, there is still large lacking of design for firefighting and rescue. Most codes and regulations for structure design mainly regulate the requirement of evacuation for trapped people, but do not consider the needs of firefighters independently. Those lack largely result from the limited understanding and research of firefighters' behavior, and this phenomenon has been improved with more research on firefighters and their behaviors on wildfire firefighting safety [45, 46].

4.2 Safe Firefighting Time (SFT)

By referring to the aforementioned safe egress time (SET) for evacuation design, we herein propose the Safe Firefighting Time (SFT). The concept of safety covers various aspects, e.g., no instantaneous injury or no long-term health issue, either in physical or mental. In this chapter, safety emphasizes the physical aspects, that is, no burning on the skin and no injury to the respiratory tract or other organs or tissues. Note that the concept of SFT should be updated for different fire situations and eventually include mental health.

SFT differs significantly from SET in terms of residents' evacuation since there are significant disparities between firefighting and evacuation behaviors (summarized in Table 2). SFT must include the firefighting mission, timetable, moving direction, and path, as well as the impact of actions on the fire and the need to free trapped residents.

Task: Firefighters reach the fire scene, control the fire, delve deep into the fire to look for trapped or injured people, help them move to a safe area while the residents' mission is to escape the burning structure. Also, firefighters must also leave the fire scene before the fire becomes dangerous or the firefighters are dehydrated.

Timeline: It typically takes a few minutes for residents to begin the evacuation after they become aware of the fire through broadcast, smoke, or other people's warnings [47]. Within the next few minutes, firefighters show up to the structure and dive headfirst into the blaze when the majority of occupants flee. The fire is rather intensive, and the smoke's temperature and toxicity are also very high. Generally,

Table 2 A brief summary of differences between SET and SFT

Aspects	Safe egress time (SET)	Safe firefighting time (SFT)
Tasks	Evacuation	Firefighting, search, rescue
Timeline	A couple of minutes	A few to dozens of minutes
Moving path	From inside to outside	From outside to inside, back-and-forth search, and then from inside to outside
Evacuation	By themselves	Post firefighting with loads and injured occupants
Influence on fire	Little influence	Big influence on firefighting operations

the fire scene is hazardous to stay and move through without any personal protective equipment.

Moving path: Occupants trapped in a building evacuate from inside to outside of the building. In contrast, firefighters must enter the burning building firstly, which is exactly the opposite of evacuation direction. The search operation requires a lot of back and forth movement inside the building while firefighters may lack a thorough understanding of the layout of the building. After they find and carry trapped occupants, they must leave the structure promptly.

Evacuation: The occupants leave the fire scene on their own. Firefighters, on the other hand, are also vulnerable entering, moving through and leaving the dangerous fire scene, even though they have completed professional training and have maintained a good athletic capability, especially after a prolonged stay at the fire scene. Additionally, they frequently have to help or free the injured individuals, which adds to their already hefty load.

Influence on fire: In most cases, while people are being evacuated to safe places, their movement won't affect the spread of the fire, because they don't frequently utilize fire extinguishers when a fire gets out of control and are discouraged from doing so. However, controlling or extinguishing the fire is one of the firefighters' main tasks. These actions will alter how a fire develops, the level of danger, and corresponding firefighting tactics.

4.3 Design Values of SFT: ASFT and RSFT

Referring to ASET and RSET (Fig. 2), the **Available Safe Firefighting Time (ASFT)** and **Require Safe Firefighting Time (RSFT)** should be proposed correspondingly, as illustrated in Fig. 5 [20]. **ASFT** estimates the amount of time after a fire is first ignited until smoke, heat, and toxic gases make circumstances for firefighters wearing protective gear and facilities intolerable. **RSFT** refers to the amount of time needed for the firefighters to finish their tasks and carry trapped people to reach a safe area.

With reference to comparable computation techniques, Table 3 examines and highlights the differences between ASET and ASFT in terms of their motivation, objects, and tenable criteria. Similar to how ASET is calculated, ASFT can be obtained by using CFD fire modeling.

Temperature, visibility, CO concentration, and thermal radiation (Table 1) are the four parameters utilized in the ASET calculation to determine the level of hazard in the evacuation environment. For calculating the ASFT, the following four parameters can also be utilized. According to the tenable criteria such as temperature, thermal radiation, visibility, and CO, four thresholds could be obtained, and the minimum value will be chosen as ASFT, see Eq. (3). However, different values should be chosen based on the corresponding scenarios. For instance, CO is not a deciding factor if firefighters are wearing the SCBA (self-contained breathing apparatus). Furthermore, there is disagreement over whether visibility should be included as a criterion. Most

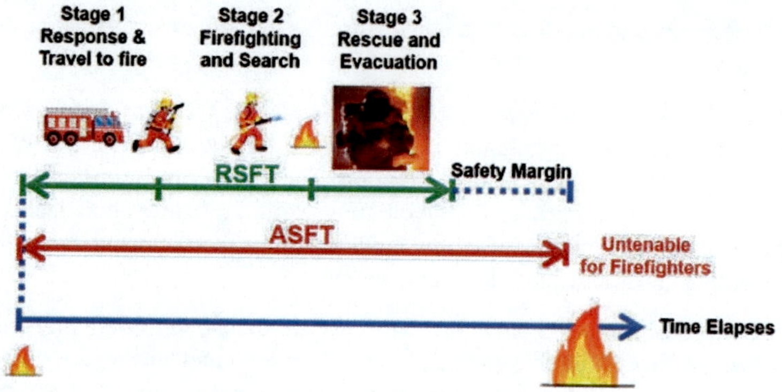

Fig. 5 Diagram of Available Safe Firefighting Time (ASFT) and Require Safe Firefighting Time (RSFT)

Table 3 Comparison of ASET and ASFT

	ASET	ASFT
Objective	To facilitate safe evacuation of trapped occupants	To support effective firefighting and rescue operations
Target users	Building occupants	Firefighters
Primary approach	CFD modelling	CFD modelling
Tenable criteria	• Widely accepted by researchers and engineers	• Studied by limited researchers and lacks recognized criteria
	• Fixed criteria	• Vary from firefighters' suits
	• All criteria are vital	• Classified as principle and secondary
	• Consider CO	• Do not consider CO with masks

firefighters will enter the low-light, smoke-filled fire scene with personal protective equipment (PPE). Commonly, there are several firefighters and fire engines waiting outside the building in fire. Thus, it is safe to assume that limited visibility near a building's exit will not provide a serious threat to firemen or the occupants who have been saved when they arrive at the outlets. Therefore, temperature and thermal radiation are two primary factors, and the CO and visibility thresholds should be taken into consideration secondarily.

$$\text{ASFT} = \min\{T_T, T_R, T_V, T_{CO}\} \tag{3}$$

where

T_T, T_R T_V and T_{CO} refer to the amount of time till the temperature, thermal radiation, visibility and CO that reach the threshold, respectively.

RSFT (Required Safe Firefighting Time) refers to the amount of time needed for the firefighters to complete their jobs and leave the fire scene carrying trapped people. RSFT is the sum of three separate parts, (1) the response and travel time to the burning building, (2) the time spent fighting the fire inside the structure, which includes putting out the flames and looking for people who may be trapped, and (3) the time spent on rescue and evacuation, when firefighters help injured people to leave the building.

RSFT could be computed using firefighting movement models and the RSFT is a function of the fire circumstances, the firefighters' skills and tactics, and the inhabitants. In practice, each firefighter's RSFT is a dynamic number that changes in response to rapidly evolving fire scenarios and firefighting activities. The differences between RSET and RSFT in terms of methodologies, movement phases, and deciding variables are compared and summarized in Table 4.

Fire drills with firefighters and numerical modeling of firefighting movement could be used to determine the RSFT, but there has been relatively little study on firefighting movement modeling or targeted modules in evacuation software like Pathfinder or FDS + Evac compared to the vast majority of RSET evacuation models. Due to the distinctions between firefighting and evacuation, RSFT should take into account the various responsibilities, collaboration, and moving paths, as displayed in Fig. 5. RSFT could then be written as Eq. (4) [20]:

$$\text{RSFT} = \sum_{i=1}^{n} f(M_t, D_j, C, B) \tag{4}$$

where

$n = 3$ *and* $i = 1, 2, 3$ refer to the firefighting tasks following the timeline. For $i = 1$ (Task 1), it refers to when firefighters get noticed and arrive at the fire scene; for $i = 2$ (Task 2) refers to when firefighters execute their assigned firefighting and searching

Table 4 Comparison of RSET and RSFT

	RSET	RSFT
Method	Evacuation modelling	Firefighter modelling
Software	Pathfinder, FDS+Evac, etc.	Not available
Models	Social force model Cellular automaton model Etc	No wide-accepted physics-based model
Research	Extensive research since the 1990s	Limited research
Components	• Recognition time • Response time • Movement time	• Recognition time Response time Travelling time to the fire • Firefighting and searching time • Rescue and evacuation time

trapped people in the building; and for i = 3 (Task 3) refers to when firefighters move out of the building and rescue trapped occupants to the outside;

M_t refers to the movement time for firefighters, and the detailed parameters refer to Eq. (5);

D_j refers to firefighters' different duties and tasks;

C refers to the cooperation among groups such as communication and support between members;

B refers to the burden the firefighters carry during the rescue (firefighting facilities, injured people);

$$M_t = g(S, F, v, C_i^t, C_N^t) \quad (5)$$

where

S refers to the building structures, including floors, facility layout, and technical installations etc.;

F refers to the fire conditions, including the fire location, comburent etc., and it will be influenced by firefighters' behavior dynamically;

v refers to the velocity of firefighters considering their starting velocity, velocity decay and velocity distribution;

C_i^t is the condition of firefighter i at time t;

C_N^t is the condition of all other firefighters at time t.

These equations provide a methodology for determining RSFT and motivate further study of firefighters' movement models. Further investigations are in urgent need on firefighters' speed, preferred trajectory, collaboration, and mental states, among other factors, in order to effectively capture their behavior and movement. Additionally, it is urged to develop models and software based on the traits of the firefighters.

4.4 Operation Values of SFT: Smart Firefighting and Rescue

SFT is a significant factor in evaluating the safety of firefighters in firefighting, both in fire safety design and smart firefighting. Apart from the proposed ASFT and RSFT to be involved in fire safety PBD, the element of SFT also contribute to conducting smart firefighting and rescue. Before the framework of smart firefighting and rescue is proposed, we made comparisons of SFT's role and value in design period and operation period. Some significant differences between these two aspects are listed as below (see Table 5).

Smart firefighting is a critical approach that relies on the acquisition and utilization of real-time information on fire dynamics and evacuation processes through the employment of IoT and digital twin systems. This real-time information facilitates the development of intelligent firefighting strategies and operations. As firefighters' behaviors are significantly influenced by real-time fire dynamics and evacuation progress, obtaining precise SFT via smart firefighting systems is essential and

Table 5 Comparison between SFT application on fire safety PBD and smart firefighting

	Fire safety PBD	Smart firefighting
Principles	The approach involves modeling fire scenarios and firefighters' behavior to design and optimize firefighting settings	The approach involves selecting the best or optimal firefighting strategies from a vast number of possibilities
Aim	To ensure the structures and facilities for firefighting are sufficient	To ensure firefighters' safety and their decisions are wise and feasible
Users	Engineers	Engineers, firefighters, decision makers
Process time	This approach can take hours to days	This approach can take seconds to minutes
Prerequisite	Knowledge of fire dynamics and firefighters' behavior	Database of thousands of cases with fire dynamics and firefighters' behavior
Development	The approach is mature and commercialized	The approach is in its early stages of development

supportive in developing more efficient and explicit firefighting strategies and vice versa.

The newly proposed smart firefighting systems suggest that the modified smart firefighting system embedded with SFT could comprise a sensor network, a cloud data server, an AI engine, and user interfaces [20, 47]. The intricate details of this modified system are elaborated below as Fig. 6.

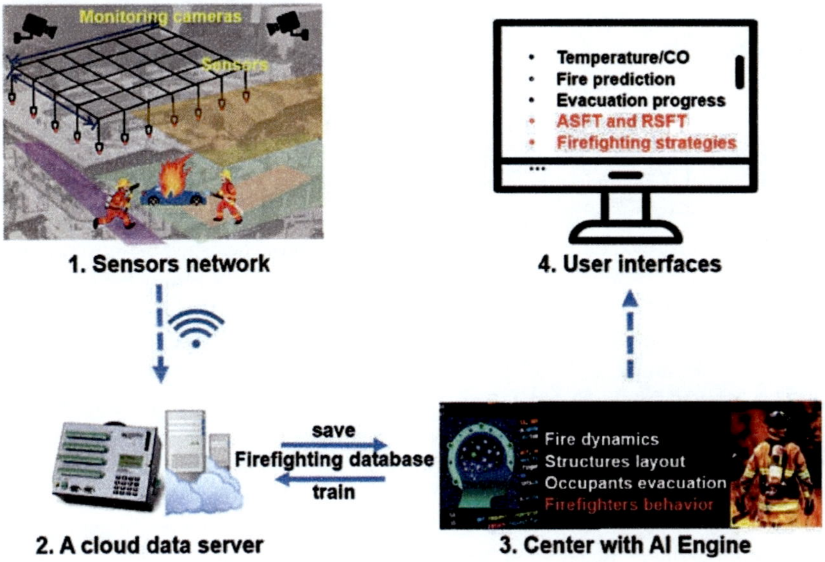

Fig. 6 The framework of SFT in smart firefighting design and operation

(1) Sensors network. Sensors are installed in buildings to monitor various environmental factors, such as temperature and CO concentration. Along with CCTV camera monitoring systems, the sensors provide real-time data on fire processes and occupants' movements, which are then uploaded to the cloud server.
(2) A cloud data server. The server plays a crucial role in the smart firefighting system by storing and processing the massive amounts of data generated by the sensors network. The server reads the data from the sensors remotely, filters out noise using algorithms, and stores the data in standard formats. The cloud database is then formed by organizing the data from previous and potential fire scenarios, which is used by the AI engine to develop predictive models.
(3) AI engine. The AI engine utilizes machine learning algorithms to analyze the data from the cloud database. By training on previous and potential fire and evacuation scenarios, the AI engine can predict and forecast the dynamic ASFT and RSFT based on various factors, such as the fire development, trapped occupants, and the time stamp when firefighters arrive at the fire scene. This predictive capability provides firefighters with critical information for developing effective firefighting strategies and tactics.
(4) User interface. The user interface is the final component of smart firefighting systems, which displays the fire scene, evacuation scene, and rescue scene in various modules and elements. The interface presents the ASFT and RSFT to firefighters and provides commanders with suggestions for adjusting firefighting strategies dynamically. This allows firefighters and commanders to make informed decisions based on real-time data and forecasts, resulting in more efficient and effective firefighting and rescue operation.

5 Summary

Fire is a vital threat to both occupants inside the building and firefighters during their firefighting and rescue operations. Once a fire occurs, the building environment changes rapidly, becoming hot, toxic, low luminance and invisible. The chapter provides an overview of the fire safety design progress for evacuation, firefighting and rescue. For evacuation designs, conventional approaches such as prescriptive-based codes and performance-based design are introduced. Driven by the emerging technologies and new understanding of human behavior, smart evacuation design approaches with intelligent evacuation modelling and real-time guidance systems are discussed and summarized. Nevertheless, very few design and operation considerations are given to firefighters who enter fire scenes and are exposed to more dangerous environment. For firefighting and rescue design and operation, we proposed the concept of safe firefighting time and introduced the application both for the aspect of design and real-time firefighting. From the study and discussion in this chapter, in the future, key areas for advancing fire safety include IoT integration, real-time human behavior modeling, collaboration among stakeholders, and sustainable design. IoT integration from different administrations will enhance communication and response coordination among different departments and roles. Accurate behavior modeling will improve evacuation planning and guidance. Collaborative efforts among firefighters, staff in the building and stakeholders will optimize emergency response and enhance the efficiency of evacuation and firefighting. Sustainable design will prioritize environmental impact reduction while ensuring fire safety and life safety. As such, the chapter provides some summary and insights to new research and application of fire safety design in the future.

Acknowledgements This work is funded by the National Natural Science Foundation of China (52204232), the MTR Research Fund (PTU-23005), and the Hong Kong Research Grants Council Theme-based Research Scheme (T22-505/19-N).

References

1. S.T. McKenna, N. Jones, G. Peck, K. Dickens, W. Pawelec, S. Oradei, S. Harris, A.A. Stec, T.R. Hull, Fire behaviour of modern façade materials—understanding the Grenfell Tower fire. J. Hazard. Mater. **368**, 115–123 (2019). https://doi.org/10.1016/j.jhazmat.2018.12.077
2. K. Fridolf, D. Nilsson, H. Frantzich, Fire evacuation in underground transportation systems: a review of accidents and empirical research. Fire Technol. **49**, 451–475 (2013)
3. Y. Zhang, Z. Yan, H. Zhu, Y. Shen, Q. Guo, Q. Guo, Experimental investigation of pedestrian evacuation using an extra-long steep-slope evacuation path in a high altitude tunnel fire. Sustain. Cities Soc. **46**, 101423 (2019). https://doi.org/10.1016/j.scs.2019.101423
4. A. Cowlard, A. Bittern, C. Abecassis-Empis, J. Torero, Fire safety design for tall buildings, in *Procedia Eng, Elsevier Ltd.* (2013), pp. 169–181. https://doi.org/10.1016/j.proeng.2013.08.053
5. S.E. Magnusson, H. Frantzich, K. Harada, Fire safety design based on calculations: uncertainty analysis and safety verification. Fire Saf. J. **27**, 15–334 (1996)
6. A.H. Buchanan, Implementation of performance-based fire codes. Fire Saf. J. **32**, 377–383 (1999)
7. Y.E. Kalay, Performance-based design. Autom. Constr. **8**, 395–409 (1999)
8. C.M. Fleischmann, Is prescription the future of performance-based design? in *Fire Safety Science* (2011), pp. 77–94. https://doi.org/10.3801/IAFSS.FSS.10-77
9. G. Spinardi, Fire safety regulation: prescription, performance, and professionalism. Fire Saf. J. **80**, 83–88 (2016). https://doi.org/10.1016/j.firesaf.2015.11.012
10. A.A. Sheeba, R. Jayaparvathy, Performance modeling of an intelligent emergency evacuation system in buildings on accidental fire occurrence. Saf. Sci. **112**, 196–205 (2019). https://doi.org/10.1016/j.ssci.2018.10.027
11. M.P. Manuel, M. Faied, M. Krishnan, M. Paulik, Robot platooning strategy for search and rescue operations. Intell. Serv. Robot. (2021). https://doi.org/10.1007/s11370-021-00390-7
12. X. Zhang, Z. Tang, Z. Fang, L. Zhang, Assessment of emergency evacuation in tunnel fire environment, in *2016 International Conference on Robots & Intelligent System (ICRIS)* (2016), pp. 128–131. https://doi.org/10.1109/icris.2016.27
13. Y. Zhang, W. Li, Y. Rui, S. Wang, H. Zhu, Z. Yan, A modified cellular automaton model of pedestrian evacuation in a tunnel fire. Undergr. Space Technol. 130 (2022) https://doi.org/10.1016/j.tust.2022.104673.
14. E. Duarte, F. Rebelo, J. Teles, M.S. Wogalter, Behavioral compliance for dynamic versus static signs in an immersive virtual environment. Appl. Ergon. **45**, 1367–1375 (2014). https://doi.org/10.1016/j.apergo.2013.10.004

15. H. Yin Wong, Y. Zhang, X. Huang, A review of dynamic directional exit signage: challenges and perspectives (2022). www.nfpa.org/foundation
16. E. Vilar, F. Rebelo, P. Noriega, E. Duarte, C.B. Mayhorn, Effects of competing environmental variables and signage on route-choices in simulated everyday and emergency wayfinding situations. Ergonomics **57**, 511–524 (2014). https://doi.org/10.1080/00140139.2014.895054
17. Y. Zhang, X. Huang, A review of tunnel fire evacuation strategies and state-of-the-art research in China. Fire Technol. (2022). https://doi.org/10.1007/s10694-022-01357-5
18. A. Haghighat, K. Luxbacher, Tenability analysis for improvement of firefighters' performance in a methane fire event at a coal mine working face. J. Fire Sci. **36**, 256–274 (2018). https://doi.org/10.1177/0734904118767066
19. L. Bel-Latour, M.A. Granié, The influence of the perceived masculinity of an occupation on risk behavior: the case of firefighters. Saf Sci. **150** (2022). https://doi.org/10.1016/j.ssci.2022.105702
20. Y. Zhang, X. Zhang, X. Huang, Design a safe firefighting time (SFT) for major fire disaster emergency response. Int. J. Disaster Risk Reduct. **88** (2023). https://doi.org/10.1016/j.ijdrr.2023.103606
21. F. Tao, Q. Qi, L. Wang, A.Y.C. Nee, Digital twins and cyber–physical systems toward smart manufacturing and industry 4.0: correlation and comparison. Engineering **5**, 653–661 (2019). https://doi.org/10.1016/j.eng.2019.01.014
22. L. chu Su, X. Wu, X. Zhang, X. Huang, Smart performance-based design for building fire safety: prediction of smoke motion via AI. J. Build. Eng. **43** (2021). https://doi.org/10.1016/j.jobe.2021.102529
23. P. Reneke, C. Grant, N.P. Bryner, A.W. Jones, G.H. Koepke, in *Research Roadmap for Smart Fire Fighting, Gaithersburg, MD* (2015). https://doi.org/10.6028/NIST.SP.1191
24. N. Naraghiaraghi, Z. Feng, R. Lovreglio, S. Wilkinson, Combining BIM and VR for future earthquake damage assessment training tools A REVIEW OF THE IMPACTS OF COVID-19 ON THE CONSTRUCTION INDUSTRY IN AUCKLAND, NEW ZEALAND View project Building Quake & People-a serious game platform for informing life saving strategies View project Combining BIM and VR for future earthquake damage assessment training tools (2022). https://www.researchgate.net/publication/358808115
25. Bateman, Edwards, Gender and Evacuation_ A Closer Look at Why Women Are More Likely to Evacuate for Hurricanes (n.d.)
26. M. Kobes, I. Helsloot, B. de Vries, J.G. Post, N. Oberijé, K. Groenewegen, Way finding during fire evacuation; an analysis of unannounced fire drills in a hotel at night. Build. Environ. **45**, 537–548 (2010). https://doi.org/10.1016/j.buildenv.2009.07.004
27. S.L. Poon, A dynamic approach to ASET/RSET assessment in performance based design, in *Procedia Eng, Elsevier Ltd.* (2014), pp. 173–181. https://doi.org/10.1016/j.proeng.2014.04.025
28. Y. Zhu, T. Chen, N. Ding, M. Chraibi, W.C. Fan, Follow the evacuation signs or surrounding people during building evacuation, an experimental study. Phys. A: Stat. Mech. Its Appl. **560** (2020). https://doi.org/10.1016/j.physa.2020.125156
29. J. Norén, M. Delin, K. Fridolf, Ascending stair evacuation: what do we know? Transp. Res. Procedia **2**, 774–782 (2014). https://doi.org/10.1016/j.trpro.2014.09.087
30. E. Ronchi, Testing the predictive capabilities of evacuation models for tunnel fire safety analysis. Saf. Sci. **59**, 141–153 (2013). https://doi.org/10.1016/j.ssci.2013.05.008
31. PIARC, Safety in tunnels: transport of dangerous goods through road tunnels, OECD (1999)
32. H. Hou, L. Wang, Measuring the rationality in evacuation behavior with deep learning (2021). https://doi.org/10.3390/e24020198
33. Y. Chen, S. Hu, H. Mao, W. Deng, X. Gao, Application of the best evacuation model of deep learning in the design of public structures. Image Vis. Comput. **102** (2020). https://doi.org/10.1016/j.imavis.2020.103975
34. D. Xu, X. Huang, J. Mango, X. Li, Z. Li, Simulating multi-exit evacuation using deep reinforcement learning. Trans. GIS **25**, 1542–1564 (2021). https://doi.org/10.1111/tgis.12738
35. H.H. Yen, C.H. Lin, H.W. Tsao, Time-aware and temperature-aware fire evacuation path algorithm in IOT-enabled multi-story multi-exit buildings. Sensors (Switzerland). **21**, 1–24 (2021). https://doi.org/10.3390/s21010111

36. T. Zhang, Z. Wang, Y. Zeng, X. Wu, X. Huang, F. Xiao, Building artificial-intelligence digital fire (AID-fire) system: a real-scale demonstration. J. Build. Eng. **62**, 105363 (2022). https://doi.org/10.1016/j.jobe.2022.105363
37. X. Wu, X. Zhang, X. Huang, F. Xiao, A. Usmani, A real-time forecast of tunnel fire based on numerical database and artificial intelligence. Build. Simul. **15**, 511–524 (2022). https://doi.org/10.1007/S12273-021-0775-X/METRICS
38. Z. Wang, T. Zhang, X. Huang, Predicting real-time fire heat release rate by flame images and deep learning. Proc. Combust. Inst. (2022). https://doi.org/10.1016/j.proci.2022.07.062
39. G. Zhang, D. Lu, H. Liu, IoT-based positive emotional contagion for crowd evacuation. IEEE Internet Things J. **8**, 1057–1070 (2021). https://doi.org/10.1109/JIOT.2020.3009715
40. S. Li, L. Tong, C. Zhai, Extraction and modelling application of evacuation movement characteristic parameters in real earthquake evacuation video based on deep learning. Int. J. Disaster Risk Reduct. **80** (2022). https://doi.org/10.1016/j.ijdrr.2022.103213
41. G. Cosma, E. Ronchi, D. Nilsson, Way-finding lighting systems for rail tunnel evacuation: a virtual reality experiment with Oculus Rift®. J. Transp. Saf. Secur. **8**, 101–117 (2016). https://doi.org/10.1080/19439962.2015.1046621
42. G.-Y. Jeon, J.-Y. Kim, W.-H. Hong, G. Augenbroe, Evacuation performance of individuals in different visibility conditions. Build. Environ. **46**, 1094–1103 (2011). https://doi.org/10.1016/j.buildenv.2010.11.010
43. I. Rodríguez-Garzón, M. Martínez-Fiestas, A. Darmohraj, A. Delgado-Padial, R. Chumpitaz, Voluntary and involuntary risk acceptance: a case study of firefighters. Saf. Sci. **142** (2021). https://doi.org/10.1016/j.ssci.2021.105394
44. M. Martínez-Fiestas, I. Rodríguez-Garzón, A. Delgado-Padial, Firefighter perception of risk: a multinational analysis. Saf. Sci. **123** (2020). https://doi.org/10.1016/j.ssci.2019.104545
45. B.W. Butler, J.D. Cohen, Firefighter safety zones: a theoretical model based on radiative heating. Int. J. Wildland Fire **8**, 73–77 (1998). https://doi.org/10.1071/WF9980073
46. P.E. Dennison, G.K. Fryer, T.J. Cova, Identification of firefighter safety zones using lidar. Environ. Model. Softw. **59**, 91–97 (2014). https://doi.org/10.1016/j.envsoft.2014.05.017
47. X. Wu, X. Zhang, Y. Jiang, X. Huang, G.G.Q. Huang, A. Usmani, An intelligent tunnel firefighting system and small-scale demonstration. Tunn. Undergr. Space Technol. **120**, 104301 (2022). https://doi.org/10.1016/j.tust.2021.104301

Advanced Fire Identification

Fire Database and Cybersecurity

Tianhang Zhang, Yishuo Jiang, and Ray Y. Zhong

1 Introduction and Background

In science, the terminology "data" is used to describe a gathered body of facts, which represents the information obtained from observing and testing experiments. The database refers to an organized collection of structured data, which needs to store, organize, index, query, analyze, maintain, and mine such heterogeneous scientific data. High-quality data allows researchers to interpret findings accurately, act as a foundation for future studies, and give credibility to their findings. Over the last decades, to understand fire hazards and study fire behaviors, numerous full-scale and reduced-scale fire tests, and numerical simulations have generated a sea of fire data. The mysteries behind various fire phenomena, e.g., spontaneous ignition [1], flashover [2], backdraft [3], smoke back-layering [4], and ghosting flame [5] are well revealed with clear mechanisms by fire data and nice theoretical analysis.

In recent years, Artificial Intelligence (AI) and Deep learning (DL) tools are becoming popular to achieve the super-fast fire modelling and fit the complex fire scenarios [6]. For example, Hodges et al. [7] used a transpose convolutional neural network (TCNN) and simulated results conducted by FDS to predict the temperature distribution inside compartment rooms. Ghoreishi [8] compared the performance of various models and demonstrated the feasibility of their faster regional CNNs in detecting fire, meanwhile limiting the false positive. Cao et al. [9] proposed an enhanced bidirectional long short-term memory (LSTM) model to predict wildland

T. Zhang (✉)
Research Centre for Fire Safety Engineering, Department of Building Environment and Energy Engineering, The Hong Kong Polytechnic University, Hong Kong, China
e-mail: tianhang.zhang@connect.polyu.hk

Y. Jiang · R. Y. Zhong
Department of Industrial and Manufacturing Systems Engineering, The University of Hong Kong, Hong Kong, China

X. Huang and W. C. Tam (eds.), *Intelligent Building Fire Safety and Smart Firefighting*, Digital Innovations in Architecture, Engineering and Construction,
https://doi.org/10.1007/978-3-031-48161-1_11

fires with video images. Zhang et al. [10, 11] proposed an AI-driven building fire digital twin which can predict the fire scenario with a 20-s lead time. Wang et al. [12, 13] established a vision-based algorithm to inverse the fire Heat Release Rate (HRR) with fire and image images in real time. It was reported that this method could provide a more accurate prediction since it can take tempo-spatial features into account to detect fire. These studies explored the potential application of AI methods in the detection and forecasting of fire [14–17].

It should be pointed out that all AI models need to be fine-tuned before being applied to practical problems. Often this is realized through sufficient training iterations on a database containing a large volume of data. A well-structured database is thus inevitable for the development of the AI model. The establishment of the database could be a challenging task since a sufficient amount of data need to be extracted from the literature, and then these data should be organized in a format to be conveniently used for further training. Nevertheless, most of the literature data (both experimental and numerical) are organized in very different ways, based on the habit and taste of researchers, so they are impossible to train by AI directly [4].

The task of fire data organization has to be done by a fire researcher or expert, because it requires expertise and knowledge in fire dynamics and modelling know-how, such as the fire size, ventilation conditions, and burning material. For example, Naser et al. [15] built up a database by collecting large amounts of data on material components of timber structures and correlated equations derived from fire tests. The fire resistance of timber structures was predicted by an AI model and the established database. For tunnel fires, Wu et al. [18] established a big tunnel-fire database of numerical simulations for varying fire locations and sizes, and ventilation conditions, and then demonstrated the use of AI and deep learning to identify the fire source and forecast the temperature field and evolution of tunnel fire.

To date, fire-safety scientists and engineers have understood more about the fire dynamics by analyzing massive fire data and establishing fire models. However, to transfer that knowledge to the AI model is a big challenge, that is, the absence of a solid database that can be fed to AI for training. To be more specific, most of the data presented in the literature are fine-processed for highlighting the major findings of authors' work, rather than allowing other researchers to further utilize the data. Very often, many data features are ignored in the data-precessing to make easy comparison with other data or even partially "cleaned" to avoid challenges. Morever, many authors are unwilling to share the raw data or make them open access online, probably because there is a lack of motivation, time, and paradigm of organizing and opening the data. All of these create a big barrier for both human and AI to use the data. Instead of re-using the literature data, many researchers prefer or have no choice but repeatedly devoting massive time and energy to developing their databases and analysis.

Meanwhile, cybersecurity issues, such as cybercriminals to exploit vulnerabilities and gain unauthorized access, and manipulation or misrepresentation of data to serve specific interests or agendas raise public doubts. Clients are worried about the cybersecurity of open API for fire sensors. Even if we set the single-directional or read-only assess to the sensor, there is still a risk of getting hacked. Thus, a database containing sufficient and precious experimental data and relevant cybersecurity technologies would be most useful and preferable for AI applications and smart firefighting.

This chapter aims to propose a paradigm of creating a fire database that can support the training of the AI-based fire model. Different approaches to collecting and generating the fire database are reviewed. Meanwhile, the cyber vulnerabilities caused by the interconnectedness of the fire database, and derivational threats are also discussed and prospected. The proposed methodology to establish and guard fire database provides new directions for future fire data collection and management and lay a foundation to promote the application of AI and smart firefighting in fire protection engineering reliably and safely.

2 Overview of Fire Database

2.1 *The Framework of Fire Database*

Data is the bloodline in the era of Artificial Intelligence [19]. To enable AI/DL model to learn the hidden features of fire phenomena, the quantity and diversity of the database should be sufficiently large and representative [20]. This could be achieved by considering the influences of different parameters and scenarios in fire tests and simulations. The database can be constructed based on previous data from fire tests and numerical simulations. In general, the establishment of the database is a multi-step task, including:

(1) **Data collection** by searching all available literatures and extracting all useful data from these documents;
(2) **Data preprocessing** by data quality check, outliers and noises removal, and filtering, and
(3) **Data mining** by extracting valuable information.

The fire phenomenon is a non-linear synamic system with strong trubulence so that it is challenging to properly store, manage, maintain, and analyse a big fire database. Most of the fire test data, especially for commercial products, are not open to the public. Moreover, the current fire test and research community has a poor habit of unwilling to share data, making the establishment of a big fire database even harder.

Table 1 compares the feature of "current fire data" and the "ideal fire data for AI". As shown, for instance, most of the literature only reports constant or steady-state quantities, such as the time-average of fire HRR, temperature distribution, pre-set ventilation condition, and features of smoke layers, because these steady-state values are easy to report, analyze, and extrapolate. However, any real fire process is a transient process, and the steady state conditions are rarely reached. Although most of these transient raw data are recorded during the tests, these data were rarely documented, analyzed, and presented to the fire community. It is mainly because the amount of these transient data is tremendous, and they are difficult to analyze and visualize and cannot provide useful engineering correlations.

Therefore, it is urgent to establish a global standard and guideline for organizing, presenting, and sharing fire research data as a community effort. There have been some initiatives on collecting experimental and numerical data for the fire modelling validation and verification, such as the IAFSS working group of Measurement and Computation of Fire Phenomena (MaCFP) [20]. New initiatives for establishing fire database for AI applications are needed in the future.

To exploit the potentialities of valuable existing data and facilitate the database establishment of future fire tests, a **framework and guideline** for data collection in fire tests is given as a baseline [4]. Taking the tunnel fire experiment as an example, the following information should be contained in the data collection step:

(1) ***Structure and test information***. Basic settings of the fire tests, including the data source (e.g., experiments, simulation, or from real fire incident), scaling factor, tunnel dimensions, data and time, ventilation system, boundary material, altitude, and on-site weather conditions (ambient temperature, humidity, and wind).
(2) ***Fire information***. Fire location, fire geometrical dimension, components of combustible material, ignition method, amount of burning materials should be given.
(3) ***Sensor information***. Layout of the sensors and measuring devices, information on sensor type, measuring range, delay time, and frequency or interval of data collection need to be provided.

Table 1 Comparison of "current fire data" and "ideal fire data for AI"

Item	Current fire data	Ideal fire data for AI
Geometric information	Represented by characteristic length or 2-D layout	3-D digital model established by CAD, BIM, and relevant platform
Ambient condition	Constant or time-averaged value	Transient time-dependent variation of ambient temperature, pressure, and wind velocity
Fire information	Fixed fire position, area with a time-averaged HRR	Transient time-dependent variation of fire HRR, area and location with fuel type and ignition method
Sensor data	Well-processed time-average value or dimensionless value	Raw data with time stamp and spatial information
Visual data	Limited photos on some critical phenomena without detailed camera setting	Various videos from different angles, including a detailed description of the shooting location, exposure settings, environmental and personal information
Imperfection data	Not provided in most of publications	Should be included for failure detection and abnormal data identification

(4) ***Spatial–temporal sensor data***. The measured data obtained from sensors during whole stages of fire, including pre-fire ignition, fire development, decay, and firefighting should be provided with a time stamp.
(5) ***Visual data***. The fire phenomena should be recorded by various video cameras from different angles. All video data should include a detailed description of the shooting location, exposure settings, environmental and personal information, and time sequence of different fire processes and critical events, such as the ignition, fire spread, ventilation, and suppression.
(6) ***Infrared Ray (IR) data (if possible)***: Compared with visual light, the infrared ray has a longer wavelength so that performs a stronger smoke penetration ability. *IR* data is supposed to be collected with a time stame which can help reveal a clearer fire scenario pattern.
(7) ***Imperfection data***. Some researchers prefer not to provide imperfect data due to the unnormal measurement, missing, or other reasons. However, it is highly encouraged to provide all data, because those data not only could be treated with technics, such as filling and filtering, but also provide crucial experiences for other researchers and future tests.

Once the raw data are collected and documented, it is crucial to analyze the data by utilizing preprocessing methods. Measuring devices, human factors, and fire impacts may cause noises, redundancies, and outliers to the data collected. Thus, those collected data need to be inspected and processed. With the limited amount of experimental data in the literature, the experimental data were directly extracted from the available documents, and no further processing was conducted to guarantee the authenticity of the database. Figure 1 shows a flowchart of the proposed procedure for the database generation:

(a) Investigate all available literatures and extract basic information based on the proposed data collection framework (step 1 in the Fig. 1),
(b) Extract testing results in the form of a table, figure, and supplemental material, and pre-process the data by quality check, outliers, and noises removal (step 2 in the Fig. 1), and
(c) Organize extracted data of various sources into tables with consistent data format and metadata (step 3 in the Fig. 1).

2.2 Establishment of Fire Database

Different from the AI applications in other research field, e.g., traffic flow prediction, which has a mature data generation system (GPS [21] or Beidou) can collectthe location of vehicles continuously, the data collection for fire is more difficult. It's almost impossible to obtain data from a real fire since the fire harzard itself is a low-probability event during the life cycle. Therefore, fire experiments and numerical fire simulation become the often-used methods to generate fire data. Two case studies to

Fig. 1 Flowchart of the proposed database generation procedure

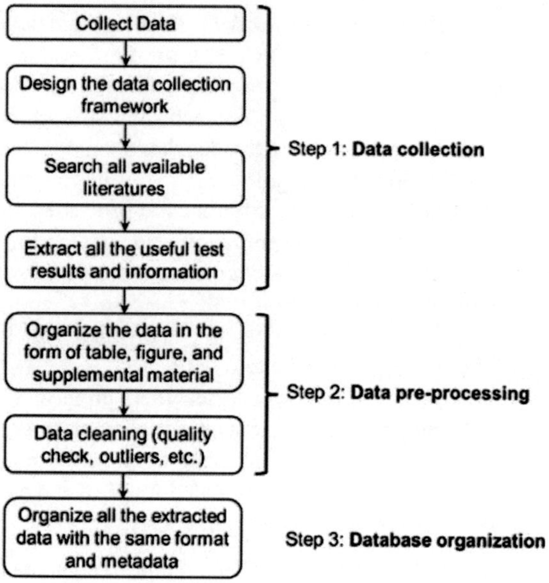

organize a fire database with experiments and numerical simulation are conducted to demonstrate the proposed framework, respectively.

To generate an experimental fire database, a number of parameters are commonly collected, and the data can be classified into two main categories. One category is sensor data, such as the temperature by thermocouples or other heat detectors, the presence of smoke by detectors, CO and CO_2 concentration by gas detectors, and heat flux by a radiometer or plate thermometer. Sensors are often installed at one or many locations for long-term measurement. In other words, sensor data are often time sequences of point or line distribution, which have both spatial and temporal dimensions. The other category is the visual data, such as the video from closed-circuit television (CCTV) and infrared cameras that are commonly installed inside the building. The visual data also include airborne cameras carried by airplanes and drones and the processed satellite image [22], particularly for large-scale fires in urban, wildland, WUI [23], and informal settlement [24].

Currently, the database can only be established with all available sensor data that can be accessed online, such as journal publications and technical reports. Nevertheless, because most of the existing fire test data initially are not produced and presented under the principle of sharing, it is extremely challenging to extract all the necessary information and data from the accomplished and ongoing tests, as well as to make a fair comparison among existing data. Figure 2 illustrates the process to construct a database for underwater tunnel fire based on a set of full-scale experiments [25]. Through a thorough literature review, the literature can first be categorized by key parameters in tunnel fire, such as the fire size, smoke layer thickness, critical velocity, back-layer length, and plug-holing. Afterwards, all related experimental data were extracted from these documents. To enforce a convenient search in the big database, all raw data and images should be well named, organized and indexed with a detailed description. Since the fire process is generally complex and influenced by multiple factors, it is essential to provide adequate test information before the data process.

It would be ideal if adequate sensor data can be obtained from full-scale fire experiments or real fire incidents. However, the existing tunnel fire data from the literature are not large enough or sufficiently well organized to support deep learning. Alternatively, a database formed by **massive CFD-based numerical simulation**

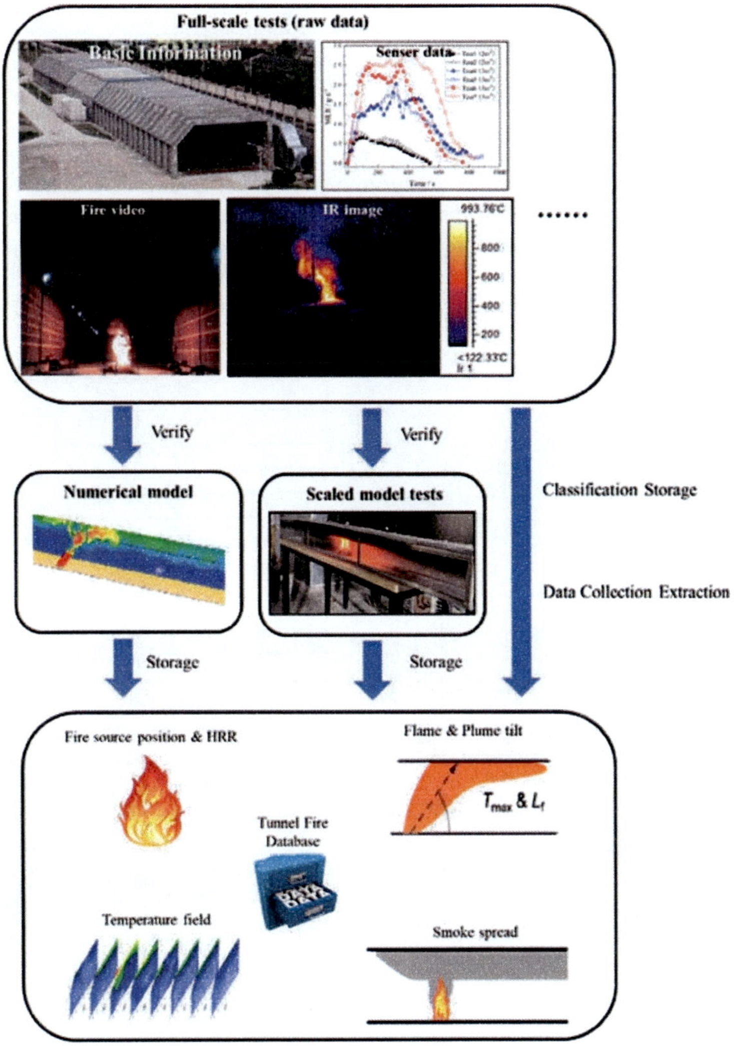

Fig. 2 Framework of full-scale underwater tunnel fire database [25]

could enable AI-based fire forecasting. Particularly, the open-source programme of FDS [26] has been extensively verified and validated and used as a standard tool for performance-based fire safety design and fire research. Moreover, the CFD modelling results can provide much more detailed temperature information than any experiment, which are ideal for the verification of the established AI method for the fire forecast.

For numerical simulations, the visual data refer to the contours and videos generated from computational results. These time-sequence data (e.g., images and videos) can directly show the real-time scene and scale of the fire, evacuation process, distributions of smoke and temperature, and firefighting activities. Compared to sensor data in fire tests, the 2-D or 3-D numerical data are several orders of magnitude larger, which can show more details about the fire process.

In contrast, the video data in fire tests are quite rare and mostly unpublic. Even if there are fire test videos, they are less uniform because of various angles of the shot, and their large size and complexity prohibit further data processing and scientific analysis. Figure 3 shows a typical construction process of a numerical database for tunnel fire [27], where the point sensor data matrix and 2-D temperature contour data are coupled. The point sensor data are used because in the fire test and real application, the point temperature data by thermocouples are most common. Such a coupling enables a database that combines limited experimental data and more abundant and detailed numerical simulations.

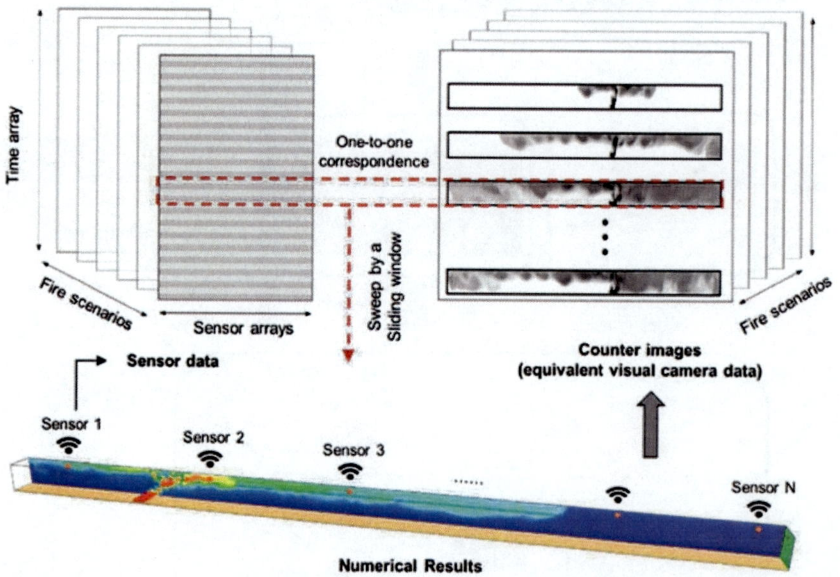

Fig. 3 Demonstration of a numerical simulation-based fire database [27]

2.3 Existing Fire Data Reporsities

Against all odds mentioned above, there are still many attempts to develop the fire data for different usage, such as material burning characteristics, vision-based fire identification, temperature patterns evoluation. Several famous and widely-used fire data repositories are summarized below for reference:

(1) **Corsican Fire Database (CFDB)** developed by the University of Corsica (Fig. 4) [28]. The database contains fire and smoke images with various environmental conditions, and shooting parameters captured by visual cameras and IR cameras. The fire zone in each figure is labeled at the pixel level. This database is widely adopted in the training/validation of visual-based fire detection/identification AI models. (https://cfdb.univ-corse.fr/index.php?newlang=english&menu=1).

(2) **University of Queensland Cladding Materials Library** [29]. The Cladding Materials Library comprises an extensive database of cladding materials based on their composition and flammability as individual components, and which may be used to perform hazard analyse. The database is a tool for qualified engineers to enable an adequate fire hazard identification and quantification of the potential fire spread of cladding materials. The flammability of cladding materials (aluminum composite panels, insulation, and any other materials such as sarking) is defined based on well-established testing frameworks widely accepted in the fire safety engineering community. These frameworks have been applied and peer-reviewed within the scope of fire research studies on the fire performance of aluminum composite panels and insulation materials at The University of Queensland and the University of Edinburgh.

Fig. 4 Samples in Corsican fire database [28]

(3) **NIST Fire Calorimetry Database (FCD).** The database contains videos for fires of different burning items associated with the heat release rate of various materials measured by the calorimeter, as shown in Fig. 5. The database contains information on the thermal and combustion properties of over 1,300 materials, including plastics, textiles, and building materials. The data was obtained through experiments using cone calorimetry, The peak heat release rate varies from 50 kW to 20,000 kW, representing single burning items to fully furnished room [13]. The database has been adopted in the training of a visual-based fire calorimeter which uses the fire image/video to identify the fire heat release rate successfully. The database can be accessed at: https://www.nist.gov/el/fcd.

(4) **National Fire Information Database (NFID)** [30]. This project was graciously funded by the Canadian Safety and Security Program (CSSP), a federal program led by Defense Research and Development Canada's Centre for Security Science, in partnership with Public Safety Canada. The NFID is a demonstration database that describes three types of fires: structural, vehicular and outdoor fires, in Canada from 2005 to 2014 and is keep upgrading. The database includes important information about fire accidents and relevant victims (civilians and firefighters).

(5) **Materials and Products (MaP) Database** developed by UL's Fire Safety Research Institute (FSRI). This is an interactive, web-based repository populated with data from rigorous testing of the contemporary products and materials commonly found in the modern built environment. The MaP is divided

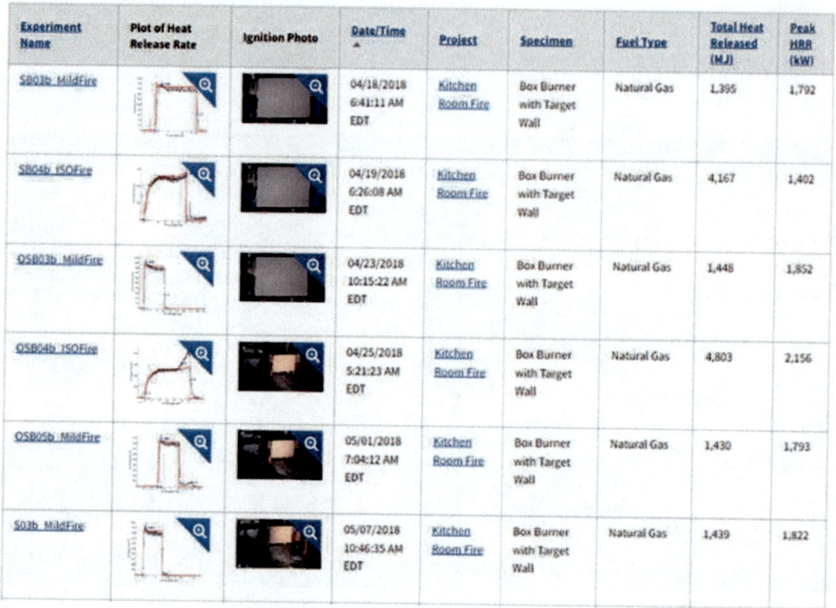

Experiment Name	Plot of Heat Release Rate	Ignition Photo	Date/Time	Project	Specimen	Fuel Type	Total Heat Released (MJ)	Peak HRR (kW)
SB03b_MildFire			04/18/2018 6:41:11 AM EDT	Kitchen Room Fire	Box Burner with Target Wall	Natural Gas	1,395	1,792
SB04b_ISOFire			04/19/2018 6:26:08 AM EDT	Kitchen Room Fire	Box Burner with Target Wall	Natural Gas	4,167	1,402
OSB03b_MildFire			04/23/2018 10:15:22 AM EDT	Kitchen Room Fire	Box Burner with Target Wall	Natural Gas	1,448	1,852
OSB04b_ISOFire			04/25/2018 5:21:23 AM EDT	Kitchen Room Fire	Box Burner with Target Wall	Natural Gas	4,803	2,156
OSB05b_MildFire			05/01/2018 7:04:12 AM EDT	Kitchen Room Fire	Box Burner with Target Wall	Natural Gas	1,430	1,793
S03b_MildFire			05/07/2018 10:46:35 AM EDT	Kitchen Room Fire	Box Burner with Target Wall	Natural Gas	1,439	1,822

Fig. 5 Samples in the NIST calorimetry fire database

Fig. 6 Demonstration of the University of Maryland's buring item database [31]

into eleven categories: roofing, exterior siding, structural materials, insulation, plumbing, cable, engineered wood, upholstered furniture, sleeping products such as mattresses and sheets, general polymers, and interior finishes such as vinyl tile, cotton, and rayon. Over 100 materials are tested to populate the categories. Additional materials will be added to the MaP as more standard fire test data and property data are measured. (https://github.com/ulfsri/fsri_materials_database).

(6) **Burning Item Database (BID)** developed by University of Maryland (Fig. 6) [31]. The database is a collection of fire test data for commonly used household/office furniture (i.e. chairs, sofas, mattresses, bookcases, etc.) The original data were collected during furniture calorimeter tests and the BID database is constructed from a review of different sources presenting test results, including books, articles published in technical journals, and technical websites on the internet. The BID database includes information on the material make up of the object, the material properties relevant to ignition, the heat release rate history, the mass loss rate history (if available), the initial and final mass (if available) and a description of, or link to the source of the original data.

(7) **PolyU Fire database** developed by The Hong Kong Polytechnic University [4]. Specifically built to facilitate the application of AI algorithms in field of tunnel fire safety, different tunnel fire accidents and fire test data (both the full-scale experimental and numerical data) are included, as well as a framework to adopt the comprehensive database to train the AI fire model to predict the critical events in tunnel fire scenarios, such as the temperature rise, critical velocity, smoke back layering length, plug-holing, etc. The PolyU fire database can be found at: https://github.com/PolyUFire/Tunnel_Fire_Database.

The key information of the above-mentioned fire database is summarized in Table 2 for intuitive demonstration. As shown, fire videos are not included in most of the existing fire databases, except for the NIST Fire Calorimetry Data-base. For the one hand, the volume of video data is typically significantly larger than that of sensor data, resulting in various challenges and greater resource consumption when

Table 2 Summary of the existing fire database

Database	Data type	Video	Website	Upgrading	Free/pay
Corsican fire database	Image for wildfire	No	Yes	Yes	Licensed for free use
NIST fire calorimetry database	Video data with transient HRR	Yes	Yes	No	Free
University of Queensland cladding materials library	Parameters of cladding materials	No	Yes	No	Licensed for free use
National fire information database	Fire incidents and victims	No	Yes	Yes	Licensed for free use
Materials and products database	Sensor data	No	Yes	Yes	Free at Github
Burning item database	Sensor data	No	Yes	No	Yes
PolyU fire database	Fire incident data and sensor data	No	Yes	Yes	Free at Github

it comes to collection, management, storage, and sharing. On the other hand, the traditional paradigm of fire research primarily relies on sensor data as they provide the most direct physical parameters for quantifying fire behavior. However, image and video data have not received sufficient attention in this context. In the realm of AI, computer vision technology enables the extraction of features from fire images and videos, opening new possibilities. Therefore, it is encouraged to expand future fire databases to encompass a broader range of video data.

In summary, although there is still a large gap between the current fire data repositiories to an ideal fire database for AI, these efforts provide vital experience of constructing a comprehensive database for fire and pioneer the roadmap for the standard fire data sharing platform.

3 Cybersecurity

3.1 Significance of Cybersecurity

Cybersecurity has become a major issue in the application of information technology for building maintenance, such as fire safety. In the internat -of-thing (IoT)-based data sharing environment, it becomes particularly sensitive because the data is collected in different places even in all the globe [32]. Security is the combination of *confidentiality* (the prevention of the unauthorized disclosure of information), *integrity* (the prevention of the unauthorized amendment or deletion of information), and *availability* (the prevention of unauthorized withholding of information) [33].

Firstly, a business or a project always involves multiple stakeholders with different expectations, interests, and motives, which causes poor information transparency, inefficient transactions, and even frauds. Multiple stakeholders normally use their own proprietary and centralized Enterprise Information System (EIS). However, stakeholders are facing potential security and privacy challenges to share information throughout projects, where "Islands of Information" issue is commonly existing and creates information fragmentation and discontinuity.

Secondly, automatic data collection, processing, and storage can save an organization's time and money, but trusting the system is more important because the real asset of any organization is the data which they share in the cloud to use the needed services by putting it either directly in the relational database or eventually in a relational database through an application.

Thirdly, the trust of the entire system depends on the data protection and prevention techniques. It is hard to develop a secure and transparent sharing environment to ensure the smooth transactions, trustable data storage, and data retrieval in projects. Numerous disputes and quality frauds have hindered clients and contractors to verify and validate the true information, such as professional qualification, building material certification, and module deliverable timing and quality.

Thus, data security becomes more important for the future application of advanced information technology in government, industry, and business. To make information technologies being adopted by more users and enterprise, the security concerns of users should be rectified first to make the data sharing environment trustworthy.

3.2 *Blockchain Technology for Cybersecurity*

Blockchain is a distributed ledger to establish a secure and reliable source of cryptographically secured time-stamped transaction records with several advantages, such as immutability, traceability, and reliability, etc. [34]. Transactions are grouped into blocks, which are then added to the chain of previously validated blocks in a secure and immutable manner. Each block contains a cryptographic hash of the previous block, creating a chronological and tamper-evident record of all transactions on the network. This distributed system provides a high level of security, as each node in the network has a copy of the entire ledger, making it difficult for any one party to manipulate the data [35]. The blockchain core constructs along with its mechanism are described as below:

(1) **Shared Ledger**. Ledger is a series of consistent blocks stored among a distributed network with two key special properties. On the one hand, a distributed ledger is a peer-to-peer network where the stakeholders can share information. On the other hand, ledger works as a distributed database. Instead of storing information in one database, ledger provides multiple scattered space to keep the reliability and immutability of data. Block is a cryptographically secured package as a fundamental element, which carries transactions generated from the construction process and is retrieved by explorers forming a bidirectional data flow, as shown in Fig. 7. A ledger contains the immutable

and sequenced records in each block, as well as a state database. Each ledger generally has one channel and several organizations, allowing a subgroup of participants in a single ledger;

(2) **Consensus**. Consensus is the core blockchain backend function, which is used to keep all the data consistent in blockchain nodes and directly related to blockchain performance. There are two types of consensus algorithms: trust-based consensus and non-trust-based consensus. Typical trust-based consensus algorithm contains Proof of Authority (PoA), Raft, etc., which has high-performance and cost-efficiency advantages. Non-trust-based consensus algorithm contains Proof of Work (PoW), Proof of Stake (PoS), practical Byzantine Fault Tolerance, which has the advantages in data transparency and integrity;

(3) **Smart Contract**. Smart contract is a series of code to execute the predefined logic. Typically, there are primarily three types of smart contracts used for computation, management, and storage. The application of smart contract could

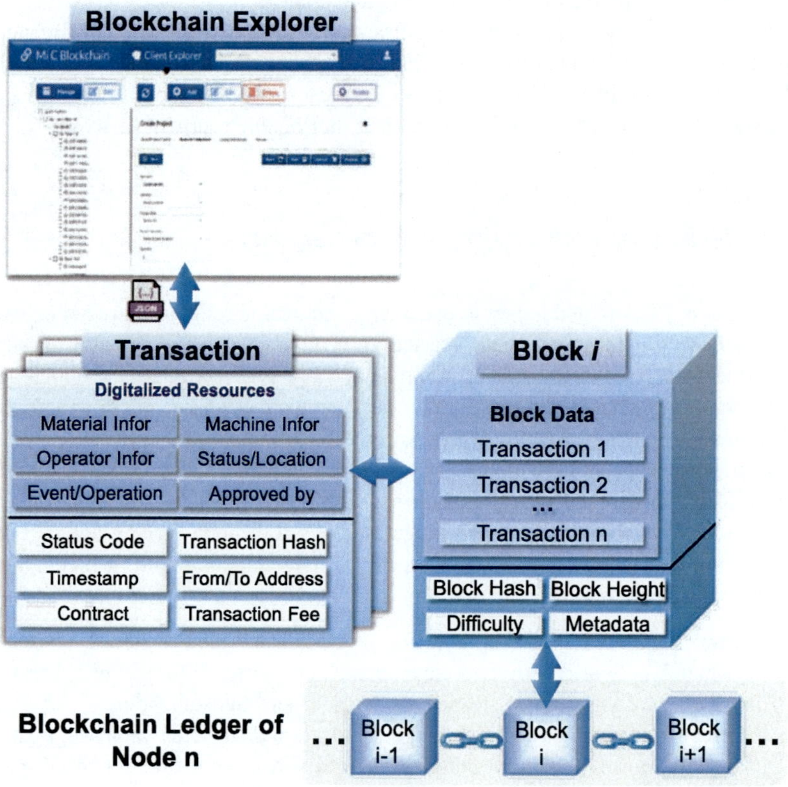

Fig. 7 Data flow through explorer/transaction/block/ledger under blockchain

be streamlined into three aspects, including data storage, computation, and function operation. Storage contract is used to store the data transactions. Computational contract is used to cope with the complex transaction transfer, evaluation, and execution. Functional contract is defined by the users to facilitate the workflow management and control, which is mostly applied in different project scenarios [36];

(4) **Tokens/Cryptocurrency**. Tokens are often used to represent assets such as digital currencies, commodities, or even access rights to a particular platform or service. They can be bought, sold, and traded on cryptocurrency exchanges, and their value can fluctuate based on market demand and other factors. Cryptocurrencies are digital assets that represents a form of currency in digital form. They use cryptographic techniques to ensure transaction security and control the creation and management of the currency. Unlike traditional fiat currencies issued and controlled by central authorities like governments or banks, cryptocurrencies operate independently of any centralized institution, which rely on distributed ledger technology to record and verify transactions. Both tokens and cryptocurrencies are important components of the blockchain ecosystem and are used to facilitate transactions, secure data, and enable new forms of economic activity.

As a novel information and communication technology in recent years, blockchain provides unified standards and protocols for information sharing with decentralized P2P framework, and guarantees the security and privacy based on the cryptographic mechanism, as shown in Fig. 8. Key data in projects are always owned and managed by large organizations and companies, and these large bodies are self-guarded of business interest, who cannot completely share their data or open their systems with others. But with blockchain network, all the transactions data can be simultaneously shared with all the stakeholders and each participant has the authority to manage the data. Due to the long-term lifecycle of building management, historical data can be traced through blockchain for sustainability and maintenance [37]. Each stakeholder can easily trace any previous transactions by accessing any node, which can also facilitate the root cause determination and accountability.

In addition, data could be absent in some situations due to the short-term relationships among participants. If an outsourcing company withdraws from the project or goes bankrupt, the data owned by the company is lost and cannot be inquired anymore. But blockchain can permanently and securely store all the data in projects. Most importantly, the data recorded cannot be tampered with, deleted, altered, concealed, or falsified, which strongly guarantees the data security.

Furthermore, the use of blockchain technology can help address the network security issues of open APIs for fire sensors. Some instructions on how to use blockchain to enhance the network security of fire sensor APIs are listed as follows:

1. **Data integrity**: Blockchain can ensure the integrity of sensor data by using Cryptography hash function and distributed consensus mechanism. Every time sensor data changes, it is added to the blockchain in the form of blocks, forming an immutable record. Therefore, if someone attempts to tamper with sensor data

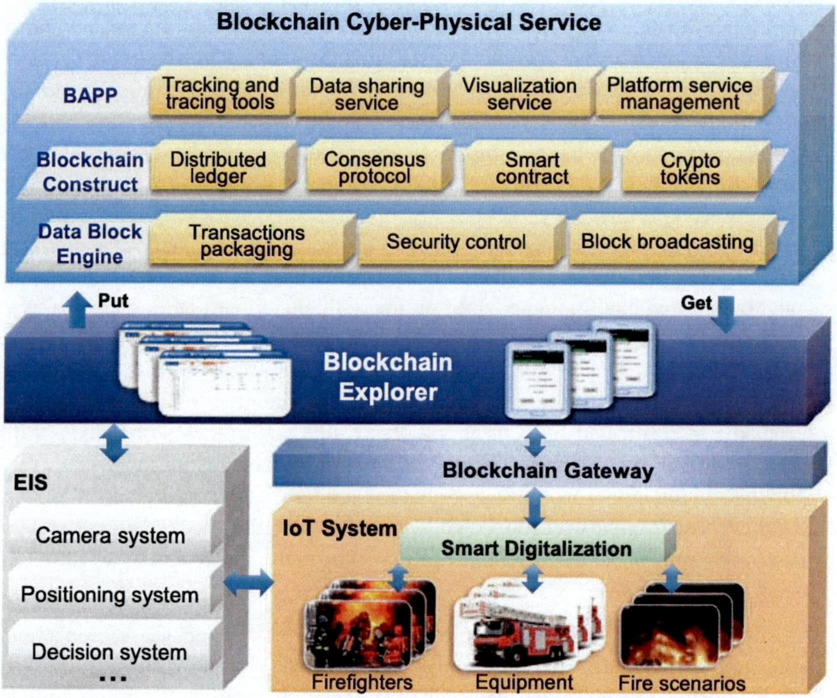

Fig. 8 A typical framework of blockchain-enabled cyber-physical system

or access unauthorized data, the consistency mechanism of blockchain will make this operation very hard to achieve.

2. **Authentication and Access Control**: Blockchain can provide authentication and access control mechanisms to ensure that only authorized entities can access sensor data. By registering and verifying the identities of devices, users, and other participants on the blockchain, a decentralized trustworthy framework can be established to ensure that only legitimate users can access sensor APIs.
3. **Decentralized architecture**: The decentralized nature of blockchain means that no single centralized control point exists, thereby reducing potential attack targets. Even if a node or device is attacked, other nodes can still continue to verify and record data, ensuring the security and reliability.

Before applying a specific blockchain network, we compare two mainstream blockchain platforms: Ethereum and Hyperledger Fabric. The aim of this comparison is to explore the advantages and disadvantages of two types of blockchain. Table 3 presents a comparative analysis between Ethereum and Hyperledger Fabric. Ethereum is a decentralized platform that runs smart contracts: application that runs exactly as programmed without any possibility of downtime, censorship, and fraud or third-party interference. Hyperledger Fabric is an open-sourced community to benefit an ecosystem that will work across a variety of industrial sectors. The following

Table 3 Comparison of Ethereum and Hyperledger fabric blockchain

No	Attribute	Hyperledger fabric	Ethereum
1	Token	None	Ether
2	State	Key-value database	Account data
3	Consensus	Pluggable: pBFT/Raft	Proof of work/Proof of authority
4	Network	Permissioned	Public
5	Transactions	Public or confidential	Anonymous or private
6	Privacy	Open to private	Open
7	Smart contract	Chaincode (Go, Java, Node.js)	Smart contract (Solidity)

seven parameters (token, state, consensus, network, transactions, privacy, and smart contract) are considered since they are typical attributes to describe a blockchain platform [38].

According to the comparison, it is suggested that Ethereum-based blockchain is chosen as the development platform for building fire security management due to the public network. Public network means that the stakeholder could flexibly registry without centralized authorities. This advantage matches the building fire prevention features, such as openness, sharing, etc. Many facilities can update their structure data for fire prevention to the platform with improved security and reliability. Admittedly, data security might be challenging under public network. Therefore, a combined solution of Ethereum blockchain and private IPFS cluster is designed and achieved to protect the data security through the access control and hash function [39].

3.3 IoT-Based Blockchain Development and Implementation

To develop the frontend of IoT-based blockchain in Fig. 9, Visual Studio Code is used as the Integrated Development Environment (IDE). React.js is used as frontend programming language to develop the interfaces of blockchain explorers. Web3.js provides the APIs for blockchain explorers to access the blockchain ledgers. The information revealed in blockchain explorers include text, number, time, location, user ID, checkbox, URL, image, and BIM & CAD file. Four blockchain-based applications are developed, including tracking and tracing tools, visualization service, data sharing service, and platform service management. The frontend interfaces of these applications are developed using react.js programing language. REST APIs, SDKs, and HTTP are used to bridge the interactions with blockchain backend.

The main users of tracking and tracing tools, data sharing service, and visualization service, are the stakeholders in projects. The main user of platform service management is the system administrator, who could also be the project owners/ representatives. For the development of visualization service, high-fidelity 3D models of machines, workers, materials, and building structure are firstly established using Revit, Maya, or Solidworks. Then these 3D models are imported into a game engine,

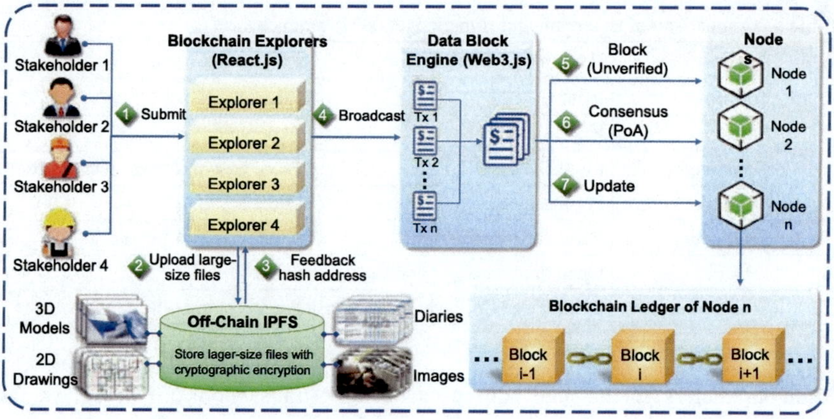

Fig. 9 Technical roadmap of IoT-based blockchain development and deployment

Unity 3D, to build the digital twin functions, such as free-moving perspective, multi-view transferring, and collision and gravity simulation [40]. The C# based construction digital twin is converted into JavaScript language through Web Graphics Library (WebGL) and uploaded into InterPlanetary File System (IPFS), which could be operated on the web-based platform. The JSON-based restored data in blockchain can be interoperated with the visualized models through Data Block Engine (DBE).

Two key components in blockchain backend are developed, including Data Block Engine (DBE) and InterPlanetary File System (IPFS). *Data Block Engine (DBE)*: DBE packs the collected transactions into a block and uploads the block to blockchain ledger. The collected transactions include order transactions, machine transactions, and material transactions. DBE serves as the middleware to connect, manage, and maintain the physical nodes with blockchain. Specifically, DBE is used to achieve the basic blockchain functions using Web3.js, such as transaction ordering and packaging, broadcasting, and consensus. Web3.js is a collection of libraries that allow the users to interact with the local/remote blockchain node using a HTTP connection. Web3.js is applied to achieve the interaction between blockchain explorers and blockchain backend through calling the predefined smart contracts.

InterPlanetary File System (IPFS): IPFS is a peer-to-peer file system used to store off-chain large-size files. Because the size of the construction design files and the semantic recording files in projects is large, which needs to be stored off-chain, such as BIM models, CAD drawings, UAV (Unmanned Aerial Vehicle) inspection images, and site diaries, etc. There are three steps to build the private IPFS cluster. The swarm key is firstly applied to achieve the local private building [41]. Secondly, the bootstrap IPFS node is used by client nodes to connect to the private IPFS network. Finally, it needs to start the private IPFS network and limit the node access by configuring the environment variable "LIBP2P_FORCE_PNET". Thus, private IPFS cluster is built to store the large-size files, while blockchain is used to store its permalinks/hash address with the timestamp, metadata, and version information. This combined

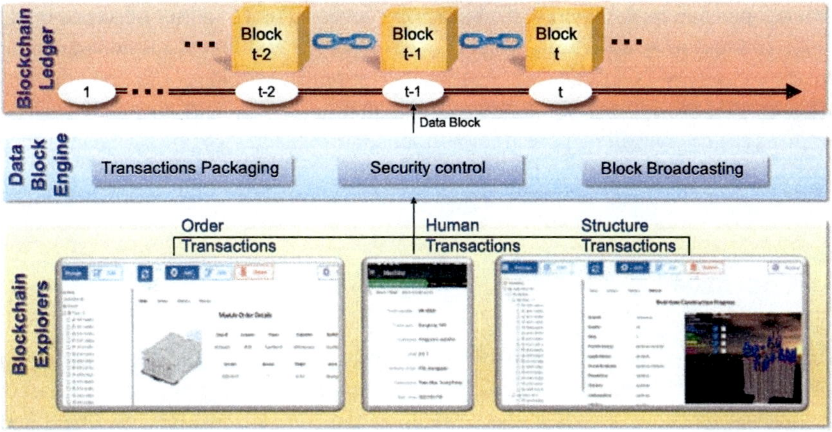

Fig. 10 Block generation process

solution not only saves vast blockchain storage space but also increases the platform performance.

The process of block generation has three stages, as shown in Fig. 10. The first stage is that blockchain explorers receive data from EISs. All these EISs contain the project information, such as tasks, workers, machines, and building structure and layout. Some critical information is transferred to blockchain explorers through pre-defined APIs. The second stage is that blockchain explorers update the critical information to DBE. Blockchain explorers could be a browser-based or mobile app-based interfaces. The third stage is that DBE packs the collected transactions into a block and uploads the block to the blockchain ledger. DBE is used to collect transactions from the blockchain explorers.

Three key components are included, namely transactions packaging, security control, and block broadcasting. Transactions packaging is used for packaging transactions in a certain time interval to form a block. Security control can guarantee data privacy and safety through the native cryptographic approaches. Block broadcasting is used to send the generated block to the blockchain network. After consensus, the uploaded block is stored in blockchain ledger. The connection between blockchain and EISs is achieved through blockchain explorers and DBE. It not only provides an easy way for users to upload information to blockchain network, but also facilitates the information sharing for stakeholders. In the Blockchain ledger, three types of information are stored, including planned program, adjusted program, and real progress. A planned program for a project is formulated by contractors and then uploaded to the blockchain ledger. A master program defines activities and their key dates, and outlines the scope of activities. In addition, some critical events are recorded and shown in a dashboard. Usually, the statuses of activities are recorded as well as the key dates. With the implementation of the system, an adjusted program is developed and represented with another line as the planned master becomes infeasible. The real progress of the project is also recorded and visualized in the project progress chart. With KPI tracking and visualization, stakeholders can track the task progress effectively, due to the shared information on the real progress, and the progress can be controlled more efficiently since an adjusted program can be informed timely.

4 Conclusion

This chapter aims to propose a paradigm of fire database that can support the training of the AI-based fire model and future smart firefighting practices. Different approaches to collecting data and generating the fire database are reviewed and demonstrated. The investigation reveals that the current fire data lacks the desired level of richness and detail when compared to the ideal fire data. By summarizing the existing fire data repository, we found the importance of fire video data have not received enough attention whereas they are a vital compontent of computer vision-based smart fire modelling. Moreover, the methods and guidelines of constructing a reliable database with experimental and numerical results for training AI-based fire model are introduced.

The interconnection of devices, networks, and systems brings great convenience and efficiency, but it also introduces cyber risks. The cyber vulnerabilities caused by the interconnectedness, and derivational threats are also discussed and prospected in Cybersecurity is crucial for maintaining trust in digital systems and safeguarding the integrity of data. In an increasingly interconnected world, where data is exchanged and stored across various platforms and devices, ensuring the trustworthiness and accuracy of information is vital. Cybersecurity measures protect against unauthorized access, data breaches, and malicious activities, allowing individuals, organizations, and societies to have confidence in the reliability and authenticity of digital interactions and information.

The proposed methodology to establish and guard fire database provides new directions for future fire data collection and management and lay a foundation to promote the application of AI and smart firefighting in fire protection engineering reliably and safely.

References

1. V. Babrauskas, Ignition Handbook, Fire Science Publishers/Society of Fire Protection Engineers (Issaquah, WA, 2003). https://doi.org/10.1023/B:FIRE.0000026981.83829.a5
2. B. Karlsson, J. Quintiere, *Enclosure Fire Dynamics*, 1st edn. (CRC Press, 1999). https://doi.org/10.1201/9781420050219

3. W.G. Weng, W.C. Fan, Critical condition of backdraft in compartment fires: a reduced-scale experimental study. J. Loss Prev. Process Ind. **16**, 19–26 (2003). https://doi.org/10.1016/S0950-4230(02)00088-8
4. X. Zhang, X. Wu, Y. Park, T. Zhang, X. Huang, F. Xiao, A. Usmani, Perspectives of big experimental database and artificial intelligence in tunnel fire research. Tunn. Undergr. Space Technol. **108**, 103691 (2021). https://doi.org/10.1016/j.tust.2020.103691
5. Y. Utiskul, J.G. Quintiere, A.S. Rangwala, B.A. Ringwelski, K. Wakatsuki, T. Naruse, Compartment fire phenomena under limited ventilation. Fire Saf. J. **40**, 367–390 (2005). https://doi.org/10.1016/j.firesaf.2005.02.002
6. X. Huang, X. Wu, A. Usmani, Perspectives of using artificial intelligence in building fire safety, in: M.Z. Naser (ed.), *Handbook of Cognitive and Autonomous Systems for Fire Resilient Infrastructures* (Springer, New York, 2022). https://doi.org/10.1007/978-3-030-98685-8_6
7. J.L. Hodges, B.Y. Lattimer, K.D. Luxbacher, Compartment fire predictions using transpose convolutional neural networks. Fire Saf. J. **108**, 102854 (2019). https://doi.org/10.1016/j.firesaf.2019.102854
8. M. Ghoreishi, A. Bagchi, M. Sultan, Review of the punching shear behavior of concrete flat slabs in ambient and elevated temperature. J. Struct. Fire Eng. (2013)
9. Y. Cao, F. Yang, Q. Tang, X. Lu, An attention enhanced bidirectional LSTM for early forest fire smoke recognition. IEEE Access. **7**, 154732–154742 (2019). https://doi.org/10.1109/ACCESS.2019.2946712
10. T. Zhang, Z. Wang, H.Y. Wong, W.C. Tam, X. Huang, F. Xiao, Real-time forecast of compartment fire and flashover based on deep learning. Fire Saf. J. **130**, 103579 (2022). https://doi.org/10.1016/j.firesaf.2022.103579
11. T. Zhang, Z. Wang, Y. Zeng, X. Wu, X. Huang, F. Xiao, Building artificial-intelligence digital fire (AID-Fire) system: a real-scale demonstration. J. Build. Eng. **62**, 105363 (2022). https://doi.org/10.1016/J.JOBE.2022.105363
12. Z. Wang, T. Zhang, X. Wu, X. Huang, Predicting transient building fire based on external smoke images and deep learning. J. Build. Eng. **47**, 103823 (2022). https://doi.org/10.1016/j.jobe.2021.103823
13. Z. Wang, T. Zhang, X. Huang, Predicting real-time fire heat release rate based on flame images and deep learning, in *Proceedings of the Combustion Institute* (2022)
14. N.K. Kim, K.M. Jeon, H.S. Kim, Convolutional recurrent neural network-based multiple microphones. Sensors **19** (2019)
15. M.Z. Naser, Fire resistance evaluation through artificial intelligence—A case for timber structures. Fire Saf. J. **105**, 1–18 (2019). https://doi.org/10.1016/j.firesaf.2019.02.002
16. S. Mahdevari, S.R. Torabi, Prediction of tunnel convergence using artificial neural networks. Tunn. Undergr. Space Technol. **28**, 218–228 (2012). https://doi.org/10.1016/j.tust.2011.11.002
17. X. Wu, Y. Park, A. Li, X. Huang, F. Xiao, A. Usmani, Smart detection of fire source in tunnel based on the numerical database and artificial intelligence. Fire Technol. **57**, 657–682 (2021). https://doi.org/10.1007/s10694-020-00985-z
18. X. Wu, X. Zhang, X. Huang, F. Xiao, A. Usmani, A real-time forecast of tunnel fire based on numerical database and artificial intelligence. Build. Simul. **15**, 511–524 (2022). https://doi.org/10.1007/s12273-021-0775-x
19. C. Grant, Integrating modern technologies to realize fire-resistant infrastructures, in *Handbook of Cognitive and Autonomous Systems for Fire Resilient Infrastructures* (Springer, 2022), pp. 61–92
20. M. Chi, A. Plaza, J.A. Benediktsson, Z. Sun, J. Shen, Y. Zhu, Big data for remote sensing: challenges and opportunities. Proc. IEEE **104**, 2207–2219 (2016). https://doi.org/10.1109/JPROC.2016.2598228
21. L. Pun, P. Zhao, X. Liu, A multiple regression approach for traffic flow estimation. IEEE Access **7**, 35998–36009 (2019)
22. R.S. Allison, J.M. Johnston, G. Craig, S. Jennings, Airborne optical and thermal remote sensing for wildfire detection and monitoring. Sensors (Switzerland) **16** (2016). https://doi.org/10.3390/s16081310

23. S.E. Caton, R.S.P. Hakes, D.J. Gorham, A. Zhou, M.J. Gollner, Review of pathways for building fire spread in the wildland urban interface part I: exposure conditions. Fire Technol. **53**, 429–473 (2017). https://doi.org/10.1007/s10694-016-0589-z
24. A. Cicione, M. Beshir, R.S. Walls, D. Rush, Full-scale informal settlement dwelling fire experiments and development of numerical models. Fire Technol. **56**, 639–672 (2020)
25. Y. Jiang, T. Zhang, S. Liu, Q. He, L. Li, X. Huang, Full-scale fire tests in the underwater tunnel section model with sidewall smoke extraction. Tunn. Undergr. Space Technol. **122**, 104374 (2022). https://doi.org/10.1016/j.tust.2022.104374
26. K.B. McGrattan, G.P. Forney, J. Floyd, S. Hostikka, K. Prasad, Fire Dynamics Simulator (version 4)—User's Guide. US Department of Commerce, Technology Administration, National Institute of … (2005)
27. X. Wu, X. Zhang, X. Huang, F. Xiao, A. Usmani, Xiqiang, Wu, Xiaoning, Zhang, Xinyan, Huang, Fu, Xiao, A. Usmani, A real-time forecast of tunnel fire based on numerical database and artificial intelligence. Building Simulation (2021). https://doi.org/10.1007/s12273-021-0775-x
28. T. Toulouse, L. Rossi, A. Campana, T. Celik, M.A. Akhloufi, Computer vision for wildfire research: An evolving image dataset for processing and analysis. Fire Saf. J. **92**, 188–194 (2017). https://doi.org/10.1016/j.firesaf.2017.06.012
29. J.P. Hidalgo, M.S. McLaggan, A.F. Osorio, M. Heitzmann, C. Maluk, D. Lange, J. Carrascal, J.L. Torero, Protocols for the material library of cladding materials–part I: framework (2019)
30. P. Maxim, D. Plecas, L. Garis, Report on the Feasibility of a Canadian National Fire Information Database, Centre for Public Safety and Criminal Justice Research, University of the … (2013)
31. T.E. Minnich, Thermal Properties Database (n.d.)
32. R.V. Rao, K. Selvamani, Data security challenges and its solutions in cloud computing. Procedia Comput. Sci. **48**, 204–209 (2015)
33. Y. Sun, J. Zhang, Y. Xiong, G. Zhu, Data security and privacy in cloud computing. Int. J. Distrib. Sens. Netw. **10**, 190903 (2014)
34. Y. Jiang, X. Liu, K. Kang, Z. Wang, R.Y. Zhong, G.Q. Huang, Blockchain-enabled cyber-physical smart modular integrated construction. Comput. Ind. **133**, 103553 (2021)
35. D. Mahmudnia, M. Arashpour, R. Yang, Blockchain in construction management: applications, advantages and limitations. Autom. Constr. **140**, 104379 (2022)
36. S. Ahmadisheykhsarmast, R. Sonmez, A smart contract system for security of payment of construction contracts. Autom. Constr. **120**, 103401 (2020)
37. K. Kang, X. Liu, Y. Jiang, K. Lee Kam Hang, S.K.W. Wan, G.Q. Huang, R.Y. Zhong, Blockchain opportunities for construction industry in Hong Kong: a case study of RISC and site diary. Constr Innov. **23**, 443–466 (2023)
38. X. Liu, Y. Jiang, Z. Wang, R.Y. Zhong, H.H. Cheung, G.Q. Huang, imseStudio: blockchain-enabled secure digital twin platform for service manufacturing. Int. J. Prod. Res. 1–20 (2021)
39. X. Liu, Y. Yang, Y. Jiang, Y. Fu, R.Y. Zhong, M. Li, G.Q. Huang, Data-driven ESG assessment for blockchain services: a comparative study in textiles and apparel industry. Resour. Conserv. Recycl. **190**, 106837 (2023)
40. Y. Jiang, M. Li, D. Guo, W. Wu, R.Y. Zhong, G.Q. Huang, Digital twin-enabled smart modular integrated construction system for on-site assembly. Comput. Ind. **136**, 103594 (2022)
41. Y. Jiang, X. Liu, Z. Wang, M. Li, R.Y. Zhong, G.Q. Huang, Blockchain-enabled digital twin collaboration platform for fit-out operations in modular integrated construction. Autom. Constr. **148**, 104747 (2023)

Sensor and Modelling Driven Real-Time Fire Forecast

Wolfram Jahn

1 Introduction and Motivation

Firefighters often face uncertain conditions when entering a building to attack a fire, and have to make decisions based solely on their experience. They would greatly benefit from an improved and more detailed knowledge of the overall state of the scenario they are about to encounter. Ideally, they could even use short term forecasts of fire development in order to adjust their course of action accordingly.

1.1 Characterising Fire Events

Modern buildings commonly feature sensors to some extent, e.g. for the use in building management. The measurements obtained from these sensors include temperature readings (e.g. heat detectors or temperature sensors in smoke exhaust ducts), particle density or obscuration (from continuously monitoring smoke detectors), CCTV camera recording, etc. Additionally, sensors in modern buildings may inform about opening of windows and doors, and air supply or exhaust through HVAC and smoke control systems. All of these sensors could provide useful information to the emergency response services, thus introducing a paradigm shift in the way fires are attacked.

W. Jahn (✉)
Department of Mechanical and Metallurgical Engineering, Pontificia Universidad Católica de Chile, Santiago, Chile
e-mail: wjahn@uc.cl

Centro Nacional de Excelencia Para la Industria de la Madera (CENAMAD), Pontificia Universidad Católica de Chile, Santiago, Chile

X. Huang and W. C. Tam (eds.), *Intelligent Building Fire Safety and Smart Firefighting*, Digital Innovations in Architecture, Engineering and Construction,
https://doi.org/10.1007/978-3-031-48161-1_12

The large amount of sensor readings, however, would rapidly lead to information overload if the data were not adequately synthesised. Additionally, due to operational reasons, sensors can usually only be installed at compartment boundaries (walls and ceilings), so that spatially resolved information of the interior necessarily will result from extrapolation of sensor data. The highly non-linear nature of fire dynamics requires the use of computational tools to protect sensor data from the compartment boundaries into the compartment interiors. Such tools usually correspond to sophisticated computer models, such as computational fluid dynamics (CFD) solvers combined with pertinent complementary models (combustion, radiation, pyrolysis models etc.).

Synthesised information about the fire and the fire development could eventually be available at different levels of insight. The lowest level of insight into a fire scenario is the knowledge of its existence. Since fire detection necessarily precedes any fire service intervention, detection itself does not really form part of firefighting. The next level of insight corresponds to the state of the fire scenario at the time of arrival of the fire services. This could include simple information on the number of compartments involved in the fire, or average smoke and fire spread velocity based on consecutive detector readings. It could, in principle, also be a complete characterisation of the fire scenario in terms of spatially resolved room temperatures or probability of occurrence of extreme events such as flashover or window breakage (using CFD simulation tools as indicated above). The ultimate level of insight into a fire scenario would feature the possibility of forecasting the fire development under different intervention strategies. This would give the Incident Commanders (IC) the possibility of deciding an effective and efficient intervention strategy, tailored to the specific fire scenario.

Characterising the fire event consists of integrating and complementing the sparse data with a mathematical model of the system, so that the complete state of the system can be reconstructed, and a useful subset can then be extracted and presented in a comprehensible manner to the emergency response services. The integration of data and mathematical model is done by estimating a set of previously unknown parameters, that, together with the governing equations, adequately describe the system. These parameters include the initial state of the system, parameters related to the driving forces that move the system, thermo-physical properties of the materials, and/or parameters regarding the geometric layout of the system that is analysed (ventilation conditions, compartmentation etc.).

Several studies have been conducted to estimate fire characteristics (parameters of fire models), feeding fire models with data from measurements taken during the fire. Early studies on this topic use zero-dimensional, two-zone fire models to prove the concept of data driven fire modelling [1–3]. Jahn et al. expand the methodology to the use of CFD type fire models, estimating the spread rate and mass loss rate of the fire [4]. Overholt and Ezekoye estimate a probability distribution for the fire parameters in order to account for uncertainties in measurements and fire model [5]. Beji et al. separate the assimilation process from the forecast, and use simple correlations to estimate the size of the fire [6]. Most of these studies use temperature measurements at the boundaries or within the fire compartment, or visual estimations of the smoke layer height, and infer the parameters using some sort of inverse modelling mechanism. Jahn et al. recognise the fact that most existing buildings do not provide dense temperature measurements, and explore the use of commonly present detection devices for parameter estimation [7]. Generally only the (constant) fire size is estimated. The estimation of other parameters, such as fire location, is addressed by Overholt and Ezekoye [5], but not simultaneously with the fire size. A later study by Chu et al. addresses this issue, and provides multidimensional fire parameter estimation using Bayesian inference together with temperature measurements at ceiling level [8]. Jahn et al. also attempt simultaneous estimation of fire size and location, and conclude that the exact fire location is not of utmost importance [7]. A crude estimate based on the location of the earliest device to detect the fire is enough for a satisfactory estimation of the growth rate of the fire. Given the high complexity of fire dynamics and the multiple physical scales involved, real-time data synthesis as described above is still beyond the current computational capabilities, even for simplified fire scenarios [9]. The parameter estimation requires several iterations, with a full CFD simulation to be performed in each iteration, since the process is based on a numerical optimisation that sequentially updates informed guesses on the parameter composition (for details of the iteration process refer to [7] or [4]).

1.2 Forecasting Fire Events

The ultimate goal in terms of assisting the fire services would be to not only present the measured data in a manageable format, but to produce short- or medium-term forecasts of future fire development based on the current state. Ideally this forecast could respond to different firefighting strategies (e.g. opening certain doors to enter a compartment), thus allowing for an informed decision-making process [10].

Forecasting a physical phenomenon (such as fire) relies on the existence of a mathematical model with a well characterised set of parameters that properly represents the underlying physical processes. For complex phenomena, many parameters are normally previously unknown, and will have to be estimated from measurements of some kind. In the case of fire scenarios, where measurements vary in time and space, the model is used to estimate model parameters from measured data by some inverse modelling approach (as in the case of fire characterisation). Once all parameters are characterised the model can be used to extrapolate the current state of the system into the (near) future. Blind predictions, i.e. predictions of the future state from first principles, without any input from local measurements of current conditions, have shown not to be possible (see e.g. [11]). Too many uncertainties, some inherent to the model, others related to the nature of fire scenarios (e.g. unknown arrangement of fuel packages), arise. While the integration of fire models and data from a fire scenario had been proposed before (see e.g. [12]), Cowlard et al. [10] recognize the inherent need for combining both data and models in order to enable reliable forecasts of fire development.

Forecasting of fire development is conceptually very similar to numerical weather prediction (NWP) [13]. However, there is a subtle yet fundamental difference in the nature of the forecast exercise of NWP compared to fire development predictions. Atmospheric flows are determined mainly by the past state of the system and the (known) energy input of the sun. NWP is thus essentially a problem of initialization. To estimate an evenly distributed initial state of the system, numerical weather models are used to interpolate physical quantities (such as temperature, pressure and humidity) between measurement locations (which are unevenly distributed around the globe). The process of fitting weather models to measured data is called data assimilation and is used to produce an initial condition for the set of differential equations that forecast the state of the system. The future state of the system is determined completely by its initial state, and can be forecast as far into the future as the theoretical predictability limit allows [14].

Fire development prediction posts a somewhat different challenge. Virtually all operational fire models are actually "fire consequence" models, in the sense that they reproduce smoke movement and temperature distributions for a given fire size and/or growth rate (or time curve of fire size). Due to the high complexity of fire spread and the divergent spatial scales associated with its modelling, it is not currently possible to simulate fire growth from first principles within full scale fire simulations [15]. Other phenomena, such as ignition of secondary items are equally challenging [16]. The fire size (i.e., the actual fire) is thus not an output of the model, but an input in the form of a (time changing) boundary condition to the fire consequence model.

This difference, while subtle in nature, is very important to consider in future efforts of fire forecasting, as it has a direct effect on the forecast horizon. For actual predictions of the future state of a fire system, it is of utmost importance to focus on predicting the behaviour of the fire source (rather than the consequence only). The fundamental aspects of fire source behaviour, such as fire spread over a solid surface, are well beyond the current capabilities of even the most powerful computers, or would take a prohibitive amount of computational time to simulate a few moments of the phenomenon (e.g. [15, 17]). Other features, such as flashover, are highly sensitive to unknown parameters, and despite some attempts in the past (e.g. [18, 19]), it has not been possible to agree on a universally validated numerical approach to simulate these phenomena. To complicate matters further in terms of simulating extreme events, changing environmental conditions (window breaking, door opening, etc.), including those triggered by the fire services, will significantly influence the course of the fire, and make predictions all the more challenging.

Forecasts of any event are only useful if they can be delivered in super-real-time, i.e. before the event takes place (or at least before it finishes). The iterative nature of the data-model integration, together with the extended run time of CFD type fire models result in a large computational effort that cannot be completed in real-time (let alone super-real-time) with current technology. The alternative of using a two-zone fire model to complement the data is not suitable, since spatial resolution is required. A third possible approach is to replace the full Navier–Stokes formulation by a faster running reduced order model (ROM) that takes advantage of the often localised nature of smoke transport in a fire [20]. Regions of the computational domain away from the fire or below the smoke layer are only marginally affected, and present only minor disturbances that could be ignored in the numerical solution.

2 Fire Dynamics and Their Modelling

The reason for the rather pessimistic statement of the previous paragraph is illustrated in the following. Firstly, an overview of fire dynamics is given, and then the different modelling approaches are briefly introduced.

2.1 Complexity of Fire Dynamics

Fire results from the complex, non-linear interaction of several physical processes; for flaming combustion to take place, (gaseous) fuel and oxidiser need to coincide under certain conditions of concentration and temperature. The heat released from the flame radiates onto the surface of solid objects in the vicinity, inducing thermal decomposition (pyrolysis) as the temperature of the objects increases. The resulting combustible vapours that emanate from the objects mix with the surrounding air and feed the flame, thus closing the feedback loop. In the early stages of a fire the pyrolysis occurs close to the flaming region, and as unburned material undergoes pyrolysis, the fire area grows (i.e. the fire spreads). The heat release rate (HRR) during that stage is controlled by the availability of gaseous fuel. In later stages of the fire radiative heat fluxes from walls and accumulated smoke can cause pyrolysis in objects further away, resulting in secondary ignition. Eventually, the oxygen available in the compartment is not enough to burn all the combustible vapours, and the supply of oxidiser into the compartment will become the limiting factor for the HRR.

Figure 1 shows a schematic representation of a typical compartment fire. During the early stages of the fire, the mechanisms that control the fire spread are the mixing of air and combustible vapours on one hand, and the thermo-chemical degradation of the fuel on the other [21]. The modelling of fire spread from first principles would thus requires the simultaneous solution of equations of fluid flow (for the mixing of fuel and oxidiser), heat transfer within the flame and from the flame onto the solid, and heat conduction with chemical decomposition in the solid phase. There are considerable challenges in the numerical modelling of all of these processes [22]: The relatively high flow velocities result in turbulent flow, and its solution requires an adequate turbulence model with corresponding boundary conditions. If the mixing process is not captured correctly, the flame structure and geometry is affected. Since radiative heat transfer depends on the geometry of the radiating object (in this case the flame surface), this would alter the incident radiative heat flux on the receiving objects. The heat transfer within the gas phase is strongly coupled to fluid flow, and its main result is the flame sheet temperature. Due to the large temperature gradients in the reaction zone, a fine numerical mesh (grid size comparable to flame thickness)

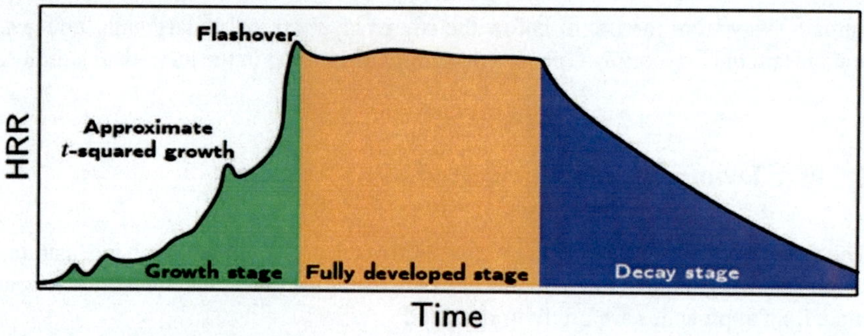

Fig. 1 Typical compartment fire development

is necessary in order to correctly reproduce the flame temperature. Since the radiative heat transfer from the flame onto the solid fuel is proportional to the fourth power of temperature, small errors can result in relatively large errors in the incident radiative heat flux. The main challenges of the modelling of the solid phase lie in the chemical decomposition of the material, which will result in a change of the thermal properties as well as the geometry of the solid.

The positive feedback loop that maintains the burning process of a solid fuel results in the propagation of modelling errors when coupling the numerical solutions. An under-estimation of flame height and/or temperature, for example, results in an under-estimation of the incident heat flux. The incident heat flux in turn is the boundary condition for the pyrolysis model, and the mass flow of combustible gases computed by this model is thus affected by the gas phase model output. As the combustible gases emanating from the solid feed the flame, its structure and temperature are again conditioned by the solid phase model output.

In the later stages of a fire (the fully developed stage shown in Fig. 1) the ventilation rate into the compartment controls the burning process. During that stage the gaseous fuel production due to pyrolysis in solid fuel objects is of less importance (as sufficient fuel is available), and the interaction between solid and gas phase reduces to the estimation of heat losses through compartment boundaries. This conjugated heat transfer can thus be represented by simplified models, making the fluid dynamics somewhat more manageable. But unlike in the early stage, where the combustion is controlled by the availability of fuel and the mixing of fuel vapours and air, the chemical kinetics start playing an important role during the ventilation controlled stage of a fire. The chemical composition of combustion products (i.e. smoke), and thus the concentration of toxic species and soot, as well as oxygen within the compartment is determined by kinetics [23]. Flames are much less localised during the advanced stages of a fire, and since chemical kinetics are strongly dependant on temperature (the reaction rate follows Arrhenius' law, which postulates an exponential dependence on temperature), the numerical grid would require high resolution in large parts of the fire compartment in order to include all flaming regions. The transition from the earlier fuel controlled stage to the posterior ventilation controlled stage is not smooth, and is associated with thermal instabilities and fundamental changes in the flow and combustion dynamics.

These preceding paragraphs illustrate the complexity of fire dynamics, and the significant challenges it posts to any modelling effort. It has been empirically shown that the state-of-the-art of fire modelling cannot currently reproduce phenomena like fire spread and fire growth in a satisfactory manner [11, 24]. In practice, fire dynamics models are thus used to evaluate quantities such as smoke movement and temperature distributions within a compartment, given a certain time curve of heat release. In other words, they are used as fire-consequence models, rather than fire-growth prediction models. This is of paramount importance in the context of the fire fighting effort, since it limits the usefulness of fire modelling tools for the Emergency Response Services when attacking a fire.

2.2 *Numerical Modelling of Fire Dynamics*

There are essentially two types of computational fire models available, one at each extreme of the complexity spectrum. While so-called *zone fire models* run extremely fast (within a few seconds), they only yield very limited output, and are thus not well suited for comparison to measured data (and consequently for parameter estimation). *Fire field models* based on CFD, on the other hand, report detailed output, but require much longer run times (several minutes or even hours). All fire models need additional sub-models, through which combustion characteristics, heat transfer to the compartment boundaries etc., are evaluated.

2.2.1 Zone Models

Zone type fire models are based on the observation that fire compartments form two very distinct layers, an upper smoke layer and a lower layer of air, as a consequence of the elevated temperature of the smoke. A schematic illustration of this is shown in Fig. 2. Simple mass and energy balance equations can be cast to model the thermal equilibrium between the hot upper layer and the cool lower layer. Since there is no explicit spatial resolution, the resulting model is zero-dimensional. While some details can be added through additional sub-models (e.g. heat transfer to walls, flow through openings or ceiling jet considerations), this approach essentially results in a set of ordinary differential equations that can be efficiently solved in a traditional manner using simple numerical techniques. The following set of equations, for example, results from mass conservation of the upper and lower layer, that-together with a energy balance in the fire plume-can be solved to obtain the upper layer height and temperature [2].

$$(P_{\text{zone}})\begin{cases} \rho_u \frac{dV_u}{dt} + V_u \frac{d\rho_u}{dt} = \dot{m}_p \\ \rho_a \frac{dV_l}{dt} + \dot{m}_e + \dot{m}_p = 0, \end{cases} \tag{1}$$

where ρ_u and V_u represent the upper layer density and volume, respectively. The lower layer volume is denoted by V_l, while ρ_a represents the lower layer density (in this case the lower layer is assumed to be air at ambient temperature). The mass fluxes $\dot{m}_e$ and $\dot{m}_p$ correspond to air flowing into the compartment, and the plume, respectively. More sophisticated zone models (e.g. CFAST [25]) include flow through openings (windows, doors or cracks), heat losses to compartment boundaries. The additional features are mainly based on semi-empirical correlations, and do not add significantly to run-time. The main output of *zone models* is the hot layer height together with the average hot layer temperature as a function of time, although more complete models can include lower layer temperature and species concentrations in the upper and lower layers. Note that the main input to a zone model is the evolution of the HRR of the fire (which essentially determines the plume mass flux $\dot{m}_p$).

Due to the averaged characteristic of the temperature output of zone models, it cannot be directly compared to actual measurements, since these can only be obtained at compartment boundaries (even if the location within the hot layer where the temperature corresponds to the average temperature were known, it would not be practical to have a thermocouple installed in the middle of the room).

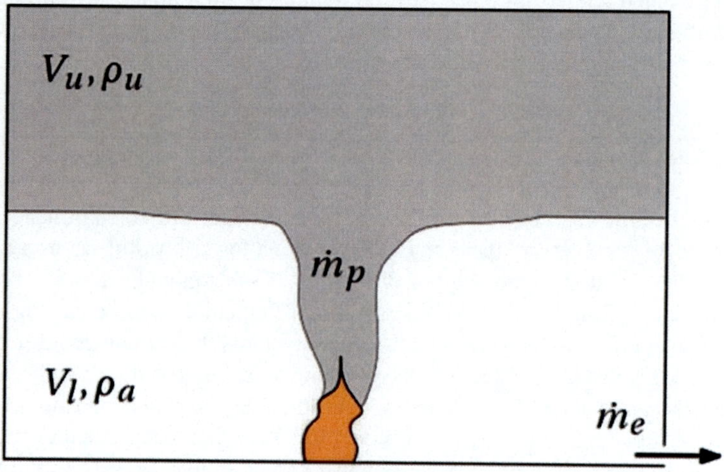

Fig. 2 Two layer configuration in a compartment fire

2.2.2 Fire Field Models

Fire field models yield a spatial distribution of outputs of interest, such as temperature, flow velocities, species concentration etc. They are based on computational fluid dynamics, complemented with fire specific submodels (such as combustion, as detailed in the following sub-section). Due to the flow characteristics, normally a low-Mach number solver for buoyancy driven flow is used [26]. As a result of the inherent complexity of the underlying physics, field models typically take large amounts of time to run, and require detailed specification of boundary conditions.

Computational fluid dynamics consist of numerically solving the conservation equations for mass (continuity), species (mixture fraction), momentum (Navier–Stokes equations) and energy. The full system of equations that govern fluid flow is as follow:

$$(P_{\text{CFD}})\begin{cases} Find\, u,\ v,\ w,\ \rho,\ p,\ f \text{ and } T : [t_0, t) \times \Omega \subset \mathbb{R}^3 \to \mathbb{R}^6, \text{ such that:} \\ \frac{\partial \rho}{\partial t} + \nabla \cdot (\rho \boldsymbol{u}) = 0 \\ \rho \frac{D\boldsymbol{u}}{Dt} = -\nabla p + \nabla \cdot (\mu \nabla \boldsymbol{u}) + \bar{\mathbf{S}}_M \\ \rho c_v \frac{Df}{Dt} = \nabla \cdot \mathcal{D} \nabla f \\ \rho c_v \frac{DT}{Dt} = -p\Delta \cdot \boldsymbol{u} + \nabla \cdot k \nabla T + \Phi + S_{Ei} \\ p = \rho R T \\ B.C. \quad \text{on } \partial\Omega \\ I.C \quad \text{in } t = t_0 \end{cases} \tag{2}$$

For the implementation of the combustion and turbulence models, additional conservation equations for species and turbulent energy must be solved. Computational fluid dynamics solvers apply some discretisation scheme to convert these partial differential equations into a set of algebraic equations that can be solved numerically.

2.2.3 Simulation of Reactive Flow

Simulating reactive flow, such as flames in a fire, poses a special challenge for computational fluid dynamics. For numerical reasons, flow solvers are normally categorised into compressible solvers (for high flow velocities) and incompressible solvers (for flows with a low Mach number). Attempting to allow for compressible effects in low Mach number flows produces stability problems in the numerical implementation, since information within the flow can travel at two different velocities that are orders of magnitude different, the velocity of sound and the velocity of the flow. Incompressible solvers thus essentially assume that the density of the gas remains constant in space and time (for a more detailed account the reader is referred to [27]). While this solves the stability problems, it poses a new problem for buoyancy driven flows, since they are moved precisely by changes in density. To overcome this problem, the Boussinesq approximation is often invoked, where a temperature dependent density is introduced in the gravity term of the Navier–Stokes equation, maintaining the constant density in the inertial term. This allows for the simulation of a buoyancy driven flow with incompressible characteristics. The Boussinesq approximation works well for small temperature variations, but struggles when larger temperature gradients are present (as is typically the case in reactive flow). As a consequence, flow simulation of combustion processes requires special numerical treatments with considerations for the aforementioned challenges.

In addition to the intrinsic complications of simulating low Mach number flow with large temperature gradients that impulse the gases, the simulation of reactive flow requires a series of auxiliary models that have to be implemented in parallel to the main flow solver. Of special importance is the combustion model, i.e. the way the chemical reaction is modelled and how it relates to the flow solver.

The energy that moves a buoyancy driven flow ultimately stems from the combustion reaction of volatile fuel with air. The simplest approach to model the release of this energy is to ignore the combustion process altogether, and inject hot air into the computational domain. The flow of hot air is dimensioned in a way such that a given HRR is reached. Since plume temperatures are relatively well known, the flow of hot air can finally be expressed as

$$\dot{m}_a = \frac{HRR}{c_p(T_{plume} - T_{amb})} \tag{3}$$

The fire diameter, through which $\dot{m}_a$ is injected, is calculated in a way such that the overall flow field in the computational domain resembles that of a fire driven flow (using Froude scaling). This approach decouples the fire development from the fire consequences (flow field, temperatures, species concentrations, heat feedback etc.), and is thus not suitable for predictive simulations of fire dynamics.

A more sophisticated approach is to allow for localised heat release of the reacting gases, whenever they meet under favourable conditions. For this an infinite rate combustion is typically assumed (e.g. [26]). Under Burke-Schuhmann's assumption of stoichiometric combustion, this model only requires the additional solving of a single conservation equation for a quantity called the mixture fraction, and thus results in a relatively minor computational cost. This infinite rate combustion model produces satisfactory results as long as enough oxygen is available to allow for complete combustion.

If more advanced features of combustion, such as flame extinction in vitiated conditions, are to be captured, the infinite rate combustion model is not suitable, and a finite rate approach must be implemented. This necessarily requires the solving of additional conservation equations, one for each species to be tracked. Considering even simple combustion mechanisms, this quickly becomes unfeasible for practical applications (especially in the context of real time forecasting of fire dynamics).

3 Inverse Modelling and Fire Forecast

The parameter estimation procedure, that allows for fire characterisation (which can then be used as an parting point for a fire forecast) consists of assuming an initial set of parameters in order to perform the computational simulations. The output of the simulations is compared to the sensor data at measurement locations, and a measure of error is defined in order to determine the quality of the set of parameters. Some optimisation procedure is then performed that seeks for a better set of parameters. This is repeated iteratively until the simulation output matches the recorded data to an acceptable degree. Once the parameters are estimated, the current state of the fire development is fully spatially characterised, and synthesised information for the emergency response services can be prepared. A fully characterised fire environment could then be used to highlight conditions leading to critical events in the development of the fire. This would enable the Fire Services to take mitigating measures and to attack the fire in a more informed manner. This fully characterised fire environment could then also be used as initial condition for a forecast of future fire development.

The forecasting of fire development is thus a sequential process, where governing parameters of a fire model are estimated from sensor measurements by inversion of a computational fire model, as explained in the previous paragraph. Those parameters are assumed to be constant until a critical event (window breakage, flashover, door opening, etc.) occurs, and it is in principle possible to make a forecast based on these invariant parameters, using the computational model in forward mode. Once a critical event has occurred, the parameters may no longer be valid, and a new set of parameters needs to be estimated. This process is schematically illustrated in Fig. 3.

The underlying premise of this approach is that in order for a forecast to be possible, the system must be completely characterised (at least for a limited time) by

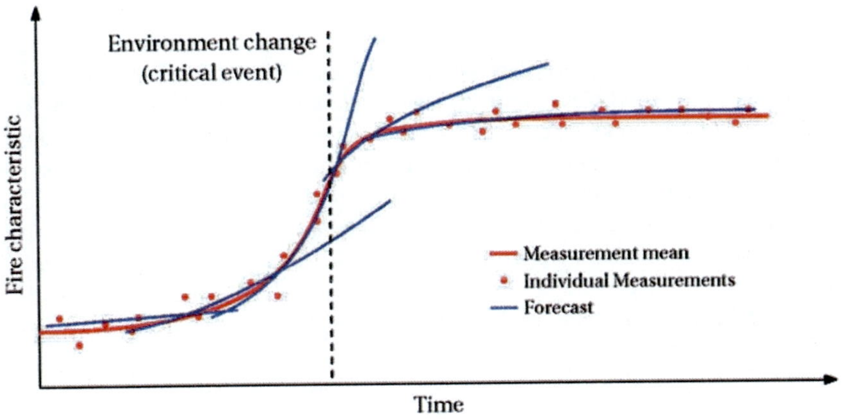

Fig. 3 Schematic illustration of a fire forecast process (after [2]). As more data come in, simulation parameters are adjusted until a forecast can be made. After the occurrence of a critical event, the cycle must start over

the combination of model and model parameters. In other words, the estimated parameters must converge to their physical value. Otherwise it cannot be guaranteed that the forecast will be representative of the true development of events. Koo et al. proposed a Monte-Carlo based approach to steer fire simulation and forecast fire development, launching a large number of zone model simulations with different input parameters [1]. Simulation output is continuously compared to measured temperature data, and the parameter space of the model is adjusted based on a Bayesian inference procedure until a good match between measurements and simulation output is achieved. While the convergence of parameters is monitored, it is not verified whether the converged values correspond to their physical counterpart. This disconnection between the model and the physics that govern the fire development somehow limits the forecast capabilities of the approach (especially given the low resolution of the forecast model). A similar methodology, based on Ensemble Kalman Filtering was presented by Lin and Wang [28]. They also used a zone model for parameter estimation but ensured convergence of the estimated parameters towards their true values. Ji et al. follow a similar setup (Ensemble Kalman Filtering and two zone model), and claim achieving real-time forecasts [29]. In both studies (Lin and Wang, and Ji et al.) only the constant HRR is estimated, and it remains open whether a more realistic scenario could also be forecast in real-time. The use of simplified fire models (such as zone models) for forecasting the fire development not only limits the level of detail that the forecast delivers, but also limits the parameter estimation process itself.

3.1 Decoupling of Fire Dynamics

Section 2 clearly shows that simulating fire growth and fire development from first principles, including all aspects and processes, is not feasible with current computational capacities, let alone in super-real-time. One of the most challenging aspects is the interaction between the gas phase and the solid, which is paramount in the effort of simulating fire growth, and which includes heat exchange between phases, volatile production in the solid phase, charring and heterogeneous combustion at the char surface. Beyond the inherent difficulty of modelling all these processes numerically, there is an additional, and perhaps more fundamental complication, corresponding to the strong coupling between processes in both phases. The resulting heat and mass fluxes in each phase are the boundary conditions to the other phase. In order to solve the gas phase, it is necessary to have solved the solid phase, which, in turn, depends on the results of the gas phase. This necessarily leads to an iterative process that is prohibitively time consuming, and thus not suitable for super-real-time simulations. It is important to recall that the data integration step is also iterative, and requires the solving of both gas and solid phase in each iteration. Data integration with a fully gas-sold coupled CFD based fire model would thus require a nested iterative process of computationally highly intensive simulations.

The solution that has been proposed to this problem is to decouple the gas phase processes from the solid phase processes and replace the latter with a simplified model with a fixed set of parameters that can be easily estimated. The simplest approach that has been proposed in this sense is to replace the fire source (the solid phase processes) with a constant HRR (e.g. [3]). The value of the HRR (the single parameter that characterises the fire source in this case) is estimated from data measurements using a fire model to transport the information from the source to the sensor location. But more complicated parametrisation can, of course, be used. It has been shown, for instance, that the spread rate of a fire is, at least under certain circumstances, relatively constant. A constant spread rate translates into a fire area growing as time squared (which is the origin of the t-square fire so extensively used in fire safety engineering). The fire source could thus be parametrised according to the following relation,

$$HRR = \pi r^2 t^2 \dot{m}_b'' \Delta h_c, \tag{4}$$

where r denotes the (constant) spread rate, t the time (so that $\pi r^2 t^2$ corresponds to the fire area), $\dot{m}_b''$ is the burning rate and Δh_c is the heat of combustion of the fuel. This spread model consists of three parameters that need estimating based on data at the compartment boundaries (such as temperatures, heat fluxes, etc.). In general, decoupling of gas and solid phase in fire modelling, requires an enhanced understanding of the underlying physical processes, so that suitable fire growth and flame spread models can be developed and integrated into predictive fire models.

3.2 *Inverse Modelling Approaches*

The process of parameter estimation from measured data using a mathematical model that represents the physical process is usually called inverse modelling. While different inverse modelling techniques exist, the general idea is common to them all. The model output, generated with an assumed set of parameters, is compared to measurements, and a cost function based on some measure of distance between data and model output is evaluated. A new set of parameters is then selected, with the objective of decreasing the cost function. This is repeated until the cost function is sufficiently small by some criterion. The difference between techniques arise in the way the updated set of parameters is selected. Two groups of techniques can generally be identified: gradient based and gradient free minimisation methods.

3.2.1 Gradient Based Methods

In gradient based methods, not only the cost function (the "distance" between model output and data) is evaluated, but also the rate of change in the cost function with respect to a change in the set of parameters, i.e. the gradient of the cost function (the partial derivatives with respect to each parameter). This provides information as to how the set of parameters should be adjusted. The cost function may be more sensitive to changes in certain parameters and less to changes in others. With this information a systematic search of the parameter space can be conducted (generally using some algorithm similar to or based on the Newton–Raphson method [30]). A typical cost function is as follows,

$$J(\theta) = \sum_i \left(y_i - \hat{y}_i(\theta)\right)^T W_i \left(y_i - \hat{y}_i(\theta)\right) \tag{5}$$

where θ is the set of parameters, y_i are the variables that are measured, and $\hat{y}_i(\theta)$ is the model output for a certain set of parameters. The subindex i represents the time instances of comparison, and W_i is a weight matrix. The gradient of the cost function is as follows,

$$\nabla J(\theta) = -2\sum_i \nabla\hat{y}_i(\theta)^T W_i \left(y_i - \hat{y}_i(\theta)\right). \tag{6}$$

It is important to note that the evaluation of the cost function involves the running of the numerical fire model as shown in Eq. (6). In order to evaluate the gradient of the cost function, which represents the change of the cost function with respect to each of the parameters, the numerical fire model needs to be differentiated with respect to its input parameters, i.e. $\nabla\hat{y}(\theta)$ is needed.

The differentiation of the numerical fire model can be obtained in several, yet related, ways. The simplest approach is a finite difference approach. Each parameter in θ is perturbed individually, maintaining all others the same, and the model is run with this modified set of parameters (one run for each perturbed parameter), as shown in Eq. (7),

$$\nabla\hat{y} = \begin{pmatrix} \frac{\partial\hat{y}}{\partial\theta_1} \\ \vdots \\ \frac{\partial\hat{y}}{\partial\theta_i} \\ \vdots \\ \frac{\partial\hat{y}}{\partial\theta_N} \end{pmatrix} = \begin{pmatrix} \frac{\hat{y}(\theta+\epsilon_1)-\hat{y}(\theta)}{\epsilon_1} \\ \vdots \\ \frac{\hat{y}(\theta+\epsilon_i)-\hat{y}(\theta)}{\epsilon_i} \\ \vdots \\ \frac{\hat{y}(\theta+\epsilon_N)-\hat{y}(\theta)}{\epsilon_N} \end{pmatrix}. \tag{7}$$

To obtain this, as many runs of the numerical fire models have to be performed, as parameters are to be estimated. This is computationally expensive, although it is relatively straightforward to parallelise the gradient estimation, by running the numerical fire model simultaneously with different, slightly perturbed, sets of input parameters.

An alternative approach is to use automatic (or algorithmic) differentiation [31]. This technique relies on the fact that any CFD solver and/or fire model is ultimately a function evaluation, i.e. it is a computational code that receives a certain number of input arguments, and returns an output. Each line of the code performs a relatively simple algebraic operation. The derivative of the function (the code) with respect to one of the input parameters can be thus obtained by adding to each line of the code an automatically produced line that contains the differentiation of the original line. Note that running the differentiated code produces the derivative of all output quantities with respect to one input variable. Automatic differentiation can also be used to obtain the adjoint of a system, if run in backwards mode (applying the chain rule). In that mode, the differentiated code produces in one run the derivative of one output quantity with respect to all input variables. Depending on the number of input variables and output quantities, it can be beneficial applying automatic differentiation in forward or backward mode. If the number of measurement comparisons is larger than the number of parameters to be estimated, the forward mode is to be preferred. If, however, the number of parameters is higher than the amount of measurement points, the backward or adjoint mode requires less evaluations of the fire model (and its derivative).

3.2.2 Gradient Free Methods

Gradient free methods rely on holistic approaches to seek for a minimum of a cost function. Generally they recreate some procedure encountered in nature. Prominent examples are genetic algorithms and swarm optimisation. Genetic algorithm based optimisation approaches mimic the evolutionary process of species by defining a set of traits (related to the parameters of the optimisation problem) that are passed on to following generations. An initial population of instances is defined, with different combinations of these traits. Instances then re-combine producing a new generation of instances, with random mutations in the traits. The fitness of the instances in each generation is evaluated, and more successful (fitter) instances are assigned a larger number of offsprings. In the long term those mutations that have proven more successful will prevail and eventually an optimal set of traits (parameters) is found.

In general terms, gradient free methods tend to require a larger number of function evaluations than gradient based methods, since large populations and many generations are necessary to find an optimum (each instance corresponds to a function evaluation). On the other hand, gradient free methods cover the entire search space and thus do not generally get caught in local optima as gradient based methods tend to do.

3.3 Assimilation Window and Rollback

Time is of the utmost importance when it comes to super-real-time fire characterisation or fire forecast. Unfortunately there are some time constraints inherent to the fire characterisation or forecast process that have to be considered. First, information of an incipient fire needs to arrive at some sensor location in order to be detected. Smoke or combustion products need to travel from the fire to the sensor. Depending on the sensor density, this can take several seconds, since sensors will typically be located at the compartment boundaries (ceiling or walls). But more importantly, if parameters relating to fire growth are to be estimated, it is necessary to allow for a sequence of measurements in order to capture the dynamics of the process. To illustrate this, one can think of a single signal of a smoke detector. From this signal it would be, in theory, possible to establish that there is a fire. But it would not be possible to make an estimate of the size of the fire. Once a neighbouring smoke detector provides a second signal, the fire size (HRR) could, in principle, be estimated based on the time difference between signals (because the HRR relates to smoke travel velocity). A third signal, from yet another neighbouring smoke detector, could give information about the growth rate of the fire following a similar logic. If a continuous signal were available (e.g. a temperature reading), it would, of course, be possible to extract the information from sequential readings of the signal. In that case an assimilation window, i.e. a period of time during which signal readings are recorded, would have to be defined. Intuitively it is acceptable to assume that the narrower this assimilation window is, the less precise is the parameter estimated based on the data from this window. As the assimilation window widens, parameter estimation becomes more precise, but valuable time for the model inversion and forecast is lost, since the inversion process can only start once all the data is available. Additionally, a wider assimilation window requires longer running simulations of the model for comparison.

The graph of Fig. 4 shows the estimated value of the spread rate in a compartment fire, for different assimilation windows. For narrow assimilation windows (less than 200 s in this case), there is considerable variability in the estimated value. As the window grows the value converges to the true spread rate (taken from [4]).

Another aspect, related to the width of the assimilation window, that needs to be considered is the amount of time for rollback. When the estimated parameters change due to some critical event (e.g. window breakage), so that a new estimation needs to be performed, it has to be decided how far back the simulations need to be restarted. A critical event is only indirectly noted when the estimated parameters are no longer valid, and simulation outputs and measurements start to diverge again. The time of occurrence of the critical event thus becomes a parameter of the model, and its estimation is necessary to keep the energy and mass balance of the system. This remains an open issue that has been given very limited attention in the literature.

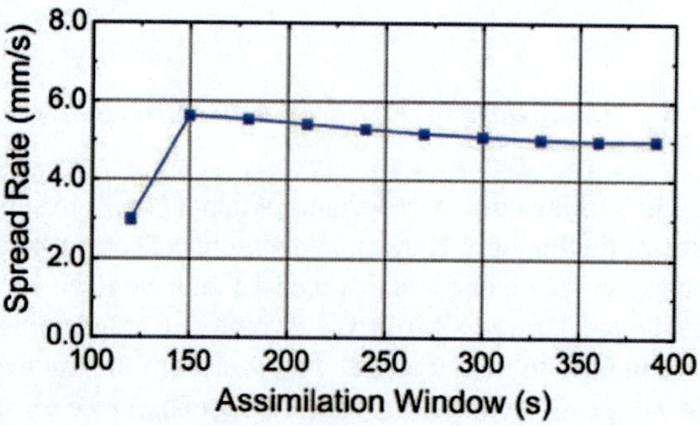

Fig. 4 Estimated spread rate as a function of the growing assimilation window

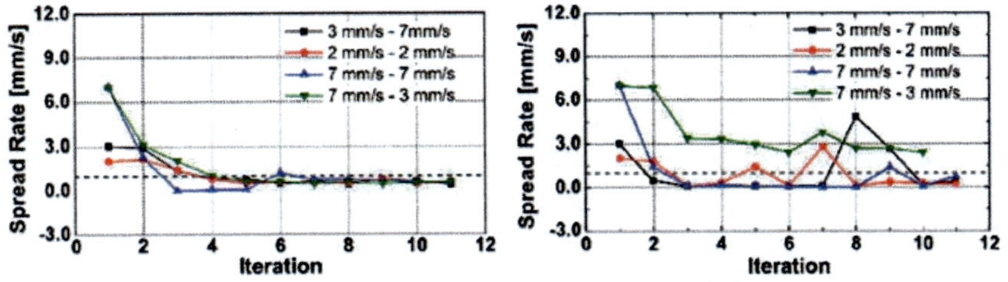

Fig. 5 Convergence of spread rate estimation for different sensor densities

3.4 Data Density

While modern buildings do feature a relatively large number of sensors of all kinds, the overall data density is generally quite low. Data useful for fire characterisation might include smoke sensors, some thermocouples and perhaps CCTV cameras. But even if thermocouples are present, they are not expected to systematically cover all ceilings with several sensors per unit area.

Intuitively, the more data are available, the more precise should the estimation of the spread rate be. The graphs of Fig. 5 show an example of multi-parameter estimation in a compartment fire with different sensor densities (all other conditions are the same), with the graph on the right having a lower sensor density. The true spread rate in this case is 1 ms^{-1}. Different lines in the graph show different first guesses for the parameters. While all first guesses converge to the true value in about 6 iterations for the high data density, it takes significantly more iterations in the case with lower data density, with worse overall performance in terms of convergence (taken from [4]). Other studies have shown, however, that for larger compartments single parameters, such as the growth rate of the fire, can be estimated satisfactorily from data of commonly used smoke detectors (installed according to NFPA standard) [7].

3.5 Speeding Up Fire Simulations

A key aspect of operational forecasting of fire dynamics is maximizing the lead time, i.e., the time into the future the forecast is valid at the moment it is delivered. The iterative process of merging data into the forecast model (by model inversion, to estimate the parameters that characterize the current state of the fire) requires several simulations runs before an actual forecast can be made. This means that a fire model to be used in a forecast framework needs to be able to deliver outputs with an acceptable degree of precision in super- or hyper-real time (i.e. faster than real time), so that it can be run several times and still yield a positive lead time. Table 1 summarises the speed levels necessary for fire forecast simulations. Fully resolved CFD simulations of fire dynamics have shown to take considerably more time to complete than the simulated time, even with coarse numerical meshes (see for example [7]). Given the current computing capabilities, this model type would thus not be eligible for forecasting purposes. Zone models, on the other hand, can deliver output in super-real time, but the produced temperature time series cannot be directly compared to measurements, as they represent averaged quantities.

The need of simplifying fully resolved fire field models in order to speed up the simulations was recognised by several authors. One way of speeding up the models is to use coarse numerical meshes. While this affects the accuracy of the output, in many occasions the impact is acceptable and, if properly assessed, can lead to significantly shorter computing times [7]. Additionally, to the decrease in output accuracy, geometric problems may arise when using coarse grids. The dimensions of openings or obstructions that need to be included in the simulations pose an upper limit to the grid size. Kallada, Janardhan, and Hostikka showed that for fire spread simulations the geometric aspect results in under-estimation of the HRR if not treated adequately [32]. Large grid cells can also lead to under-estimation of local temperatures due to the averaging over a larger volume. Marino et al. proposed to use a sub-grid correction to enable soot formation calculations in coarse grid fire simulations [22].

An interesting approach to solving the problem of computing time in CFD fire simulations is proposed by Würzburger and Arnold [33], who introduce dynamic domain expansion for the numerical computations. The flow field is only solved where velocity or temperature variations occur, thus saving CPU (or GPU) time. As smoke travels, the computational domain is expanded accordingly. They report that dynamic domain simulations take only about 65% of the simulation time compared to the static domain.

An alternative approach is to start from a zone type fire model and to include additional physics with the objective of achieving higher precision in the output. The first attempt towards that direction was made by Davis and Cooper, who included a ceiling jet algorithm in a zone model [34]. Zhang et al. proposed a sub-model to account for heat and mass transfer (additional to balance) between the upper and lower layer [35]. With this they were able to reproduce pre- and post-flashover fires with improved accuracy compared to previous zone models. A multi-zone model approach, presented by Johansson, resulted in increased spatial resolution (compared

Table 1 Speed levels of fire simulations

Speed level	Definition
Real time	The wall time (time the simulation runs on the computer) is less or equal to the simulated time
Super-real time	The wall time is less than the simulated time, and enough time is available to interpret the results of the simulation
Hyper-real time	The wall time is a fraction of the simulated time, so that several model runs can be performed in order to assimilate data, and a usable forecast with enough lead time for effective decision making is available

to a traditional zone model) and good agreement with experimental fire data (comparable to CFD based fire simulations) [36]. The computational cost remains low (super-real time simulations are possible) but approaches the order of magnitude of the total extension of the simulations.

When fire scenarios in large spaces with multiple compartments (entire buildings) need to be simulated it has been proposed to use network models that are based on similar considerations as zone type models (heat and mass transfer between nodes). Bilyaz and Ezekoye present such a model, and report significant speed ups in terms of computation time when compared to a full CFD model (seconds compared to days of runtime) [37]. They consider uniform temperature in the fire origin room and put emphasis on the flow calculations of adjacent rooms. If flow and temperature details of the fire compartment are of interest in similar scenarios, a multiscale approach has been proposed, where the simulation domain is divided into a far field and a near field (relative to the main flaming region). A network type model is then applied in the far field, which solves for average temperature and smoke density, and acts as a dynamic boundary condition to the fully resolved CFD simulation in the near field. This approach has been implemented by Colella et al. for tunnels, and by Ralph et al. for buildings [38, 39].

A fundamentally different approach to speeding up fire simulations was envisioned by Watts [40], who predicted a paradigm shift in the modelling of fire dynamics, foreseeing early on the use of big data and neural networks [40]. An operational implementation is presented by Hodges et al. and Lattimer et al. [41, 42], who propose to use concepts of Machine Learning to simulate fire behaviour. Two separate methodologies of ML are introduced by Lattimer et al.: supervised and unsupervised learning [42]. Reduced-order modelling (ROM) is an example of unsupervised learning. Here, a new, reduced set of mathematical equations is developed, and the full-order equations (a complete CFD based fire model) are projected into the reduced-order space by minimising the misfit when compared to data. The training of artificial neural networks is an example of supervised learning. Here, a collection of interconnected nodes is used to represent the relationship between input and output. This leads to a series of linear equations with weights that are determined by comparison to data. The latter approach is used by Hodges et al., who present a simulation framework consisting of a zone model and an ANN. The zone model receives the compartment geometry, ventilation condition and fire intensity as input, and produces output consisting of mean smoke layer temperature and smoke layer height. This is then fed into a previously trained ANN, which outputs spatially resolved thermal fields [41].

3.6 *Predictability of Fire Scenarios*

There is an upper limit to the horizon of predictability for dynamical systems [14]. The non-linear nature of these systems results in large uncertainties of the output for only small perturbation in the initial conditions, effectively turning the system chaotical.

Fire scenario forecasts are no exception. But there is an even more restrictive limit of predictability of fire development forecasts, which is related to the fire dynamics and their modelling itself.

During the early stages of a fire, there is little interaction between the fire and the surroundings (walls, ceiling, enclosures in general), which makes it relatively easy to simulate using CFD based tools (at least when considering some approximations, as indicated in Sect. 3.1). The problem in terms of predictability is that there is generally no prior information regarding fuel load, fuel distribution and fuel geometry. The predictability is thus limited not so much by the model, as by the large uncertainty in the input parameters, since it is impossible to keep track of how furniture and other combustible items will be located during normal operation of a building.

Fire dynamics change fundamentally between the fuel controlled and ventilation controlled stages, as indicated in Sect. 2.1. Modelling this transition is beyond the capabilities of any existing tool, due to the complexity of the underlying physics, and the involved time and length scales. It has further been repeatedly shown that ventilation controlled and under-ventilated fire scenarios are notoriously difficult to recreate using numerical simulations [43], due to intrinsic limitations of the combustion and flame extinction models. Since in the ventilation controlled stage the geometry of fuel packages (related to usage of the compartment) is essentially irrelevant, in principle the predictability horizon of this stage of the fire could extend further. However, as stated in previous sections, current fire modelling tools are not capable of reproducing under-ventilated fire conditions to a satisfactory level, due to intrinsic problems in the combustion and flame extinction models.

4 Conclusions

The forecasting of fire events is a complicated endeavour. The physical processes involved in fire dynamics are too numerous, complicated and interlinked to allow for any attempt of predicting their development from first principles, based solely on numerical integration of the mathematical expressions that constitute the fire model. Data from the ongoing event necessarily have to be considered and merged into the model. Merging data with numerical simulations is done through an iterative procedure, where each iteration involves a forward run of the numerical fire model. In order to be useful for decision making, fire development forecasts need to be delivered in super- or hyper-real time, with at least some level of spatial distribution. This means that both, the data integration process and the actual forecast need to finish a significant amount of time before the events take place, thus requiring fire models much faster than those presently available. While fire models and computing hardware are continuously being improved, so that significant steps forward in speed and resolution have been made, there is an ongoing need for more research in all aspects related to this topic, in order to enable true forecast capabilities.

Acknowledgements The author would like to recognise the financial support provided by the Timber Innovation Center UC and ANID BASAL FB210015 (CENAMAD).

References

1. S. Koo, J. Fraser-Mitchel, S. Welch, Sensor-steered fire simulation. Fire Saf. J. **45**(3), 193–205 (2010). https://doi.org/10.1016/j.firesaf.2010.02.003
2. W. Jahn, G. Rein, J.L. Torero, Forecasting fire growth using an inverse zone modelling approach. Fire Saf. J. **46**(3), 81–88 (2011). https://doi.org/10.1016/j.firesaf.2010.10.001
3. K.J. Overholt, O.A. Ezekoye, Characterizing heat release rates using an inverse fire modeling technique. Fire Technol. **48**, 893–909 (2012). https://doi.org/10.1007/s10694-011-0250-9
4. W. Jahn, G. Rein, J.L. Torero, Forecasting fire dynamics using inverse computational fluid dynamics and tangent linearisation. Adv. Eng. Softw. **47**(1), 114–126 (2012). https://doi.org/10.1016/j.advengsoft.2011.12.005
5. K.J. Overholt, O.A. Ezekoye, Quantitative testing of fire scenario hypotheses: a Bayesian inference approach. Fire Technol. 1–33 (2014). https://doi.org/10.1007/s10694-013-0384-z
6. T. Beji, S. Verstockt, R. Van de Walle, B. Merci, On the use of real-time video to forecast fire growth in enclosures. Fire Technol. **50**(4), 1021–1040 (2014). https://doi.org/10.1007/s10694-012-0262-0
7. W. Jahn, Using suppression and detection devices to steer CFD fire forecast simulations. Fire Saf. J. **91**, 284–290 (2017). https://doi.org/10.1016/j.firesaf.2017.03.062
8. Y.Y. Chu, V.K.R. Kodur, D. Liang, A Probabilistic inferential algorithm to determine fire source location based on inversion of multidimensional fire parameters. Fire Technol. (2016). https://doi.org/10.1007/s10694-016-0620-4
9. W. Jahn, F. Sazunic, C. Sing-Long, Towards real-time fire data synthesis using numerical simulations. J. Fire Sci. 1–16 (2021). https://doi.org/10.1177/0734904121993449
10. A. Cowlard, W. Jahn, C. Abecassis-Empis, G. Rein, J.L. Torero, Sensor assisted fire fighting. Fire Technol. **46**(3), 719–741 (2010). https://doi.org/10.1007/s10694-008-0069-1
11. G. Rein et al., Round-robin study of a priori modelling predictions of the Dalmarnock fire test one. Fire Saf. J. **44**(4), 590–602 (2009). https://doi.org/10.1016/j.firesaf.2008.12.008
12. W.D. Davis, G. Forney, A sensor-driven fire model, in *12th International Conference on Automatic Fire Detection*, Gaithersburg, MD: NIST, pp. 494–504 (2001)
13. E. Kalnay, *Atmospheric modeling, data assimilation and predictability* (Cambridge University Press, Cambridge, UK, 2003)
14. E.N. Lorenz, The predictability of a flow which possesses many scales of motion. Tellus **21**(3), 289–307 (1969). https://doi.org/10.3402/tellusa.v21i3.10086
15. D. Zeinali et al., Study of the importance of non-uniform mass density in numerical simulations of fire spread over MDF panels in a corner configuration. Combust. Flame **200**, 303–315 (2019). https://doi.org/10.1016/j.combustflame.2018.11.020
16. S. Sazegara, M. Spearpoint, G. Baker, Benchmarking the single item ignition prediction capability of B-RISK using furniture calorimeter and room-size experiments. Fire Technol. **53**(4), 1485–1508 (2017). https://doi.org/10.1007/s10694-016-0642-y
17. E. Markus, A. Snegirev, E. Kuznetsov, L. Tanklevskiy, Application of the thermal pyrolysis model to predict flame spread over continuous and discrete fire load. Fire Saf. J. **108**, 102825 (2019). https://doi.org/10.1016/j.firesaf.2019.102825
18. M. Lazaro, H. Boehmer, D. Alvear, J.A. Capote, A. Trouve, Numerical simulation of fire growth, transition to flashover, and post-flashover dynamics in the dalmarnock fire test. Fire Saf. Sci. **1**, 1377–1388 (2008). https://doi.org/10.3801/IAFSS.FSS.9-1377
19. V. Novozhilov, Nonlinear dynamical model of compartment fire flashover. J. Eng. Math. **67**(4), 387–400 (2010). https://doi.org/10.1007/s10665-009-9333-8

20. F. Sazunic, W. Jahn, C. Sing-Long, Interpretable reduced order modeling for fast fire dynamics simulation. Appl. Math. Comput. vol. submitted (2021)
21. D. Drysdale, *An Introduction to Fire Dynamics*, 3rd edn. (Wiley & Sons, New York, 2011). ISBN 0-470-31903-1
22. O. Mariño, F. Muñoz, W. Jahn, Soot production modelling for operational computational fluid dynamics fire simulations. J. Fire Sci. (2020). https://doi.org/10.1177/0734904120905579
23. I. Glassman, R. Yetter, N. Glumac, *Combustion*, 5th edn. (Academic Press, 2014). https://doi.org/10.1016/C2011-0-05402-9
24. J.-W. Kwon, N.a. Dembsey, C.W. Lautenberger, Evaluation of FDS V.4: upward flame spread. *Fire Technol.* 43(4), 255–284. https://doi.org/10.1007/s10694-007-0020-x
25. W. Jones, R. Peacock, G. Forney, P. Reneke, CFAST—Consolidated Model of Fire Growth and Smoke Transport, Version 6 (2005)
26. R.G. Rehm, H.R. Baum, The equations of motion for thermally driven, buoyant flows. J. Res. Natl. Bur. Stand. **83**(3), 297 (1978). https://doi.org/10.6028/jres.083.019
27. H.K. Versteeg, W. Malalasekera, *An Introduction to Computational Fluid Dynamics* (Pearson Education, 2007)
28. C.C. Lin, L. Wang, Forecasting smoke transport during compartment fires using a data assimilation model. J. Fire Sci. **33**(1), 3–21 (2015). https://doi.org/10.1177/0734904114548837
29. J. Ji, C. Liu, Z. Gao, L. (Leon) Wang, Real-time forecasting of fire in a two-story building using ensemble Kalman filter method. *Fire Saf. J.* **97**, 19–28 (2018). https://doi.org/10.1016/j.firesaf.2018.02.007
30. J. Nocedal, S. Wright, *Numerical Optimization*, 2nd edn. (Springer, USA, 2006). ISBN 978-0387-30303-1
31. A. Griewank, *Automatic Differentiation.* Princeton Companion to Applied Mathematics, Nicolas Higham Ed., (Princeton University Press, 2014)
32. R. Kallada Janardhan, S. Hostikka, Predictive computational fluid dynamics simulation of fire spread on wood cribs. Fire Technol. **55**(6), 2245–2268 (2019). https://doi.org/10.1007/s10694-019-00855-3
33. M.L. Würzburger, L. Arnold, Dynamic domain expansion in smoke spread simulations with ARTSS: speedup and overhead. Fire Saf. J. **120**(June 2020), 103168 (2021). https://doi.org/10.1016/j.firesaf.2020.103168
34. W.D. Davis, L.Y. Cooper, A computer model for estimating the response of sprinkler links to compartment fires with draft curtains and fusible link-actuated ceiling vents. Fire Technol. **27**(2), 113–127 (1991). https://doi.org/10.1007/BF01470863
35. L. Zhang, Y. Wang, H. Shi, L. Zhang, Modeling and analyzing 3D complex building interiors for effective evacuation simulations. Fire Saf. J. **53**, 1–12 (2012). https://doi.org/10.1016/j.firesaf.2012.06.008
36. N. Johansson, Evaluation of a zone model for fire safety engineering in large spaces. Fire Saf. J. March, 103122 (2020). https://doi.org/10.1016/j.firesaf.2020.103122
37. S. Bilyaz, O.A. Ezekoye, Fire smoke transport and opacity reduced-order model (FireSTORM): a new computer model for high-rise fire smoke simulations **55**(3). Springer US (2019). https://doi.org/10.1007/s10694-019-00815-x
38. F. Colella, G. Rein, R. Borchiellini, J.L. Torero, a novel multiscale methodology for simulating tunnel ventilation flows during fires, **47**(1) (2011). https://doi.org/10.1007/s10694-010-0144-2
39. B. Ralph, R. Carvel, J. Floyd, Coupled hybrid modelling within the fire dynamics simulator: transient transport and mass storage. J. Build. Perform. Simul. **12**(5), 685–699 (2019). https://doi.org/10.1080/19401493.2019.1608304
40. J.M. Watts, Future of fire modeling. Fire Technol. **27** (1991). https://doi.org/10.1007/BF01470860
41. J.L. Hodges, B.Y. Lattimer, K.D. Luxbacher, Compartment fire predictions using transpose convolutional neural networks. Fire Saf. J. **108**(November 2018), 102854 (2019). https://doi.org/10.1016/j.firesaf.2019.102854

42. B.Y. Lattimer, J.L. Hodges, A.M. Lattimer, Using machine learning in physics-based simulation of fire. Fire Saf. J. **114**(July 2019), 102991 (2020). https://doi.org/10.1016/j.firesaf.2020.102991
43. B. Lafdal et al., Numerical study of the combustion regimes in naturally-vented compartment fires. Fire Saf. J. 131(August 2021), 103604 (2022). https://doi.org/10.1016/j.firesaf.2022.103604

Fire and Smoke Image Recognition

Yoon Ko, M. Hamed Mozaffari, and Yuchuan Li

1 Introduction

In combustion reactions, organic fuels (containing Carbon compounds) generally release a combination of signatures of heat, light, gases, and soot particulates. Many of these signatures are observable or could be visualized using imaging technologies functioning as volume sensors in various emission spectrums (e.g., visible range, infrared range) [1]. Image-based solutions provide data on the fire's direction, growth, and size in a wide coverage area, overcoming the limitations of point sensor-based solutions. In addition, image-based solutions analyze the image patterns of smoke and flame, which vary with different lighting and background conditions [2]. Regarding accuracy, efficiency, and robustness, image-based solutions rapidly advance by embracing deep learning technologies utilizing high-level feature analyses [3].

Application. Image-based solutions have the potential to provide unique advantages over typical systems in various applications, including the following:

- Detection of smoke and flame
- Fire development prediction
- Assisting in firefighting and rescue

Y. Ko (✉) · M. Hamed Mozaffari · Y. Li
Fire Safety Unit, Construction Research Centre, National Research Council Canada, Ottawa, ON K1A 0R6, Canada
e-mail: yoon.ko@nrc-cnrc.gc.ca

M. Hamed Mozaffari
e-mail: mhamed.mozaffarimaaref@nrc-cnrc.gc.ca

Y. Li
e-mail: yuchuan.li@nrc-cnrc.gc.ca

X. Huang and W. C. Tam (eds.), *Intelligent Building Fire Safety and Smart Firefighting*, Digital Innovations in Architecture, Engineering and Construction,
https://doi.org/10.1007/978-3-031-48161-1_13

With the advancement of computer processing and image analysis technologies, VID (Video Image Detection) systems have been developed and used for fire and smoke detection [4]. The VID systems often employ surveillance video cameras and are for use mostly in large spaces, outdoor facilities, and tunnels. For example, the VID systems are installed for smoke and fire detection in some historic buildings, warehouses, power plants, automotive plants, nuclear research facilities, automotive plants, road/rail tunnels and oil platforms [5].

While image-based solutions are being used to detect smoke and fire in a broad range of applications, they are also considered to assist firefighting and rescue activities [6, 7]. For example [8–12], smart firefighting solutions, including robotic firefighting, are being studied by bringing in image processing technologies for analyzing smoke/flame signatures. New algorithms have been developed not only to capture key visual signs of fire but also to predict an impending sudden fire propagation, such as a room flashover. The new vision-based approach is promising to enhance the understanding of fire dynamics, overcoming the limitations of experimental and numerical studies that have been conducted for decades.

With advanced vision-based image and video processing approaches developed in the past two decades, smoke and flame images can be effectively analyzed using Machine Learning (ML) and Deep Learning (DL) techniques. The feasibility of applying image processing techniques to fire safety has been verified in vision-based flame and smoke detection systems recently developed [13]. However, how these vision-based techniques can be applied to assist firefighting and to predict sudden fire propagation/flashover has not yet been thoroughly verified.

Advantages. The Video Image Detection for Smoke and Fire (VID_SF) systems have demonstrated unique advantages in the application areas where conventional systems cannot be effective for fire/smoke detection.

Conventional smoke detectors monitor the presence of smoke using the characteristics of smoke particulates in scattering/obstructing a light source; and/or interfering with ions and reducing the ion mobility [14]. Since these technologies are usually a spot type, beam type or aspirating air sampling type, they are not practical for monitoring a large space as the space requires a number of detectors and air sampling points/lines [15]. Unlike traditional smoke detection technologies requiring local air sampling, the VID_SF systems detect smoke by real-time processing of the image data within the field of view. Thus, the VID_SF systems are capable of monitoring a large space with a relatively high ceiling height without many limitations, within the verified field of view. Figure 1 shows an example of a smoke and flame detection facility application [16]. This detection system is equipped with multiple video image cameras and a central analysis and decision-making system to help detect fire and smoke patterns quickly and accurately.

With the ability to monitor a large space, the number of devices required to be installed is much less for the VID_SF systems than the conventional smoke detectors. In addition, the VID systems can use the image data from basic camera devices, including common video surveillance cameras. Hence, the VID_SF systems are often integrated with video surveillance devices, as shown in Fig. 1. Consequently,

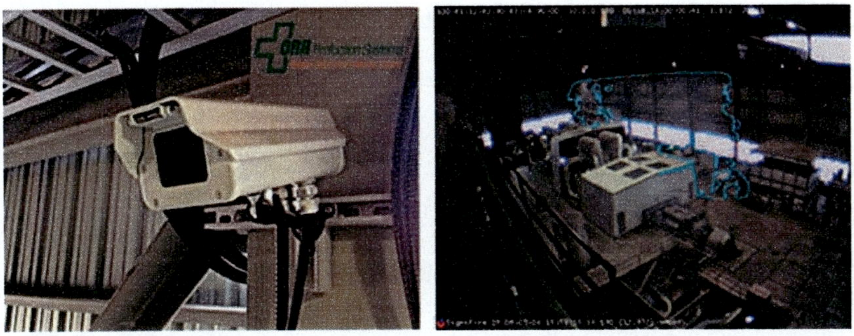

Fig. 1 An application of large scale smoke and flame detection by video image detection from ORR® Protection Co. [16]. Left: An example of the equipped video image cameras in the detection system. Right: An example of real-time analysis and processing of potential smoke and fire patterns in the interested region, circled by cyan dashed curve

the VID_SF systems have advantages over easy installation, simple maintenance, and high system reliability due to frequent use [17]. Furthermore, unlike the conventional smoke detectors, VIS_SF systems can be used for outdoor fire and smoke detection tasks (e.g., oil and gas industries or wildfires).

In detecting flames, conventional flame detector technologies detect light emitted by flames in the wavelength designed for target fires/fuels. While they deliver robust flame detection performance in the field of view with the ability to locate the flame, their detection ability is limited to the designed wavelength. Thus, early smoke generation, heat buildup, and temperature increase cannot be fully detected [18]. On the other hand, the VID_SF systems provide flame detection with an algorithm capable of recognizing flame's light, color, visual patterns and features, and motion. Also, the VID_SF systems have the advantage of processing live video and providing alarm immediately [19], which also allows fast situational awareness with an accurate fire location in the monitored area.

The recent advancement in image processing methods implemented by the power of deep learning techniques, including Convolutional Neural Networks (CNNs) enables accurate and speedy detection of smoke and flame. Also, the advanced solutions could be used as means to diagnose/predict fires and potential problems when properly trained with the data from various scenarios involving different fire sources and nuisance sources.

This chapter will discuss fire and smoke image signatures and sensors/devices capable of capturing those signatures. Then, computer vision technologies will be discussed for processing smoke and flame using traditional computer vision and deep learning approaches. Image-based solutions for firefighting and prediction of sudden fire propagation will be also discussed briefly.

2 Fire and Smoke Image Signatures and Sensors

2.1 *Smoke and Flame*

Fire or flame starts when fuel vapor is dispersed and reacts with oxygen in the surrounding atmosphere. The chemical oxidation reaction emits gaseous products, including carbon dioxide, carbon monoxide, water vapor, and unburned hydrocarbon. Heat is generated in most combustion conditions of smouldering (327 °C–627 °C), and flaming (927 °C–1427 °C) [20]. The heat is an important signature enabling to differentiate fire from the ambient surrounding. In particular, when the chemical oxidation reaction of organic fuels occurs rapidly, a flame is generated together with heat and light, which are visible gases undergoing the highly exothermic chemical reaction.

From the combustion reaction of organic fuel, radiative energy is also released at approximately 30–40% of the total combustion energy, and the radiative energy is dissipated in the form of electromagnetic radiation in broad spectral ranges, covering Ultraviolet (UV), visible (VIS), and infrared (IR) bands [21]. Likewise, the combustion reaction and the combustion gas products (e.g., carbon dioxide, carbon monoxide, water vapor, and unburned hydrocarbon) are detectable through photon emission at specific wavelengths.

Smoke containing visual soot particulates is produced when organic materials go through most combustion processes of pyrolysis, smouldering and flaming [22]. Light-colored smoke will be formed in the pyrolysis condition while the fuel decomposes and emits fuel vapors. Smouldering combustion generates smoke, generally with large particle sizes, while there is no visible flame. The amount of smoke released generally depends on the type and the amount of fuel involved in the reaction. The smouldering combustion generates smoke much larger than the flaming combustion, although fuels burn slowly with low temperatures. When an organic material burns in the flaming combustion, fine smoke particulates are usually generated. The smoke includes black soot particles, so the color is darker. Smoke is considered as a significant signature sought for early fire detection, using smoke detection technologies, including optic sensors. These signatures of heat, light, gases, and soot particulates are visual or could be visualized using cameras and imaging devices.

2.2 *Image Characteristics of Smoke and Flame*

Smoke image characteristics

Smoke is a sooty aerosol containing light absorbing carbon (i.e., elemental carbon) and light scattering carbon (i.e., organic carbon) [23]. Due to these characteristics, smoke is visible to the human eye with a light source, and smoke is gray in color with various color saturations and opacities. In addition to uneven smoke color density distribution, the contour/shape of smoke is also irregular, yet smoke streamline is generally from a fixed source. To detect smoke, its motion is also analyzed in many of the VID_SF systems, which include the motion of smoke dispersion in vertical and horizontal directions; wavelet/oscillation of smoke optical flows.

Flame image characteristics

Flame is visible combustion gases undergoing a highly exothermic reaction. The flame color varies depending on the fuel and separates the flame from the background [24]. The light and heat generated from flame are also visible to the human eye, and they could be visualized and detected, even when obstructed, using appropriate vision and thermal IR imagers. Flame shape/contour usually changes over time as the fire grows. Another distinct characteristic of flame is its flickering and puffing motion. In detecting flame, many VID_SF systems focus on the motion/movement, dynamic texture of flame as well as intensity changes of flame.

2.3 Image Sensor Technologies

Vision sensors and cameras

The visible image features (e.g., color, shape, motion, transparency, pattern) of smoke and flame could be well captured by regular vision/RGB cameras [25]. However, their application is limited due to the challenging environment of fires (e.g., in heavy fog or in dark conditions) could disguise the presence of smoke and flame. Also, the systems relying on vision cameras should be designed considering the fact that the prominent image features of smoke and flame, such as motion, shape, transparency, patterns and colors, change with the environmental conditions. Thus, these systems often require techniques beyond traditional image processing methods to sufficiently reduce the rate of false alarms.

IR sensors and cameras

Optical flame sensor and camera technologies are based on radiation produced by the flames in the spectrums of UV, VIS, and IR. IR data provide information that could hardly be captured from regular RGB cameras. IR imagers provide a view of environmental temperature, including sharp imaging of hot objects or moving humans hidden behind smoke and flames [26].

When imaging flame/fire, generally, MWIR (Mid-Wave IR) cameras [27] are used because the intensity differences between a flame and its background are much more distinct in the MWIR band of 3–5 μm than in the LWIR (Long-Wave IR) band 8–14 μm. Also, a MWIR camera is usually considered as the most suitable range to study flame and combustion since many combustion products, such as water vapor, carbon dioxide, carbon monoxide and hydrocarbons, are infrared-active in the MWIR band. A LWIR camera is also used for firefighting [28], because in the LWIR band, the radiance absorbance by soot and water vapor are relatively low while the energy from solid surfaces is radiated in the band.

The advantages of using thermal images come from their measurement, that is, the temperature/energy intensity of each pixel [29]. IR cameras are capable of providing IR radiation/thermal data even in challenging environments (in the dark and on fireground). This means that feature extraction/classification of images of smoke and fire is less affected by the environment with thermal image data than regular vision image data.

2.4 Other Vision-Based Sensors and Cameras

Radar, Sonar, Laser sensor

Radio (e.g., Radio Detection and Ranging or Radar), Sound (e.g., Sound Navigation and Ranging or Sonar), and laser (e.g., Light Detection and Ranging or Lidar) sensors are used for navigation in various applications, such as wildfires [30, 31]. The physics behind these technologies is the time (called time-of-flight (ToF)) for the signal (radio wave, sound wave, or laser beam) to be emitted by a sensor, reach the target object, and reflected, and received by a sensor. The data recorded can be visualized in the form of a three-dimensional (3D) image (i.e., 3D point cloud or 3D map). The recorded image can then be visualized and used for the detection of fire and smoke [32]. However, utilization of these imaging and mapping technologies is still in its infantry stage, and a significant investigation [33, 34] must be undertaken to be adopted as vision-based technologies for fire and smoke detection purposes.

Pan Tilt Zoom (PTZ) camera

PTZ cameras are a certain type of camera that are built with mechanical parts that allow for movements like swiveling from left to right, tilting from up and down, and zooming in and out of a scene [35]. A typical application for them is to monitor wide open areas requiring a more than 180-degree view. It could also be deployed in guard stations and be operated by a remote camera controller from far away. They can be set to either automatically follow activity triggered by motion or adhere to a pre-set schedule on a certain region of interests. Nowadays, PTZ cameras are generally implemented in tandem with a larger surveillance system, which tracks movements instead of detailed image shots.

Unmanned Aerial Vehicle (UAV)

Cameras can be assembled and installed on Unmanned Aerial Vehicles, such as drones, airplanes, satellites, etc., and sometimes called UAV cameras. Unlike most stationary cameras, the UAV camera allows for a great expansion of capabilities in video and image capture [36]. Indeed, they have severely affected fire safety engineering in the last few years. They provide enhanced flexibility in aerial shot setup, access to hard-to-reach spaces, and the possibility of novel visual effects and shot types. Also, UAV cameras offer features like easy target tracking and active following, which are crucial to smoke or flame pattern detection and recognition in large areas. Besides, they also dramatically reduce the operation cost, which makes them far more accessible than other options like cranes and spider cams.

Stereo vision and stereo camera

A stereo camera is usually a system of multiple cameras that are designed to reconstruct objects in images from multiple 2-dimensional captures [37]. A simple example is a system of two cameras placed horizontally, and the captured images are then processed for the recovery of visual depth information. It could be generally divided into 2 types, depending on the solution of stereo reconstruction. The first one is algorithm-based, which means the captured information is 2D features, such as depth, and 3D image data is estimated by the algorithm by matching patterns of 2D information. The second type of stereo camera systems contains a dedicated sensor to acquire depth information, such as a ToF or structured light sensor. Compared with the algorithm-based stereo camera system, sensor-based models are more efficient and accurate in 3D image reconstruction.

3 Image Recognition of Smoke and Fire

3.1 Introduction to Computer Vision

Computer Vision (CV) is an active research area of Computer Science, which typically refers to a subject of research in enabling the machine to process and analyze visual surroundings. Those visual understandings, enabled by the processing (e.g., image thresholding, contrast enhancement, image denoising, morphological methods, methods in frequency domain, brightness normalization, etc.) and analysis (e.g., object localization, detection, identification, recognition, segmentation, etc.), are the ability to recreate the entire surroundings, including the properties, like shape, illumination, and color distribution, of the surroundings from one or more images [38].

Image Processing (IP) is a crucial and flexible technique for data processing in CV, which is typically conducted by transforming images using mathematical functions. IP transforms images in a wide variety of applications in industries and academies [39], such as Colorization, Texture Synthesis and Inpainting, HDR and Tone Mapping, Intrinsic Images, Deblurring, Denoising, Warping and Morphing.

After IP technique is applied, the processed image is further analyzed by various techniques to provide image scene interpretation, which involves many tasks, such as classification, segmentation, localization, recognition (detection), reconstruction, etc. [40]. Figure 2 shows an example of comparing fire and smoke image segmentation, localization, classification, and detection tasks in CV. It could be divided based on the level of features that it focuses on, such as pixel level, single object level, and

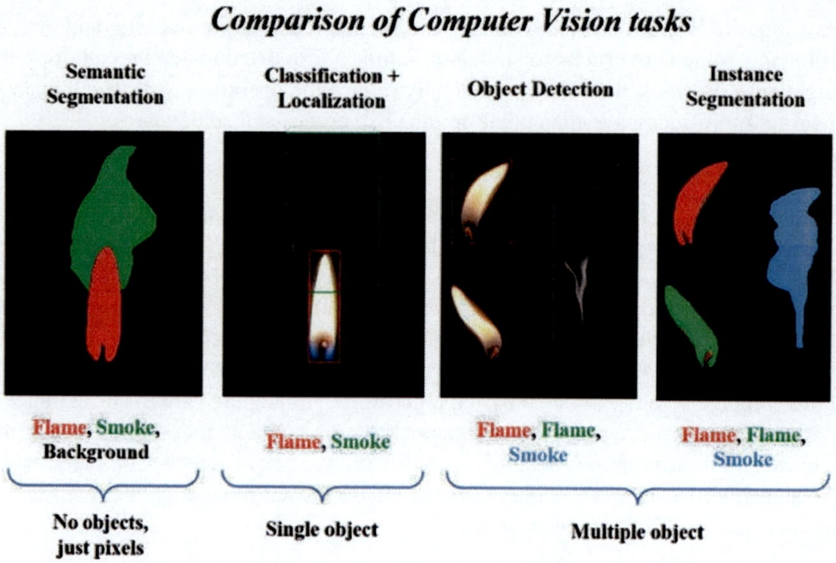

Fig. 2 A comparison of different computer vision tasks, including semantic segmentation, classification + localization, object detection, and instance segmentation. They could be generally divided into 3 categories, depending on the level of feature, such as pixel or object

multiple objects level. In fact, a complex CV task, such as smoke and fire detection, usually requires a collaboration of multiple approaches. These CV techniques are briefly defined as below:

- Segmentation aims to label each pixel in the image with a single category label within or without different instances, depending on the segmentation task [41].
- Localization is a task of locating a single instance of a particular category of object in an image, which is typically done by specifying a tightly cropped bounding box centered on the instance [42].
- Classification is a task that assigns a label to an input image from a known set of categories [38]. It is also one of the core problems in CV that, despite its simplicity, has a large variety of practical applications.
- Object detection is a process of finding certain interested instances of target objects in images or videos [43].

3.2 *General Methods of Smoke and Fire Image Detection*

In general, there are two main approaches, (a) Conventional image processing methods (b) Deep Learning (DL) methods (the most common technique is Convolutional Neural Networks (CNNs)) in CV for flame and smoke image detection tasks.

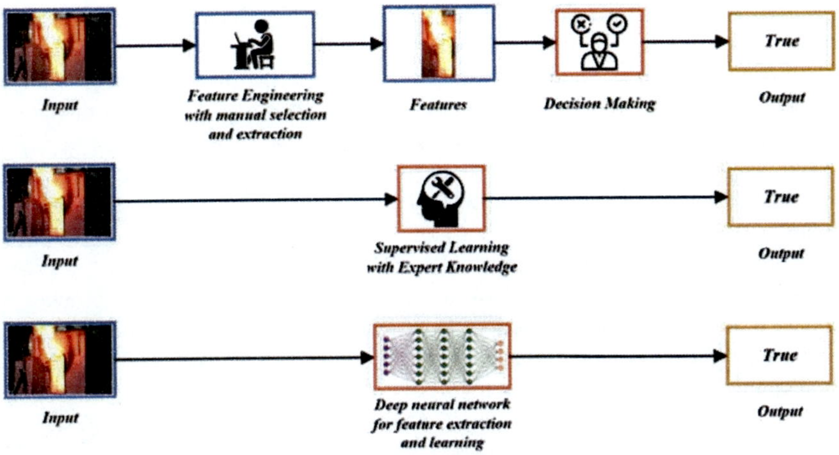

Fig. 3 The overall structures for vision-based image processing approaches on an example of a fire detection task. Upper: Conventional image processing method with manual feature extraction and decision making. Middle: Supervised learning with expert knowledge. Lower: Convolutional Neural Networks (CNNs) based methods

Figure 3 shows a comparison between 3 levels of solution for fire detection tasks: the conventional manual feature engineering method, the supervised learning with expert knowledge, and automatic feature extraction and learning by CNNs. Unlike the conventional method, DL enables a faster and more efficient analysis process with even better performance [44].

As for the conventional method with techniques like image processing, the entire task includes several steps of feature extraction/object localization and object classification/detection sub-tasks. And the most important task is to understand the characteristics of smoke and flame images, which include features like color, shape, dynamic texture, motion, and flickering/oscillation behavior. Besides, the conventional image processing approach usually requires domain expert knowledge for designing a certain feature extractor to transform input image data into a feature vector [45], which will be fed in the decision making step for further smoke and fire detection result decision.

For approaches with expert knowledge involved, it becomes a supervised learning method compared with conventional approaches. The feature extraction and decision making steps are combined here, and is solely executed by experts. They key factor is the previous experience on similar tasks and inputs within the experts' brain, which could be regarded as a supervised training and learning procedure to help dealing with the current task. In addition, this kind of approach usually includes a majority voting part to make sure that the randomness and probability of making a mistake from experts are reduced as much as possible.

In CNN-based approaches, the basic concept and process of detecting smoke and flame images remain the same as the conventional image processing method.

However, those tasks are done in a different manner this time: end-to-end fashion, which means the output is directly available after the processing of CNN. CNN-based methods are able to automatically capture important image features that are signatures of smoke and flame [46], so there is no need to do handcraft operation for the features extraction and classifier design. Therefore, an insight into the CNN design is required in applying the approach to the problem of smoke and fire detection since there is a lack of feature interpretability and information visualization over the entire internal processes.

In additional, there is hybrid method in CV that combines the above approaches together, traditional and modern ones [47]. Hybrid techniques could produce a universal model with versatile options to perform accurate fire and smoke detection. One foreseeable advantage is that a significant amount of computation and training time could be saved when inputs are processed by traditional methods. Thus, it is still valuable to understand techniques for feature extraction and classification.

Feature extraction

Most of the traditional image processing methods start from Feature Engineering, which processes and selects image features in raw data, to better represent the problems and directly affect the accurate computation of results [45]. In smoke and flame image recognition, features are usually small informative patches of smoke and flame patterns in images, which include:

- Smoke/flame pixel intensity in RGB color space (Red, Green, Blue coordinate)
- Smoke/flame pixel intensity in HSI (Hue, Saturation, Intensity coordinate), YUV (Brightness, Blue projection, Red projection), YCbCr (Y: luminance, CbCr: chrominance) color spaces.
- Smoke/flame image texture descriptors
- Localized or delineated smoke/flame by texture segmentation
- Spatial analysis and frequency domain analysis for edge and corner detection
- Threshold foreground and background using image thresholding methods
- Temporal variation of image features (e.g., color, luminance) to recognize wavelets of smoke, and flame flickering or oscillation that related to motion features
- Dynamic texture analysis for the temporal evolution of pixel intensities to detect flame and smoke.

The feature extracted from raw input can be done either manually or automatically using more advanced, well-established image processing models such as SIFT, SURF, BRIEF, etc. [48].

Classification

After the feature extraction, the second step is to set up a classifier made of a neural network to find out the extent that the extracted features match the target criteria of smoke and fire images. In traditional CV methods, the classification is conducted commonly using machine learning approaches, such as a decision tree [48] or support vector machine (SVM) [49], which fundamentally mimic the process of human understanding and decision making. When it comes to the task with multiple images and features, including spatial, temporal, and spectral smoke and flame data, those features could be handpicked to form either joint feature vectors for further analysis or a single neural network classifying the data for smoke and fire detection automatically. In addition, the classification and detection task for smoke or fire could be conducted at different scale levels, such as pixel level or image block level, depending on the ultimate target of the task.

3.3 Traditional Computer Vision Approaches for Smoke and Fire Recognition

Traditional methods include feature engineering, image processing, and naïve machine learning algorithms, and they have advantages in research and practice over the rising CNN methods due to their high interpretability [49].

Smoke detection

Traditional CV approaches for smoke detection use multiple image features of smoke, including spectral, spatial, and temporal features. Among many image features, a review of the smoke detection techniques [50] reported that the disorder and growth characteristics are good indicator of smoke. Relatively high detection rates were reported with the techniques focusing on smoke motion and wavelet as well as analyzing space–time features of smoke. The motion features and history analysis using invariant moment methods (for shape recognition and identification) resulted in high smoke detection rates in tunnels as reported in Han and Lee [51]. The space–time features of smoke texture were analyzed in Yuan [51] using LBP (Local Binary Pattern) and LBPV (Local Binary Pattern Variance), which resulted in high detection rates with their high discrimination power for recognizing smoke texture and movement direction. High detection rates were also reported by Zhou et al. [51] in which smoke wavelet was analyzed using texture descriptors (e.g., angular second moment, entropy, contrast, image pixel correlation and inverse different moment). However, they reported that their method is computationally costly. In [50], it is reported that the flickering feature could be excluded for smoke detection, which does not supply quite an improvement to the overall performance but bring computational costs.

Overall, the reported detection rate for smoke is lower than flame since open flame has many prominent image features, distinct from the background. Also, [50] discussed that the rate of false positives is high for smoke detection, and the false positive could be reduced by increasing the scope for both smoke and flame detection.

Flame detection using traditional approaches

For flame detection, the traditional CV approaches considered multiple image features of flame, including spectral, spatial, and temporal features. Discriminating fire and non-fire objects by the color-based approach using various color coordinates (e.g., RGB, YUV and YCbCr) resulted in good detection rates. Concerning flame luminance, the statistical model analyzing input images in YCbCr color mode proposed by Celik et al. [51] resulted in a high flame detection rate. However, to reduce false positives, the features of the shape change and disordered movement of flame as well as flickering are also included. Spatiotemporal feature and dynamic texture analysis were used in the real-time flame detection model proposed by Dimitropoulos et al. [52]. Chromatic, dynamic texture and contour features of flame are analyzed to extract the moving regions and check the colors of moving fire pixels in Rui et al. [53]. The motion history information is also analyzed for checking flame flickering frequency, and LBP features are extracted and used to train the SVM classifier.

Compared with the detection of smoke patterns, flame patterns are easier as most flame patterns are in brighter scenes that provide better contrast and overall quality to the input image. As a result, most of the techniques and approaches for smoke detection are also suitable for flame detection with a small amount of parameter tuning.

3.4 Deep Learning Approaches for Video Image Detection for Smoke and Fire

As briefly discussed in Sect. 3.2, Convolutional Neural Network (CNN) [54] computes the steps of feature extraction/object localization and object classification/detection as one big task in end-to-end manner. The entire process of the CNN involves deep layers to automatically make spatial correlations between those image pixels that are likely to share the same features in convolutional layers, which is also the core part of CNN, in finding out the more precise feature for the task. In the convolutional layer, the most crucial element is kernels (also known as filters or called weights) [55], which are essentially a matrix of values that could be trained and altered to help detect specific features. Therefore, CNNs spatially convolve the kernel matrix on an input image to detect its features matching to those trained in the kernel matrix, and the kernel matrix could also be updated by the feedback from the last layer of CNN and the target, using an error function. Subsequently, the richness and diversity of the training dataset is an essential factor in a successful CNN approach for smoke and fire detection tasks, as the image features of smoke and flame might vary dramatically in different fire scenarios and environments.

In the field of flame and smoke detection, many studies have achieved good results in some general tasks for fire research, such as fire detection [56], smoke detection [57], and fire prediction [58]. Some of them even greatly surpass the performance of traditional methods, which demonstrates the potential superiority of deep learning-based approach over conventional ones.

Smoke Detection

Smoke detection is regarded as an object detection/segmentation task in CV by deep learning approaches. In fact, there are many successful and efficient approaches being proposed for general detection tasks outside of the fire safety area. However, researchers have found that those successes could not be easily adopted due to several reasons [59], such as the relatively lower quality of training smoke image samples due to equipment factors, varying environment and background, and the nature of smoke patterns itself, which has a lower contrast and darker images, leading to a much harder detection task without pre-processing.

As a result, researchers have been trying to propose an additional mechanism or modify previous deep learning models. A few researchers claimed that they solved the challenging problem of smoke detection by employing a Dual-Channel Convolutional Neural Network (DC-CNN) using transfer learning for detecting smoke images [60]. Their network achieved a high detection rate on a publicly available dataset. Besides, Hu et. al. also reported a high detection rate by using Spatio-Temporal CNN for a video smoke detection in [61]. Recurrent Neural Networks (RNNs) are also considered for smoke detection in a study by Yin et al. [62]. Their model is able to capture the spatial and dynamic motion context in smoke videos, which works well in video data.

Nevertheless, there are some open problems that remain to be solved [63]. For example, most of the existing approaches are only validated on a private dataset, which makes it almost impossible to compare different approaches. Furthermore, many previous studies used pre-processing methods to enhance images before training and testing, but the pre-processing cannot be implemented for real-time fire image processing tasks.

Flame Detection

Flame detection is a sub-direction of the object detection/segmentation tasks in the general CV area with more focus on the flame patterns. Unlike smoke detection task, flame patterns are relatively easier to be recognized than smoke patterns with a dark background [64]. As a result, most of the proposed deep neural networks for flame detection are less complicated than those for smoke detection.

There are many existing works on flame detection by deep learning. In [65], Xu et al. proposed a fire detection method based on color features using wavelet analysis and CNN. It is the first appearance of CNN in fire flame detection, and it was only used as a complementary method for decision making. After that, Zhong et al. in [66] proposed a novel flame detection algorithm using CNN. Their proposed method achieved the outstanding performance with a high accuracy. Kim et al. [13] brought in Faster R-CNN (Region-Based CNN) for the same task and achieved a comparable accuracy with a much higher detection speed. Filonenko et al. proposed a combination of a CNN and RNN to detect flame benefiting of using temporal information in parallel to special data [67]. The CNN automatically builds the low-level features, whereas the RNN finds the relation between the features in different frames of the same event.

Figure 4 shows examples of flame and smoke detection by segmentation techniques on various popular algorithms. Instead of image detection techniques, several researchers focused to use benefits of image segmentation methods for the task of flame detection. PSPNet (Pyramid Scene Parsing Network) was proposed by Zhao et al. in [68] based on a pyramid pooling module, which is able to process information in different scales and achieve a better segmentation performance. DeepLab v3 [69] is another popular segmentation architecture that is equipped with several modifications and improvement. Modules in the network are employed with atrous convolution in cascade or in parallel to capture multi-scale context by adopting multiple atrous rates, which is able to handle the problem of segmenting objects at multiple scales. Also, there are image-level features encoding global context to boost the performance. Besides, they also change the process of down-sample and up-sample within the network to help reduce computational complexity and improve feature extraction performance at the same time. The last algorithm in the figure is TD-Net [70], which is different from the previous two approaches. It is designed for video segmentation, which utilizes a mechanism that takes advantage of not only each image in the video, but also the motion flow between them. It is equipped with several attention extraction modules to help focus on the key part of each frame. The extraction results would be aggregated by a motion-based attention mechanism to help combine and compare the semantic information across different images. Comparing the segmentation results for the boundaries, smoothness, and the accuracy with GT, we could observe from Fig. 4 that TD-Net provides the best segmentation results for both flame and smoke patterns for three different inputs.

Despite all the advantages listed in the above paragraphs of flame detection over smoke detection, there are certain challenges for flame detection tasks due to the

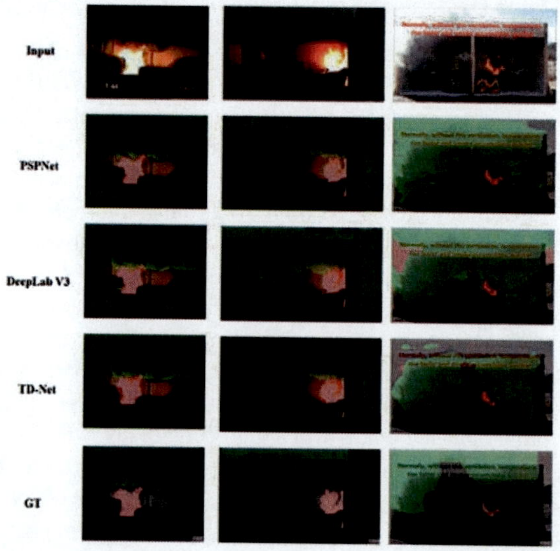

Fig. 4 Examples of flame and smoke detection by semantic segmentation techniques on various popular algorithms, including PSPNet, DeepLab V3 and TD-Net. The input images are from videos of firefighting experiments and demonstration. The label at left denotes the source of each row of images. Red patterns refer to flame, and green patterns refer to smoke. 'GT' refers to the ground truth, which is manually annotated by human

nature of those flame patterns. First of all, high detection rates would likely result in high false positive rate [66]. It means that deep learning models tend to segment all the bright patterns in the images, like light and reflections from other objects as flame. While it is difficult to reduce the false positives, inter-image features, such as optical flow and motion flow, could help in reducing this false segmentation issue. Besides, as the flame usually appears to be bright, it could also affect the dynamic range of capture devices resulting in a brighter image and loss of darker features from the background. This is a limitation of capture hardware, which means that image pre-processing would be helpful for this issue.

3.5 Thermal Image Data Processing

Smoke and flame are hard to be detected by vision-based solutions in dark or obstructed conditions. There are a few solutions based on regular cameras that are effective for both daytime and nighttime detection of smoke and flame. The use of thermal cameras is considered to overcome the drawback and adopt the technique for image fusion utilizing vision and thermal devices [71].

Thermal image data

Thermal images of a space with a heat source of fire are captured and generated by InfraRed (IR) cameras, which visualize the dispersion of electromagnetic radiation from the heat source and the heated surrounding surfaces in space [72]. There are many types of IR cameras with different specifications suited to different applications, including fire detection and firefighting, which are dependent on the designed wavelength (i.e., short, mid or long spectrum) of the camera. In adopting IR cameras to smoke and fire recognition, the understanding of IR imaging principles and technical specifications (e.g., temperature range and sensitivity) are necessary to ensure a fluent and accurate thermal image capture. Also, calibration/training is required to understand a wide range of setting parameters and tailor them to each application and potential fire scenarios since the measurement data are generally sensitive to the target and environment of image shooting [73].

Thermal image data processing

The advantages of using thermal images come from their measurement, that is, the temperature/energy intensity of each pixel. IR cameras can provide thermal data even in challenging environments (in dark and on fireground). This means that feature extraction/classification of images of smoke and fire is less affected by the environment with thermal image data than regular vision image data. Consequently, thermal image data enable us to overcome the application limitation and false-alarm likelihood of regular vision cameras [74].

Adopting the image processing and computer vision technologies used with regular vision data (e.g., RGB data), thermal video image detection technologies have been enhanced for wide applications, including object detection, object classification and scene reconstruction. Thermal images are also computed for segmenting the region of interest and object detection (smoke/fire) using feature extraction [75].

The use of thermal data is discussed in smoke detection during the night [76–79]. Thermal images from an LWIR camera are analyzed by Verstockt et al. [80] for the spatial and temporal features of the thermal images. They discussed the application of LWIR thermal images to hot spot detection by looking through the smoke environment. For forest fire detection, thermal images are used by Bosch et al. [81, 82] for analyzing forest fire features of persistence and growth. They reported that thermal data could also be affected by environmental factors (e.g., wind, precipitation, and cloud).

Data fusion/image transformation with RGB

The image fusion of visible and thermal data is explored for effective smoke and fire detection. Verstockt et al. [83] used both vision and thermal data to detect smoke and flame, which enabled fire localization and fire size estimation in three dimensions. Dios et al. [84] used both visible and infrared cameras placed on ground stations and mounted on UAS (unmanned aerial systems) to acquire forest fire information on the location, shape, height, spread rate. In combining the information obtained from different cameras, a statistical data fusion approach is used to focus on the direct view of fire origin.

Deep neural networks have shown their dominant performance against traditional feature engineering algorithms and even human abilities in various applications, including image translation. The recent success of deep learning models demonstrated another possibility of data transformation between thermal and vision images [85] by learning the mapping of certain objects between the pixels of thermal image and vision images. In a recent work [86] conducted by NRCC, the authors introduced a deep learning network that is designed for transforming RGB (vision) images to thermal images. They showed a good preliminary result on compartment fire tests achieving a good performance in both high and low temperature region thanks to its attention mechanism. The data conversion between RGB and thermal images shows the possibility of assisting in the analytics of flame and fire while compensating for the limitations of RGB and thermal images.

3.6 Performance of Deep Learning Approach

In this subsection, techniques to enhance the accuracy and speed of deep learning methods will be explained briefly. By solving existing problems in CV and deep learning field, we can pave the way for implementing fully automatic and real-time systems for the field of fire safety to better study fire and save lives.

Non-stationary Camera

Almost all aforementioned techniques considered stationary camera images, and the stationary images were uncomplicated in the analysis of the image features of smoke and flame motion and their wavelets that are prominent in detection. The analyses of images from non-stationary cameras are also in demand for the smoke and fire detection of forest fires and smart/robot firefighting solutions.

One approach used in the traditional image processing is to use a covariance descriptor and Background Subtraction Method (i.e., detect objects by eliminating non-stationary background in complex environment). Using this approach, Habiboglu et al. [87] converted input video into spatiotemporal blocks and developed the technique that allows the use of non-stationary cameras mounted on UAVs to detect forest fires. Zhang et al. [88] proposed a method for forest fire detection working with non-stationary cameras by localizing fire area and analyze them at the patch level. The main idea of these methods is to highlight region of interest, and work on processed images.

Movements of a camera can be formulated as image transformation comprising of translation, rotation, and scaling. Fortunately, in deep learning literature, CNN models are popular in CV because they are translational invariance due to the existence of local and global pooling layers in CNN network structures [89, 90]. At the same time, the application of data augmentation (i.e., applying various random transformations to each training image) [91, 92] provides CNN architectures with a semi-invariance to image rotation and scaling. Therefore, by employing these strategies in designing and training of CNN models, they can detect, segment, and even predict behavior of smoke and flame in images with different transformations. In other words, movement of camera is not a significant problem for study of fire using non-stationary camera setups.

Computational cost

For accurate smoke and fire detection, algorithms should combine various image features including dynamic texture descriptor analysis and deep learning layers. New deep learning techniques are promising in the accurate detection, but computational costs could be high, which inhibit deployment of the new solutions in real-time smoke and fire detection on fire ground. We are still unable to accurately predict the time it will take to train a deep learning network to solve a given problem [93, 94]. Computational cost of training and validation of deep learning models comes from two factors: (1) size of the network where more than hundreds of million parameters should be tuned in each iteration, and (2) the number and the size of training images (i.e., approximately million images with large pixel sizes are used in a typical CV application). Employing data enhancement techniques (e.g., data augmentation) and combining more than one deep learning method, exacerbate this problem.

Various techniques have been proposed to decrease computational cost of deep learning methods [95, 96]. However, as they employ deeper networks, advancement in this field will require dramatically more computationally-efficient methods [94]. Based on the application, a new advanced CNN based approach could maintain the performance while reducing the computational burden by adopting the traditional image pre-processing [97], such as using new training strategies (e.g., transfer learning and domain adaption [96]), and employing lighter and smaller network structures [98] for specific tasks. Other alternatives are utilizing more efficient hardware [99] and software [100] facilities.

Evaluation Metrics

Various evaluation metrics have been proposed for computer vision and image processing tasks [101, 102]. For instance, accuracy of fire and smoke detection using one CV technique can be assessed by a comparison between the output of the technique and manual detection results. Using mathematical comparisons, we can evaluate the performance of deep learning models during training, validation, and testing using a criterion (e.g., objective, loss, or error function between labeled or ground truth data and predicted or output data). Common assessment criteria for deep learning models are confusion matrix [101], logarithmic losses (e.g., SoftMax or Sigmoid), statistical (e.g., absolute error and mean square errors) and entropy-based losses (cross-entropy) [103] for classification and regression tasks and similarity-based losses such as dice or intersection over union (IoU) [104, 105] for object detection and segmentation tasks. There are also custom-designed loss functions for specific tasks or specific deep learning model (e.g., min–max loss for GAN).

Validation Dataset

One reason for the success of deep learning models in CV tasks is the advance of new digital cameras and sensors available for collecting higher quality and larger image datasets for better training and validation [105]. Detecting fire and smoke in image data is not exceptional, and deep learning models require a large training dataset. One of the problems for vision-based techniques for fire and smoke analysis is lack of real datasets [106]. The reason is that conducting actual fire tests is expensive and laborious. Furthermore, datasets should be balanced, meaning that the number of images with specific object (i.e., fire and smoke) should be equal between each object, otherwise, deep learning model could be biased for one object over another [107]. Techniques such as employing simulation data, data augmentation, and using small-scale fire tests can alleviate this issue, but still, it is a burden for progress of vision-based techniques. Moreover, even if validated on one fire scenario, a model might not work well on another scenario with different fire scenario/conditions. Current technique to increase the generalization ability of deep learning models are transfer learning and domain adaptation [106, 108–110]. Also combining data from various datasets, even mix of synthetic and real images, can be another alternative [110].

4 Image-Based Prediction of Sudden Fire Propagation Using Deep Learning Approaches

Sudden fire propagation phenomena

More than 30,000 firefighters get injured each year during firefighting operations [11]. Therefore, the prediction of the fire phenomena is a crucial part of fire analysis, evacuation planning, and fire service's decision-making. There are many sudden fire propagation phenomena [111], such as backdraft [112], fire gas ignition [113], smoke explosion [114], flash fire [115], flameover [111], backdraft and flashover [116]. In this chapter, we focus on the most common enclosure fire phenomenon, "Flashover".

Flashover might happen in an enclosure/room fire when all the combustible materials in the room are suddenly ignited near-simultaneously since the hot smoke layer accumulated near the ceiling rapidly heats the exposed surface of all the items in the room. Figure 5A–D shows a few images from a room fire test. It takes less than 3 min from ignition to flashover onset. Failing to recognize signs of the impending flashover could cost the lives of occupants in the building. While there are no absolute criteria for flashover prediction, the onset of flashover generally coincides with the room ceiling temperature reaching around 600 °C, and the heat flux (HF) rate on the floor reaching approximately 20 kW/m^2 [117, 118] as well as the growth of the Heat Release Rate (HRR).

Traditional monitoring techniques

Various investigations have been conducted to develop a new method for flashover detection and prediction, employing HRR, HF, and temperature data [117–119]. The national institute of standards and technology (NIST) generated several sets of big synthetic data using the zone fire model, CFAST [120]. They utilized Graph Neural Network (GNN) [121] as a binary classifier to determine the possibility of flashover impending on two fixed time spans. Artificial neural network (ANN) models using fuzzy and regression models [122, 123] and probabilistic mapping with maximum entropy [124] were also proposed to predict the occurrence of the flashover based on the HRR, temperature, and the geometry of a fire compartment.

Data recorded from reduced-scale room fire experiments were used in [125–127] for training of machine learning models. Researchers used a convolutional

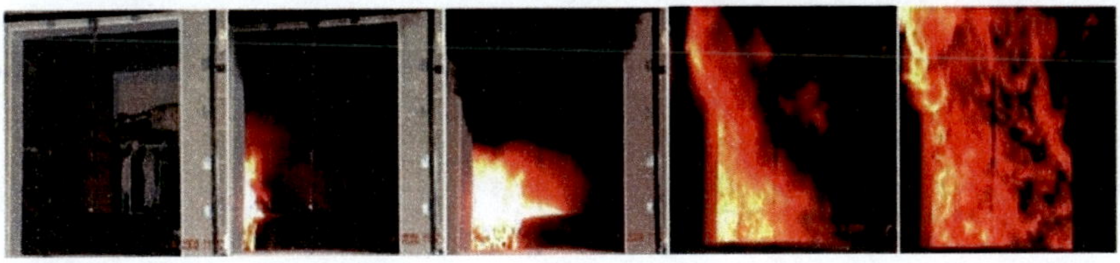

Fig. 5 The rapid growth of fire in a test room with modern upholsteries. From left to right, it takes less than 4 min from ignition to flashover

neural network accompanied by long short-term memory layers (CNN-LSTM) as a regression model [122, 124, 128, 129], which detected abrupt and extreme thermal measurement changes to predict flashover. The main problem of these techniques is that several sensors for thermocouples, HF and HRR with sufficient high-temperature resistance should be fixed readily in specific places in a fire compartment. Therefore, these approaches cannot be easily utilized by firefighters in actual fire ground.

Image-based flashover detection

These days, firefighters have been equipped with different types of RGB and IR cameras. Also, flashover onset comes with a few visual signs, such as "extrusive flame" and "fire rollover" [106, 125, 130].

Based on vision data rather than traditional sensor-based data, recently few studies have attempted to develop deep learning techniques for flashover detection. Figure 6 shows a typical CNN model for flashover detection. For instance, some studies [12, 86] employed RGB and IR cameras to predict flashover in room fires [12], and the authors in [131] proposed a preliminary method for prediction of flashover time based on the dynamic change of measured smoke area in RGB video data. A popular CNN model Mask RCNN, often used for object detection and segmentation, was employed for the segmentation of smoke areas in [131]. Then, flashover was detected from the trend of the smoke area in recorded videos of room fires.

In [8], the authors used a binary CNN model (VGG16 [132] an optimized object recognition model with 16 layers) to detect the occurrence of flashover 30 s in advance. They trained VGG16 over a significantly large dataset of smoke images in two view angles generated by CFD modelling. Their model reached a detection accuracy of ~80% with the CFD simulated data. Limited studies also attempted to use thermal infrared (IR) data from actual room fires [133] or transforming RGB videos into thermal images using a generative adversarial network (GAN) [86, 134]. A new CNN model proposed recently [106] as a multi-label classifier. This technique was the first sample of training a CNN model, all on actual full-scale RGB room fire test data. This work proved that the onset of flashover phenomena has unique visual

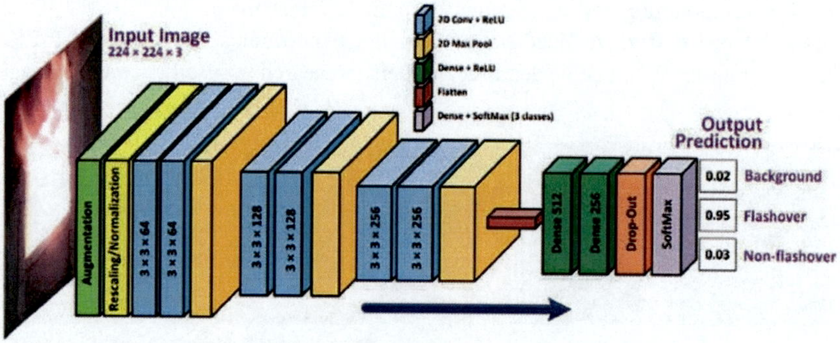

Fig. 6 FlashoverNet architecture for detection of flashover images [106]

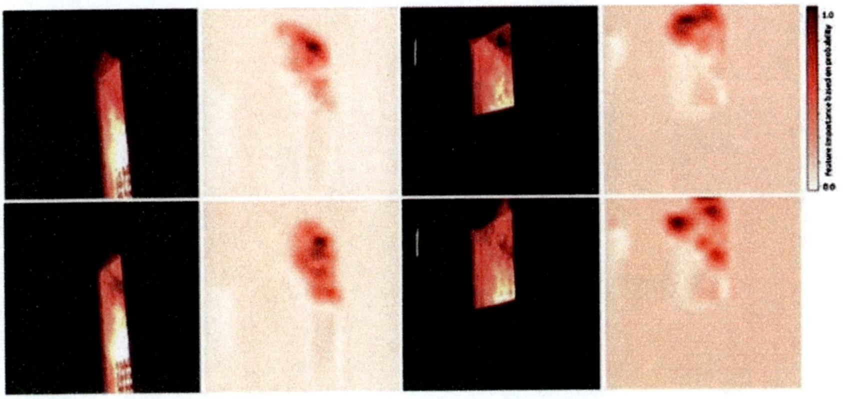

Fig. 7 Random features extracted from sample flashover images in room fire test experiments. The left columns are input test images, and the right columns are corresponding class activation maps (CAM) [106]

features, which can be found as extrusive flame and smoke crossing the head of a vent (Fig. 7).

Results of recent investigations in vision based deep learning approaches for detection and prediction of flashover [135, 136] revealed that these methods can be a promising alternative to the traditional techniques. In comparison, these methods are repeatable, generalizable, and most importantly capable of using for real-time applications. Based on the availability of data type and field of view in vision data, it is possible to utilize RGB and/or IR data [12, 137]. In general, IR data, if available, can be used to find common flashover signs, such as temperature of hot upper smoke layer, and RGB data could be added to provide new features for detection and prediction of flashover. As a comparison example, Fig. 8 is the result of the CNN model proposed by [11] on IR data. Figure 9 shows one example of image conversion between thermal and RGB data, which was made for analysis of fire and smoke. Figure 10 is a comparison between the performance of different deep learning techniques for RGB data obtained from a room fire test [106]. Figure 10 shows the probability of flashover onset resulted from the image-based CNN models in comparison to the HRR, HF, and temperature measured in the fire tests. The experimental results indicated that the CNN models are capable of extracting distinctive flashover attributes, and based on those features, they can detect flashover onset even a few seconds in advance.

5 Conclusion and Future Works

Since fires release heat, light, gases and particles, image-based detection/recognition of smoke and flame relies on cameras and imageries acting as volume sensors in various emission spectrums (e.g., visible range, infrared range). In this chapter

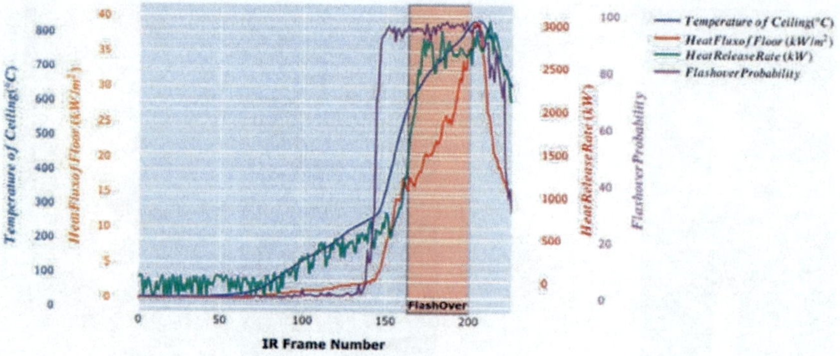

Fig. 8 Results of flashover detection by ResNet18 model trained from scratch, which is compared with the flashover onset verified by the sensor data from a room fire test [11]

Fig. 9 One example of data conversion from RGB to IR data [12]

we discussed how the image data are processed for extracting and analyzing the visual features of flame and smoke, such as shape, shade, motion, flickering, color/temperature, emission intensity, and behavior characteristics such as rapid growth behavior, which vary depending on fire scenarios. We explained methods of image recognition in the field of fire safety for the task of fire and smoke detection, recognition, and segmentation. By enumerating and summarizing these methods, we categorized these into conventional and state-of-the-art techniques. Conventional vision-based methods are iterative algorithms based on feature engineering, require human manipulations and manual processing. On the other side of the spectrum, state-of-the-art approaches benefit from pattern recognition and automatic feature extraction ability of artificial intelligent for aforementioned tasks. The discussion will include the localization/recognition of smoke and flame at the ignition stage as well

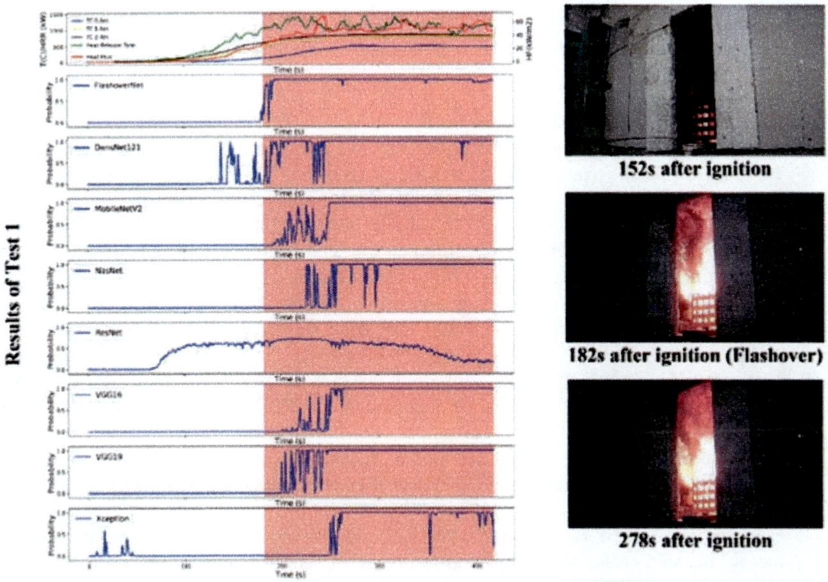

Fig. 10 Comparison of different deep learning models for predicting a flashover occurrence with image data from a fire test. On the left, the performance of CNN models of flashover detection is plotted. On the right, three frames of the fire test are shown [106]

as the detection/prediction of sudden flame propagations (e.g., room flashover) in the growth stage. We also reviewed advanced technologies adopting image processing (e.g., image fusion) and artificial intelligence (e.g., deep learning) methods in this chapter which have demonstrated notable potential to overcome the weaknesses of conventional image-based flame and smoke detection systems. Although deep learning computer vision approaches are powerful enough in automatic detection and recognizing fire and smoke even in real-time, yet lack of image data recorded from actual fire ground is the main challenge in this field. By increasing the use of imagery devices in fire incidents and growing the usage of smart and IoT devices in buildings, we will see more vision-based methods and tools in this field for firefighting purposes.

References

1. Z. Wang, X. He, J.-X. Wang et al., Combustion visualization and experimental study on spark induced compression ignition (SICI) in gasoline HCCI engines. Energy Convers. Manag. **51**, 908–917 (2010)
2. Y. Wang, Y. Yu, X. Zhu, Z. Zhang, Pattern recognition for measuring the flame stability of gas-fired combustion based on the image processing technology. Fuel **270**, 117486 (2020)

3. S. Sarkar, K.G. Lore, S. Sarkar et al., Early detection of combustion instability from hi-speed flame images via deep learning and symbolic time series analysis, in *Annual Conference of the PHM Society* (2015)
4. D.T. Gottuk, J.A. Lynch, S.L. Rose-Pehrsson et al., Video image fire detection for shipboard use. Fire Saf. J. **41**, 321–326 (2006)
5. T.X. Tung, J.-M. Kim, An effective four-stage smoke-detection algorithm using video images for early fire-alarm systems. Fire Saf. J. **46**, 276–282 (2011)
6. Z. Wang, T. Zhang, X. Huang, Predicting real-time fire heat release rate by flame images and deep learning, in *Proceedings of the Combustion Institute* (2022)
7. T. Zhang, Z. Wang, Y. Zeng et al., Building artificial-intelligence digital fire (AID-Fire) system: a real-scale demonstration. J. Build. Eng. **62**, 105363 (2022)
8. Z. Wang, T. Zhang, X. Wu, X. Huang, Predicting transient building fire based on external smoke images and deep learning. J. Build. Eng. **47**, 103823 (2022). https://doi.org/10.1016/j.jobe.2021.103823
9. J. Wang, W.C. Tam, Y. Jia et al., P-Flash–A machine learning-based model for flashover prediction using recovered temperature data. Fire Saf. J. **122**, 103341 (2021). https://doi.org/10.1016/j.firesaf.2021.103341
10. M.Z. Naser, C. Lautenberger, E. Kuligowski, Special Issue on "Smart Systems in Fire Engineering." Fire Technol. **57**, 2737–2740 (2021). https://doi.org/10.1007/s10694-021-01196-w
11. H. Mozaffari, Y. Li, Y. Ko et al, Detecting flashover in a room fire based on the sequence of thermal infrared images using convolutional neural networks. Canadian AI 2022 (2022)
12. Y. Li, Y. Ko, W. Lee, RGB image-based hybrid model for automatic prediction of flashover in compartment fires. Fire Saf. J. **132**, 103629 (2022)
13. B. Kim, J. Lee, A video-based fire detection using deep learning models. Appl. Sci. **9**, 2862 (2019)
14. F.N. Simon, G.D. Rork, Ionization-type smoke detectors. Rev. Sci. Instrum. **47**, 74–80 (1976)
15. K. Lee, Y.-S. Shim, Y.G. Song et al., Highly sensitive sensors based on metal-oxide nanocolumns for fire detection. Sensors **17**, 303 (2017)
16. (2023) ORRProtection Co. https://www.orrprotection.com/detection/video-smoke-detection
17. K. Muhammad, J. Ahmad, I. Mehmood et al., Convolutional neural networks based fire detection in surveillance videos. IEEE Access **6**, 18174–18183 (2018)
18. G. Yadav, V. Gupta, V. Gaur, M. Bhattacharya, Optimized flame detection using image processing based techniques. Indian J. Comput. Sci. Eng. **3**, 202–211 (2012)
19. N. Rahmatov, A. Paul, F. Saeed, H. Seo, Realtime fire detection using CNN and search space navigation. J. Real-Time Image Proc. **18**, 1331–1340 (2021)
20. J.A. Milke, Monitoring multiple aspects of fire signatures for discriminating fire detection. Fire Technol. **35**, 195–209 (1999)
21. W. Kim, Y. Sivathanu, J.P. Gore, Characterization of spectral radiation intensities from standard test fires for fire detection. NIST Special Publication SP 91–106 (2001)
22. B.C.H.R. Hagen, J.A. Milke, The use of gaseous fire signatures as a mean to detect fires. Fire Saf. J. **34**, 55–67 (2000)
23. L. Chen, P. Mao, Y. Zhang et al., Experimental study on smoke characteristics of bifurcated tunnel fire. Tunn. Undergr. Space Technol. **98**, 103295 (2020)
24. Y. Yong, Q. Tian, L.U. Gang et al., Recent advances in flame tomography. Chin. J. Chem. Eng. **20**, 389–399 (2012)
25. C. Yuan, Z. Liu, Y. Zhang, Vision-based forest fire detection in aerial images for firefighting using UAVs, in *2016 International Conference on Unmanned Aircraft Systems (ICUAS). IEEE* (2016), pp. 1200–1205
26. J.-H. Kim, S. Jo, B.Y. Lattimer, Feature selection for intelligent firefighting robot classification of fire, smoke, and thermal reflections using thermal infrared images. J. Sensors (2016)
27. R. Méndez-Rial, Souto-Lpez , Garca-Daz A, MWIR infrared gating imaging with uncooled PbSe FPAs for surveillance applications, in *Unconventional Optical Imaging. SPIE* (2018), pp. 564–569

28. J.-H. Kim, B. Keller, B.Y. Lattimer, Sensor fusion based seek-and-find fire algorithm for intelligent firefighting robot, in *2013 IEEE/ASME International Conference on Advanced Intelligent Mechatronics*. IEEE (2013), pp. 1482–1486
29. K. Jindal, A. Wang, D. Thakur et al, Design and deployment of an autonomous unmanned ground vehicle for urban firefighting scenarios (2021). arXiv preprint arXiv:210703582
30. J.W. Starr, B.Y. Lattimer, Evaluation of navigation sensors in fire smoke environments. Fire Technol. **50**, 1459–1481 (2014)
31. J.W. Starr, Rangefinding in fire smoke environments (2016)
32. J.-H. Kim, J.W. Starr, B.Y. Lattimer, Firefighting robot stereo infrared vision and radar sensor fusion for imaging through smoke. Fire Technol. **51**, 823–845 (2015)
33. A.M. Fernandes, A.B. Utkin, A.V. Lavrov, R.M. Vilar, Neural network based recognition of smoke signatures from lidar signals. Neural Process. Lett. **19**, 175–189 (2004)
34. A.B. Utkin, A. Fernandes, F. Simes et al., Feasibility of forest-fire smoke detection using lidar. Int. J. Wildland Fire **12**, 159–166 (2003)
35. A.S. Olagoke, H. Ibrahim, S.S. Teoh, Literature survey on multi-camera system and its application. IEEE Access **8**, 172892–172922 (2020)
36. V. Sherstjuk, M. Zharikova, I. Sokol, Forest fire-fighting monitoring system based on UAV team and remote sensing, in *2018 IEEE 38th International Conference on Electronics and Nanotechnology (ELNANO)*. IEEE (2018), pp. 663–668
37. N. Lazaros, G.C. Sirakoulis, A. Gasteratos, Review of stereo vision algorithms: from software to hardware. Int. J. Optomechatronics **2**, 435–462 (2008)
38. R. Szeliski, *Computer Vision: Algorithms and Applications* (Springer Nature, 2022)
39. R.C. Gonzalez, *Digital Image Processing* (Pearson Education India, 2009)
40. N. O'Mahony, S. Campbell, A. Carvalho et al., Deep learning vs. traditional computer vision. in*Advances in Computer Vision: Proceedings of the 2019 Computer Vision Conference (CVC)*, vol. 1 1. (Springer, 2020), pp. 128–144
41. A. Garcia-Garcia, S. Orts-Escolano, S. Oprea et al., A review on deep learning techniques applied to semantic segmentation (2017). arXiv preprint arXiv:170406857
42. T.-Y. Lin, M. Maire, S. Belongie et al., Microsoft coco: common objects in context, in *European Conference on Computer Vision* (Springer, 2014), pp. 740–755
43. Z.-Q. Zhao, P. Zheng, S. Xu, X. Wu, Object detection with deep learning: a review. IEEE Trans. Neural Netw. Learn. Syst. **30**, 3212–3232 (2019)
44. Y. LeCun, Yoshua bengio, and geoffrey hinton. Deep Learn. Nat. **521**, 436–444 (2015)
45. T.-H. Chen, P.-H. Wu, Y.-C. Chiou, An early fire-detection method based on image processing, in *2004 International Conference on Image Processing, 2004. ICIP'04*. IEEE (2004), pp. 1707–1710
46. S. Dara, P. Tumma, Feature extraction by using deep learning: a survey, in *2018 Second International Conference on Electronics, Communication and Aerospace Technology (ICECA)*. IEEE (2018), pp. 1795–1801
47. H. Ali, M. Sharif, M. Yasmin et al., A survey of feature extraction and fusion of deep learning for detection of abnormalities in video endoscopy of gastrointestinal-tract. Artif. Intell. Rev. **53**, 2635–2707 (2020)
48. L. Juan, O. Gwun, A comparison of sift, pca-sift and surf. Int. J. Image Process. (IJIP) **3**, 143–152 (2009)
49. M.A. Chandra, S.S. Bedi, Survey on SVM and their application in image classification. Int. J. Inf. Technol. **13**, 1–11 (2021)
50. A. Gaur, A. Singh, A.A. Kumar et al., Video flame and smoke based fire detection algorithms: a literature review. Fire Technol. **56**, 1943–1980 (2020). https://doi.org/10.1007/s10694-020-00986-y
51. D. Han, B. Lee, Flame and smoke detection method for early real-time detection of a tunnel fire. Fire Saf. J. **44**, 951–961 (2009)
52. K. Dimitropoulos, P. Barmpoutis, N. Grammalidis, Spatio-temporal flame modeling and dynamic texture analysis for automatic video-based fire detection. IEEE Trans. Circuits Syst. Video Technol. **25**, 339–351 (2014)

53. R. Chi, Z. Lu, Q. Ji, Real-time multi-feature based fire flame detection in video. IET Image Proc. **11**, 31–37 (2017)
54. K. O'Shea, R. Nash, An introduction to convolutional neural networks (2015). arXiv preprint arXiv:151108458
55. L. Alzubaidi, J. Zhang, A.J. Humaidi et al., Review of deep learning: concepts, CNN architectures, challenges, applications, future directions. J. Big Data **8**, 1–74 (2021)
56. Q. Zhang, J. Xu, L. Xu, H. Guo, Deep convolutional neural networks for forest fire detection. in *2016 International Forum on Management, Education and Information Technology Application* (Atlantis Press, 2016), pp. 568–575
57. S. Khan, K. Muhammad, S. Mumtaz et al., Energy-efficient deep CNN for smoke detection in foggy IoT environment. IEEE Internet Things J. **6**, 9237–9245 (2019)
58. F. Abid, A survey of machine learning algorithms based forest fires prediction and detection systems. Fire Technol. **57**, 559–590 (2021)
59. L. He, X. Gong, S. Zhang et al., Efficient attention based deep fusion CNN for smoke detection in fog environment. Neurocomputing **434**, 224–238 (2021)
60. F. Zhang, W. Qin, Y. Liu et al., A dual-channel convolution neural network for image smoke detection. Multim. Tools Appl. **79**, 34587–34603 (2020)
61. Y. Hu, X. Lu, Real-time video fire smoke detection by utilizing spatial-temporal ConvNet features. Multim. Tools Appl. **77**, 29283–29301 (2018)
62. M. Yin, C. Lang, Z. Li et al., Recurrent convolutional network for video-based smoke detection. Multim. Tools Appl. **78**, 237–256 (2019)
63. A. Filonenko, L. Kurnianggoro, K.-H. Jo, Comparative study of modern convolutional neural networks for smoke detection on image data, in *2017 10th International Conference on Human System Interactions (HSI)*. IEEE (2017), pp. 64–68
64. D. Shen, X. Chen, M. Nguyen, W.Q. Yan, Flame detection using deep learning, in *2018 4th International Conference on Control, Automation and Robotics (ICCAR)*. IEEE (2018), pp. 416–420
65. Z. Xu, W. Wanguo, L. Xinrui et al., Flame and smoke detection in substation based on wavelet analysis and convolution neural network, in *Proceedings of the 2019 3rd International Conference on Innovation in Artificial Intelligence* (2019), pp. 248–252
66. Z. Zhong, M. Wang, Y. Shi, W. Gao, A convolutional neural network-based flame detection method in video sequence. SIViP **12**, 1619–1627 (2018)
67. A. Filonenko, L. Kurnianggoro, K.H. Jo, Smoke detection on video sequences using convolutional and recurrent neural networks, in *International Conference on Computational Collective Intelligence* (Springer, 2017), pp. 558–566
68. H. Zhao, J. Shi, X. Qi et al., Pyramid scene parsing network, in *Proceedings of the IEEE Conference on Computer Vision and Pattern Recognition* (2017), pp. 2881–2890
69. L.-C. Chen, G. Papandreou, F. Schroff, H. Adam, Rethinking atrous convolution for semantic image segmentation (2017). arXiv preprint arXiv:170605587
70. P. Hu, F. Caba, O. Wang et al., Temporally distributed networks for fast video semantic segmentation. in *Proceedings of the IEEE/CVF Conference on Computer Vision and Pattern Recognition* (2020), pp. 8818–8827
71. W. Krüll, R. Tobera, I. Willms et al., Early forest fire detection and verification using optical smoke, gas and microwave sensors. Procedia Eng. **45**, 584–594 (2012)
72. R. Gade, T.B. Moeslund, Thermal cameras and applications: a survey. Mach. Vis. Appl. **25**, 245–262 (2014)
73. K. Ribeiro-Gomes, D. Hernndez-Lpez, J.F. Ortega et al., Uncooled thermal camera calibration and optimization of the photogrammetry process for UAV applications in agriculture. Sensors **17**, 2173 (2017)
74. W.H. Maes, A.R. Huete, K. Steppe, Optimizing the processing of UAV-based thermal imagery. Remote Sens. **9**, 476 (2017)
75. L. Deng, Q. Chen, Y. He et al., Fire detection with infrared images using cascaded neural network. J. Algorithms Comput. Technol. **13**, 1748302619895433 (2019)

76. L. YunChang, Y. ChunYu, Z. YongMing, Nighttime video smoke detection based on active infrared video image, in *2010 International Conference on Electrical and Control Engineering*. IEEE (2010), pp. 1359–1362
77. B.U. Treyin, Fire detection algorithms using multimodal signal and image analysis (2009)
78. O. Günay, K. Taşdemir, B.U. Töreyin, A.E. Çetin, Video based wildfire detection at night. Fire Saf. J. **44**, 860–868 (2009)
79. S. Verstockt, R. Dekeerschieter, A. Vanoosthuyse, et al., Video fire detection using non-visible light, in *Proceedings of the 6th International Seminar on Fire and Explosion Hazards* (2010)
80. S. Verstockt, A. Vanoosthuyse, S. Van Hoecke, et al., Multi-sensor fire detection by fusing visual and non-visual flame features, in *International Conference on Image and Signal Processing*. Springer (2010), pp. 333–341
81. I. Bosch, S. Gomez, L. Vergara, J. Moragues, Infrared image processing and its application to forest fire surveillance, in *2007 IEEE Conference on Advanced Video and Signal Based Surveillance*. IEEE (2007), pp. 283–288
82. I. Bosch, S. Gomez, R. Molina, R. Miralles, Object discrimination by infrared image processing, in *International Work-Conference on the Interplay Between Natural and Artificial Computation*. Springer (2009), pp. 30–40
83. S. Verstockt, N. Tilley, B. Merci et al., Future directions for video fire detection. in *10th International Symposium on Fire Safety Science*. International Association for Fire Safety Science (IAFSS) (2011), pp. 529–542
84. J.R. Martnez-de Dios, L. Merino, F. Caballero, A. Ollero, Automatic forest-fire measuring using ground stations and unmanned aerial systems. Sensors **11**, 6328–6353 (2011)
85. R. Zhang, J. Bin, Z. Liu, E. Blasch, WGGAN: a wavelet-guided generative adversarial network for thermal image translation, in *Generative Adversarial Networks for Image-to-Image Translation* (Elsevier, 2021), pp. 313–327
86. Y. Li, Y. Ko, Development of a hybrid algorithm to predict room fire flashovers based on vision data, NRC Report A1-020368.1. 53 (2021)
87. Y.H. Habiboğlu, O. Günay, A.E. Çetin, Covariance matrix-based fire and flame detection method in video. Mach. Vis. Appl. **23**, 1103–1113 (2012)
88. L. Zhang, M. Wang, Y. Fu, Y. Ding, A forest fire recognition method using UAV images based on transfer learning. Forests **13**, 975 (2022)
89. J.C. Myburgh, C. Mouton, M.H. Davel, Tracking translation invariance in CNNs, in *Southern African Conference for Artificial Intelligence Research* (Springer, 2021), pp. 282–295
90. C. Mouton, J.C. Myburgh, M.H. Davel, Stride and translation invariance in CNNs, in *Southern African Conference for Artificial Intelligence Research* (Springer, 2021, pp. 267–281
91. L. Taylor, G. Nitschke, Improving deep learning with generic data augmentation, in *2018 IEEE Symposium Series on Computational Intelligence (SSCI)*. IEEE (2018), pp. 1542–1547
92. M.H. Mozaffari, W.-S. Lee, Encoder-decoder CNN models for automatic tracking of tongue contours in real-time ultrasound data. Methods **179**, 26–36 (2020)
93. D. Justus, J. Brennan, S. Bonner, A.S. McGough, Predicting the computational cost of deep learning models, in *2018 IEEE International Conference on Big Data (Big Data)*. IEEE (2018), pp. 3873–3882
94. N.C. Thompson, K. Greenewald, K. Lee, G.F. Manso, The computational limits of deep learning (2020). arXiv preprint arXiv:200705558
95. V. Monga, Y. Li, Y.C. Eldar, Algorithm unrolling: interpretable, efficient deep learning for signal and image processing. IEEE Signal Process. Mag. **38**, 18–44 (2021)
96. M. Hamed Mozaffari, W.-S. Lee, Domain adaptation for ultrasound tongue contour extraction using transfer learning: a deep learning approach. J. Acoustical Soc. Am. **146**, EL431–EL437 (2019)
97. M.H. Mozaffari, W.-S. Lee, Semantic segmentation with peripheral vision, in *International Symposium on Visual Computing* (2020), pp. 421–429
98. A.G. Howard, M. Zhu, B. Chen et al., Mobilenets: efficient convolutional neural networks for mobile vision applications (2017). arXiv preprint arXiv:170404861

99. A. Saiyeda, M.A. Mir, Cloud computing for deep learning analytics: a survey of current trends and challenges. Int. J. Adv. Res. Comput. Sci. **8** (2017)
100. H. Dong, A. Supratak, L. Mai et al., Tensorlayer: a versatile library for efficient deep learning development, in *Proceedings of the 25th ACM International Conference on Multimedia* (2017), pp. 1201–1204
101. A. Akbari, M. Awais, M. Bashar, J. Kittler, How does loss function affect generalization performance of deep learning? Application to human age estimation, in *International Conference on Machine Learning*. PMLR (2021), pp. 141–151
102. Q. Wang, Y. Ma, K. Zhao, Y. Tian, A comprehensive survey of loss functions in machine learning. Ann. Data Sci. **9**, 187–212 (2022)
103. M.H. Mozaffari, L.-L. Tay, Convolutional neural networks for Raman spectral analysis of chemical mixtures. in *2021 5th SLAAI International Conference on Artificial Intelligence (SLAAI-ICAI)*. IEEE (2021), pp. 1–6
104. H. Harkat, J. Nascimento, A. Bernardino, Fire segmentation using a DeepLabv3+ architecture. in *Image and signal processing for remote sensing XXVI*. SPIE (2020), pp. 134–145
105. P. Barmpoutis, K. Dimitropoulos, K. Kaza, N. Grammalidis, Fire detection from images using faster R-CNN and multidimensional texture analysis, in *ICASSP 2019–2019 IEEE International Conference on Acoustics, Speech and Signal Processing (ICASSP)*. IEEE (2019), pp. 8301–8305
106. M.H. Mozaffari, Y. Li, Y. Ko, Real-time detection and forecast of flashovers by the visual room fire features using deep convolutional neural networks. J. Build. Eng. **64**, 105674 (2023)
107. A. Mikołajczyk, M. Grochowski, Data augmentation for improving deep learning in image classification problem, in *2018 international interdisciplinary PhD workshop (IIPhDW)*. IEEE (2018), pp 117–122
108. S. Geetha, C.S. Abhishek, C.S. Akshayanat, Machine vision based fire detection techniques: a survey. Fire Technol. **57**, 591–623 (2021)
109. G. Xu, Y. Zhang, Q. Zhang et al., Deep domain adaptation based video smoke detection using synthetic smoke images. Fire Saf. J. **93**, 53–59 (2017)
110. G. Xu, Q. Zhang, D. Liu et al., Adversarial adaptation from synthesis to reality in fast detector for smoke detection. IEEE Access **7**, 29471–29483 (2019)
111. (2007) Rapid fire phenomena. fire tactics magazine
112. J. Wu, Y. Zhang, X. Gou et al., Experimental research on gas fire backdraft phenomenon. Procedia Environ. Sci. **11**, 1542–1549 (2011)
113. A.C. Fernandez-Pello, On fire ignition. Fire Saf. Sci. **10**, 25–42 (2011)
114. N. Chen, Smoke explosion in severally ventilation limited compartment fires (2012)
115. W.P. Behnke, Predicting flash fire protection of clothing from laboratory tests using second-degree burn to rate performance. Fire Mater. **8**, 57–63 (1984)
116. D. Mackay, T. Barber, E. Leonardi, CFD model of a specific fire scenario (2007)
117. V. Babrauskas, Estimating room flashover potential. Fire Technol. **16**, 94–103 (1980)
118. R.D. Peacock, P.A. Reneke, R.W. Bukowski, V. Babrauskas, Defining flashover for fire hazard calculations. Fire Saf. J. **32**, 331–345 (1999). https://doi.org/10.1016/S0379-7112(03)00027-4
119. G.E. Gorbett, R. Hopkins, P. Kennedy (2007) The current knowledge and training regarding backdraft, flashover, and other rapid fire progression phenomena, in *Annual Meeting of the National Fire Protection Association*, Boston, MA
120. W.C. Tam, E.Y. Fu, R. Peacock et al., Generating synthetic sensor data to facilitate machine learning paradigm for prediction of building fire hazard. Fire Technol. (2020). https://doi.org/10.1007/s10694-020-01022-9
121. W.C. Tam, E.Y. Fu, P. Reneke et al., A generic flashover prediction model for residential buildings using graph neural network, in *Proceedings of the 12th Asia-Oceania Symposium on Fire Science and Technology*. AOSFST December (2021), pp. 7–9
122. E.W.M. Lee, R.K.K. Yuen, S.M. Lo et al., A novel artificial neural network fire model for prediction of thermal interface location in single compartment fire. Fire Saf. J. **39**, 67–87 (2004). https://doi.org/10.1016/S0379-7112(03)00092-4

123. E.W.M. Lee, Y.Y. Lee, C.P. Lim, C.Y. Tang, Application of a noisy data classification technique to determine the occurrence of flashover in compartment fires. Adv. Eng. Inform. **20**, 213–222 (2006)
124. E.W.M. Lee, R.K.K. Yuen, S.M. Lo, K.C. Lam, Probabilistic inference with maximum entropy for prediction of flashover in single compartment fire. Adv. Eng. Inform. **16**, 179–191 (2002)
125. A. Dexters, R.R. Leisted, R. Van Coile et al., Testing for knowledge: application of machine learning techniques for prediction of flashover in a 1/5 scale ISO 13784–1 enclosure. Fire Mater. **45**, 708–719 (2021)
126. R.R. Leisted, M.X. Srensen, G. Jomaas, Experimental study on the influence of different thermal insulation materials on the fire dynamics in a reduced-scale enclosure. Fire Saf. J. **93**, 114–125 (2017)
127. A. Dexters, R.R. Leisted, R. Van Coile et al., Testing for knowledge: maximising information obtained from fire tests by using machine learning techniques, in *Interflam 2019*. Egham, United Kingdom (2019), pp. 515–527
128. S.A. Yusuf, A.A. Alshdadi, M.O. Alassafi et al., Predicting catastrophic temperature changes based on past events via a CNN-LSTM regression mechanism, in *Neural Computing and Applications* (2021), pp. 1–16
129. E.Y. Fu, W.C. Tam, J. Wang et al., Predicting flashover occurrence using surrogate temperature data, in *Proceedings of the AAAI Conference on Artificial Intelligence* (2021), pp. 14785–14794
130. J. Francis, A.P. Chen, Observable characteristics of flashover. Fire Saf. J. **51**, 42–52 (2012)
131. A.L. Huyen, K. Yun, S. De Baun et al., Dynamic fire and smoke detection and classification for flashover prediction **1173502**, 1 (2021). https://doi.org/10.1117/12.2588175
132. K. Simonyan, A. Zisserman, Very deep convolutional networks for large-scale image recognition (2014). arXiv preprint arXiv:14091556
133. M.H. Mozaffari, Y. Li, Y. Ko, Detecting flashover in a room fire based on the sequence of thermal infrared images using convolutional neural networks, in *Proceedings of the Canadian Conference on Artificial Intelligence* (2022). https://doi.org/10.21428/594757db.7c1cd4e1
134. K. Yun, J. Bustos, T. Lu, Predicting rapid fire growth (flashover) using conditional generative adversarial networks, in *IS and T International Symposium on Electronic Imaging Science and Technology 2018* (2018), pp. 2751–2757.https://doi.org/10.2352/ISSN.2470-1173.2018.09.SRV-127
135. D. Cortés, D. Gil, J. Azorn et al., A review of modelling and simulation methods for flashover prediction in confined space fires. Appl. Sci. **10**, 5609 (2020)
136. H. Mozaffari, Y. Li, Y. Ko, Real-time assistance to firefighters using convolutional neural networks, in *Celebrating the Success of Women in STEM Symposium: Pushing the Frontiers of Research Through Collaboration 2016* (2022)
137. K. Yun, K. Yu, J. Osborne et al., Improved visible to IR image transformation using synthetic data augmentation with cycle-consistent adversarial networks, in *Pattern Recognition and Tracking XXX*. SPIE (2019), p. 1099502

Internet of Things and Digital Twin in Fire Safety Management

Xiaoning Zhang, Tianhang Zhang, Yifei Ding, and Xinyan Huang

Abstract IoT and Digital Twin are emerging cutting-edge technologies in the building and construction industry. There is a considerable amount of research has been conducted on the applications of IoT in building fire safety. However, the research and applications of Digital Twin in building fire safety are quite limited, and its definition and scope are not well-defined. This chapter first clarifies the concepts of IoT and digital twin and highlights their unique features in fire safety and firefighting. Then, it systematically reviews the studies related to Digital Twin in fire safety, including those that do not contain the keyword of "digital twin" but meet its definition. Afterwards, the framework of Fire Digital Twin is proposed, and some case studies are presented to facilitate future research and development. The enabling technologies and tools for the fire digital twin will also be introduced and discussed. Finally, we discuss the main challenges and future research areas in fire digital twin.

1 Introduction

Fire that occurs in buildings and infrastructure poses a significant risk and hazard to human and property. Effective fire safety management and timely emergency response are essential to minimize fire hazard. Emerging technologies are widely used in firefighting operations and fire safety management to improve decision-making. The Internet of Things (IoT) and Digital Twin are two promising technologies that will be widely used in the near future to improve fire safety.

IoT is regarded as an ecosystem that integrates data collection, data transmission and data processing technologies to connect smart devices and end users. The common goal of IoT applications is to promote the quality of human life through smart services. The concept of IoT was initially proposed to describe objects connected with radio-frequency identification (RFID) technology [1]. Later, more

X. Zhang · T. Zhang · Y. Ding · X. Huang (✉)
Research Centre for Fire Safety Engineering, Department of Building Environment and Energy Engineering, The Hong Kong Polytechnic University, Hong Kong, China
e-mail: xy.huang@polyu.edu.hk

X. Huang and W. C. Tam (eds.), *Intelligent Building Fire Safety and Smart Firefighting*, Digital Innovations in Architecture, Engineering and Construction,
https://doi.org/10.1007/978-3-031-48161-1_14

technologies, such as smart sensors, machine learning, wireless communication and cloud computing are involved in IoT. With these latest technologies, IoT can connect physical and digital, and provide communication and automatic system control. One of the most important applications of the IoT is real-time monitoring and subsequent decision-making to facilitate effective management. In firefighting and fire safety management applications, the IoT system is intended to support the onsite fire information collection, visualization and decision-making [2].

Digital twin is defined as virtual representations of physical objects, processes, or systems, which are dynamic, bidirectional, interactive, and executable. Digital twin enabled with the latest data communication and modelling techniques can be applied to describe the geometric and physical information and behavior patterns of physical entities [3]. It has been successfully implemented in a variety of fields, like manufacturing, construction, smart cities, smart traffic, and health care. In fire safety engineering, the digital twin is one feasible way to improve building fire safety by developing the digital representation of buildings and fires and connecting the virtual model with physical buildings and building facilities [4, 5].

This chapter will include the introduction of IoT and digital twin, and several applications of building fire digital twin. Section 2 will introduce the Fire-Safety IoT systems; Sect. 3 will include the information of digital twin; Sect. 4 will provide three applications of digital twin in fire safety; and Sect. 5 will summarize this chapter.

2 Sensor Network and IoT Systems

2.1 IoT Architecture

Today, there is still no single definition accepted by the whole world that describes the architecture of IoT. Different IoT architectures (from 3 to 7 layers) have been proposed based on the complexity and application scenario. For example, the sensing layer, network layer and application layer are the main components of a three-layer IoT architecture. And one proposed seven-layer IoT architecture comprises layers of environment, hardware, network, communication, service, ASM, and application [6].

Typical IoT mainly comprises four layers, namely, the sensor layer, networking layer, service layer, and application layer, and these four layers, correspond to four functions of IoT, which are data collection, data transmission, data processing, and data visualization, respectively. The four layers and their corresponding technologies or applications are demonstrated in Fig. 1.

The sensing layer, also known as the perception layer or smart things, is used to capture and collect the data from physical world by IoT sensors. It is regarded as the interface between the digital and physical environments. As shown in Fig. 1, there are diverse data types, such as temperature, humidity, speed, pressure, and position, that can be collected by sensory devices in the sensing layer. The collected physical parameters are converted to digital signals by these sensors that are then transmitted and processed for decision-making. In terms of IoT in fire safety management, temperature and smoke information of fire can be collected by fire detectors, sensors, or cameras that are largely available in modern buildings, as required by fire safety regulations and codes.

The networking layer is also called the gateway or transport layer, which connects sensor networks and communication networks. It plays an important role in connecting devices and integrating different IoT technologies and protocols. The communication technologies of sensing layering include wireless, wired or hybrid networks. Each communication technology has its advantages and disadvantages. For example, wireless communication technologies have strengths in mobility, scalability, and cost-effectiveness, but perform poorly in coverage area and power consumption. In contrast, wired networks have advantages in reliability, security, and bandwidth. Currently, Cellular, Wi-Fi, Bluetooth, ZigBee and LoRa networks are widely used in IoT and smart city applications.

The service layer, also named the processing layer or platform layer, provides the platform and interface for managing and using IoT data. It acts as the intermediary

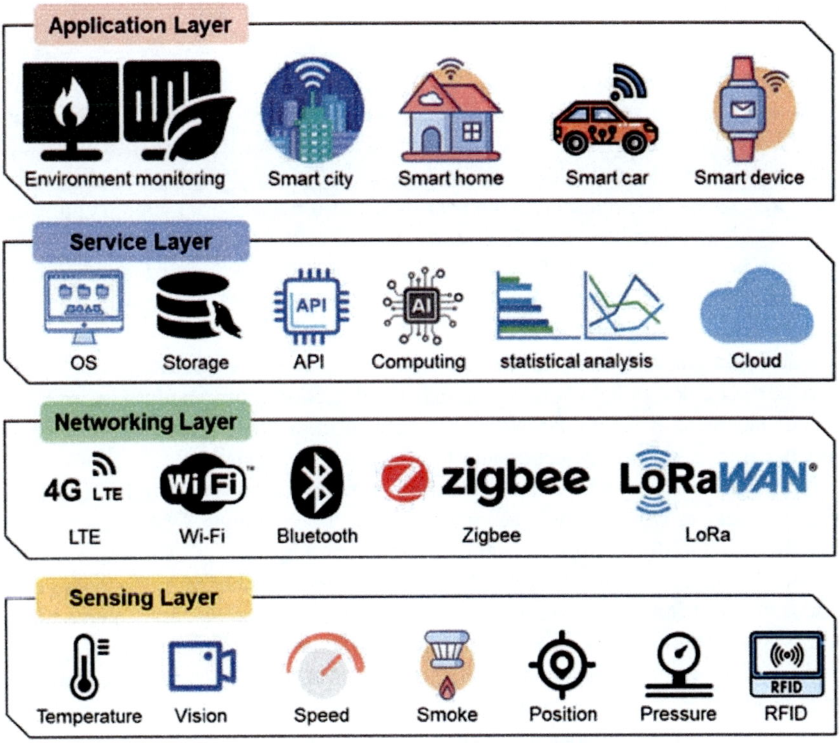

Fig. 1 Four layers of IoT and related technologies (revised based on [7])

that connects the data collected from the sensing layer and the end-user applications, enabling the transmission of useful information and the development of various solutions. The service layer contains a wide range of functionalities, such as data storage, cloud computing, real-time processing, and analytics. Besides, this layer provides interfaces and APIs (Application Programming Interfaces) that facilitate the development of IoT applications and the integration of existing systems. In Fire-Safety IoT, the service layer enables real-time fire detection, fire monitoring and fire prediction.

The application layer is the top layer of the IoT architecture and focuses on delivering services and functionalities to end users. A wide range of domains can be facilitated by this layer, including manufacturing, transportation, environmental monitoring, health care and so on. For example, in the application of smart transportation, vehicle monitoring, and traffic control are applied to improve the transportation management coefficient. In Fire-Safety IoT systems, the fire information can be visualized in the application layer in the forms of charts, 2D maps, 3D models and VR/AR models.

2.2 *Sensor Network*

Sensing is the first layer of a typical IoT system, and sensors can work independently or form a wireless sensor network (WSN). WSN is one foundational technology of IoT, which is mainly used for sensing and monitoring by numerous interconnected intelligent sensor nodes [8]. Each node is usually composed of the power unit, processing unit, transceiver, and sensing unit. Physical or ambient conditions like vibration, sound, speed, temperature, humidity, gas concentrations, and pressure can be perceived by WSNs to build a variety of real-time IoT applications. The latest fire detection research by sensing technology can be divided into two categories, single sensors with complicated algorithms, and multiple sensors with appropriate algorithms. Fire detection based on multiple sensors usually has higher detection accuracy and lower false rate than the detection using a single sensor. With the advancement of Neural Networks and data fusion technologies, WSN has been widely deployed for fire detection and fire monitoring purposes.

WSN has different architectures for different scenarios, such as buildings, forests, roads, tunnels, as well as emergency management. Two common WSN architectures are introduced in this section, including Layered Architecture and Clustered Architecture. The layered architecture WSN includes a large number of sensor nodes and base stations. It follows the architecture of the common Open Systems Interconnection (OSI) model and has five layers including the physical layer, data link layer, network layer, transport layer and application layer. Above the five layers, there are three cross layers planes, namely the task management plane, connection management plane, and power management plane. With these layers and planes, the sensor nodes can work together with good network efficiency [9].

The advantage of layered architecture is that sensor nodes have lower power consumption because each node only communicates with neighboring nodes at low power. In clustered architecture network, sensor nodes are divided into different groups, also known as clusters. Clustered architecture flows specific protocol, such as "LEACH Protocol" which stands for "Low Energy Adaptive Clustering Hierarchy". Based on the protocol, the distributed algorithm can divide sensor nodes into clusters, and the cluster head in each cluster will construct and maintain a Time-division multiple access (TDMA) schedule [10]. The data fusion concept is used to improve energy efficiency by sharing the data through the cluster head and base station.

2.3 IoT-Based Fire Systems

The Fire-Safety IoT aims to connect different things over the networks related to fire [11]. A typical building fire-safety IoT system includes a network of fire sensors or detectors, communication networking, event-detection model, and fire visualization and management platform, which corresponds to four layers of the IoT architecture mentioned above. Besides the different layers of the fire-safety IoT system, there are different applications regarding different firefighting targets. In the present chapter, four applications or areas of Fire-Safety IoT systems are considered and investigated, including detection, monitoring, forecast, and response, as shown in Fig. 2. These four areas have been widely investigated in previous research.

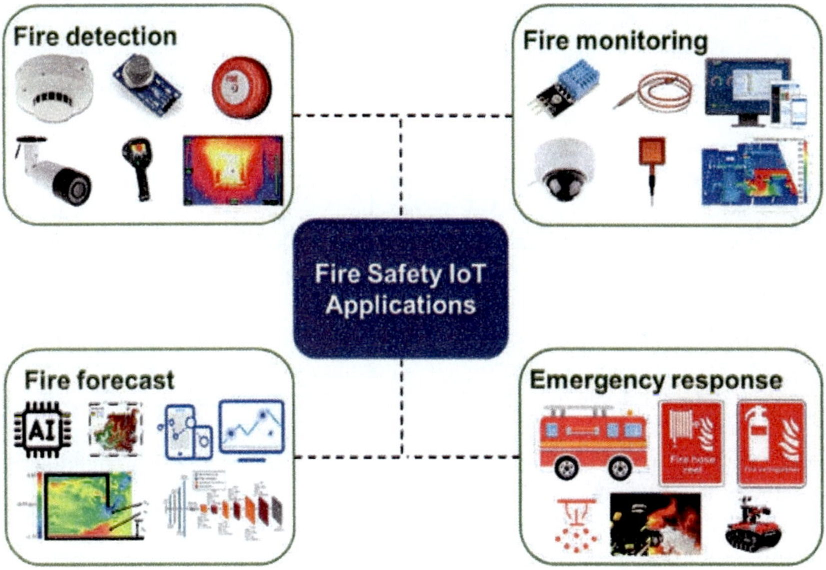

Fig. 2 Four main applications of fire-safety IoT systems (CC-BY Xiaoning Zhang)

The first application is fire detection, which should be detected as early and accurately as possible to minimize the consequences. Building fire detection usually includes sensors-based detection and video-based detection. Flame, gas, smoke, and temperature sensors are widely used fire detection sensors. Early detection and timely warning of fire events are important for the elimination of fire consequences. Therefore, data collection frequency and energy consumption rate (e.g., the duration of the battery) should be considered when designing portable wireless fire sensors. For example, sleep scheduling techniques can be used to prolong the lifespan of batteries and sensors. Video fire detection is realized by detecting flame or smoke separately or simultaneously based on their features. The algorithms of video fire detection are mainly based on handcraft features and deep learning methods [12]. Once the fire is detected, the building fire protection system will be activated, and the fire information will be sent to the building manager and fire service department.

The monitoring of fire is also crucial to provide critical fire information and better decision-making in evacuation and firefighting operations. Due to the dynamic nature of building fires, the lack of real-time fire information often leads to casualties or property damage [2]. Fire monitoring aims to obtain real-time transient fire information based on real-time fire data and event-detection models. The real-time fire data is collected from sensors (e.g., fixed sensors, and sensors carried by firefighters or firefighting robots). This information can be exchanged between firefighters during the firefighting operation to improve situation awareness. The measured raw data need to be further processed by the event-detection model to obtain intuitive fire information. For example, the tunnel fire location and fire size can be identified by a fire detection model based on deep learning and real-time temperature sensor data [13].

The goal of fire prediction is to estimate the likelihood of fire events occurrence or the future development of fires, such as flashover and backdraft in compartment fires [14] and extreme fire events in wildfires [15]. In building fires, the fire prediction takes input data from sensors or cameras inside the building or carried by firefighters, as well as the building information (e.g., building materials, geometry, and building systems) and environmental conditions. For example, the occurrence of flashover can be predicted by the pre-flashover fire characteristics, including color, size, shape, and temperature of fire smoke [16]. Compared with fire monitoring, building fire prediction focuses on fire risk assessment and implementation of preventive measures.

Finally, fire emergency response plays the most important role in extinguishing fire hazards. Applications in detection, monitoring and prediction are used to obtain fire information for decision-making and responses, to extinguish or slow down fire development, as well as reduce fire casualties. Fire response includes the evacuation process inside the building, response of fire protection systems, firefighting by firefighters, and firefighting by robots.

There are still some issues and challenges that need to be addressed when implementing IoT systems for firefighting, including extreme temperature, long-term sensing, high accuracy, and reliability, requirements of fire codes and standards, etc.

3 Digital Twin

3.1 Concepts and Evolution of Digital Twin

The digital twin is regarded as the digital representation of a physical object, process, or system. It is a digital copy of whole physical assets (or systems) that capture the entities, behaviors, and relations in physical, enabling real-time processing, monitoring, simulation, and bidirectional interactions. A broad definition of digital twins is the digital representation of everything, including human beings and things. People can become a part of a digital twin, being monitored, simulated, and represented in real-time. With the advances in sensory technology, big data analytics, as well as machine learning and AI, digital twin is widely used in manufacturing, smart cities, healthcare, etc.

Digital twin is not a new concept, the origins of digital twin can be traced back to the Apollo space program in the 1960s when engineers used digital models to simulate conditions on the moon's surface. Two aircraft are built, the one on Earth can simulate and predict the aircraft in space [17]. Then in 1991, the concept of digital twins was mentioned in a book called *Mirror Worlds* written by David Gelernter [18]. It refers to software models that mimic reality with the input of the physical world information. The model of digital twins was first introduced and applied in manufacturing by Michael Grieves, which was proposed as a conceptual model in product lifecycle management (PLM). He also published a white paper in 2014 to further the model, which consists of three parts: physical products, virtual products, and the connection between physical and virtual products [19]. The detailed definition of digital twin was given by NASA in 2010. NASA defines a digital twin as "*A Digital Twin is an integrated multi-physics, multiscale, probabilistic simulation of an as-built vehicle or system that uses the best available physical models, sensor updates, fleet history,* etc., *to mirror the life of its corresponding flying twin*" [20]. In 2016, Siemens applied digital twin in Industry 4.0 [3]. In 2017, the digital twin was listed as the top 10 Strategic Technology Trends. They believed billions of things would be represented by digital twin within five years [17]. Since then, numerous studies have been conducted on digital twin, and digital twin has been applied in different fields. The evolution of digital twins is shown in Fig. 3.

There is no clear definition of digital twin in the architecture, engineering, and construction (AEC) industry or civil engineering sector, some studies have confused digital twin with other similar concepts [21], such as digital model, digital shadow, cyber-physical systems (CPS) [22] and building information modelling (BIM, see Chap. 3). It is important to distinguish digital twin from other concepts, as the unclear definition may lead people to regard it as just hype [23].

Compared with digital twin, digital model has no data interaction between physical objects and digital objects. Most fire simulations run by fire engineers and researchers to study a fire process are a typical digital model. The digital shadow only has a one-way data flow from physical space to digital space, while the digital twin integrates bidirectional data flow between digital and physical. The BIM model is a digital

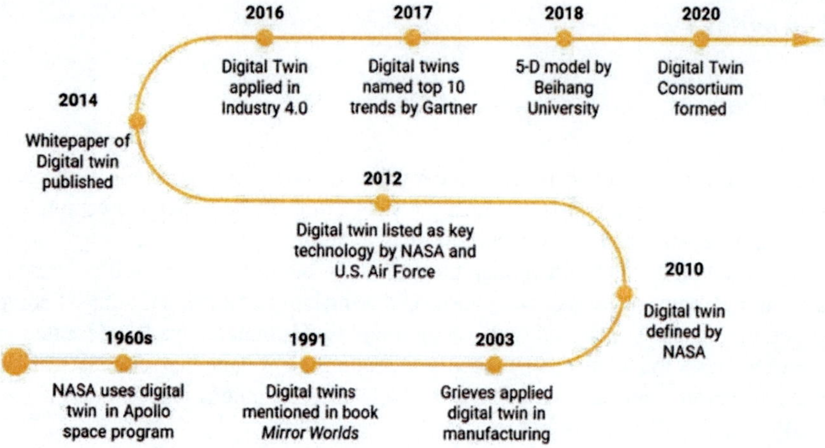

Fig. 3 History and evolution of digital twin technology (revised based on [3])

representation of the physical and functional characteristics of a building, which can be used to represent something that does not exist. While digital twin must describe something that exists, and emphasize the connections between physical and digital spaces. CPS is the integration of computer and physical systems, which also enables data transmission and connections between the physical system and the virtual model. However, the digital twin focuses more on the virtual model and the relationship between physical and virtual entities, while CPS emphasis more on cyber and the virtual model is not necessary in CPS [21].

3.2 Digital Twin Model and Application Fields

A whitepaper defines the digital twin as physical entity, virtual entity, and connection between physical and virtual. This three-dimensional digital twin model is mostly used nowadays [19]. Based on the three-dimensional model, various definitions and models of digital twin were proposed. For example, Tao et al. added two more dimensions to the three-dimension model and proposed a five-dimension digital twin model, which includes physical entities, virtual models, services, digital twin data, and connections in digital twin [24]. Later, a "Digital Twin 8-dimension model" was developed to describe the behaviors and the context of the digital twin. The eight dimensions were divided into two groups, focusing on digital twin context and environment, as well as behavior and capability richness, respectively [25].

As shown in Fig. 4, the classical digital twin model comprised the physical twin, the digital twin, and the connections between the physical and digital twin. Digital twin technology is to create a virtual model to simulate the behavior of physical objects or systems, so the physical world is the basis of the digital twin model. The digital models are designed as replicas of physical entities, which should represent the physical world from different aspects, such as geometries, rules, behaviors, and properties [27]. Digital twin data is the key element of the system and is usually transmitted bi-directionally between physical and virtual spaces. Some data is collected from physical space, like environmental data or static data, to enable the evolution and simulation of virtual models. These data are analyzed and managed in virtual spaces, and some data (e.g., information and decisions) is generated to reflect the simulation results and affect the physical spaces.

Over the past few years, the concept of digital twin has been successfully implemented in many fields, such as manufacturing, smart city, health care, construction, and supply chain, as shown in Fig. 4. In modern manufacturing, the communications and interactions between physical systems and virtual spaces are important. Digital twin is applied to help with real-time monitoring of manufacturing processes [28], production planning [27], human–robot collaboration and interaction [29], etc. The digital twin also plays a more and more important role in the AEC industry, such as city planning and design [30], disaster response and emergency management [31], construction progress monitoring [32], and construction safety monitoring and management [33]. Digital twin is still demonstrating its application potential.

Fig. 4 Digital twin model and application fields of digital twin (revised based on [26])

3.3 *Enabling Technologies for Building Digital Twin*

To fulfill the functions of the digital twin in different applications, a variety of tools and technologies are required. These tools and technologies can be divided into five groups, i.e., physical world cognizing and controlling, connections, data management, modelling, and services [3]. These five aspects of digital twin technologies also correspond to the five dimensions of the 5-dimensional digital twin model. The digital twin shares a lot of enabling technologies with IoT, such as sensing, controlling, data transmission, data management, and visualization, but the digital twin focuses more on modeling than the IoT does. As the digital twin is defined as the digital representation of the physical entity, modeling technologies are the basis for the digitization of physical entities. In this section, enabling technologies and tools for sensing and controlling the physical world, data management, and digital twin modeling are introduced and illustrated in Fig. 5 (contents of connections and services are not included but shown in Fig. 5, as they share similar technologies with data management and modeling).

A clear understanding of the physical world is the foundation for building a digital twin, which involves multidisciplinary knowledge, like electronics, structural mechanics, fluid mechanics, aerodynamics, thermodynamics, material science, computer science, control theory, etc. With the technologies of sensing and measurement, the information of objects in the physical world is reflected in the virtual world. To create a high-fidelity model in virtual space, it is significant to accurately cognize the physical world. The geometry, materials, and characteristics of physical entities

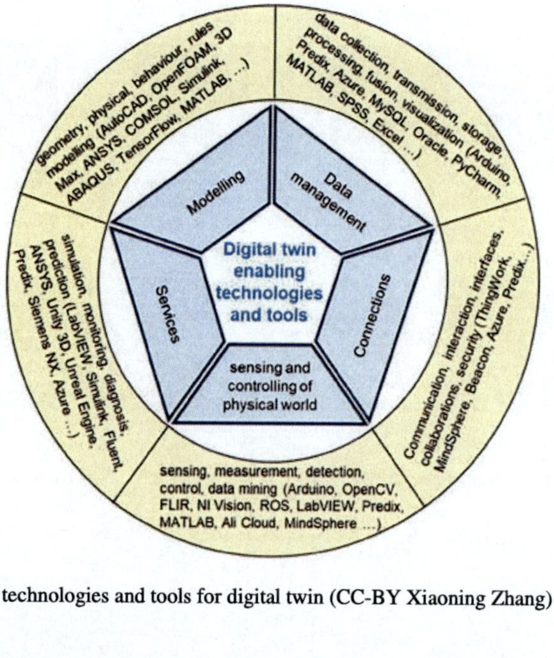

Fig. 5 Enabling technologies and tools for digital twin (CC-BY Xiaoning Zhang)

are measured to build the static model. Real-time data (e.g., temperature, speed, pressure, voltage, etc.) are collected to build the dynamic model or synchronize the virtual model. The world can also be perceived through data mining technologies like big data analytics. In turn, physical entities and systems can be controlled from the virtual world based on control theories and technologies. To control the physical world, control technologies like embedded control, electrical control, and programmable control can be adopted. Some tools for physical world perceptions and controlling are listed in Fig. 5.

The digital twin modelling technologies include geometric modelling, physical modelling, behavioural modelling, and rules modelling, as illustrated in Fig. 5. The Geometric model characterizes the geometric shape, appearance, and element relations of physical entities with the proper data format. The geometric shapes including the lines, curves, angles, and surfaces, can be modelled by wireframe, surface, and solid modelling. The appearance modelling aims to enhance the reality of the model with texture techniques (e.g., textures, materials, and skins in 3D graphics). The features of physical entities are modelled by physical modelling. The information such as heat conductivity, strength of materials, and ignition temperature of combustibles are defined and added to the model. Simulations of behaviours are complicated, may include the state model, dynamic model, and probability model, which describe the model capacity in functions, response, updating, interactions, and operations, etc. The rule model is to help the virtual model in reasoning, prediction, and optimization, which is developed from historical data, simulation data, and expert knowledge. For example, the rule model can be defined using XML-based data. For building fire digital twin, the modelling of building fire includes building modelling, fire dynamics simulation (flame, smoke, gases, visibility, temperature, heat flux, etc.), fire system simulation and human evacuation modelling.

Data management of the digital twin describes the entire lifecycle of data from collection, processing to application. Data management technologies include data acquisition, data transmission, data storage, data fusion, and data visualization. Data is usually collected from the environment by hardware (e.g., sensors, cameras and IoT devices), and can also be acquired from software and network. The collected data is transmitted by wireless or wired transmission. Wireless communications have short-range and long-range technologies, which are used in different scenarios or purposes. For example, Wi-Fi and Bluetooth can be used in short-distance transmission with high data capacity. The target of data storage is to store the collected data for further processing. SQL database is the most common tool for data storage. New data storage methods such as NoSQL and cloud storage are becoming popular with the development of big data technologies. Data processing is to extract useful features from the collected raw data. Some procedures like data cleaning, data transformation and data reduction are required before processing. Then the data is analysed by quantitative, statistical, or machine learning methods. Data fusion methods are required to get better results when multisource data are collected. Kalman filtering is a commonly used data fusion technique. Visualization technology is then applied to display the processed data. Visual analytics aims not only to view the data from different perspectives through data mining, statistics, and visualization tools but also to enable human interaction allowing user control during the data analysis process [34]. Currently, different types of charts, 2D/3D rendered graphics, and VR/AR/MR are widely used visualization methods. Some tools for data management can be found in Fig. 5.

4 Digital Twin of Fire and Applications

The digital twin of fire is still a very new concept, as fire occurrence is unpredictable and real-time simulation and monitoring of fire behaviors are challenging. The establishment of fire digital twin may require the combination of many technologies, such as Building Information Modeling (BIM), Computational Fluid Dynamics (CFD) simulations, IoT sensors, IoT platforms, deep learning models, and game engines. In this section, three fire digital twin applications are demonstrated, including building fire digital twin, tunnel fire digital twin and human evacuation digital twin.

4.1 Digital Twin of Tunnel Fire

This section presents a tunnel fire digital twin framework for the simulation and connection between real tunnel and virtual tunnel model, especially during tunnel fire events [35].

Digital twin framework. The overall framework of the proposed tunnel fire digital twin, each component, and their relationships are shown in Fig. 6. The system includes a real tunnel, a digital tunnel, and the data center and hardware module. The real tunnel is the physical entity in the Digital Twin model, and the digital tunnel is the virtual model. The data and information between the real tunnel and virtual tunnel are the connection between physical and virtual spaces. Compared with past systems [36, 37], this Digital Twin not only can recognize and forecast real-time tunnel fire but also support daily facility management and fire safety emergency response. The digital twin model can be integrated with other facility management systems, such as tunnel ventilation and lighting systems, to control their operation and urgent use during a fire emergency.

The data transfer between the physical tunnel and the digital tunnel has to be bidirectional; otherwise, it is only a Digital Shadow rather than a Digital Twin [38]. From left to right, the data is transferred from the physical tunnel to the digital tunnel, and the real fire scenario is projected into the digital space. At the same time, the data (e.g., the facility control signal or evacuation/rescue instructions) can be transmitted from the digital tunnel to the physical tunnel. For example, the user can operate in the digital tunnel to control the smoke ventilation and fire suppression system in the physical tunnel.

Data management. The proposed digital twin system can evolve with the changing tunnel environment, with the feed of data collected from installed sensors or devices

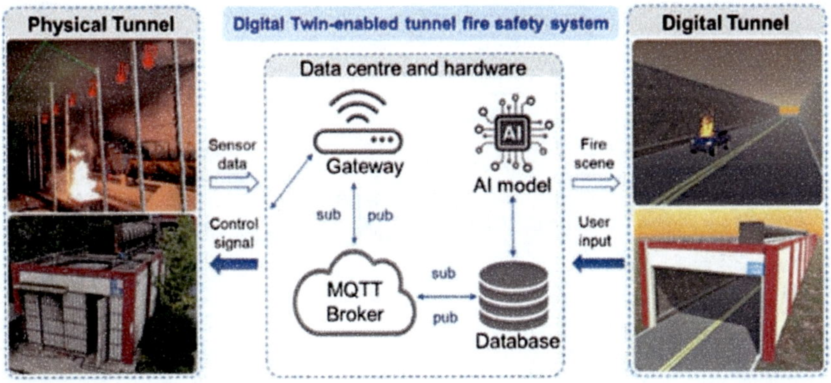

Fig. 6 The overall digital twin framework for tunnel fire safety management (CC-BY Xiaoning Zhang)

to form a dynamic data-driven system. The whole lifecycle of the proposed digital twin system includes data collection, data transmission, data storage, data processing, and data visualization, as shown in Fig. 7. The data is initially collected from the IoT sensors, cameras, other facilities inside the tunnel, or generated by AI and simulation. Then the data is transferred by wired or wireless methods, such as Wi-Fi and 5G. Database technology is widely used to store the collected data, MySQL is used in the present study. Then the data (e.g., temperature, and video) is processed to extract useful information (e.g., fire size and location) for decision-making. Finally, the obtained fire information can be visualized using appropriate visualization techniques. In the present work, the fire information is simulated and visualized in the game engine Unity 3D.

The proposed tunnel fire digital twin system is designed for all tunnel fire safety aspects, including daily fire risk monitoring, facility management and control, fire identification and prediction, as well as evacuation and rescue guidance. Among these applications, real-time fire recognition and rendering are the most crucial during a fire, such as the fire size and fire location. Therefore, the focus of this section is to development of a fire recognition model and real-time fire rendering platform. The fire recognition model is developed based on a deep learning algorithm and a large amount of numerical simulation data.

Fire cognitive model. In the present work, the Fire Dynamics Simulator (FDS) developed by NIST is adopted to conduct the fire simulation of different scenarios and generate the training dataset. The 3D tunnel model is simulated using FDS 6.7 and built based on the real fire test tunnel in Sichuan Fire Research Institute (SCFRI), with 140 m in length, 8 m in width and 6 m in height. 29 thermocouple devices with an interval of 5 m were distributed along the tunnel centerline to record the ceiling gas temperature every 1 s. Fire location and fire heat release rate (HRR) are varied to simulate different fire scenarios, 13 fire locations and 10 fire sizes are considered. All the scenarios are simulated for 120 s to reach the steady stage. After the simulation,

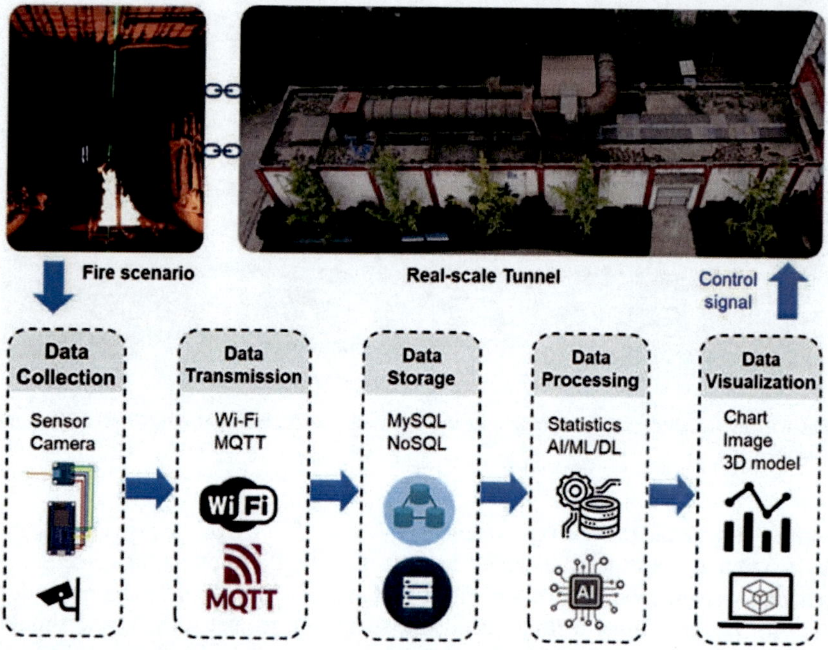

Fig. 7 Data management and interoperation of tunnel fire digital twin system (CC-BY Xiaoning Zhang)

the results of all cases are extracted and processed to form the training database. Figure 8 illustrates the process of database generation.

Transformer is a deep learning architecture for sequential data processing based on the self-attention mechanism [39]. In the present case, the sensors far away from the fire source may not get any information, and the sensors near the fire should have heavier weights than other sensors. Therefore, the Transformer model is suitable for fire identification using distributed sensors. The architecture of the proposed Transformer model is shown in Fig. 8. Temperature data along the tunnel are input to the Transformer model, and fire information, including fire size and location, is predicted. Before the input and after the output, normalization and inverse normalization processes are conducted, respectively, to optimize the training process. The model includes 4 encoder blocks and 4 decoder blocks, and each encoder-decoder architecture has a multi-head attention layer and a fully connected layer.

Digital twin modeling. Based on the real tunnel in SCFRI, a virtual tunnel model is constructed using *SketchUp* with a scale of 1:1, and some unnecessary components or facilities are simplified. The fire and smoke are modelled in *Unity 3D (V2020.3.25)* using the Particle System. The fire and smoke in virtual space are synchronized with the real fire scene based on the collected real-time sensor data and fire cognitive model. The proposed tunnel fire digital twin aims to provide an immersive virtual

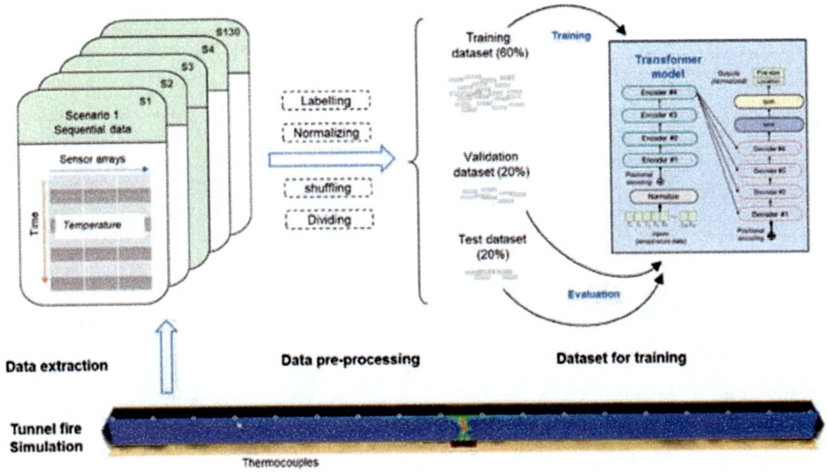

Fig. 8 Generation of the database and architecture of the Transformer model (CC-BY Xiaoning Zhang)

environment for different firefighting purposes, so it is important to make the fire rendering realistic enough. After the fire size and fire location are confirmed, the fire can be represented in a virtual space with suitable dimensions and location. The location of fire can be mapped directly, and the dimension of fire can be calculated using empirical equations [40].

Implementation of Digital twin. The user interface of the digital twin is developed using *Unity 3D*, as shown in Fig. 9a, b. Then the digital twin is demonstrated in real-time fire recognition using numerical experiments and full-scale tunnel fire experiments. The numerical experiment is conducted based on the CFD simulation data, and the full-scale demonstration is conducted based on real-scale tunnel fire test data. Numerical data with changing fire sizes are input to the digital twin platform, which has not been seen in the training process of the fire recognition model. The simulation fire size and prediction fire size are compared in Fig. 9c. A full-scale tunnel fire test is conducted in a large-scale experimental tunnel at SCFRI, which is the basis for building the digital tunnel model. Then the experimental data is used to demonstrate the digital twin in real fire recognition. The predicted fire size and actual fire size from measurement are compared and shown in Fig. 9d.

4.2 *Digital Twin of Room Fire*

Driven by an AI engine and the hardware, the framework of the building Artificial Intelligence Digital Fire (AID-Fire) is proposed in Fig. 10 [41]. To achieve the target of fire identification and prediction, the heat detectors (could be an existing fire

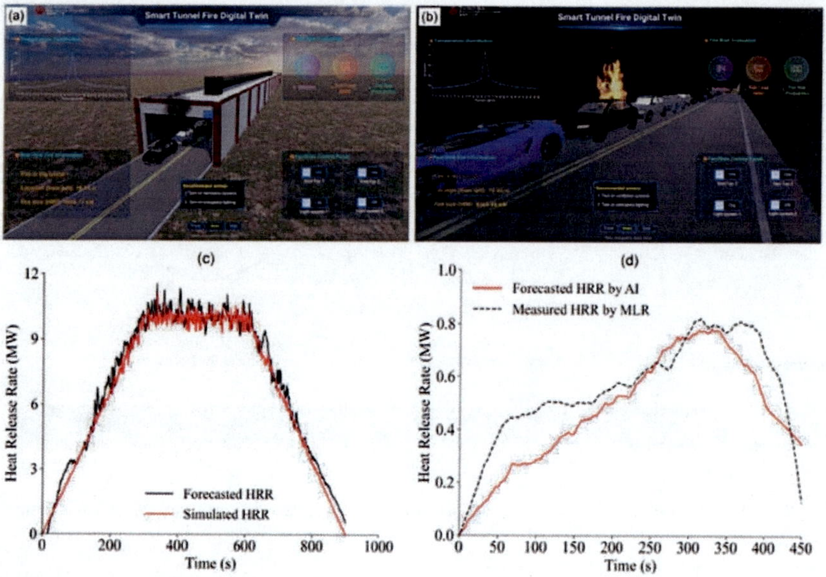

Fig. 9 Demonstration of tunnel fire digital twin in simulation and experimental data: **a** main view, **b** inside view, **c** simulation data, and **d** experimental data (CC-BY Xiaoning Zhang)

alarm system or pre-installed thermocouple network) will first collect temperature data using an IoT sensor network and then transmit data to a cloud server. The server can manage and store the data following a pre-set and standard format, which is accessible by the AI engine. The AI engine can make use of the data to identify fire state and provide early warning on the potential fire risks to occupants and firefighters in real-time. Finally, a User Interface (UI) is developed to display the fire information (measured data and AI prediction) and allows cyber-physical interaction. To demonstrate the proposed system, a fire-test room with the dimensions of 7.5 (L) × 3.4 (W) × 5.4 (H) m^3 was selected as the target physical twin. The four main sections are introduced in detail subsequently.

IoT Sensor network. Many indicators (e.g., temperature, heat flux, gas concentration, etc.) can be used to analyze fire intensity. Among them, the temperature is one of the most feasible indicators for many reasons (e.g., direct, convenient, and relatively reliable). Commercial temperature sensors (essentially thermocouples) are cheap, fire-resistant, and pre-installed. Therefore, the heat-detector network and temperature data are adopted as the data collected to identify the fire scenario in the AID-Fire system. Alternatively, a matrix of smoke detectors can also be used for building the sensor network and forming a database of smoke visibility.

The temperature evolution is collected by a wired local data logger that is protected from the high temperature and irradiation from fire and smoke. The data collection controller is equipped with a Wi-Fi module to achieve wireless data transmission to the cloud, as shown in Fig. 10. To extend the system lifetime, the data collecting

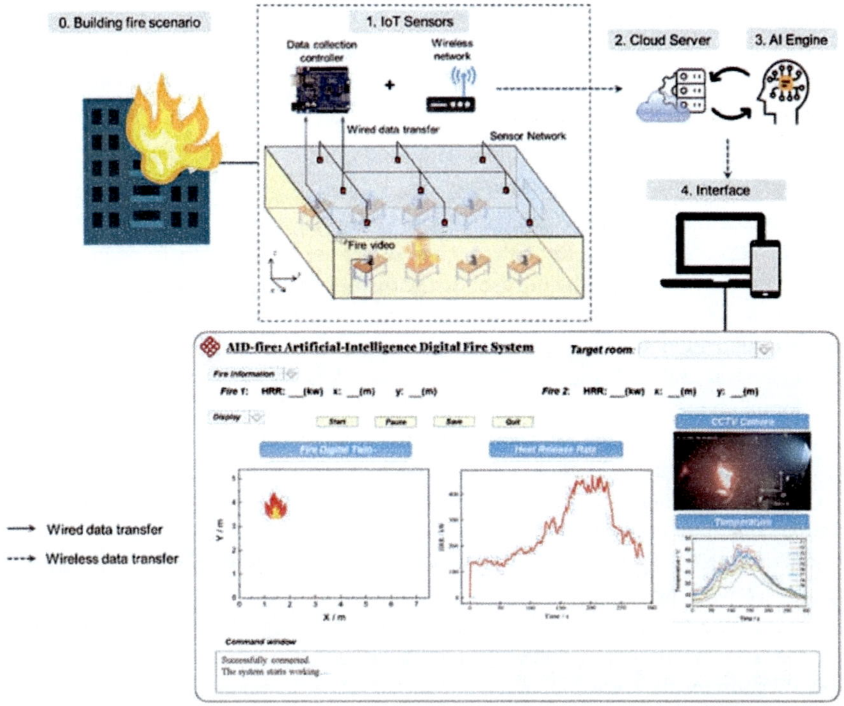

Fig. 10 The framework of the intelligent fire digital twin (CC-BY Tianhang Zhang)

frequency is set to be low (e.g., 1/60 Hz) in daily use [42, 43] and can switch to the high-frequency mode (e.g., >1 Hz), once any sensor temperature exceeds a pre-set threshold (e.g., 60 °C in this study) in a fire [36].

Cloud Server. To store, manage, and process the sensor data transferred from the fire field, a remote cloud database is built in a server. There are many advantages to adopting a cloud server in the AID-Fire system, such as global access and stability. The data in the database are stored as multiple sheets organized by MySQL [44]. Currently, there is no universal standard to organize fire data. Therefore, the data structure refers to our previous tunnel fire database [36, 45]. The database is formed by two sheets, namely, (a) measurement sheet, which records the temperature data from each thermocouple in the associate column; (b) forecast sheet, which stores the real-time prediction results by the AI model. The real-time video captured by the CCTV camera (for the validation purchase) is also stored in the server as video streaming. The data interaction uses the MySQL Connector [46].

AI Engine. The AI engine builds the correlation between the spatial–temporal temperature distribution and the fire information (fire HRR, location, etc.), so it is the core of the AID-Fire system. All raw temperature data in the database are pre-processed by a data-cleaning algorithm, which can avoid the influences of potential invalid data (e.g., "missing," "out of range," "burn out," etc.) caused by broken sensors. Fire Dynamic Simulator (FDS) is adopted to generate the building fire scenarios, and this model is calibrated by experimental data. Fire scenarios with various fire locations, HRRs, and fire source number are included in the synthesis database.

User interface. The User Interface (UI) is developed to realize the function of information visualization and cyber-physical interaction, as shown in Fig. 10. The UI can connect to the target database on the remote server according to the IP address, username, and password input by the user. In the display area, the measured raw temperature data are plotted and fed to the AI engine. The AI predictions are rendered in the digital twin showing the number, size, and 2-D location of the virtual fire. The historical and real-time fire HRRs are also plotted to give an overview of fire development.

The CCTV video is also shown to verify the AI predictions before it is blocked by smoke, but in reality, it is usually not available inside the room for privacy reasons. On the bottom, a dialogue box presents the system operation status, e.g., successful actions, failed actions, and possible reasons. Moreover, the occurrence of a critical event, such as fire spread, flashover, backdraft, etc., will be shown in this dialogue box with a warning signal.

To examine the performance of the proposed AID-fire system in a realistic fire environment, a sofa fire is tested. The experimental procedure and the associated identification results are summarized in Fig. 11. A small amount of propanol (around 30 ml) is used to ignite the sofa at the beginning of the test. During the test, a small fire hose is used to suppress the fire when the fire becomes too large to test the performance of the proposed system when interacting with the firefighting system (i.e., sprinkler), as well as for safety concerns.

Figure 11a shows a good overall agreement between AI's identification and the observed fire evolution phenomenologically. The predicted fire HRR increases at the development stages (the light-yellow regions) and decreases when the fire is discontinuously suppressed by a fire hose (the purple regions). Before the sofa fire is suppressed by a fire hose, the AI engine predicted a peak HRR of 470 kW at around 220 s, where half of the sofa is burning. Such a value is roughly 1/2 of the peak HRR of 992 kW found in previous literature [47] so the AI's identification is reasonable.

The fire position can also be accurately identified at the beginning stage when the fire develops. However, when the fire hose suppresses the fire at around 220 s, the AI model predicts a wrong result, as shown in the rightmost blue point in Fig. 11b. To identify the fire position, the AI model analyzes the input temperature data and conducts non-linear interpolation by searching the temperature patterns with high similarity and correlation. However, when using the fire hose to suppress the fire, a large amount of vapor is rapidly generated. Such a vapor plume disturbs buoyancy-driven smoke layer and changes the ceiling temperature distribution. Thus, the AI model tends to predict multiple fire sources in the suppressing stage, but such an abnormality quickly disappears after stopping suppressing. This issue could be addressed by adding more fire scenarios with different fire suppressions to the

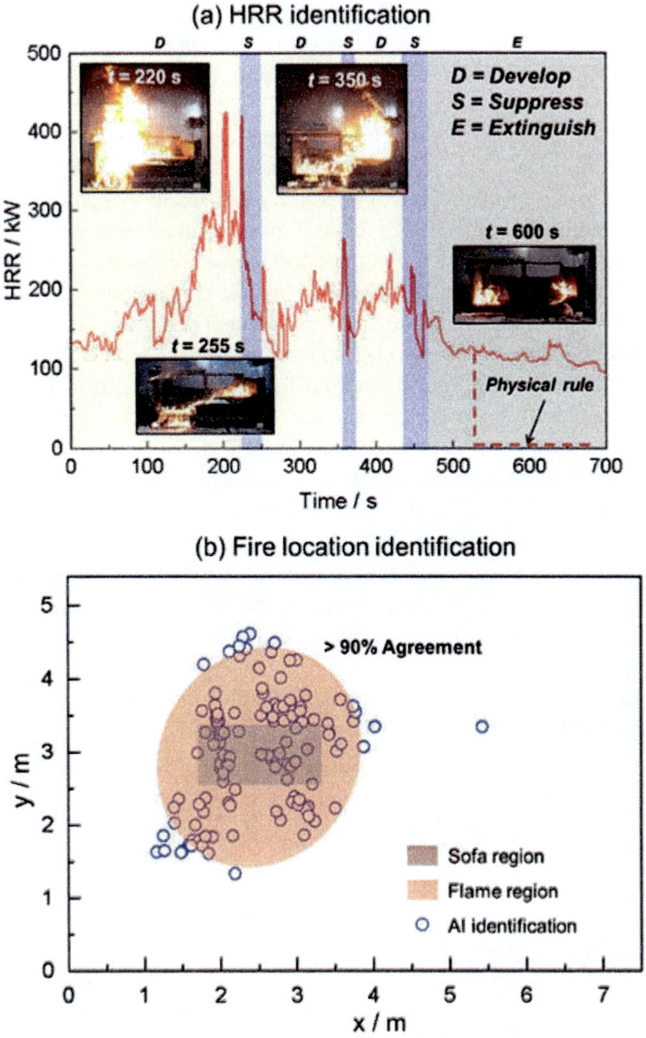

Fig. 11 AI identification results of the sofa burning test **a** fire heat release rate, and **b** fire location (CC-BY Tianhang Zhang)

database. The overall accuracy of fire location identification is more than 90% (excluding suppressing stages), as shown in Fig. 11b.

Another interesting phenomenon is found after the fire is extinguished, where the AI model still predicts a fire with a certain HRR after 540 s. This is expected because the upper layer of gas can still maintain a relatively high temperature after the fire burns out. In fact, the AI's prediction is not just reasonable but also much safer. That is, although the flame is extinguished, the fire hazard (e.g., smoke and high temperature) still remains. Nevertheless, if perfect fire identification is required, some physical rules can be applied to the outputs of the AI model. For example, we can set the output HRR to be 0 when the ceiling temperature data meet the two conditions: (a) the temperature keeps decaying in 10 s, and (b) the average temperature is lower than 40 °C. Then, the overall prediction becomes more reasonable, as shown by the dashed line in Fig. 11a.

Compared to the results in 1-D tunnel fire scenarios, in which the fire location is easier to identify than the HRR [36], it is more difficult to locate the fire in the 3-D building fire scenario. The different performances are caused by the temperature distribution characteristics. In the 1-D tunnel fire, the temperature of the smoke layer decays dramatically under the cooling effects of tunnel ceiling and ventilation airflow during spreading along the tunnel [48], in other words, the temperature distribution has a large gradient. Therefore, it is easier for the AI model to identify the high-temperature region close to the fire and the low-temperature region far from the fire.

However, in the building fire scenario, the space is relatively closed, and the smoke flow is restricted by the boundaries. Therefore, the smoke generated from the fire will accumulate and form a relatively stable smoke layer below the ceiling. Since there is no strong heat and mass transfer process in the smoke layer, the spatial gradient of the temperature is not that large, increasing the difficulties in identifying the fire source location. Based on this assumption, it can be concluded that the AID-fire system should perform better in buildings with vast spaces, e.g., large underground garages, malls, and stadiums, where there is enough space for smoke to spread, and cause a large smoke temperature-gradient.

4.3 Digital Twin of Fire Evacuation

The provision of real-time and detailed evacuation information feedback is vitally significant for the formulation and adaptation of the onsite evacuation strategy. The conventional surveillance system relying on surveillance cameras is limited to processing video and fails to extract human behaviour or provide privacy protection. In this section, an Intelligent Emergency Digital Twin system [49] is introduced based on computer vision and deep learning. The proposed system enables fusion of evacuation information and privacy protection of occupants and lay the foundation for the implementation of smart human monitoring in fire scenarios for buildings.

The framework of the proposed intelligent fire emergency digital twin monitoring system [49]. The system mainly contains four components (see Fig. 12): (1) CCTV network, (2) YOLOv4 [50] detector, (3) DeepSORT [51] tracker, (4) Perspective transformer and (5) Digital twin interface, which is responsible for three tasks: data obtaining, data processing and digital presentation.

Data obtaining. CCTV devices are pre-installed in critical positions such as in the ceilings and handrails of the staircase along egress routes to record original videos of

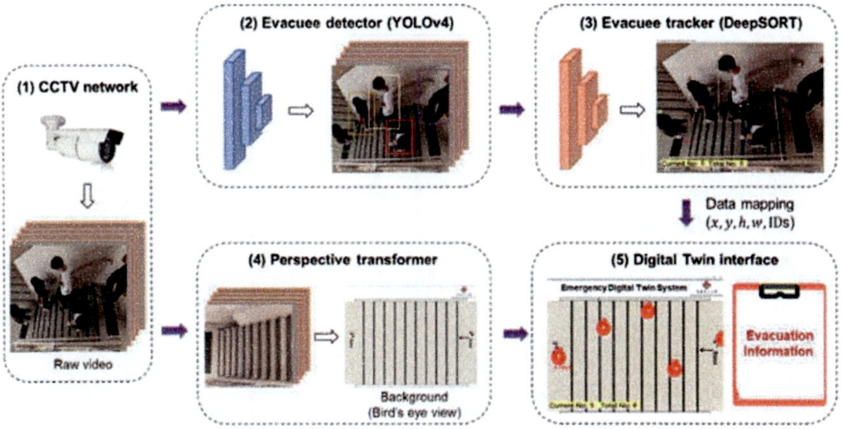

Fig. 12 Summarised structure of fire emergency digital twin system for monitoring fire evacuation (CC-BY Yifei Ding)

evacuation. The cameras and corresponding LAN-based wireless connectivity form the CCTV networks to obtain raw video data. The raw data will be stored in the central server of the emergency digital twin and the real-time video stream will be read by the YOLOv4 detector and the DeepSORT tracker for further processing and analysis. The low latency data acquisition and transmission is a key prerequisite to ensure Fire emergency digital twin outputs evacuation information in real-time.

Data processing. Data processing is conducted by an AI engine comprised of the YOLOv4 detector and the DeepSORT tracker. The YOLOv4 detector is a deep learning object detection algorithm for positioning the evacuees with bounding boxes in the real-time video and analysing the distribution information of evacuees as well as the pixel coordinates of locations. The number of evacuees detected in each frame will also be recorded by the YOLOv4 detector. During the frame-by-frame detection analysis of the input video, the DeepSORT tracker is introduced to ensure that the evacuees' assigned IDs are constant in every frame. The speed estimator will calculate the passage time and speed of each evacuee based on the corresponding coordinates under each unique ID. After real-time data processing, the integrated evacuation information including the distribution of evacuees, the passing time and average speed of each evacuee and the count of evacuees is mapped and presented on the digital twin interface.

Digital twin presentation. Digital twin interface is the core component of the proposed system to show the processed evacuation information. The background of the interface is designed as a virtual scene of an egress component in the building, for example, the staircase (see Fig. 13). Not only the dynamic traces of evacuees can be presented in the virtual interface, but it also demonstrates diverse crucial evacuation data, which helps engineers, policymakers, firefighters, etc., to acquire

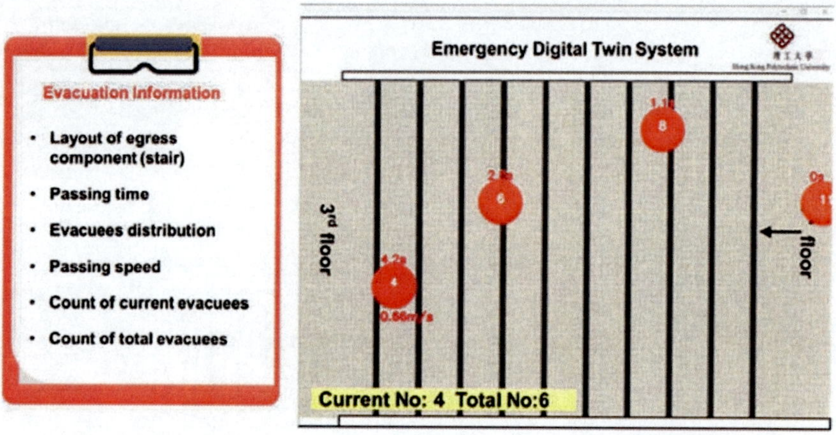

Fig. 13 Illustration of the digital twin interface of the fire emergency digital twin system (CC-BY Yifei Ding)

real-time information in evacuation sites and benefits to further evacuation and rescue decision-making. Overall, the digital twin interface offers the following strengths:

(1) Real-time evacuation information assembly: Traditional surveillance interfaces can only transmit and record video without any data analysis capabilities. Important evacuation data and parameters of evacuee behaviors can only be processed at a later stage and are not available in providing sufficient dynamic evacuation guidance on site. In response to this issue, the proposed Fire emergency digital twin system combines deep learning algorithms and computer vision techniques to realize the real-time evacuation data acquisition and processing and presents the critical information in the digital twin interface, see Fig. 13. The dynamic information on the DT interface contains the distribution and trace of each evacuee, the time required to pass the building key egress component, the average passing speed as well as the count of escaping evacuees. The monitor can visualize this information through the DT interface and use this information to inform the subsequent evacuation strategy development.

(2) Privacy protection: Compared to traditional monitoring systems, the evacuees' appearance would be overlooked in the proposed digital twin interface, and uniformly represented as many orange balls (see Fig. 13). Moreover, only the virtual interface would be presented as the visualization output of the proposed Fire emergency digital twin system and the physical edge of the original video captured by CCTV would be stored in the server rather than being played. The monitor can only observe the evacuees' traces and distribution through the virtual interface and personal information such as individual appearance would be invisible. Therefore, the occupants do not need to worry that their behaviors and privacy will be exposed if the proposed system is used during the emergency escape.

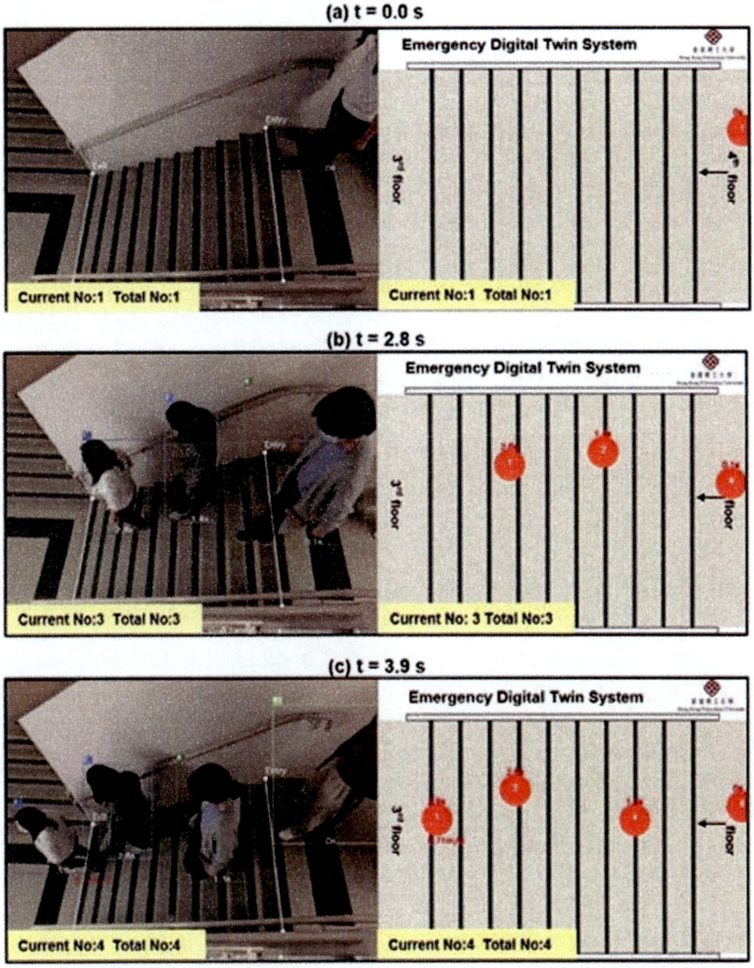

Fig. 14 Demonstration of the flowchart of the Fire emergency digital twin monitoring system for multiple objects test (CC-BY Yifei Ding)

Demonstration. To test the performance of this intelligent fire emergency digital twin system, a fire drill test was performed in a staircase of an office building located at the Hong Kong Polytechnic University. This demonstration of fire drill shows how the four critical components of the Fire emergency digital twin system work to realise occupant detection and tracking as well as speed estimation, as shown in Fig. 14. Firstly, the video stream was generated by a CCTV camera and transited to an AI model. A combination of the YOLOv4 detector and DeepSORT multiple objects tracker can solve the problems of object loss and ID change during the multiple occupants tracking. While the method of continuous timing in the whole experiment process to estimate the traveling speed of each occupant is not suitable. To address this issue, we assign a separate timer to each participant and associate it with the respective ID of each participant.

In other words, the timer keeps track of each participant's individual traveling time, and it would not restart if their identities had not switched. As each participant passes near the exit line, their respective timers stop and record individual traveling time and average traveling speed. The values of total traveling time and average speed will also be mapped to the right visual interface, as shown in Fig. 14. For example, when t = 3.9 s, the first participant crosses the exit line with 3.9 s, and the speed estimator calculates the traveling speed as 0.71 m/s. When t = 5.1 s, the second participant crosses the exit line with 3.8 s, and the speed estimator calculates the traveling speed as 0.74 m/s. The performance of the proposed system for the multiple occupant's test was high with a 93% detection ratio, 84% tracking ratio and 75% speed estimation ratio. In the future developments of the Fire emergency digital twin system, solving the overlapping problem and increasing the accuracy of tracking and speed estimation would be the primary points.

5 Summary

In this chapter, the architectures, and key technologies of IoT systems, sensor networks, and digital twins are introduced, and three applications of fire digital twin are demonstrated. Typical IoT mainly comprises four layers, namely, sensor layer, networking layer, service layer, and application layer. WSN is an important technique in IoT, which can be used to better detect and monitor fire with the development of Neural Networks and data fusion technologies. WSN has layered and clustered architectures for different scenarios. Then the Fire-Safety IoT is introduced, the applications of Fire-Safety IoT systems usually include detection, monitoring, prediction, and response. The digital twin is defined as the digital representation of physical entities, and it enables real-time bidirectional interactions. Digital twin originates from the Apollo space program, and nowadays different models have been proposed and applied. Enabling technologies and tools are the basic to construct digital twins, which can be divided into five groups, i.e., physical world cognizing and controlling, connections, data management, modelling, and services. Finally, the concept of fire digital twin is introduced, and three case studies of tunnel fire digital twin, building fire digital twin, and human evacuation digital twin are demonstrated.

Acknowledgements This work is funded by the Hong Kong Research Grants Council Theme-based Research Scheme (T22-505/19-N) and MTR Research Fund (PTU-23005).

References

1. L. Da Xu, W. He, S. Li, Internet of things in industries: a survey. IEEE Trans. Ind. Inform. **10**, 2233–2243 (2014). https://doi.org/10.1109/TII.2014.2300753
2. P.A. Beata, A.E. Jeffers, V.R. Kamat, Real-time fire monitoring and visualization for the post-ignition fire state in a building. Fire Technol. **54**, 995–1027 (2018). https://doi.org/10.1007/s10694-018-0723-1
3. Q. Qi, F. Tao, T. Hu, N. Anwer, A. Liu, Y. Wei, L. Wang, A.Y.C. Nee, Enabling technologies and tools for digital twin. J. Manuf. Syst. **58**, 3–21 (2021). https://doi.org/10.1016/j.jmsy.2019.10.001
4. Y. Zeng, X. Huang, Smart building fire safety design driven by artificial intelligence, in *Interpretation Learning Analysis Des. Assessment, Information Decision Civil Infrastructure*, ed. by M.Z. Naser (Elsevier, New York, 2023). https://doi.org/10.1016/B978-0-12-824073-1.00011-3.
5. Y. Jiang, M. Li, W. Wu, X. Wu, X. Zhang, X. Huang, R.Y. Zhong, G.G.Q. Huang, Multi-domain ubiquitous digital twin model for information management of complex infrastructure systems. Adv. Eng. Inform. **56**, 101951 (2023). https://doi.org/10.1016/j.aei.2023.101951
6. D.G. Darwish, E. Square, Improved layered architecture for Internet of Things. Int. J. Comput. Acad. Res. **4**, 214–223 (2015)
7. M. Kavre, A. Gadekar, Y. Gadhade, Internet of Things (IoT): a survey, in *2019 IEEE Pune Section International Conference, IEEE* (2019), pp. 1–6
8. M. Kocakulak, I. Butun, An overview of Wireless Sensor Networks towards internet of things, in *2017 IEEE 7th Annual Computing Communicatons Work Conference, IEEE* (2017), pp. 1–6
9. K. Kaur, P. Kaur, E.S. Singh, Wireless sensor network: architecture, design issues and applications. Int. J. Sci. Eng. Res. **2** (2014)
10. C.T. Sony, C.P. Sangeetha, C.D. Suriyakala, Multi-hop LEACH protocol with modified cluster head selection and TDMA schedule for wireless sensor networks, in *2015 Global Conference Communication Technolofy*. IEEE, (2015), pp. 539–543
11. S.R. Vijayalakshmi, S. Muruganand, A survey of Internet of Things in fire detection and fire industries, in *Proceedings International Conference IoT Society Mobile, Analysis Cloud, I-SMAC 2017* (2017), pp. 703–707. https://doi.org/10.1109/I-SMAC.2017.8058270
12. A. Gaur, A. Singh, A. Kumar, A. Kumar, K. Kapoor, Video flame and smoke based fire detection algorithms: a literature review. Fire Technol. **56**, 1943–1980 (2020). https://doi.org/10.1007/s10694-020-00986-y
13. X. Wu, Y. Park, A. Li, X. Huang, F. Xiao, A. Usmani, Smart detection of fire source in tunnel based on the numerical database and artificial intelligence. Fire Technol. **57**, 657–682 (2021). https://doi.org/10.1007/s10694-020-00985-z
14. T. Zhang, Z. Wang, H.Y. Wong, W.C. Tam, X. Huang, F. Xiao, Real-time forecast of compartment fire and flashover based on deep learning. Fire Saf. J. **130**, 103579 (2022). https://doi.org/10.1016/J.FIRESAF.2022.103579
15. F. Abid, A survey of machine learning algorithms based forest fires prediction and detection systems. Fire Technol. **57**, 559–590 (2021)
16. K. Yun, J. Bustos, T. Lu, Predicting rapid fire growth (flashover) using conditional generative adversarial networks, in *IST International Symposium Electronic Imaging Science Technology* (2018), pp. 2751–2757. https://doi.org/10.2352/ISSN.2470-1173.2018.09.SRV-127
17. M. Liu, S. Fang, H. Dong, C. Xu, Review of digital twin about concepts, technologies, and industrial applications. J. Manuf. Syst. **58**, 346–361 (2021). https://doi.org/10.1016/j.jmsy.2020.06.017
18. D. Gelernter, *Mirror Worlds: Or the Day Software Puts the Universe in a Shoebox, How it Will Happen and What it Will Mean* (Oxford University Press, 1993)
19. M. Grieves, Digital twin: manufacturing excellence through virtual factory replication. White Pap. **1**, 1–7 (2014)

20. E. Glaessgen, D. Stargel, The digital twin paradigm for future NASA and US Air Force vehicles, in: *53rd AIAA/ASME/ASCE/AHS/ASC Structures. Structure Dynamic Material Conference. 20th AIAA/ASME/AHS Adaptive Structure Conference. 14th AIAA* (2012), p. 1818
21. F. Jiang, L. Ma, T. Broyd, K. Chen, Digital twin and its implementations in the civil engineering sector. Autom. Constr. **130**, 103838 (2021). https://doi.org/10.1016/j.autcon.2021.103838
22. L. Sha, S. Gopalakrishnan, X. Liu, Q. Wang, Cyber-physical systems: a new frontier, in, *IEEE International Conference Sensing Networks, Ubiquitous Trust Computing Cyber-Physical* (2008), pp. 1–9. https://doi.org/10.1109/SUTC.2008.85
23. L. Wright, S. Davidson, How to tell the difference between a model and a digital twin, Adv. Model. Simul. Eng. Sci. **7** (2020). https://doi.org/10.1186/s40323-020-00147-4.
24. F. Tao, W. Liu, M. Zhang, T. Hu, Q. Qi, H. Zhang, F. Sui, T. Wang, H. Xu, Z. Huang, Five-dimension digital twin model and its ten applications. Comput. Integr. Manuf. Syst. **25**, 1–18 (2019)
25. R. Stark, C. Fresemann, K. Lindow, Development and operation of Digital Twins for technical systems and services. CIRP Ann. **68**, 129–132 (2019). https://doi.org/10.1016/j.cirp.2019.04.024
26. Science & Tech Spotlight: Digital Twins—Virtual Models of People and Objects | U.S. GAO, (n.d.)
27. F. Tao, M. Zhang, Digital twin shop-floor: a new shop-floor paradigm towards smart manufacturing. IEEE Access **5**, 20418–20427 (2017). https://doi.org/10.1109/ACCESS.2017.2756069
28. Y. Qamsane, C.-Y. Chen, E.C. Balta, B.-C. Kao, S. Mohan, J. Moyne, D. Tilbury, K. Barton, A unified digital twin framework for real-time monitoring and evaluation of smart manufacturing systems, in *IEEE 15th International Conference Automation Science Engineering*. IEEE (2019), pp. 1394–1401
29. X. Ma, Q. Qi, J. Cheng, F. Tao, A consistency method for digital twin model of human-robot collaboration. J. Manuf. Syst. **65**, 550–563 (2022). https://doi.org/10.1016/j.jmsy.2022.10.012
30. G. Schrotter, C. Hürzeler, The digital twin of the city of Zurich for urban planning. PFG–J. Photogramm. Remote Sens. Geoinf. Sci. **88**, 99–112 (2020)
31. C. Fan, C. Zhang, A. Yahja, A. Mostafavi, Disaster city digital twin: a vision for integrating artificial and human intelligence for disaster management. Int. J. Inf. Manage. **56**, 102049 (2021). https://doi.org/10.1016/j.ijinfomgt.2019.102049
32. R. Sacks, I. Brilakis, E. Pikas, H.S. Xie, M. Girolami, Construction with digital twin information systems. Data-Centric Eng. **1**, e14 (2020)
33. Z. Ye, Y. Ye, C. Zhang, Z. Zhang, W. Li, X. Wang, L. Wang, L. Wang, A digital twin approach for tunnel construction safety early warning and management. Comput. Ind. **144**, 103783 (2023). https://doi.org/10.1016/j.compind.2022.103783
34. A. Protopsaltis, P. Sarigiannidis, D. Margounakis, A. Lytos, Data visualization in internet of things: tools, methodologies, and challenges. in *ACM International Confernce Proceeding Series* (2020). https://doi.org/10.1145/3407023.3409228
35. X. Zhang, Y. Jiang, X. Wu, Z. Nan, J. Shi, Y. Zhang, X. Huang, G.G.Q. Huang, AIoT-enabled digital twin system for smart tunnel fire safety management. Dev. Built Environ. (Under Rev.) (2023)
36. X. Wu, X. Zhang, Y. Jiang, X. Huang, G.G.Q. Huang, A. Usmani, An intelligent tunnel firefighting system and small-scale demonstration. Tunn. Undergr. Sp. Technol. **120**, 104301 (2022). https://doi.org/10.1016/j.tust.2021.104301
37. X. Huang, X. Wu, X. Zhang, A. Usmani, Smart tunnel fire safety management by sensor network and artificial intelligence, in *Leveraging Artificial Intelligence in Engineering Managing Safety Infrastructures*, ed. by M.Z. Naser (New York, 2022), pp. 423–443. https://doi.org/10.1201/9780367823467-18.
38. G. Ma, Z. Wu, BIM-based building fire emergency management: Combining building users' behavior decisions. Autom. Constr. **109**, 102975 (2020). https://doi.org/10.1016/j.autcon.2019.102975

39. A. Vaswani, N. Shazeer, N. Parmar, J. Uszkoreit, L. Jones, A.N. Gomez, Ł. Kaiser, I. Polosukhin, Attention is all you need. Adv. Neural Inf. Process. Syst. **30** (2017)
40. D. Drysdale, *An Introduction to Fire Dynamics*, 3rd ed. (John Wiley & Sons, Ltd, Chichester, UK, 2011). https://doi.org/10.1002/9781119975465.
41. T. Zhang, Z. Wang, Y. Zeng, X. Wu, X. Huang, F. Xiao, Building artificial-intelligence digital fire (AID-Fire) system: a real-scale demonstration. J. Build. Eng. **62** (2022) 105363. https://doi.org/10.1016/j.jobe.2022.105363.
42. A. Sharma, P.K. Singh, Y. Kumar, An integrated fire detection system using IoT and image processing technique for smart cities. Sustain. Cities Soc. **61**, 102332 (2020). https://doi.org/10.1016/j.scs.2020.102332
43. S. Jiang, S. Zhu, X. Guo, C. Chen, Z. Li, Safety monitoring system of steel truss structures in fire. J. Constr. Steel Res. **172**, 106216 (2020)
44. D. McCarthy, U. Dayal, The architecture of an active database management system. ACM Sigmod Rec. **18**, 215–224 (1989)
45. X. Zhang, X. Wu, Y. Park, T. Zhang, X. Huang, F. Xiao, A. Usmani, Perspectives of big experimental database and artificial intelligence in tunnel fire research. Tunn. Undergr. Sp. Technol. **108**, 103691 (2021). https://doi.org/10.1016/j.tust.2020.103691
46. J.W. Krogh, G. Krogh, *Gennick* (Springer, MySQL Connector/Python Revealed, 2018)
47. G.W. Zou, Y. Huo, W.K. Chow, C.L. Chow, Modelling of heat release rate in upholstered furniture fire. Fire Mater. **42**, 374–385 (2018). https://doi.org/10.1002/fam.2502
48. T. Zhang, G. Wang, H. Hu, Y. Huang, K. Zhu, K. Wu, Study on temperature decay characteristics of fire smoke backflow layer in tunnels with wide-shallow cross-section. Tunn. Undergr. Sp. Technol. **112** (2021). https://doi.org/10.1016/j.tust.2021.103874.
49. Y. Ding, Y. Zhang, X. Huang, Intelligent emergency digital twin system for monitoring building fire evacuation. J. Build. Eng. **77**, 107416 (2023). https://doi.org/10.1016/j.jobe.2023.107416.
50. A. Bochkovskiy, C.-Y. Wang, H.-Y.M. Liao, Yolov4: optimal speed and accuracy of object detection (2020). ArXiv Prepr. ArXiv2004.10934
51. N. Wojke, A. Bewley, D. Paulus, Simple online and realtime tracking with a deep association metric, in *2017 IEEE International Conference Image Process* (2017), pp. 3645–3649. https://doi.org/10.1109/ICIP.2017.8296962

Applying Machine Learning to Evaluate the Performance of Thin-Walled Steel Members in Fire

Qi Tong, Carlos Couto, and Thomas Gernay

1 Introduction

Structural fire safety is crucial for building design and construction. Fires in buildings can affect strength and stability of structures and cause devastating consequences on human life and property. Structural fire safety plays an important role in firefighting that includes designing, constructing and maintaining a safe building to ensure the safety of occupants and firefighters. The efforts in designing and maintaining structural integrity help to prevent fires from spreading and provide more time for evacuation and firefighting activities. To ensure safety in case of a fire, structures should be designed to withstand fire exposure and maintain their structural integrity in accordance with the performance objectives or code-based prescriptions. This requires evaluating the capacity of structural members at elevated temperatures. High-fidelity finite element models (FEM) can be applied to model the structural response under different fire conditions. However, application of FEM requires expertise, and it can be time-consuming and computationally expensive.

The development of artificial intelligence (AI) has had a significant impact on various fields of engineering. AI methods such as machine learning (ML), pattern recognition, and deep learning have already been used to address various problems in structural engineering [1]. AI can be leveraged for its cognition ability and deductive reasoning to identify patterns in data and resolve problems which are either too complex or that cannot be appropriately solved in a timely or economical manner through other traditional methods. AI is especially appropriate for problems in which

Q. Tong (✉) · T. Gernay
Department of Civil and Systems Engineering, Johns Hopkins Univesity, Baltimore, USA
e-mail: qtong5@jhu.edu

C. Couto
Civil Engineering Department, RISCO, University of Aveiro, Aveiro, Portugal

X. Huang and W. C. Tam (eds.), *Intelligent Building Fire Safety and Smart Firefighting*, Digital Innovations in Architecture, Engineering and Construction,
https://doi.org/10.1007/978-3-031-48161-1_15

a large number of variables (features) leads to the phenomena (output) and the relationship between the features and the outputs are complex to describe. AI has been increasingly used to uncover relationships between features and output and to develop fast predictive models in a variety of domains.

ML refers to a class of data analysis methods that learn the pattern from existing data and predict the response given new data within the boundaries of the original dataset. It provides an opportunity to derive efficient surrogate models and identify prevailing parameters [1], including to derive low-cost models for predicting the behavior of complex members and structural assemblies subjected to fires [2]. Although such an application is still in its infancy, a handful of studies have recently been published. Naser applied AI to predict the thermal and structural response of timber at elevated temperatures [3] or the occurrence of spalling in concrete [4, 5]. Chaudhary et al. developed ML models for probabilistic analysis of a concrete column subjected to fires [6, 7]. Others have used AI to improve the ability to predict fire or smoke behavior [8–10].

This chapter summarizes two studies on the application of ML methods to develop predictive models for the elevated temperature capacity of thin-walled steel columns [11] and beams [12]. Thin-walled steel columns and beams are commonly used as load-carrying components in buildings. Steel elastic modulus and strength are reduced at elevated temperatures, which affects the resistance and possibly the failure mode. Therefore, a proper understanding of the behavior of thin-walled steel columns and beams at elevated temperatures is needed for their structural fire design. The chapter presents the generation of the dataset through the use of high-fidelity FEM, the development of the ML models, and the obtained performance of the models. Comparison against current analytical models shows the potential benefits of the approach in terms of accuracy and ease of application.

2 Scope of the Study

The study focuses on the elevated temperature capacity of thin-walled I-shaped steel columns and beams. Their behavior is influenced by the complex interaction between local, global, and distortional instability modes, combined with the steel strength and modulus reduction at elevated temperatures. This results in complex models for the structural fire design of such members.

For thin-walled steel columns, experiments showing failure from local and global buckling at elevated temperature have been reported in the literature (see Fig. 1a, b). Franssen et al. conducted tests on columns with European class 4 cross-sections at elevated temperatures and calibrated shell finite element models on the tests [13]. High-temperature experiments on thin-walled steel sections were also reported by others [14, 15]. Eurocode 3 Part 1–2 provides specifications for class 4 cross-section columns at elevated temperatures. When no explicit calculation is made, a critical temperature of 350 °C may be assumed [16], meaning that it is conservatively assumed that the loadbearing capacity of the member is lost once the steel temperature has reached 350 °C. Alternatively, a method was proposed based on the calculation of an effective width to account for the reduction of load-carrying capacity due to local buckling simultaneously with the use of the 0.2% proof strength of steel at elevated temperatures [17]. However, this design model demonstrated to lead to inconsistencies in load-carrying capacity for stub columns or individual plates [14, 18, 19], as well as for columns with thin-walled sections [20, 21]. Recently, Couto et al. showed that the analytical model proposed in the Eurocode for class 4 cross-sections was overly conservative in certain cases, and they proposed an improved method which included replacing the use of the 0.2% proof strength with the yield strength at 2% total strain [22].

For thin-walled steel beams with I-shape sections, the main failure modes at elevated temperature include exceeding their flexural capacity, instability due to

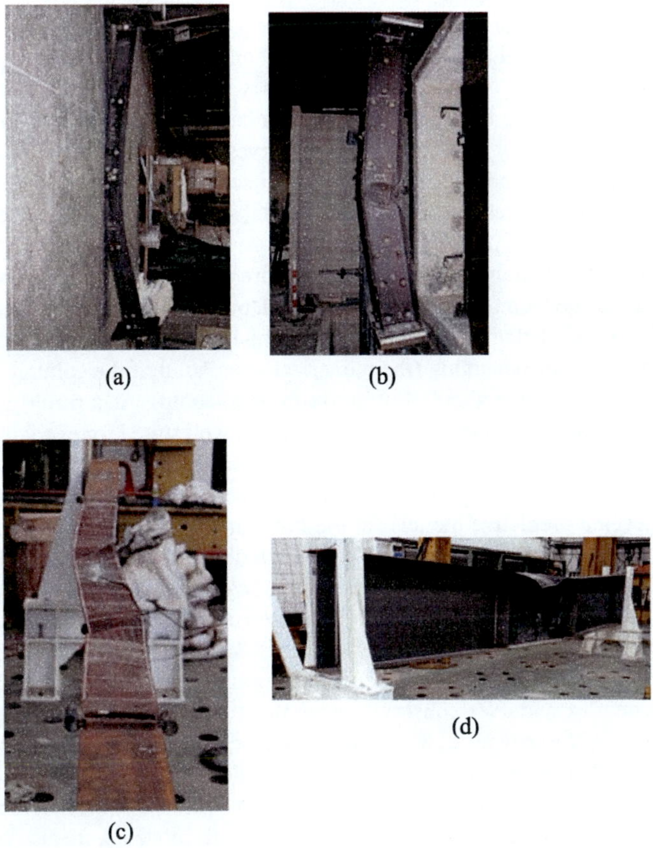

Fig. 1 Slender section showing **a** global buckling of the column, **b** global and local buckling of the column [13]. **c** lateral-torsional buckling of the beam and **d** local buckling of the beam after fire exposure [33]

lateral-torsional buckling (LTB), or local failure of the section plates under compressive stresses (i.e., local buckling) (see Fig. 1c, d). In addition, the web plate may also collapse due to the shear loading or concentrated loads that can cause web crushing, especially when its thickness is relatively small. For the case of LTB, capacity depends essentially on the beam and section sizes, the supporting conditions and the way the load is applied. The ultimate strength of beams with slender sections is mainly governed by the interaction of LTB and local buckling. Slender I-shaped beams offer benefits from the perspective of rational usage of material, but the design of slender section beams must account for the complex local–global interactive behavior that may occur [23]. The subject of LTB of laterally unsupported beams sections has been extensively studied in the past both at normal temperature (e.g. [23–26]) and for the case of fire (e.g., [27–32]). However, resistance of beams against lateral-torsional buckling still remains relatively less explored for the case of beams with slender sections and design methods for the behavior of slender steel beams in the fire situation remain complex and overly conservative.

While analytical methods derived from mechanics-based principles and experimental observations are very useful for design, they are necessarily based on assumptions. Further improving their accuracy requires increasing the level of complexity, for example in the form of additional buckling curves for local and/or global buckling, which is not practical and is error-prone to the users. Finite element model (FEM) can also be used to simulate the behavior of slender steel members subjected to fire as shown in Fig. 2. Studies published in the literature have shown that FEM with shell finite elements can capture the critical temperature and failure mode of slender steel members at elevated temperatures [22, 34–38]. However, their use requires users to have in-depth knowledge about numerical methods, and the building and running of the model is time consuming (the numerical simulation of the column subjected to fire in Fig. 2 may take about 10 min to run on a laptop). As a result, FEM with shell elements remains a specialized tool that is not well suited for everyday design, probabilistic assessments, or rapid pre-design decisions.

From the discussion above, it appears that ML methods could be particularly well suited to develop predictive models for the fire behavior of slender steel members. The capacity of thin-walled structures can be predicted by well identified features and outputs. Data can be generated from numerical models, such as FEM, which have been developed and experimentally validated over the years. ML models could provide advantages over current design methods in design codes and standards, which typically combine mechanics-based analytical equations and empirical factors in relatively complex and over conservative formulations. This chapter thus presents the application of three types of ML methods to determine the resistance of I-shaped thin-walled members at elevated temperatures. The dataset were built from shell finite element simulations in SAFIR [35] and the performance of the ML models was assessed against numerical testing data. The trained ML models were also compared with the state-of-the-art analytical models. This chapter will present the data generation from shell FE models (Sect. 3), the construction of the ML models (Sect. 4), and the results (Sect. 5) and comparison with current state-of-the-art analytical models (Sect. 6) for the prediction of the elevated temperature capacity of slender steel

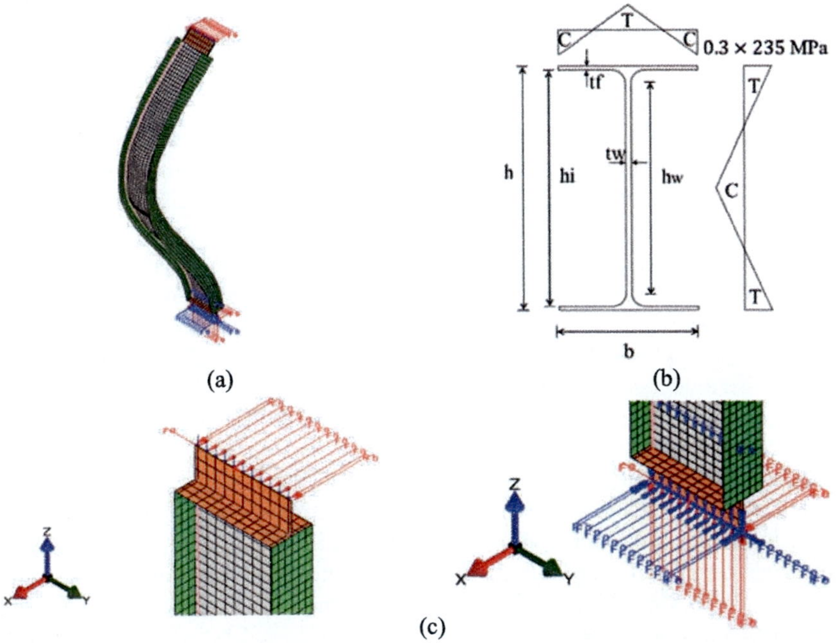

Fig. 2 Numerical model in SAFIR: **a** shell model for IPE500 column at 300 °C; **b** cross-section dimensions and residual stresses ($C = -0.3 \times 235$MPa, $T = 0.3 \times 235$MPa); **c** boundary conditions with end plates and pinned-fixed supports

columns and beams. Finite element models are applied to generate data for ML models in Sect. 3. Both thin-walled columns and beams are considered. Once the data is collected, ML models are trained and tested with the data as shown in Sect. 4. The results are shown in Sect. 5. Comparison of ML models against analytical models is provided in Sect. 6. Section 7 concludes this chapter.

3 Data Generation with Finite Element Models

3.1 Steel Columns

3.1.1 Numerical Model

Numerical modeling was used to analyze I-shaped columns with slender cross-sections subjected to axial compression at elevated temperatures. The numerical modeling was implemented with the nonlinear FE software SAFIR using shell elements. Columns with different plate slenderness and length were modeled to evaluate the effect of the width-to-thickness ratio of steel plates and the slenderness ratio of the column on the structure behavior. Webs of class 2, 3, and 4 and flanges of class 1, 2, and 3 were studied. The columns were uniformly heated on four sides. Columns with lengths of 1.5, 3.5 and 4.5 m long were analyzed with pinned or fixed supports. Residual stress and initial geometric imperfections were included in the model. As an example, Fig. 2a shows the failure mode of a column with IPE 500 cross-section at 300 °C. The failure mode consists of global buckling along the weak axis and local buckling of the web and flange.

The ultimate load-bearing capacity of the columns was calculated with SAFIR by first uniformly increasing the temperature in the section up to the target value and then progressively loading the column until failure. Steels with grades S235, S355, and S460 at ambient temperature were used in the numerical modeling. The constitutive model included in the simulations followed the non-linear stress–strain relationship and reduction factors defined in Eurocode 3 Part 1–2 [16].

Geometric imperfections in the model were obtained for local and global modes through linear buckling analyses with the software Abaqus [39]. For global imperfection, the amplitude was taken as L/1000, where L is the length of the column. For local imperfection, the amplitude was calculated as 80% of the geometric fabrication tolerances as recommended in [36, 40]. The amplitudes of local imperfections were calculated following Table 1. The global imperfection and local imperfection were combined following the recommendation of Annex C in Part 1–5 of Eurocode 3 [41]. The lowest eigenmode was the leading imperfection and the amplitude of the other eigenmode was reduced to 70%. The coordinates of the nodes with geometric imperfections were exported from Abaqus to SAFIR. Shell elements with four nodes were used. A sensitivity analysis on the mesh size was conducted, showing the convergence of the results with 120 elements on the length, 6 elements on the flange, and 10 elements on the web.

For boundary conditions, two rigid 100 mm thick endplates were added at both ends of the column, as shown in Fig. 2c. The load was axially applied on the edge of the vertical plate on the top with no eccentricity such that the load can be distributed evenly on the web and flange. The displacements of two rigid extensions were constrained except for the vertical direction on the top (Uy) which allowed free thermal expansion. The two rigid extensions were restrained as pinned or fixed.

Validation was conducted against eight experimental tests on thin-walled columns in [11] to ensure the accuracy of the numerical models. Validation against experiments is an important step in the process since the FE models were then used as the "ground truth" for training and validating the ML models. The results show that

Table 1 Amplitude of local imperfections

Flange	Web
$\frac{b}{150}\ for\ \frac{b}{tf} \leq 20$	$\frac{h_w}{200}\ for\ \frac{h_w}{t_w} \leq 80$
$\frac{b^2}{300 \times tf}\ for\ \frac{b}{tf} > 20$	$\frac{{h_w}^2}{16000 \times t_w}\ for\ 80 < \frac{h_w}{t_w} \leq 200$
	$\frac{h_w}{80}\ for\ \frac{h_w}{t_w} > 200$

the critical temperature and failure mode observed in the tests are well captured by SAFIR. The finite element model captures the failure modes observed in the test. For critical temperature, the average ratio of SAFIR/Experiment is 1.026 with a standard deviation of 0.035. Therefore, the numerical modeling approach in the software SAFIR can be used to generate the datasets on capacity of thin-walled steel columns at elevated temperatures.

3.1.2 Features and Outputs

The design parameters for the FEM are presented in Table 2. The slenderness of the plates, the slenderness of the columns, and boundary conditions influence the resistance and failure mode of the columns subjected to fire. In the parametric analysis, the column with a length of 4.5 m was studied for 24 cross-sections ranging from IPE300 to IPE600, three values of yield strength, four different boundary conditions, and six temperatures, leading to 1728 (=1 × 24 × 3 × 4 × 6) numerical models. The column with a length of 1.5 and 3.5 m was studied for cross-sections of IPE300-, IPE330-, IPE330+, and IPE600+, three values of yield strength, four different boundary conditions, and six temperatures, leading to 576 (=2 × 4 × 3 × 4 × 6) numerical models. The combination of these parameters generates 2304 (=1728 + 576) numerical models.

The ranges of values for the input parameters are listed in Table 3. Nine features were considered for the ML model development, which is discussed afterwards. The h_w/t_w and $b/2t_f$ are the adimensional web and flange dimensions, F_{y_web}/E and F_{y_flange}/E are the adimensional web and flange yield strengths. The L is column length, $N_{pl,20}$ and $N_{cr,20}$ are the plastic capacity and critical capacity at ambient temperature. The range of temperature (assumed uniform in the section and along the column) covers the typical range of critical temperatures for columns. The range of feature values considered in the training and testing dataset, which derive from the design variables in Table 2, is provided in the columns "ML training/testing" of Table 3 and was chosen to cover a common range of design parameters for slender steel columns in building structures. The output is $N_{u,T}/N_{pl,20}$, i.e., the ultimate

Table 2 Design parameters in the numerical modeling for a total of 2304 simulations

Design parameter	Range
Cross-section	24 cross-section sizes from IPE 300 to IPE600
t_w (mm)	6.1–15
b (mm)	56.4–81.1
t_f (mm)	9.2–24
f_y (N/mm^2)	235, 355, 460
Boundary condition	Pinned–pinned, pinned-fixed, fixed-pinned, fixed–fixed
Temperature (°C)	300, 400, 500, 600, 700, 800
Column length (m)	1.5, 3.5, 4.5

Table 3 Input parameters for the ML models

Notation	Feature	Input values	ML (training/testing)	
			Min	Max
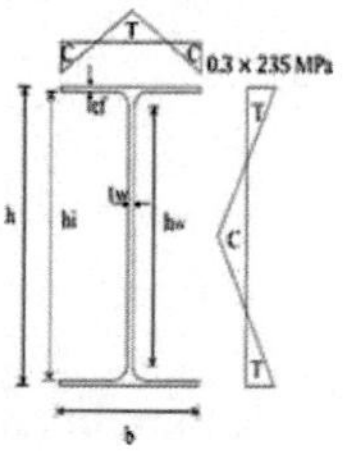	x_1	h_w/t_w	31	52
	x_2	$b/2t_f$	4.6	8.2
	x_3	h_w/b	0.56	2.34
	x_4	t_w/t_f	0.56	1
	x_5	f_{y_web}/E	(235/E)	(460/E)
	x_6	f_{y_flange}/E	(235/E)	(460/E)
	x_7	L/b_f	6.7	30
	x_8	$(N_{pl,20}/N_{cr,20})^{0.5}$	0.17	2.01
	x_9	Temperature	300 °C	800 °C

load-carrying capacity of the column at elevated temperature normalized over the plastic capacity of the section at ambient temperature.

3.2 *Steel Beams*

3.2.1 Numerical Model

Thin-walled beam is another widely used type of load-carrying steel member that is susceptible to buckling and collapse under fire. The software SAFIR was used to build the finite element model of the thin-walled steel beams. Shell elements were used. Fork-supports were used at both ends of the structural member to prevent the displacements in z-direction and y-direction as shown in Fig. 3a. To prevent the rigid body movement, the displacements in x-direction were constrained at mid-span. The loading was applied by nodal forces to produce end-moments at both ends. Additionally, end plates were included with a thickness equal to 10 times the web thickness to ensure correct load distribution. Uniform temperature along the member and cross-section was considered to allow direct comparison with existing fire design rules that were developed based on this assumption. Initial imperfections and residual stress were considered. Figure 3b, c shows two examples of the collapse shape obtained with this numerical model for two different beams.

3.2.2 Features and Outputs

In previous studies by Couto et al. [29, 30, 42], the FEM was used in extensive parametric analyses to evaluate the ultimate load of thin-walled beams at elevated temperatures. The references [29, 30] develop the background of the recently proposed

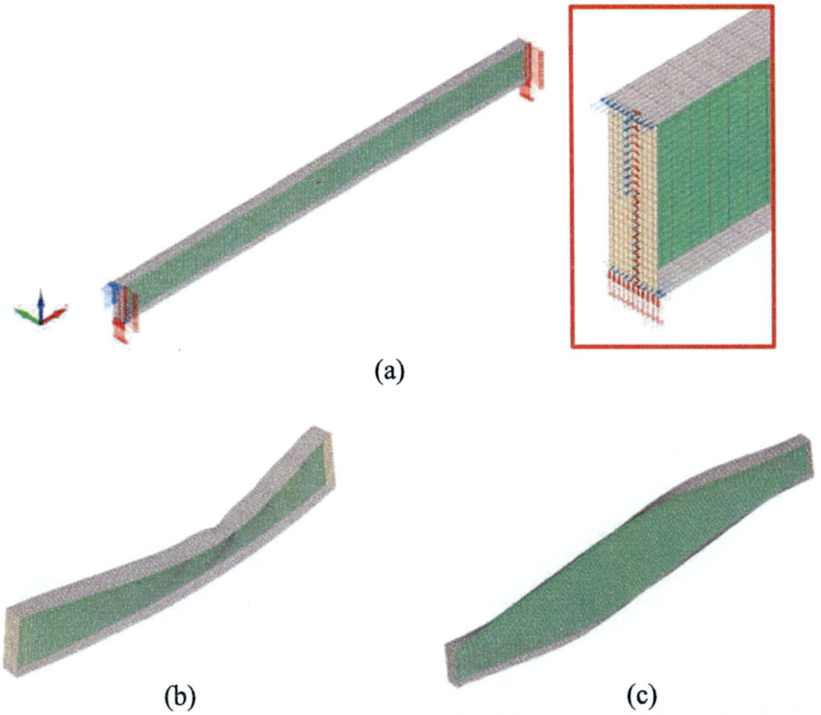

Fig. 3 a Numerical model with applied boundary conditions (loads and supports); b Examples of collapse shape of numerical models (×4) for $\psi = 0.75$ and c $\psi = -1$ showing the possible interaction between local buckling and lateral-torsional buckling

expressions that will be included in the next generation of the Eurocode 3 Part 1–2 [43] to calculate the lateral-torsional buckling of beams with slender section for the case of fire. The reference [42] contains additional results following the same principles and modelling strategy.

Herein, the calculation results from these FEM studies were considered as the dataset for the purpose of fitting, and validating, the ML models. The cases for which the collapse of the beam was identified to be the result of a shear failure were disregarded, in line with the same assumptions from [29].

Therefore, a total of 24,516 samples were considered in the dataset. 8 features were defined, as indicated in Table 4. f_y and E are the steel yield strength and Young modulus ($E = 210\,\text{GPa}$), respectively. θ is the uniform temperature along the beam. $M_{pl,20}$ and $M_{cr,20}$ are the plastic cross-section resistance and the elastic critical moment of the beam, both calculated at normal temperature. The feature x_5 was defined as f_y/E to provide a nondimensional parameter for yield strength, consistent with the slenderness-inspired feature x_7. . The table lists the minimum and maximum values of each feature within the dataset. The range of temperature (assumed uniform in the section and along the beam) covers the typical range of

Table 4 Input parameters/features of dataset used for the ML models training

Notation		Feature	Input values	Min	Max
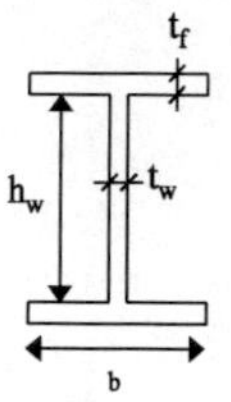	x_5	x_1	h_w/t_w	75	200
	x_6	x_2	b/t_f	6	50
	x_7	x_3	h_w/b	1.8	3.33
	x_8	x_4	t_w/t_f	0.16	0.8
		x_5	f_y/E	(235/E)	(460/E)
		x_6	θ	300 °C	700 °C
		x_7	$(M_{pl,20}/M_{cr,20})^{0.5}$	0.097	2.738
		x_8	ψ	−1	1

critical temperatures for such members. The frequency distributions of the features are shown in Fig. 8 of [12].

The output/target y of the ML models has been defined as

$$y = M_{u,T}/M_{pl,20} \tag{1}$$

where $M_{u,T}$ is the ultimate load of the beam at elevated temperature.

4 Machine Learning Models

4.1 Types of ML Models

The used ML models in this chapter include Support Vector Regression (SVR) [44], Polynomial Regression (PR) [45], Artificial Neural Networks (ANN) [46], and Random Forest (RF) [47].

The SVR is developed as an extension of the support vector machine (SVM) [48], which aims to find a hyperplane in an n-dimensional space (n is the number of features, i.e., input parameters) that classifies the training datasets in different classes. While the objective of SVM is to find a hyperplane that has the maximum margins ($\pm\varepsilon$), the extension SVR aims to find a flat hyperplane with margins ($\pm\varepsilon$) that accept the data points within or on the margins while rejecting the data points outside the margins.

The general form for PR is written as:

$$Y = X\omega + \varepsilon \tag{2}$$

in which Y is the vector of responses, X is the feature matrix, ω is the coefficient and ε is the bias. The polynomial regression extends the inputs of the linear model, which is obtained by raising the initial inputs to a power. The new inputs are created with degrees less than or equal to the specific order. The new feature matrix includes (1) bias; (2) converting the initial inputs to their higher-order terms for each degree; (3) a combination of all pairs of initial inputs. The model with higher degrees may closely fit most of the data in the training dataset, however, it may also capture the noises in the data, resulting in a larger error on the testing dataset (i.e., overfitting). To prevent over-fitting in polynomial regression, ridge regression is applied to fit the polynomial feature matrix. The ridge regression adds a regularization term to the sum of squares of residuals. The loss function of ridge regression is written as:

$$minimize\colon \sum_{i=1}^{n} \|y_i - \sum_{j=0}^{m} x_{ij} w_j\|^2 + \lambda \sum_{j=0}^{m} \|w_j\|^2 (\lambda > 0) \tag{3}$$

in which the y_i is the known observation, $\sum_{j=0}^{m} x_{ij} w_j$ is the predicted value, and λ is the tuning parameter that controls the complexity of the model. As λ grows larger, the ridge regression effectively shrinks coefficient w_j to be 0 and selects a small subset of features to build the model, which prevents training a more complex model and thus avoids over-fitting.

The RF is a learning algorithm which uses ensemble learning method for classification and regression problems. The algorithm operates by constructing a multitude of decision trees at training time and outputting the average (regression) or max voted (classification) of prediction of the individual trees. The averaging or voting improves the RF's accuracy and reduces overfitting. Besides, each tree draws a random sample from the original data set when generating its splits, adding a further element of randomness that prevents overfitting.

The ANN is a mathematical model that can be described as a group of neurons arranged in layers and their connections forming a network as shown in Fig. 4. The ANN can map a certain range of input values (features) to a specific target value (result). The most used type of ANN is the Multilayer Perceptron (MLP) [49], with an input layer, one or more hidden layers, and an output layer as shown in Fig. 4. Each neuron of a certain layer is only connected to the neurons in the next layer, and may include an extra, independent, value called the bias. The output of each neuron j can be calculated as:

$$\text{output}_j = f\left(\sum_{i=1}^{n} w_{ij} x_i + b_j\right) \tag{4}$$

where output_j is the output of the layer, w_{ij} are the weight coefficients, x_i are the input values, or features, and b_j is the bias values. The $f(x)$ is the so-called transfer or activation function as given in Table 3 of [12]. The training of ANN is achieved by exposing the network to a set of examples (input patterns) with known outputs (target output). The weights of the internal connections, and the bias values of the neurons, are adjusted to minimize errors between the network output and target output [50].

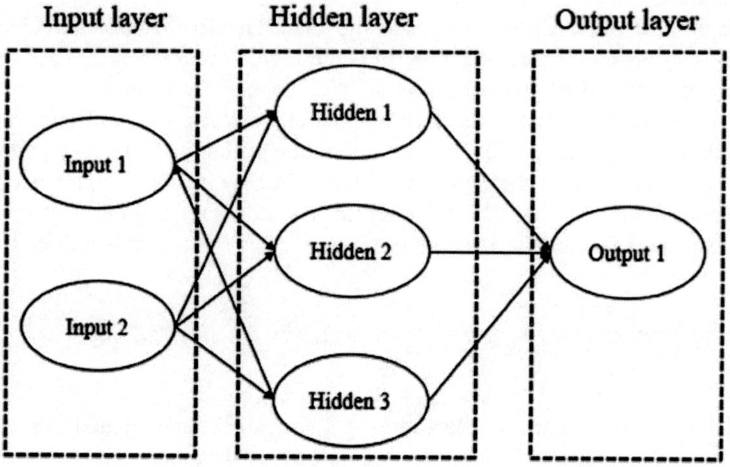

Fig. 4 Generic structure of the ANN model

4.2 Training Procedure

The numerical datasets from the FEM were randomly divided into two groups for training and testing the ML models. For the steel column, the 2304 data points were divided into two subsets: a training set with 2073 data points and a testing set with 231 data points. The training set was used to training the ML model and the testing set was used to test the robustness of the model for data that the model has never seen. For the steel beam, the dataset of 24,516 samples was divided into a training dataset and a test dataset with a ratio of 7:3.

The best hyperparameters were determined by grid search with five-fold cross-validation to avoid overfitting [51]. The Python package Scikit-learn [52] was used for the implementation of the different types of ML models discussed in the previous section.

Regarding the performance metrics, the performance of training, testing, and validation was determined by the coefficient of determination (R^2). The R^2 was calculated following Eqs. (5), (6), (7), and (8).

$$\overline{y} = \frac{1}{n}\sum_{1}^{n} y_i \tag{5}$$

$$ss_{tot} = \sum_{1}^{n} (y_i - \overline{y})^2 \tag{6}$$

$$ss_{res} = \sum_{1}^{n} (y_i - f_i)^2 \tag{7}$$

$$R^2 = 1 - \frac{ss_{res}}{ss_{tot}} \tag{8}$$

in which y_i is the ith observed data, $\overline{y}$ is the mean of observations, f_i is the predicted value, ss_{tot} and ss_{res} are the total sum of squares and residual sum of squares, respectively. The R^2 measures how well the observations are replicated by the ML models, an R^2 close to 1 is preferred.

5 Results

5.1 *Steel Columns*

Figure 5a–f plots the predicted capacity $N_{u,T}/N_{pl,20}$ using SVR, ANN, and PR against the numerical estimations from the shell finite element model in SAFIR. For the training and testing dataset, the results from SVR, ANN, and PR agree well with the capacity evaluated by SAFIR. Overall, the trained SVR, ANN, and PR models, can predict the resistance of the slender steel columns at elevated temperatures with high accuracy. The performance of the ML models can be further improved by considering a larger dataset including a greater range of inputs.

The hyperparameters for SVR are C, gamma and epsilon with values of 849, 0.354 and 0.01, respectively. The kernel used for SVR is 'rbf'. The hyperparameters for ANN are α, hidden layer sizes, initial learning rate with values of 0.37373, 8 and 0.01213, respectively. The activation function is 'Relu' and the learning rate is 'Adaptive'. The coefficients of the polynomial regression model of degree 2 are given in [11]. This model can easily be implemented in a datasheet to provide an almost-immediate prediction of the elevated temperature capacity for slender steel columns, within the range of features provided in Table 3 (column "ML training/testing").

To quantify the performance of the different models, the coefficient of determination R^2 value was evaluated for the models against the training and testing dataset. Table 5 gives the results for the SVR, ANN, and PR (degree 2) models.

5.2 *Steel Beams*

Figure 6a–h plot the predicted capacity $M_{u,T}/M_{pl,20}$ using SVR, ANN, PR, and RF against the numerical estimations from the shell finite element model in SAFIR. For the training and testing dataset, the results from the ML models agree well with the capacity evaluated by SAFIR.

The hyperparameters of the ML models are provided in [12]. The R^2 scores of the ML models plotted in Fig. 6 are, for the train and test sets, respectively:

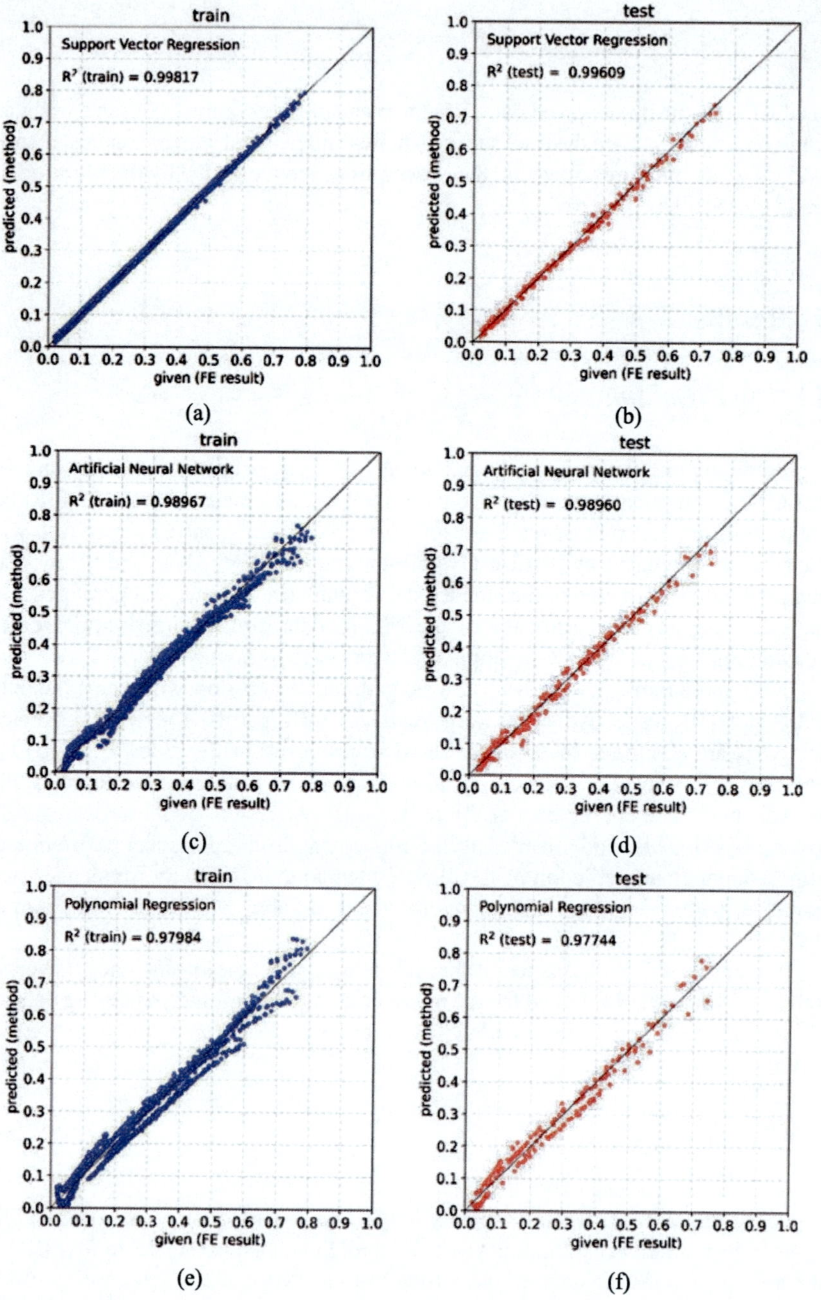

Fig. 5 Predicted capacity $N_{u,T}/N_{pl,20}$ for thin-walled steel columns at elevated temperature: Comparison between SAFIR finite element model and **a** SVR training, **b** SVR testing; **c** ANN training, **d** ANN testing; **e** PR training, **f** PR testing. Training/testing performed based on 2304 data points

Table 5 Performance of ML models and the analytical model

Regressor (R^2)	Train	Test
SVR	0.998	0.996
ANN	0.990	0.990
PR (degree 2)	0.980	0.977
Analytical [22]	0.973	

0.99812 and 0.99803 (ANN); 0. 98,739 and 0. 98,753 (SVR); 0.98113 and 0.98142 (PR); 0.99946 and 0.99813 (RF). For the PR model, the model with degree of 6 shows the best performance of both training and testing process. As the degree increases, the performance of training increases, however, when the degree reaches 6, its performance of testing starts to decrease due to overfitting. More details can be found in [12].

6 Comparison with Analytical Models from Eurocode

6.1 Steel Column

Part 1.5 of Eurocode 3 provides expressions of reduction factor for plate buckling resistance under compression [41]. Couto et al. [22] proposed an updated formula to account for the local buckling of slender steel columns (Class 3 and Class 4) at elevated temperature and replaced the use of design yield strength corresponding to the 0.2% proof strength with the yield strength at 2% total strain for Class 4. R^2 by the analytical model described in [22] with χ_{fi} from Eurocode 3 Part 1.2 [16] is 0.973 as shown in Table 1. The three ML models provide better agreement with the training and testing dataset than the analytical model. Figure 7 plots the results obtained with the analytical models with a R^2 of 0.973. It can be observed that the scatter is larger than with the ML models of Fig. 5.

6.2 Steel Beam

In the fire situation, the current version of Eurocode 3 Part 1–2 [16] recommends the same approach as for normal temperature to calculate the effective properties of Class 4 sections, but for the material properties degradation with temperature, defines that a reduced yield strength of steel based on the 0.2% proof stress must be considered at elevated temperatures instead of the stress corresponding to the 2% total strain as defined for the other classes of sections. To improve consistency of results, the new generation of Eurocode 3 Part 1–2 [43] will adopt the same yield strength for

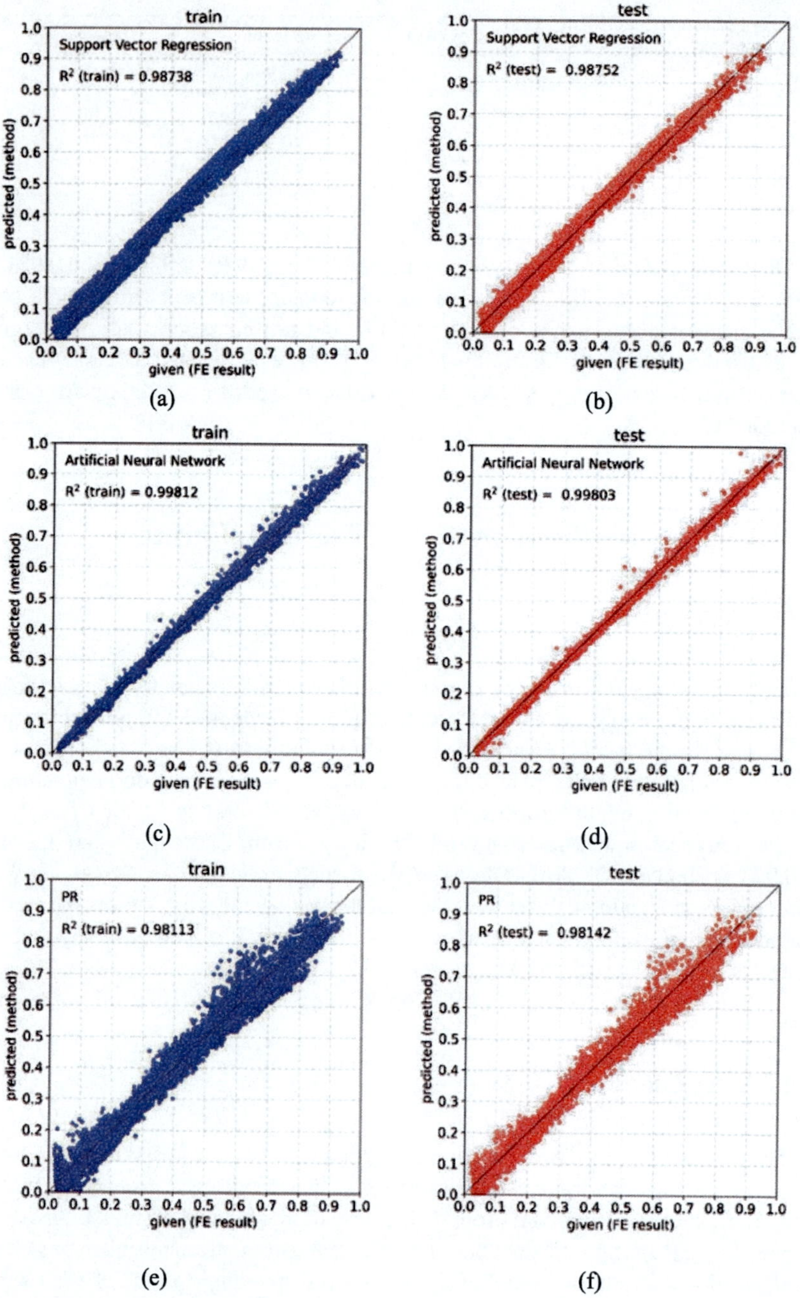

Fig. 6 Predicted capacity $M_{u,T}/M_{pl,20}$ for thin-walled steel beams at elevated temperature: Comparison between SAFIR finite element model and **a** SVR training, **b** SVR testing; **c** ANN training, **d** ANN testing; **e** PR training, **f** PR testing; **g** RF training, **h** RF testing. Training/testing performed based on 24,516 data points

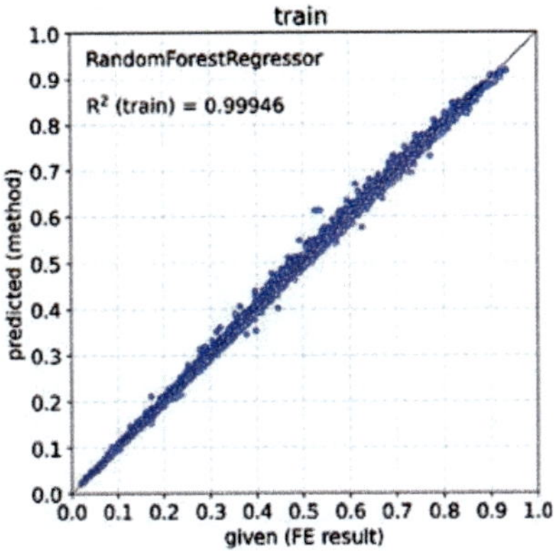

(g)

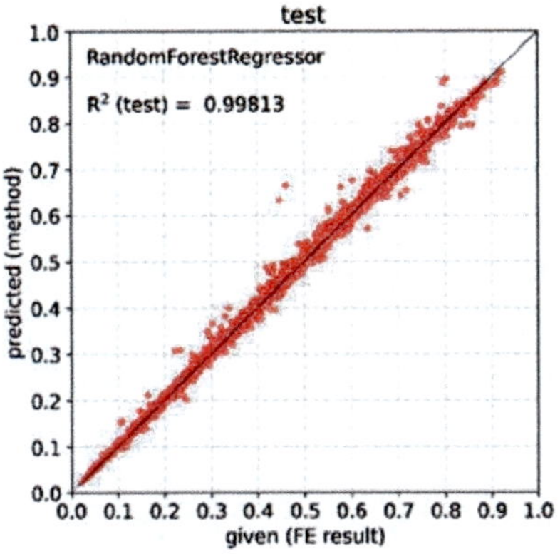

(h)

Fig. 6 (continued)

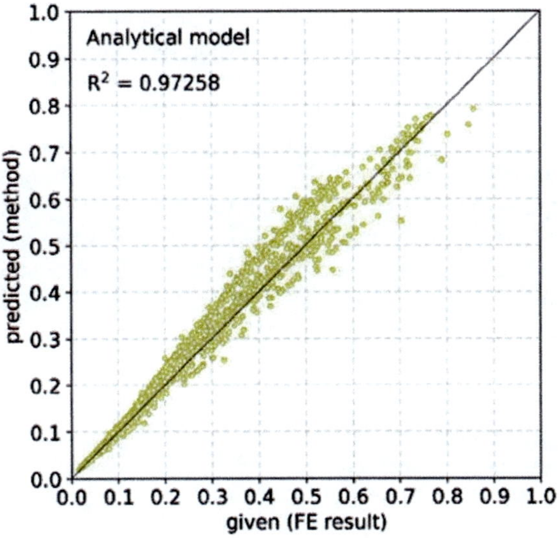

Fig. 7 Comparison between the results obtained with FE (considered as ground truth) and obtained with the analytical model

all section classes, corresponding to the stress at a 2% total strain, together with new effective width expressions developed for the elevated temperature situation [19, 22].

Table 6 summarizes the R^2 scores obtained for ML models and analytical models. The ANN and the RF both show an excellent correlation between the predicted capacities and those obtained by the numerical models. The SVR and PR models, also have a higher R^2 score when compared to the analytical models, although the results are more scattered than those obtained with the ANN and RF models. The PR is also particularly interesting considering the trade-off between the complexity

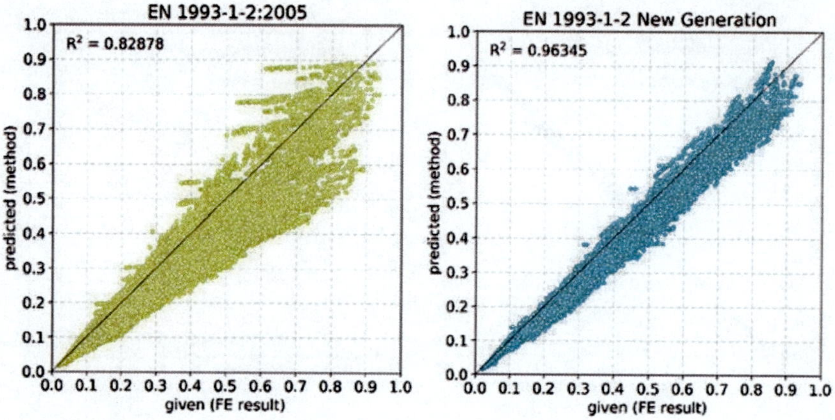

Fig. 8 Comparison between the results obtained with FE (considered as ground truth) and obtained with EN 1993-1-2:2005 and EN 1993-1-2 New Generation

of the model and the prediction ability compared to the analytical models and with much more complex models such as the ANN and the RF.

Figure 8 plots the results obtained with the analytical models. The results are normalized by the plastic section resistance of each beam, to allow direct comparison with the ML models. For the EN 1993-1-2:2005, two trends in the results are visible, namely those in the region (0.8, 0.6) stemming from the fact that EN1993-1-2 does not account for the effect of different bending moment distributions. This has been improved by the new generation version. Moreover, some results from EN 1993-1-2:2005 lay on the unsafe side, which was also addressed by the new generation version. The R^2 score is 0.829 and 0.963 for the EN 1993-1-2:2005 and the new generation, respectively, which demonstrates the improvements brought by the new version.

Table 6 Summary of the prediction capacity of the models

Model	R^2 score (train)	R^2 score (test)
EN1993-1-2:2005	0.829	
EN 1993-1-2 New Generation	0.963	
ANN (scikit-learn)	0.998	0.998
SVR	0.987	0.987
Polynomial Regression	0.981	0.981
Random forest	0.999	0.998

6.3 Discussion

For both the steel columns and steel beams, the ML models perform better than the state-of-the-art analytical models in predicting the elevated temperature capacity of the members. This demonstrates the potential of ML models to replace computationally expensive and/or sophisticated shell finite element models. Hence, ML models trained on a numerical dataset can then be used to study variations of designs within the boundaries of the considered problem, returning quasi-immediate outputs. These ML models can also serve to gain further insights into the governing parameters and identify optimum designs.

The superior performance of the ML models raises the question whether these could replace analytical models in design codes and standards, such as the Eurocode used herein for comparison. However, it should be noted that analytical models still present advantages. The equations can easily be written down in a code document and checked independently. Analytical models directly rely on the underlying mechanics of the problem, thereby inherently providing some degree of understanding on the nature of the response. In contrast, ML models have a "black box" characteristic that makes the interpretability and quality control more challenging. ML models which are not physically based present risks with extrapolation and may be unsatisfying to the engineering community, hindering adoption because of a lack of trust. While analytical equations are explicitly spelled out in standard documents, ML models would need to be shared as computer code with a clear indication of the underlying data and the limits of validity. Strict quality procedures would need to be established and implemented to ensure the transparency of the data and models. In sum, the authors' opinion is that analytical models certainly have their usefulness and value and will continue to do so. Yet, under the right conditions and with the proper procedures, ML models could progressively be incorporated into design guides and standards given their high potential to provide accurate and fast predictions.

7 Conclusion

This chapter investigated the potential of ML models to capture the capacity at elevated temperatures of thin-walled steel members. A numerical study based on validated nonlinear FEM with shell elements was conducted to build the datasets for ML models. The thin-walled steel columns exhibit failure by local and/or global buckling, while the behavior of the thin-walled steel beams is governed by the interaction between local buckling and lateral-torsional buckling. The dataset was used to train and test ML models. The performance of the ML models was compared with that of analytical models.

The results indicate that the three tested ML models can predict the resistance of the columns in the training and testing dataset with a R^2 greater than 0.990 for SVR and ANN, and greater than 0.977 for PR. The ML models agreed better with the FE data than a state-of-the-art analytical model. While for thin-walled beams, ANN, SVR, PR, and RF models trained against the numerical dataset outperformed the current best-in-class analytical model. Specifically, the R^2 scores against the test dataset were 0.999 for ANN, 0.998 for RF, 0.988 for SVR, and 0.981 for PR; to be compared with 0.963 and 0.829 for the new generation and the current Eurocode 3 Part 1–2 models, respectively. Since the analytical model tends to give overconservative predictions and FEM is time-consuming, the developed ML models can readily be applied to give quick prediction for cases within the range of inputs, which is beneficial for making quick decisions in practice.

This study complements other contemporary studies highlighting the high potential of ML methods to solve complex structural fire engineering problems, a domain which indubitably should continue to attract the attention of researchers. It is expected that the accuracy of test data could be further improved by extending the training dataset for the ML models. This work shows that the ML models can accurately predict the resistance of thin-walled members under uniform heating while also being computationally efficient. This work lays the foundation for capacity assessment of structural members subjected to fires. In future works, more complex configurations such as structural assemblies or non-uniform fire exposures will be explored. The ML models will be used to provide insights into the features influencing the fire performance of structural systems under a range of fire exposures. Looking further ahead, ML models could be developed to predict in real-time the integrity of buildings under fires to assist smart firefighting. Sensor data from the room (e.g., temperatures, heat flux) could be combined with prior data on the structure (drawings, fire design) and real-time images (e.g., from firefighter body cameras) to feed into ML models to make predictions of the integrity of buildings. Accurate and rapid assessment of capacity of structural assemblies with ML techniques would assist smart firefighting by notifying the firefighters with information of building integrity in case of fires.

References

1. H. Salehi, R. Burgueño, Emerging artificial intelligence methods in structural engineering. Eng. Struct. **171**, 170–189 (2018)
2. M.Z. Naser, Mechanistically informed machine learning and artificial intelligence in fire engineering and sciences. Fire Technol. 1–44 (2021)
3. M.Z. Naser, Fire resistance evaluation through artificial intelligence—A case for timber structures. Fire Saf. J. **105**, 1–18 (2019)
4. M.Z. Naser, Observational analysis of fire-induced spalling of concrete through ensemble machine learning and surrogate modeling. J. Mater. Civ. Eng. **33**, 04020428 (2021). https://doi.org/10.1061/(asce)mt.1943-5533.0003525
5. M.Z. Naser, H. Salehi, Machine learning-driven assessment of fire-induced concrete spalling of columns, ACI Mater. J. **117**, 7–16 (2020). https://doi.org/10.14359/51728120.
6. R.K. Chaudhary, B. Jovanović, T. Gernay, R. Van Coile, Generalized fragility curves for concrete columns exposed to fire through surrogate modelling, in *Proceeding of 11th International Conference Structures. Fire* (2020)
7. R.K. Chaudhary, R. Van Coile, T. Gernay, Potential of surrogate modelling for probabilistic fire analysis of structures. Fire Technol. (2021). https://doi.org/10.1007/s10694-021-01126-w

8. L. Su, X. Wu, X. Zhang, X. Huang, Smart performance-based design for building fire safety: Prediction of smoke motion via AI. J. Build. Eng. **43**, 102529 (2021)
9. T. Zhang, Z. Wang, H.Y. Wong, W.C. Tam, X. Huang, F. Xiao, Real-time forecast of compartment fire and flashover based on deep learning. Fire Saf. J. **130**, 103579 (2022)
10. A. Dexters, R.R. Leisted, R. Van Coile, S. Welch, G. Jomaas, Testing for knowledge: application of machine learning techniques for prediction of flashover in a 1/5 scale ISO 13784-1 enclosure. Fire Mater. **45**, 708–719 (2021)
11. Q. Tong, C. Couto, T. Gernay, Machine learning models for predicting the resistance of axially loaded slender steel columns at elevated temperatures. Eng. Struct. **266**, 114620 (2022). https://doi.org/10.1016/j.engstruct.2022.114620
12. C. Couto, Q. Tong, T. Gernay, Predicting the capacity of thin-walled beams at elevated temperature with machine learning. Fire Saf. J. **130**, 103596 (2022). https://doi.org/10.1016/j.firesaf.2022.103596
13. J.-M. Franssen, B. Zhao, T. Gernay, Experimental tests and numerical modelling on slender steel columns at high temperatures, J. Struct. Fire Eng. (2016)
14. W. Wang, V. Kodur, X. Yang, G. Li, Experimental study on local buckling of axially compressed steel stub columns at elevated temperatures. Thin-Walled Struct. **82**, 33–45 (2014)
15. J.C. Batista Abreu, L.C.M. Vieira, A.L. Moreno, T. Gernay, B.W. Schafer, Experiments on load-bearing cold-formed steel sheathed studs at elevated temperatures. Thin-Walled Struct. **156**, 106968 (2020). https://doi.org/10.1016/j.tws.2020.106968.
16. CEN, EN 1993-1-2, Eurocode 3: Design of steel structures - Part 1–2: General rules—Structural fire design (2005)
17. A. Ranby, Structural fire design of thin walled steel sections. J. Constr. Steel Res. **46**, 303–304 (1998). https://doi.org/10.1016/S0143-974X(98)00118-7
18. S.E. Quiel, M.E.M. Garlock, Calculating the buckling strength of steel plates exposed to fire. Thin-Walled Struct. **48**, 684–695 (2010). https://doi.org/10.1016/j.tws.2010.04.001
19. C. Couto, P. Vila Real, N. Lopes, B. Zhao, Effective width method to account for the local buckling of steel thin plates at elevated temperatures. Thin-Walled Struct. **84**, 134–149 (2014). https://doi.org/10.1016/j.tws.2014.06.003
20. J. Pauli, M. Knobloch, M. Fontana, On the local buckling behaviour of steel columns in fire, in *8th Fib PhD Symposium Kgs*. Lyngby, Denmark (2010)
21. J. Pauli, The behaviour of steel columns in fire: material-cross-sectional capacity-column buckling. IBK Bericht. **343** (2012)
22. C. Couto, P.V. Real, N. Lopes, B. Zhao, Resistance of steel cross-sections with local buckling at elevated temperatures. J. Constr. Steel Res. **109**, 101–114 (2015)
23. C. Couto, P. Vila Real, A proposal based on the effective section factor for the lateral-torsional buckling of beams with slender I-shaped welded sections, Thin-Walled Struct. **145**, 106389 (2019). https://doi.org/10.1016/j.tws.2019.106389
24. N.S. Trahair, Laterally unsupported beams. Eng. Struct. **18**, 759–768 (1996). https://doi.org/10.1016/0141-0296(96)00005-3
25. A. Taras, R. Greiner, New design curves for lateral–torsional buckling—Proposal based on a consistent derivation. J. Constr. Steel Res. **66**, 648–663 (2010). https://doi.org/10.1016/j.jcsr.2010.01.011
26. C. Couto, P. Vila Real, Numerical investigation on the influence of imperfections in the lateral-torsional buckling of beams with slender I-shaped welded sections. Thin-Walled Struct. **145**, 106429 (2019). https://doi.org/10.1016/j.tws.2019.106429.
27. P.M.M. Vila Real, N. Lopes, L. Simões da Silva, J.-M. Franssen, Parametric analysis of the lateral–torsional buckling resistance of steel beams in case of fire. Fire Saf. J. **42**, 416–424 (2007). https://doi.org/10.1016/j.firesaf.2006.11.010
28. P.M.. Vila Real, N. Lopes, L. Simões da Silva, J.-M. Franssen, Lateral-torsional buckling of unrestrained steel beams under fire conditions: improvement of EC3 proposal. Comput. Struct. **82**, 1737–1744 (2004). https://doi.org/10.1016/j.compstruc.2004.01.006
29. C. Couto, É. Maia, P. Vila Real, N. Lopes, The effect of non-uniform bending on the lateral stability of steel beams with slender cross-section at elevated temperatures. Eng. Struct. **163**, 153–166 (2018). https://doi.org/10.1016/j.engstruct.2018.02.033

30. C. Couto, P. Vila Real, N. Lopes, B. Zhao, Numerical investigation of the lateral–torsional buckling of beams with slender cross sections for the case of fire. Eng. Struct. **106**, 410–421 (2016). https://doi.org/10.1016/j.engstruct.2015.10.045
31. M. Prachar, M. Jandera, F. Wald, B. Zhao, Lateral torsional-buckling of class 4 steel plate beams at elevated temperature: experimental and numerical comparison. J. Struct. Fire Eng. **6**, 223–236 (2015). https://doi.org/10.1260/2040-2317.6.3.223
32. M. Jandera, M. Prachař, F. Wald, Lateral-torsional buckling of class 4 section uniform and web tapered beams at elevated temperature. Thin-Walled Struct. **146**, 106458 (2020). https://doi.org/10.1016/j.tws.2019.106458
33. FIDESC4, Fire Design of Steel Members with Welded or Hot-Rolled Class 4 Cross-Section, RFCS-CT-2011-00030, 2011-2014 (2014)
34. F. Dumont, E. Wellens, J.-M. Franssen, Fire behaviour of steel members with class 4 cross sections under axial compression with or without eccentricity (2016)
35. J.-M. Franssen, T. Gernay, Modeling structures in fire with SAFIR: theoretical background and capabilities. J. Struct. Fire Eng. (2017) JSFE-07-2016-0010. https://doi.org/10.1108/JSFE-07-2016-0010
36. C. Couto, P.V. Real, The influence of imperfections in the critical temperature of I-section steel members. J. Constr. Steel Res. **179**, 106540 (2021)
37. X. Yun, N. Saari, L. Gardner, Behaviour and design of eccentrically loaded hot-rolled steel SHS and RHS stub columns at elevated temperatures. Thin-Walled Struct. **149**, 106646 (2020). https://doi.org/10.1016/j.tws.2020.106646
38. C. Zhang, L. Choe, M. Seif, Z. Zhang, Behavior of axially loaded steel short columns subjected to a localized fire. J. Constr. Steel Res. **111**, 103–111 (2015). https://doi.org/10.1016/j.jcsr.2014.11.012
39. D. Systèmes, Abaqus 6.14: Abaqus/CAE User's Guide. http://130.149.89 (2014) v6
40. European Committee for Standardisation (CEN). EN 1090-2:2018 Execution of steel structures and aluminium structures—Part 2: Technical requirements for steel structures 2018:1–204 (n.d.)
41. E.N. CEN, 1-5: 2006-Eurocode 3: Design of steel structures-Part 1–5: Plated structural elements, Brussells Eur. Comm. Stand. (2006)
42. C. Couto, A. Sanzel, P.V. Real, N. Lopes, B. Zhao, Beam-columns with thin wall cross-sections in case of fire, in *9th International Conference Structures. Fire* (2016)
43. CEN/SC3.T6, EN 1993-1-2:20XX (E) Eurocode 3—Design of steel structures Part 1–2: General—Structural fire design—Final draft v309_22.10.2019 (2019)
44. M. Awad, R. Khanna, M. Awad, R. Khanna, Support vector regression. Effic. Learn. Mach. Theor. Concepts, Appl. Eng. Syst. Des. 67–80 (2015)
45. E. Ostertagová, Modelling using polynomial regression. Procedia Eng. **48**, 500–506 (2012). https://doi.org/10.1016/j.proeng.2012.09.545
46. B. Yegnanarayana, *Artificial Neural Networks* (PHI Learning Pvt. Ltd., 2009)
47. G. Biau, E. Scornet, A random forest guided tour. TEST **25**, 197–227 (2016)
48. W.S. Noble, What is a support vector machine? Nat. Biotechnol. **24**, 1565–1567 (2006). https://doi.org/10.1038/nbt1206-1565
49. M.W. Gardner, S.R. Dorling, Artificial neural networks (the multilayer perceptron)—A review of applications in the atmospheric sciences. Atmos. Environ. **32**, 2627–2636 (1998). https://doi.org/10.1016/S1352-2310(97)00447-0
50. M. Rafiq, G. Bugmann, D. Easterbrook, Neural network design for engineering applications. Comput. Struct. **79**, 1541–1552 (2001). https://doi.org/10.1016/S0045-7949(01)00039-6
51. S. Yuanyuan, W. Yongming, G. Lili, M. Zhongsong, J. Shan, The comparison of optimizing SVM by GA and grid search, in *ICEMI 2017—Proceedings of IEEE 13th International Conference Electronic Measurement Instruments.* 2018 January (2017), pp. 354–360. https://doi.org/10.1109/ICEMI.2017.8265815
52. F. Pedregosa, G. Varoquaux, A. Gramfort, V. Michel, B. Thirion, O. Grisel, M. Blondel, P. Prettenhofer, R. Weiss, V. Dubourg, Scikit-learn: machine learning in Python. J. Mach. Learn. Res. **12**, 2825–2830 (2011)

Smart Firefighting

Building Fire Hazard Predictions Using Machine Learning

Eugene Yujun Fu, Wai Cheong Tam, Tianhang Zhang, and Xinyan Huang

1 Introduction

In the period of 2015–2019, US fire departments responded to an estimated average of 346,800 home structure fires per year [1]. As compared to the estimate of 734,000 reported in 1980, the number of home structure fires is lower by approximately 50% (i.e., an estimate of 356,500 in 2020). The reduction in home structure fires is possibly attributed by the presence of smoke alarms [2] and/or sprinkler systems [3–5], the use of fire-resistant building materials [6, 7], better fire consumer education and awareness in fires [8, 9], and the implementation of home safety codes [10–12]. However, the current home structure fires are more hazardous. In [13], experimental burns comparing spaces containing items common to modern homes and rooms with legacy furnishings were conducted. It is clearly seen that temperature rises rapidly for relatively shorter duration in case of fires in the modern (i.e., at ~5 min of fire in the modern room versus after 29 min in case of the legacy room). This rapid temperature increase is largely due to the near-simultaneous ignition of most of the directly exposed combustible materials in the enclosed area. As seen in the figure, the temperature can exceed 800 °C in less than 20 s. This extreme fire event is known as flashover [14] and it presents an increased fire risks to the firefighters.

E. Y. Fu
Department of Rehabilitation Sciences, The Hong Kong Polytechnic University, Hung Hom, Hong Kong, China

W. C. Tam (✉)
Fire Research Division, National Institute of Standards and Technology, Gaithersburg, MD, USA
e-mail: waicheong.tam@nist.gov

T. Zhang · X. Huang
Department of Building Environment and Energy Engineering, The Hong Kong Polytechnic University, Hung Hom, Hong Kong, China

X. Huang and W. C. Tam (eds.), *Intelligent Building Fire Safety and Smart Firefighting*, Digital Innovations in Architecture, Engineering and Construction,
https://doi.org/10.1007/978-3-031-48161-1_16

Backdraft is the abrupt burning of superheated gasses in a fire, caused when oxygen rapidly enters a hot, oxygen-depleted environment [15]. The backdraft explosion often happens when firefighters open a door or window of an enclosure room when conducting firefighting operation. The backdraft can occur without warning and is one of the most dangerous threats to firefighters' lives. According to statistics of 127 fire accidents involving explosions in different countries, 109 cases were related to the backdraft phenomenon [16]. Therefore, the dangerous flashover and backdraft are major hazardous critical events in firefighting and raises the attention of fire researchers worldwide.

Currently, firefighters solely rely on their experience to detect critical fire events. There can be four warning signs [17] in a structure fire. Outside of the building structure, the presence of thick, black smoke and free-burning fires from windows and doors are two reliable signs of an impending flashover event. Inside the building structure, firefighters can try to seek for a fire phenomenon called rollover [17] near the ceiling and/or try to use their hands to feel the intense heat from the hot gas layer. However, these warning signs are generally difficult to be recognized accurately and it could take many years of experience to build up the necessary proficiency. Therefore, if the firefighters do not read flashovers correctly in time, their lives are in danger.

In the fire research community, a great deal of research efforts has been made to foster data-driven firefighting. Different methods are utilized to develop real-time fire hazard prediction models and these methods include empirical correlations [18–20], inversed modeling techniques [21, 22], and computational fluid dynamics (CFD) based approaches [23, 24]. Since the empirical correlations and the inversed modeling techniques use only closed-form mathematical expressions, these methods are easy to be implemented and are numerical efficient for quick evaluations. However, one drawback for these methods [18–22] is that they are limited to single-compartment structural settings and are not designed for multi-compartment structures. The CFD based approaches, such as [23], provide more flexibilities to account for structure variance and offer better prediction capabilities and higher prediction accuracy. Yet, the drawback of this kind of approach is that the models rely on high-performance computing machines and requires lengthy computations. As noted in [23], results from just one simulation time step takes more than five minutes to compute. Therefore, the CFD based approach at its current form will hardly be able to be used in real-time firefighting.

In addition, there is another drawback for the existing fire hazard prediction models. They do not account for realistic conditions from real fire scenarios. For example, the models [18–24] generally rely on both the continuous temperature signals from thermocouples in all compartments and the prior knowledge about the fire locations and vent opening conditions. However, fire protection devices, such as heat detectors, will stop functioning at elevated temperature [2] and the information about the fire locations and the opening conditions of windows and doors are often unknown. Since these realistic conditions (i.e., sensor temperature limit and the effect of arbitrarily fire location and vent opening conditions) have not been considered in the development process, the model performance from any of these models [18–24] is likely to be diminished. For that, a more robust data-driven approach is needed to overcome both the numerical challenge and the modeling complexity of realistic conditions from real fire scenarios.

Machine learning (ML) has made breakthroughs in various practical engineering problems, including human activity recognitions [25], recommendation systems [26], abnormal heartbeat classification [27], offline learning [28, 29], intention detections [30], cooktop ignitions [31–33], and thermal radiation analyses [34–39]. ML paradigms have also been used extensively to overcome prediction challenges for various fire forecasting tasks. Chenebert et al. [40] provided a decision tree classifier for outdoor flame detections using image data. Yin and his co-workers [41] developed a deep neural network (DNN) framework for smoke detections. Another DNN model was developed and was used to identify flames for high-resolution videos given in residential building settings [42]. Aslan et al. [43] attempted to use generative adversarial networks for fire detections in large building structures using videos from surveillance cameras. The scientific advancement in ML truly enables reliable real-time predictions. However, these ML models might not be applicable to the current fire protection devices, such as heat detectors and smoke/CO detectors, where the data are in time-series. In order to fill in this knowledge gap, the Fire Research Division (FRD) from the National Institute of Standards and Technology (NIST) have collaborated with researchers from the Hong Kong Polytechnic University (HKPU) to develop ML technologies that can be used to enhance firefighters' situational awareness in built environments, protecting them from hazardous fire environments, and to pave the way for the development of data-driven prediction systems.

This chapter highlights the 4-year research efforts and the chapter is organized as follows. Section 2 describes the problem associate with the scarcity of real-life flashover data and the solution to overcome this data challenge. Section 3 presents the ML-based model development for flashover and backdraft predictions in residential settings. Finally, Sect. 4 summarizes the conclusions of the study and provides an overlook to future research directions.

2 Flashover Prediction by Numerical Database

To develop a robust ML-based prediction model, the dataset must contain sufficiently large amounts of relevant data that cover the desired ranges of fire conditions and geometric settings. Because flashover is an extreme fire event, the data collection is challenging. Unlike the classification tasks [44, 45] and/or the regression tasks [46, 47] being well established in the AI/ML community, the study of the flashover predictions for full-scale building structures in the ML domain has not been adequately addressed. For that, sensor data, such temperature, involving flashover in multi-compartment building structures is not available in any public data repositories such as [48]. Although fire data can be obtained by physically conducting the required full-scale experiments, when more complex building structures (i.e., with increasing number of compartments) and various realistic conditions (i.e., different fire and vent opening conditions) are needed to be accounted for, the required number of physical experiments is expected to be increased dramatically. When factors, such as cost and time, are of concern, physically conducting all full-scale experiments is not feasible. Therefore, in order to facilitate the development of ML-based flashover prediction models, the learning-by-synthesis approach is utilized for data collection.

The learning-by-synthesis (LBS) approach [49] provides an alternative means for data collection. To relate this concept to fire research, this approach makes use of computer simulation programs to conduct full-scale experiments with the desired fire and vent opening settings. Since the cost and time for running the numerical experiments are considerably lower than physically conducting the full-scale experiments, LBS provides flexibility to optimize the data quality through trial-and-error and to ensure the acquisition of a large amount of relevant data.

A series of research thrusts on the development of real-time flashover prediction models for residential building structures has begun since 2019 and these efforts have also been one of the primary research focuses for the smart fire fighting (SFF) project. The objective of this section is to describe the ML-based models that are developed to account for various realistic conditions encountered from real-life firefighting and how the fire problem is being addressed systematically.

There are three main challenges that the SFF project has overcome in terms of developing functional real-time flashover prediction models for residential building structures: (1) loss of data from sensors (i.e., heat detectors) at elevated temperature in a three-compartment home, (2) unknown fire conditions and venting opening configurations in a typical single-story residential home, and (3) application at-scale with different residential homes. Consequently, three different ML-based models are developed to address the three challenges: (1) P-Flash with SVR [50], (2) P-Flash with attention-based Bi-LSTM [51], and (3) FlashNet [52] and key information is presented in Sect. 3.1, Sect. 3.2, and Sect. 3.3, respectively. Readers are suggested to seek for complete details from the literature [50–52].

2.1 *Zone-Model Fire Data Generation and Validation*

A computational tool, CFAST Fire Data Generator (CData) [53, 54], is developed to generate time series data for typical devices/sensors (i.e., heat detector, smoke detector, and other targets) in any user-specified building configurations. CData has three main modules: (1) preprocessor, (2) executor, and (3) accumulator.

Preprocessor Module: The preprocessor module is used to generate individual CFAST [55] input files. The CData input file utilizes the CFAST Namelist format. Seven different probability density functions are included to allow users to specify different simulation parameters, such as building layouts, surface materials, fire conditions, ventilation configurations, location of detector(s), and output intervals. Examples are provided in page 4 of reference [53]. Depending on the problem complexity and using approximately 200 nodes from a Linux cluster, approximately 10,000 CFAST cases can be completed in a day.

Executor Module: As part of the process of creating the individual CFAST inputs files, CData creates batch scripts for both Windows (see Pg. 25 from [54]) and Linux (see Pg. 26 from [54]) operating systems. Each batch script depends on external software to support running multiple CFAST jobs in parallel. Both of these scripts include information on the locations of these external files. With that, running the set of CFAST simulations is accomplished by running the appropriate script for Windows or Linux. By default, a maximum of 100,000 iterations are set for each run to ensure that jobs which take an extremely long time to run do not prevent the rest from running. This value can be changed in both of the batch scripts.

Accumulator Module: A summary spreadsheet containing detailed information about all input parameters for each of the CFAST cases can be generated in this module. This spreadsheet can be used to carry out data inspection to eliminate any parameters. This information is crucial to ensure the quality and the covering range of the fire data. Readers who are interested in CData can refer to [54] for additional details and examples.

CData uses CFAST as the simulation engine. In general, CFAST is validated against more than 15 other sets of full-scale experiments [55] with peak heat release rate (HRR), compartment aspect ratio (i.e., compartment length against ceiling height), and global equivalence ratio, ranging from approximately 50 kW to 15,700 kW, 0.4–4.9, and roughly 0 to a value larger than 1 for a wide range of ventilation factors, respectively.

In order to ensure data fidelity in a more complex geometric setting, additional model validation is carried out. Specifically, temperature measurements obtained from two full-scale gas burner experiments (Exp 1 and Exp 2) reported in [56] are used to benchmark the synthetic temperature data generated using CFAST. Figure 1a shows the floorplan of the single-story residential structure and it is consisted of a living room, kitchen, a dining room, a short hallway, and three bedrooms. The fire for both experiments is initiated at the living room and the heat release rate of the gas burner fires are the same for the two experiments.

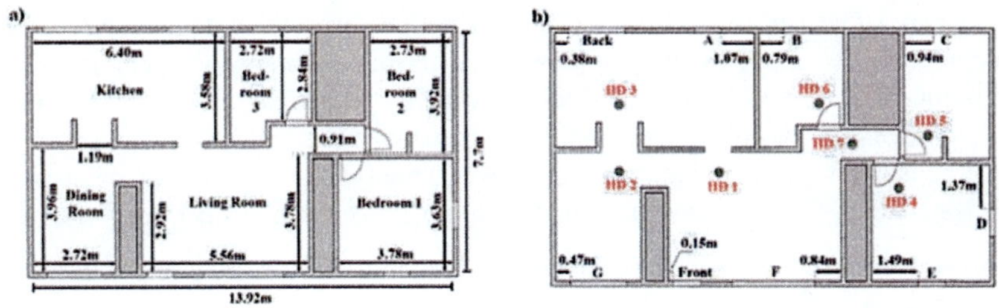

Fig. 1 Plan view dimensioned drawing of **a** the single story structure and **b** vent openings with heat sensors (HD) [56]

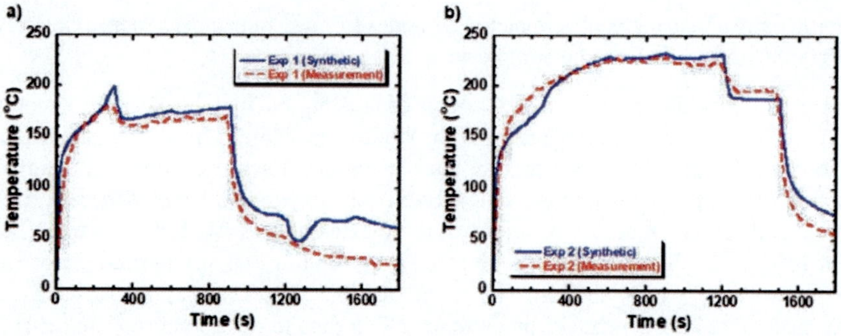

Fig. 2 CFAST validation against measurement for **a** Experiment 1 and **b** Experiment 2

Figure 2 show the upper gas layer temperature profiles for the two experiments. The blue solid lines represent the synthetic temperature data obtained from CFAST. The red dash lines are the estimated upper gas layer temperature for the experiments, and they are obtained based on the hot gas layer reduction method provided in [56]. It can be seen that the magnitudes and trends of the temperature profiles match the experimental data for different vent opening events. The window opening conditions can be retrieved from [51, 57]. This observation indicates that CFAST, the simulation engine of CData, is capable of capturing both the corresponding effect of fire and vent openings in the single story multi-compartment structure. In terms of uncertainty, the absolute root mean squared error is about 30 °C and 10 °C for Exp 1 and Exp 2, respectively. Comparison is also made for other compartments and the overall agreement is very good. Therefore, it can be expected that the generated temperature data can be reliably used for model development.

2.2 *Flashover Prediction Using Recovered Temperature Data (P-Flash with SVR)*

The primary question to be addressed in the first study is to develop a modeling framework that can be used to provide predictions even when temperature data are not available due to malfunctioning heat detectors at elevated temperature. The subsequent sections will provide discussion on: (i) numerical setup for the three-compartment home, (ii) corresponding temperature behaviors, (iii) heat detector operational temperature limit, (iv) flashover criteria, (v) methods to overcome missing data, (vi) model development of P-Flash, and (vii) model performance.

Numerical Setup: Consider a single-story building with three compartments as shown in Fig. 3a, the dimensions of Room 1 are 3.5 m × 3.5 m, and the dimensions of Room 2 and Corridor are 4.5 m × 4.5 m and 3.5 m × 1 m, respectively. The ceiling height is 2.5 m, and it is identical for all compartments. For simplicity, the

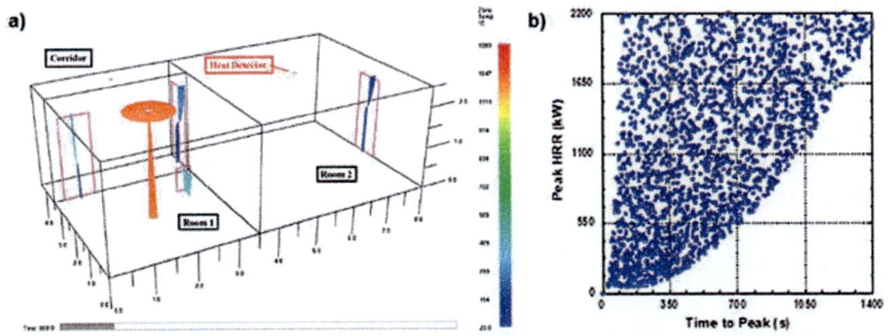

Fig. 3 **a** Schematic of the single-story three compartments with a fire in action and **b** scatter plot for peak HRR versus time to peak

material of all walls, ceilings, and floors is gypsum wallboard. As seen in Fig. 3a, there are 4 openings: (1) a window in Room 1, (2) a door between Room 1 and Corridor, (3) a door between Corridor and Room 2, and (4) an exit-door in Room 2. The openings are fully opened. There is one heat detector in every compartment, and they are all located at the center of each compartment about 4.5 cm away from the ceiling. The response time index for the heat detector is 35 $(ms)^{0.5}$.

Temperature Behaviors: CData is used to obtain 1000 different simulation runs with a t-squared fire at the center in Room 1. Based on references provided in [58], a range of fires are selected in this study. Figure 3b shows the scatter plot of peak HRR and time to peak for the 1000 cases. It can be seen that the peak HRR and the time to peak ranges from approximately 50 kW–2200 kW and from 50 to 1400 s, respectively. The selected range of peak HRR and time to peak cover various burning items from an office trash can with a slow fire growth rate to an upholstered furniture fire with an ultra-fast fire growth rate.

For all simulation runs, a fire is started in Room 1. Subsequently, the upper layer gas temperature rises, and the layer thickness increases. Some hot gases leave the building structure, and some flow through the door. Air mixing between Room 1 and Corridor occurs. Due to the mixing, the upper gas layer temperature in Corridor also increases. Similar mass transfer and heat transfer processes take place between Corridor and Room 2, and the Room 2 upper gas layer temperature gradually rises. It can be imagined that the overall temperature profile from Room 1 for a given fire is likely to be the highest and the Room 2 overall temperature is going to be the lowest (see Pg. 4 from [50]).

Heat Detector Maximum Operational Temperature Limits: Loss of heat detector (HD) temperature signal is a realistic condition in actual fire scenarios and the HDs are very unlikely to survive at elevated temperature and would fail at temperatures well below the flashover conditions. According to NFPA 72 [2], heat sensing fire detectors are categorized into seven different classes with temperature classifications ranging from low to ultra-high, and the maximum operational temperature ranging from approximately 29 °C–302 °C. In order to find a balance in between applicability and model challenge, the extra high class HD with a maximum operational temperature limit of ~150 °C is selected.

Flashover Criteria: Numerous variables can affect the transition of a compartment fire to full room involvement (i.e., flashover) [14]. Thermal influences including radiative and convective heat flux are assumed to be the driving forces and are clearly important. Ventilation conditions, compartment volume, and the chemistry of the hot gas layer can also influence the occurrence of flashover. Based on the experimental studies reviewed in [14], the onset of flashover within a compartment can be quantified by two measurable criterions: 1) heat flux and 2) temperature. Peacock and his co-workers [14] demonstrated that when the incident heat flux onto the floor surface is between approximately 15 kW/m^2 to 33 kW/m^2, there can be a potential occurrence of flashover. However, the measurement of heat flux can rarely be measured in typical building environments because heat flux gauges are usually not installed. For that, the flashover criteria associated with heat flux cannot be used and it will not be considered in this study. For the temperature criterion, the range of values obtained from nine different literature sources in [14] ranges from 450 °C to 771 °C. The wide range of values for the temperature criterion is due to the large change in temperature within a very short period of time. Yet, it can be observed from [14] that most of the values are in the 550 °C–650 °C range. To be conservative, the upper gas layer temperature of 600 °C is used as the threshold to determine the flashover moment.

Sequence Segmentation: Figure 4 depicts the temperature profiles of a medium growth fire with high peak HRR case with a temperature cut-off at 150 °C. The dash lines are the ideal temperature profiles and the solid lines are the available temperature information. It can be seen that HDs begin to fail. There are only two and one remaining temperature information from the HDs in Phase II and Phase III, respectively. This missing temperature information imposes significant challenges for the model development and it is well known that developing a ML model based on unphysical data significantly jeopardizes the model performance. For that, the sequence segmentation is applied to the temperature data. There are three benefits from using the segmented data: a) the unphysical information due to any malfunctioning detectors is eliminated, (b) the ML model can take full advantage of the available data associated with a specified phase, and (c) the new data structure provides the basis for the model development of P-Flash.

P-Flash with SVR: A traditional machine learning algorithm, namely Support Vector Regression (SVR) [59], is used to develop the Prediction model for Flashover (P-Flash) to recover the missing temperature in different phases. Since SVR does not have capabilities to extract features itself, feature extraction is needed to obtain more useful features from the raw data. Two types of features are extracted for Phase I, II, and III and they are temperature-based features and trend-based features. These extracted features are the inputs and they help to build a more robust model. It should be noted that the list of extracted features can be found from page5 from [50].

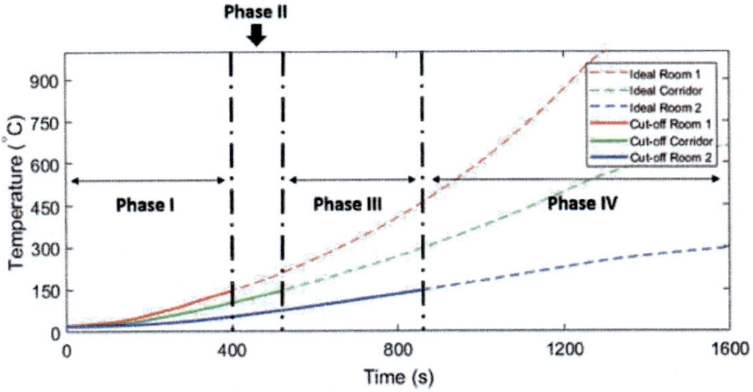

Fig. 4 Sequence segmentation and temperature profiles with 150 °C cut-off in 4 phases

Figure 5 shows the overview of P-Flash model architecture. As seen in the figure, P-Flash is consisted of two regression models (R_{cor} and R_{R2}) and a memory component. ①, ②, and ③ are the extracted features. Both regression models, R_{Corr} and R_{R2}, are executed simultaneously and three outputs are obtained. Two temperature outputs from R_{Corr} and R_{R2} using features ① and ② are averaged to yield the recovered Room 1 temperature in Phase II and the recovered Room 1 temperature output for Phase III is directed obtained from R_{R2} using features ③. The Room 1 temperature information (Phase I to Phase III) is stored the memory component (M). Since temperature information is not available in Phase IV, curve fitting is carried out to extrapolate the missing temperature in Phase IV.

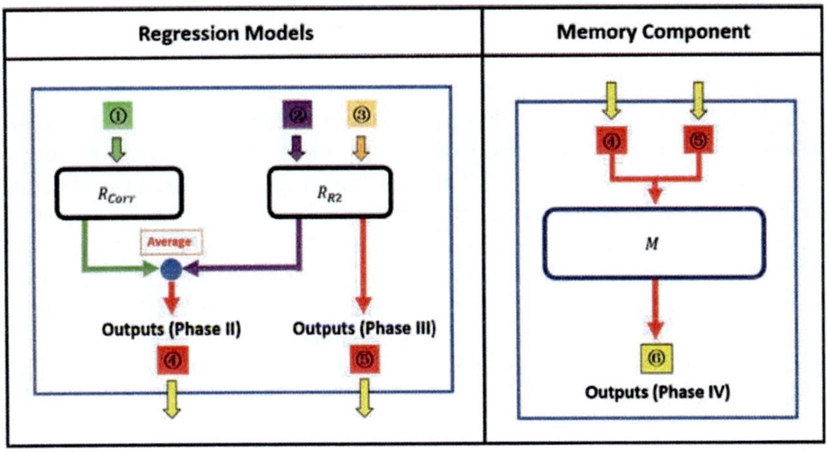

Fig. 5 Overview of model architecture for P-Flash

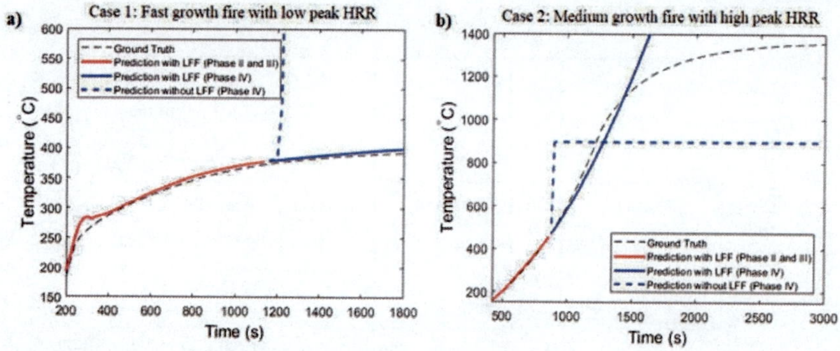

Fig. 6 Comparison between ground truth and predictions obtained from P-Flash **a** with and **b** without LFF

Results and Discussion: Figure 6 show the temperature predictions obtained from P-Flash for two selected cases: (1) a fast growth fire with low peak HRR case and (2) a medium growth fire with high peak HRR case. There are three sets of curves in each figure: (i) ground truth/Room 1 temperature, (ii) prediction with learning from fitting (LFF), and iii) prediction without LFF. For each prediction curve, it can be composed of up to two lines: (a) red line represents the Room 1 temperature predictions associated with Phase II and III and (b) blue line is for predictions in Phase IV. Since no prediction is needed for Phase I, comparison is omitted.

In Fig. 6a, it can be seen that P-Flash provides accurate temperature predictions of Room 1 in all phases, and the benefit of using LFF is noticeable. After approximately 1150 s, when all detectors are lost, P-Flash is still capable to provide predictions with similar trend and magnitude. For P-Flash without LFF, the prediction relies on the regression models, and it can be shown that the temperature prediction is unrealistic (i.e., showing a temperature increase to as high as 910 °C). This observation demonstrates that unphysical inputs will lead to unphysical outputs.

In Fig. 6b, it can be seen that the temperature of Room 1 being recovered from Phase II and III is still growing exponentially. Since P-Flash does not have any information about the change of increase of the temperature, it over-predicts the Room 1 temperature significantly after ~1250 s. However, it should be noted that although P-Flash is capable to project the temperature increase in which the determination of flashover (i.e., temperature approaching 600 °C) in Room 1 can be made, the model can only be applied for cases with fires at a fixed location and static (i.e., always open) vent conditions. In the following section, a more robust approach is presented. It is designed to account for arbitrary fire conditions and vent opening configurations in a realistic residential home and it is able to predict flashover for as many as 60 s before flashover occurs.

2.3 *Flashover Prediction Using Surrogate Temperature Data (P-Flash with Attention-Based Bi-LSTM)*

This section presents (i) the numerical setup for the fire problem (an early flashover prediction task with a more complex building structure), (ii) the unique temperature behaviors due to an arbitrary open door and the corresponding challenges, (iii) the model development of P-Flash with attention-based Bi-LSTM, (iv) results and discussion, and (v) model limitations.

Numerical Setup: Consider a single-story ranch-style home structure as shown in Fig. 1a. There are seven major compartments: a living room, a dining room, a kitchen, three bedrooms, and a hallway connecting the living room and the three bedrooms. The overall interior dimensions of the structure are about 13.92 m $\times$ 7.7 m with a ceiling height of 2.44 m. The detailed dimensions associated with each of the compartments are shown in Fig. 1a.

Figure 1b shows the relative position of the vent openings and the heat detectors in different compartments. For vent openings, there are two exterior doors (front and back), three bedrooms doors, and seven windows (from A to G). There is one heat detector (HD shown in Fig. 1b) located at each compartment and they are approximately 0.02 m away from the ceiling. The response time index of the HDs is similar to that mentioned in Sect. 2.2 and it is 35 $(\text{m-s})^{0.5}$. The reason for selecting this single-story ranch-style structure is because about 90% of residential buildings were built using this layout since the mid-1950s and this structure layout remains the most popular style of homes in 34 states across the United States [60]. For that, the research efforts with this structure are useful for the development process of real-time flashover predictions for firefighting across the U.S.

For each simulation case, there will only be one fire and it is either initiated at the center, against a wall surface, or at a corner within a compartment (except the hallway). The fire is located on the floor and this setting allows entrainment of oxygen-rich air and facilitates a more idealized condition for burning. A burning item is assumed to consist of four different fire growth stages and there is a linear growth (i.e., smoldering fire), t-squared growth (i.e., flaming fire), a peak, and a decay stage. Three items are considered and they are flaming chairs, polyurethane foam mattresses, and cotton-based mattresses. There items are found to be the leading items first ignited in home structure fires [1]. The transition HRR from smoldering to flaming fire (Q_1), peak HRR (Q_{max}), time to transition (t_1), time to peak HRR (t_2), peak time (t_3–t_3), and decay time (t_4–t_2) are summarized in Table 1. The peak HHR and time to peak HRR are obtained from [61], and the fire growth ratc, α, is determined to be in between 0.000329 kW/s^2 and 0.041387 kW/s^2 which yields fires ranging from slow to fast fire growth rate.

For vent opening conditions, all doors and windows are initially closed and they can be opened based on the following two settings: a time-trigger setting for doors and a temperature-trigger setting for windows. For the time-trigger setting, the doors in any compartments within the building structure can be opened at any given time within the t-squared fire growing stage. This setting allows mimicking door opening

Table 1 Approximate HRR parameters for chair and mattresses [61]

Items	Q_1 (kW)	Q_{max} (kW)	t_1 (s)	t_2 (s)	t_3-t_2 (s)	t_4-t_3 (s)
Chair	10–30	270–3500	150–1200	295–675	200	10
Mattress (foam)	10–30	2275–4620	150–1200	305–435	200	10
Mattress (cotton)	10–30	130–1670	150–1200	360–1240	200	10

events due to activities, such as leaving the room of the fire origin and/or leaving the building structure. In addition, the arbitrarily open conditions of doors indirectly introduce double-peak fire growth behaviors that cannot be provided by just a simple t-squared fire.

The temperature-trigger setting allows windows to be arbitrarily opened when a temperature threshold is reached. In a fire scenario, a window is subjected to heating either from the local gas temperature and/or from the fire. Due to temperature or a flux gradient [62], a window may crack and eventually break out and create an opening to the outside environment. Based on [63], breakage of a single-pane float glass is experimentally observed at temperature between 100 °C and 200 °C. For that, when the window temperature reaches these thresholds (i.e., 100 °C–200 °C), the window will then be opened. It is believed that the arbitrarily opening of doors and windows will provide more complex data behavior for development of a more robust ML-based model.

Temperature Behaviors and Challenges: Understanding the temperature data behaviors is vital to design a robust model architecture such that the ML-based prediction model is capable of learning the important relationship between temperature and flashover conditions. Figure 7a shows temperature data for a double-peak fire case. Consider a fire occurs in Bedroom 1 with all the doors and windows in the building structure being initially closed. This setting creates a "sealed" compartment for Bedroom 1. As seen in the figure, the temperature first increases and then decreases due to the lack of oxygen. At around 470 s, the door is opened and the fire is approaching its peak. The temperature increases in bedroom 1, living room, kitchen, and dining room, but the temperatures in bedrooms 2 and 3 remain at ambient temperature. Physically, it is completely understandable because the doors are closed and the heated gases cannot enter the bedrooms. However, the modeling challenge presents here is that the information about the opening of the doors in not known in a real fire scenario. So, the model needs to learn to omit data that are not important. Figure 7b shows the fire case with the heat detector maximum operational temperature limit at 150 °C. As shown in Fig. 7a, the flashover condition in the living room is met at about 500 s. Yet, the living room temperature is no longer available after ~300 s. Therefore, it can be seen that the model does not only need to discriminate data with higher importance, but it also needs to relate the temperature from other compartments for flashover conditions.

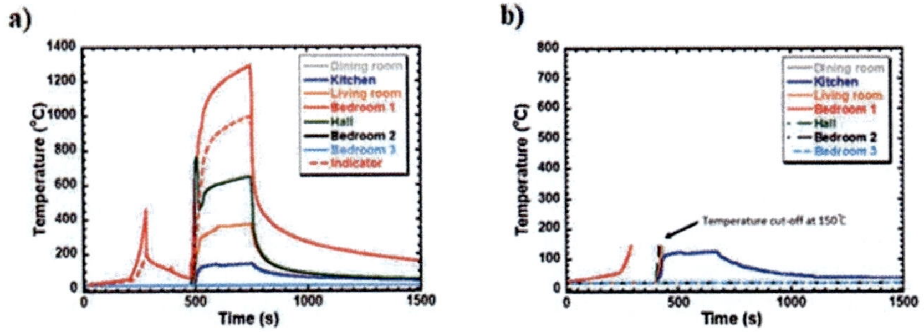

Fig. 7 **a** Temperature profiles of different compartments for a ventilation-controlled fire case and **b** temperature profiles with a cut-off value of 150 °C

P-Flash with Attention-Based Bi-LSTM: In order to overcome the two above-mentioned challenges, a bi-directional long short-term memory (Bi-LSTM) with self-attention mechanism is formulated. The Bi-LSTM algorithm [64] is to capture the complex data behaviors and the overall model structure is shown in Fig. 8a. The figure shows that for a temperature signal: $S = (s_1, s_2, \ldots, s_\tau)$ and a time step i, Bi-LSTM includes a forward hidden state $\overrightarrow{h_i}$ and a backward hidden state $\overleftarrow{h_i}$. Since the complete behavior for temperature signals is needed, only the last hidden state of $\overrightarrow{h_\tau}$ and $\overleftarrow{h_\tau}$ are extracted from the Bi-LSTM. As shown in the figure, concatenation is applied to yield $h_\tau = [\overrightarrow{h_\tau}, \overleftarrow{h_\tau}]$ to encode temperature behavior with flashover conditions.

In order to enhance the learning capability of the model in discriminating temperature signals with higher significance (i.e., neglecting bedrooms 2 and 3 temperature signal in Fig. 7b), a self-attention mechanism (see Fig. 8b) is used to model the sensor-wise relationships. With that, the contextual temperature information of all compartments within the structure can be extracted and it is believed that the

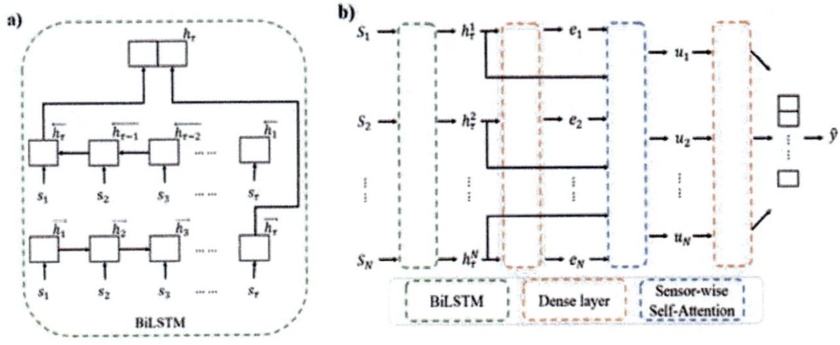

Fig. 8 **a** Model architecture of BiLSTM and **b** BiLSTM with sensor-wise self-attention

contextual information can contribute to provide more accurate flashover predictions within the coming 30 s and 60 s based on the available temperature signals. Detailed mathematical formulations of the model are provided in page 6 from [51].

Results and Discussion: Table 2 shows the model performance for flashover predictions with a lead time of 30 s and 60 s. Based on the overall accuracy, the attention-based Bi-LSTM outperforms the original Bi-LSTM. The attention-based model also yields a significantly better recall score, indicating the benefits of including the sensor-wise self-attention mechanism, and this is important because there are fewer false negatives for flashover events. As shown in Table 2, the overall performance for cases with a lead time of 60 s is generally better. One possible reason is due to the fact that it has a larger set of training samples.

Figure 9 illustrates the learned attention weights between the sensor signals in fire origin room and other compartments: kitchen (K), dining room (D), living room (L), and bedrooms 1–3 (B1, B2, B3), for two door opening conditions: all opened denoted as Open and all closed denoted as Close. As shown in the figure, the attention-based model can discover the spatial relation between sensor signals from different compartments. For example, when fire occurs in kitchen, the signal of dining room and living room are determined as the most discriminating surrogate signals (see Fig. 9a). This is expected because the sensors placed in these two rooms are very close to the kitchen. On the other hand, the signals from dining room and kitchen are barely important for our model when fire occurs in bedroom 1, regardless of the door opening conditions (see Fig. 9b). This also agrees with their spatial relations that the sensors in dining room and kitchen are farer away from that in bedroom 1. It can be seen that the learned attention weights suggest that the attention-based model can effectively learn the useful relationships between different sensor signals under different door opening conditions without the need of providing any prior information.

Limitations: One major limitation of existing prediction models such as [50, 51] lies in their generalizability across different building structures. The model architecture from these models does not support a variable number of input data and floorplans. For example, a model trained for one building structure with 3 channels of temperature data (Fig. 4) does not generalize to another with 7 channels (Fig. 7). This model limitation imposes the need of prior knowledge about the exact floorplan of the building structure. However, this kind of information is usually unknown in practical firefighting.

Table 2 Performance of P-flash

X	Model	Acc (%)	Prec (%)	Rec (%)	F1 (%)
60 s	BiLSTM	81.8	86.9	74.9	80.5
	BiLSTM-Attention	86.5	84.6	89.2	86.8
30 s	BiLSTM	78.2	77.0	80.5	78.7
	BiLSTM-Attention	81.8	79.5	85.6	82.3

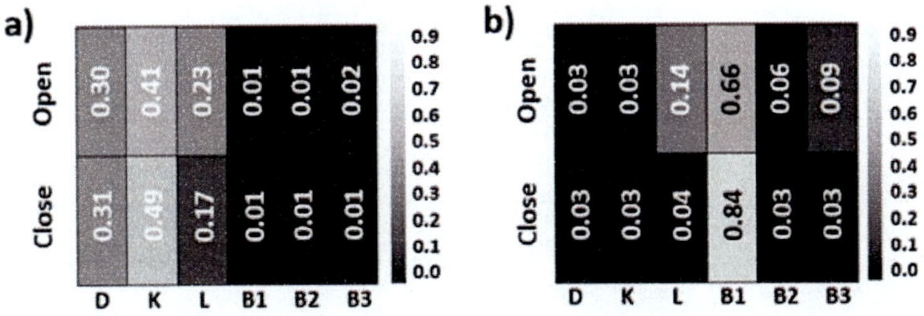

Fig. 9 Learned attention for **a** Kitchen and **b** Bedroom 1, under different door opening conditions

2.4 Generic Flashover Prediction Model (FlashNet)

This section presents the model development efforts to account for a wide range of residential homes with different floorplans and this section includes 4 subsections: (i) the home structures, (ii) the concept of graph structured data, (iii) the model formulation for FlashNet, and (iv) results and discussion.

Home Structures and Numerical Settings: Seventeen typical single-floor home structures are selected from [65], which defines 209 dwellings to represent approximately 80% of U.S. housing layouts. These 17 structures can be categorized into three types of residential buildings: (1) apartment homes, (2) attached homes, and (3) detached homes. The overall floor area ranges from 65 m^2 to 275 m^2 with three to fourteen compartments.

Graph Structured Data: The graph structured data is made up of two components: node attribute matrix which is the extracted instances for temperature and adjacency matrix. The adjacency matrix is formulated using two steps. The first step is to convert the home floorplan into a graph representation. An example is shown in Fig. 10 where each compartment is represented by a node and the corresponding opening or connection between two compartments is represented by an edge. The non-diagonal matrix elements are determined based on the size of opening between two compartments (i.e., the height and the width of a door from a bedroom to the hallway). For the diagonal elements, since the temperature information is crucial to correlate the potential occurrence of flashover, they are taken to be the maximum value of the non-diagonal elements. For the six-compartment building structure, the maximum value is 2.88 m^2. The same procedure is carried out to generate the graph representations for the rest of the sixteen building structures.

The second step is to obtain a geometric averaged adjacency matrix (GAAM) that can be used for all home structures. By visual inspection of the seventeen building structures from [52], there can be a possible of 15 × 15 compartment combinations a living room (L), kitchen (K), bedroom1 (BR1), bathroom1 (BA1), hall1 (H1), bathroom2 (BA2), bedroom2 (BR2), bathroom3 (BA3), den (Den), bedroom3 (BR3), family room (Fam), bedroom4 (BR4), hall2 (H2), hall3 (H3), or a dining room (D). The use of this information provides the dimensions of the

a)

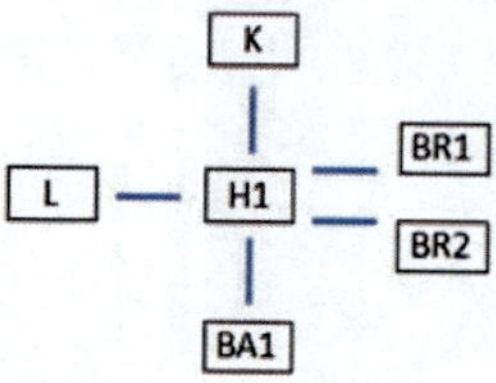

b)

	L	K	BR1	BA1	H1	BA2	BR2	BA3	...		D
L	2.88	0	0	0	2.88	0	0	0			0
K	0	2.88	0	0	1.845	0	0	0			0
BR1	0	0	2.88	0	1.845	0	0	0			0
BA1	0	0	0	2.88	1.845	0	0	0			0
H1	2.88	1.845	1.845	1.845	2.88		1.845	0	0		0
BA2	0	0	0	0	0	0	0	0			0
BR2	0	0	0	0	1.845	0	2.88	0			0
BA3	0	0	0	0	0	0	0	0			0
⋮					0				0		0
											0
D	0	0	0	0	0	0	0	0	0	0	0

Fig. 10 A schematic of **a** a graph representation and **b** an adjacency matrix for the six-compartment home

GAAM. Final GAAM matrix elements are determined using the statistical mean of the normalized adjacency matrices from the seventeen home structures. Additional explanation for the determination of GAAM and the corresponding GAAM for the fourteen-compartment detached home are provided in page 6 to 7 in [52].

Model Formulation: The overall structure of the Flashover prediction neural Network, namely *FlashNet*, is presented in Fig. 11 and the model has a single block of a spatial temporal graph convolutional network (ST-GConvN). The ST-GConvN block is used to capture the spatial and the temporal dependencies from the multivariate temperature data and consists of two temporal convolution layers and one spatial graph convolution layer. The required number of layers is determined based on numerical experiment [47]. For each temporal convolution layer, there is a 1-D convolution following by a rectified linear unit (ReLU), and this operation provides additional non-linearity to the temporal convolution layer to learn important features/information (i.e., how fast the temperature increase, how much time it takes to reach a certain temperature condition, etc.) in the time domain.

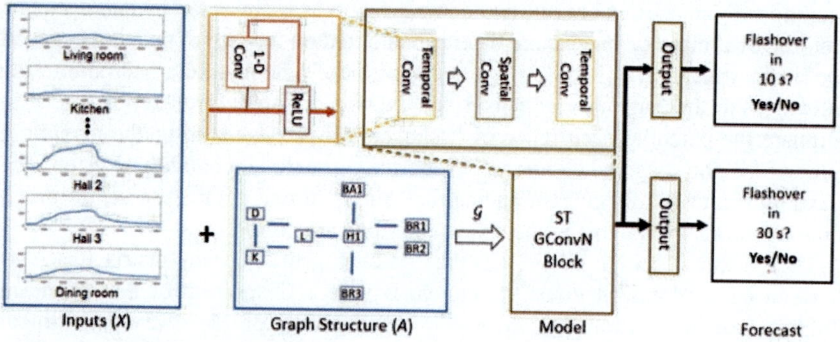

Fig. 11 Model structure of FlashNet

Table 3 Baseline comparison against five existing machine learning based flashover prediction models

Methods	Acc (%)	Prec (%)	Recall (%)	F1 (%)	Training (s)	Testing (s)
SVM	53.6	52.8	66.6	58.9	37.5	0.01
MPL	62.1	49.4	74.3	59.3	710.6	2.42
LSTM	66.6	75.7	49.7	60.1	1594.1	2.87
BiLSTM-ATT	70.0	66.6	81.2	73.2	1834.8	2.93
CNN	79.8	85.0	74.8	79.5	1351.0	2.38
FlashNet	85.2	83.5	87.8	85.6	1657.3	2.91

Model Performance and Benchmarks: Table 3 shows the results for flashover prediction with a lead time of 10 s. FlashNet is benchmarked against five state-of-the-art prediction models. The baseline models include (i) P-Flash with SVR [50], (ii) MLP—a feedforward multiple-layer perceptron [66], (iii) LSTM—a two-layer long short-term memory [67], (iv) P-Flash with attention-based Bi-LSTM [51], and (v) CNN—a three-layer standard convolutional neural network [68]. As shown in Table 5, FlashNet outperforms the existing machine learning based prediction models and achieves an overall accuracy of about 85.2%. Also, the scores from precision, recall, and F1 suggest that FlashNet is a well-balanced model which minimizes the false positive and the false negative. The main reason why FlashNet improves the model performance is that the nature for temperature data in flashover predictions accounting for different building structures with a wide range of fire and vent opening conditions is better suited for the modeling assumptions about temporal and spatial dependencies. In terms of training time, most of models, except ANN and the linear SVM, require about 30 min for convergence. For testing, models generally take about 3 s to make predictions for testing subset. On average, one prediction uses about a fraction of a second per instance.

3 Flashover and Backdraft Prediction Using Real Data

Sections 2.1–2.4 presents the machine learning model development using synthetic fire data. In this section, the development of the deep learning model uses real fire data from both small and real scale chamber fire experiments. Similar to the previous sections, descriptions for the experimental setup and the data behavior are first given. Then, the model formulations and important observation are presented.

3.1 *Flashover Forecast in Scaled Room Fire Tests*

This study [67] considers a standard 1/5th scale cubical enclosure with side lengths of approximately 0.5 m. The enclosure is primarily constructed with 4 mm thick galvanized steel. In order to observe the fire growth, a refractory glass is installed on the right side of the enclosure. In additional, there is an adjustable opening in the front side of the enclosure. Within the enclosure, three thermocouples are placed, along the centerline, on the ceiling. The first thermocouple is about 0.05 m away from the front opening and the second and third thermocouples are about 0.2 m apart. Two digital cameras are placed outside of the enclosure to visualize the fire development and to determine when there is an onset of flashover.

A series of experiments is carried out. There are 11 pool fire experiments and 13 wood crib fire experiments. Six different vent opening conditions are considered. Figure 15 shows the temperature growth from the three thermocouples for (a) a 24 cm diameter liquid pool fire with 0.4 m (height) × 0.16 m (width) opening and (b) a large wood crib fire with 0.2 m × 0.3 m opening. It can be seen in these figures that temperature data from real fire test has noticeable temperature oscillations. This kind of temperature behavior is missing from the synthetic temperature data generated by a zone model (i.e., CFAST). In Fig. 12, there are four primary fire stages: (1) growing, (2) flashover, (3) fully developed, and (4) decay. Video recordings are used to determine the onset of flashover. Specifically, when spilled flame is noted, the corresponding timestamp is marked as flashover. For example, the onset of flashover in Fig. 12a is about 320 s.

Model Formulation: The deep learning (DL) model carries out the following two tasks: (1) temperature prediction as a regression task and (2) flashover prediction as a binary classification task. Figure 13 depicts the overall workflow and the model structure of the DL model. The model inputs are the temperature profiles from the three thermocouples. A rolling time window of 20 s is used to facilitate real-time, continuous forecasts and the z-score normalization is conducted to normalize the

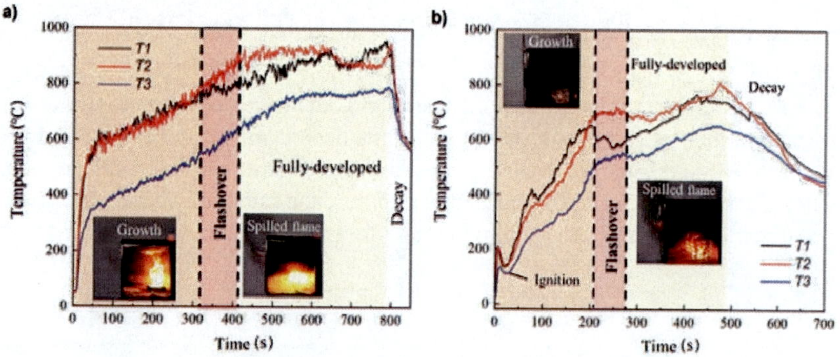

Fig. 12 Typical temperature growth for different fuels **a** 24-cm diameter liquid pool fire with 0.4 m × 0.16 m opening and **b** large wood crib fire with 0.2 m × 0.3 m opening.

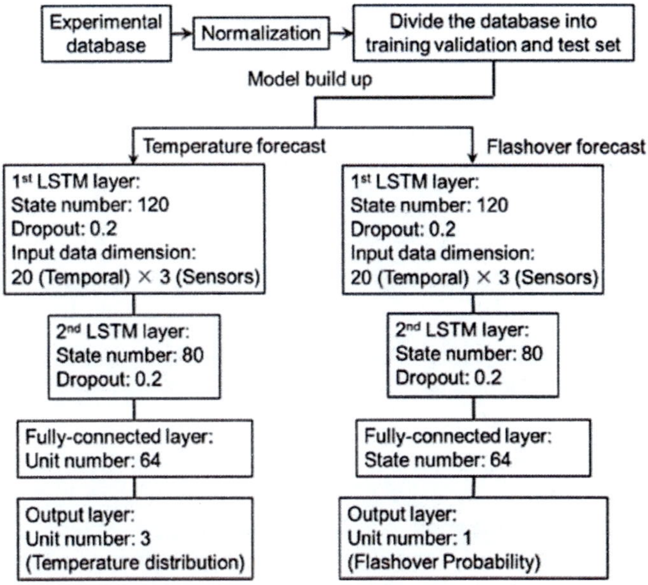

Fig. 13 The structure of the LSTM based fire forecast model

input data in the same scale. Input data is divided into three subsets and they are training, validation, and testing subsets.

As seen in Fig. 13, the DL model is consisted of a stacked Long Short-Term Memory (LSTM). The 1st LSTM has a state number of 120 and the 2nd LSTM has a state number of 80. A dropout of 0.2 is applied to both LSTM layers to facilitate training stability. The extracted features obtained from the LSTM layers are passed to a fully connected layer with a node number of 64. Relu is used to provide additional nonlinearity for mapping. Depending on the forecasting task, the output layer for the temperature prediction and the flashover prediction has a node number of 3 and 1, respectively.

Results and Discussion: Figure 14 shows the predicted results of the DL model for a large wood crib fire with 0.2 m × 0.3 m opening. Specifically, the temperature forecast and the forecasting flashover probability with a lead time of 10 s are shown in Fig. 14a and Fig. 14b, respectively. The solid lines are the ground trues and the dash lines are the model predictions. For temperature forecast, it can be seen from Fig. 17a that the DL model is capable to predict the future temperature (with a lead time of 10 s) and the temperature predictions capture the trend of the temperature growth. For this particular case, the overall discrepancy between the predictions and the ground true is about 10%. In term of flashover prediction, it can be seen that the model has relatively promising performance and it can forecast the potential occurrence of flashover correctly. The results from this study affirm a fact that machine learning paradigm can learn the important patterns for various prediction tasks such as temperature

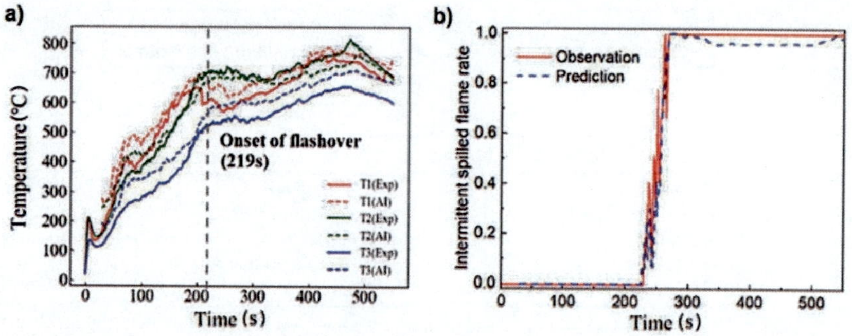

Fig. 14 Model prediction of **a** temperature and **b** flashover probability for a large wood crib fire with 0.2 m × 0.3 m opening

and flashover predictions from real fire temperature data that has various noise and temperature fluctuations.

It is hoped that these research efforts can continuously overcome limits and/or eliminate assumptions that the traditional modeling approaches fail to capture. In the next section, another thrust of research efforts is presented to highlight the current modeling advancement in other built environments.

3.2 *Predicting Backdraft in Real-Scale Room*

This section presents a full-scale fire test in a 5.5 m (length) × 2.4 m (width) × 2.4 m (height) chamber (Fig. 15). Wood planks are adopted as the fuel and arranged to simulate the furniture, room ceiling, sidewalls, and floor to simulate the fuel distribution in an actual building environment (Fig. 15a, b). At the beginning of each test, an advanced flame-thrower burner is employed to ignite the innermost fuel until the fire can sustain itself autonomously. When the fire grows to large enough, the door is then closed to create a nearly sealed room environment. After the fire further developed inside the room for a certain period, the door can be opened to generate a spilled flame or a backdraft. The operation of close and open door is repeated several times in each fire test with different internal fire scenarios (varied oxygen supply and fire HRRs) and door enclosure durations to obtain backdraft samples with various intensities.

Totally, five groups of large-scale fire tests are conducted on different dates under varied ambient and weather conditions. Each test includes multiple spilled flame and backdraft processes. The temperature profile in each experiment is measured by a set of thermocouple arrays (Fig. 15c). Two TC trees (TC#1–6 and TC#13–18) are adopted to record the vertical temperature distribution, and these data can be used to identify the smoke stratification. The horizontal TC array (TC#7–12) are set to measure the temperature distribution of the upper smoke layer. Moreover, two

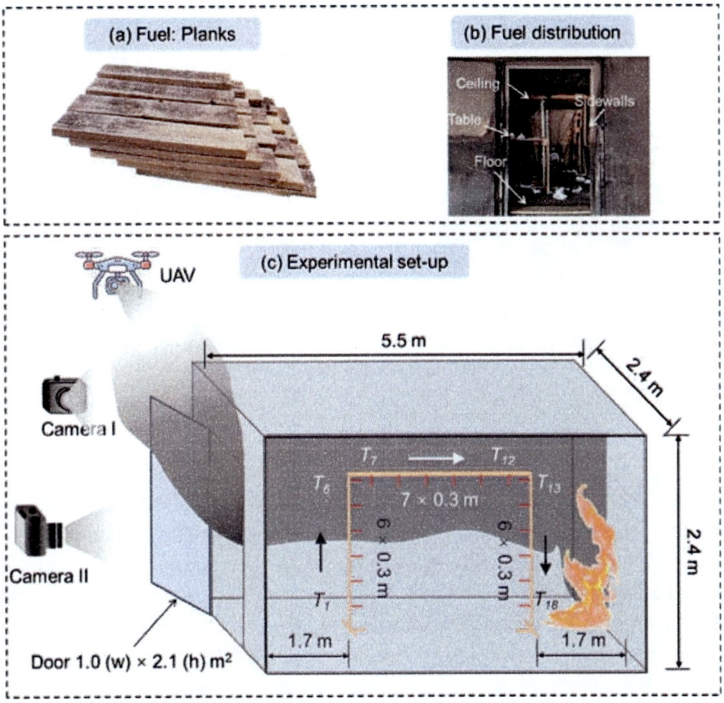

Fig. 15 Demonstration of the backdraft experimental setup (a) Fuel type, (b) Fuel distribution, and (c) Test chamber geometry and sensor layout

fixed cameras and a camera carried by portable Unmanned Aerial Vehicle (UAV) are installed a few meters away from the chamber door to record the smoke and flame videos from different view angles.

The typical temperature growth of the large-scale fire test is plotted in Fig. 16. As shown, when closing (opening) the door, the oxygen will be cut off (supplied) to induce the temperature decay (rise), respectively. When the fuel gas concentration and the internal temperature reach a critical level, an explosion will happen when the door is open. This explosion is a typical backdraft phenomenon. During the period of 1000–2200 s, the door is closed and re-opened several times. Then, several no-backdraft cases and backdraft cases with different intensities are further observed, depending on different initial states (e.g., the previous door-close duration and internal temperature) and door enclosure durations. Finally, the fire is either extinguished with a fire hose or continued until the burnout of fuel (the blue region). In total, 31 times of backdrafts have been observed in 5 tests.

Model Formulation: Unlike other critical event in the building fire, which is the spontaneous evolution of the fire system, the occurrence of the backdraft is associated with the firefighter's action, i.e., open or close the door so the state of the chamber is important information to judge the backdraft risks. Therefore, a multimodal deep

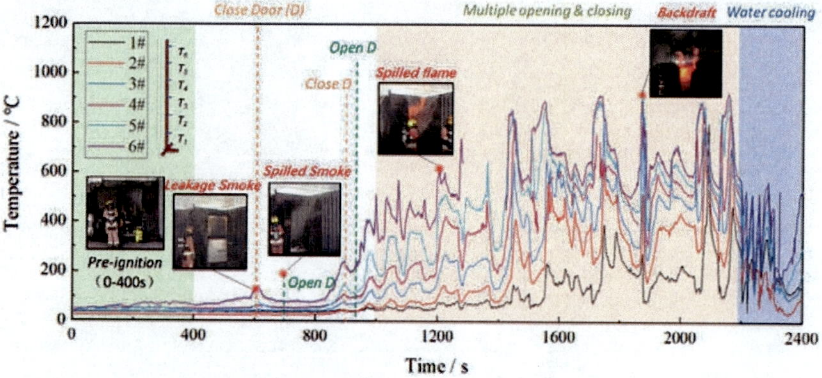

Fig. 16 Experimental setup of the full-scale backdraft test chamber

learning framework to integrate the sensor signal (temperature) and visual signal (images) are proposed to forecast the backdraft (Fig. 17). The original transformer and Vision-transformer algorithms are adopted as the encoder for the sensor data and visual data, respectively. The input image is first divided into a grid of patches and linearly embedded into a lower-dimensional vector representation. The patch embeddings are then passed through a series of transformer encoder layers (four in this work for the visual signal). In each layer, two key components, i.e., self-attention and feed-forward neural networks, are adopted to capture both local and global dependencies among the patches. Similarly, the sensor data are also processed by transformer encoder layers (two in this work) to extract the features and dependencies.

After the initial encoding of each modality, a cross-modal attention mechanism is employed in the fusion-transformer layer, in which the attention scores between the representations of different modalities, allowing them to attend and exchange information with each other. In this study, the model tries to link the temperature variation characteristics (absolute value and gradient) with the firefighter's operation (closing or opening the door) and establish potential connections among these features. Finally, the fusion-encoded information from multimodal data passes through the LSTM model to obtain the temporal features, and two fully connected layers are used to achieve the nonlinear fitting and generate the output, namely, the backdraft intensity (represented by the maximum explosion flame height).

Results and Discussion: Figure 18 presents the prediction of the deep learning model for backdraft onset with ground truth. As shown, there are eight times of close-open cycles in the whole experiment, and seven of them successfully induce a backdraft with different intensity while the last time has no backdraft due to the water-cooling effect. Accordingly, the proposed model predicts a certain backdraft intensity for the seven samples which successfully generate a backdraft and outputs a zero for the last time with the water cooling. The model performs a 100% accuracy on predicting the occurrence of the backdraft with a low delay time (less than 1.6 s). Meanwhile, the accumulation of the pyrolysis fuel gas and the gradual temperature decay can well

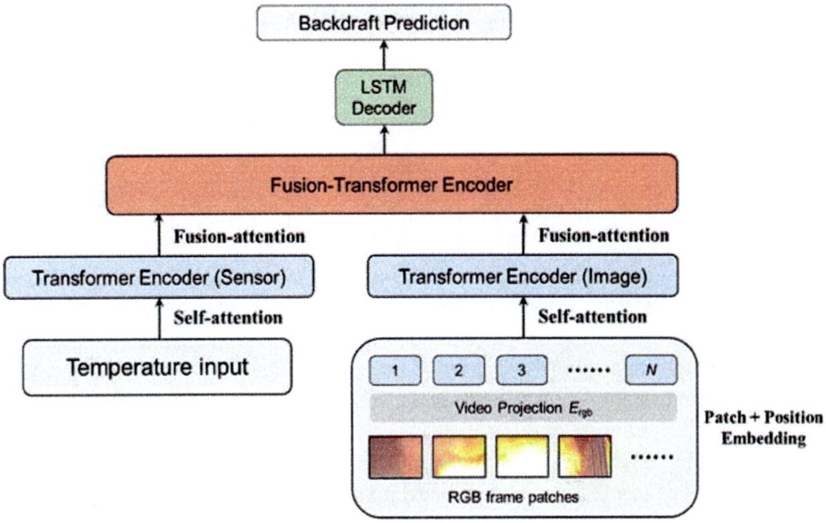

Fig. 17 Diagram of the Fusion-transformer structure

be reflected by the increase and decrease of predicted backdraft risk, respectively, indicating that the results are phenomenological.

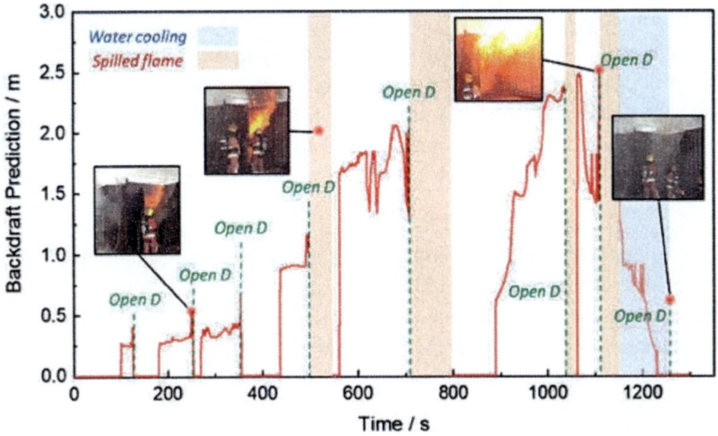

Fig. 18 Deep learning model prediction versus observed ground truth

4 Challenges and Perspectives in AI-Driven Fire Forecast

4.1 Challenges

Interpretability of the AI/ML fire modeling. Despite the numerical advantages from [50–52], there is one sufficient drawback of using ML paradigm and that is, the models are not interpretable. Often time, it is considered as a black box. It is well understood that being able to explain the model decision is important for practical engineering applications, especially when the potential applications involve lives. Also, model interpretability provides traceable information to scientists or engineers to understand the underlying reasons why a model is making such a correct or a wrong decision. In the future study, an explainable ML-based flashover prediction model is needed and model interpretability can be introduced using various techniques such as feature attention mechanism, feature visualization, and/or class activation map [69]. These techniques can help to extract discriminative information in both spatial and temporal domains for the multivariate series so that a more trust-worthy ML system can be developed to address the fire safety problems.

Lack of consistent criteria of critical events. As demonstrated above, many phenomena (critical temperature, heat flux, spilled flame, rollover, etc.) can be regarded as the flashover onset but a large difference in the predicted moment may be caused by the adaptation of those criteria. For example, according to the temperature profile in a 1/5 scaled model fire test, the average ceiling temperature can reach 600 °C in around 120 s, whereas if the visual criterion, spilled flame is utilized, the flashover occurs at around 220 s, which is around 100 s later than the temperature data-based prediction. Therefore, although many papers claimed that they are predicting the flashover onset, they are indeed forecasting different moments if there is not a consistent definition. Consequently, the prediction results with different models are not comparable, which largely impedes the communication and spread of smart forecast models.

Lack of a comprehensive database. Compared with engineering applications in other fields, data collection for fire research is always a big challenge. Full-scale fire test data are of great significance as they are closest to real fire accidents but very dangerous and costly at the same time. In contrast, the scaled model and numerical tests are friendly for both money and time, which is a better way to generate a large amount of data for AI training. However, the results from scaled and numerical models are inevitably affected by the scaling effect and simplifying assumptions. Since the prediction accuracy of the AI model relies heavily on the quality of the database, the performance of the AI model trained by the ideal database is always questionable.

4.2 Perspectives

Vision-based fire scenario recognition. Compared with sensor data, which is hard to access in real fire incidents, the visual data are more accessible as they can be captured by potable cameras, Unmanned Aerial Vehicles (UAVs), and even the cell phone of residents. Computer vision technology shows a good performance to identify the fire Heat Release Rate (HRR) based on the external smoke [70] and flame images [71, 72]. Combined with remote sensing technologies, the vision-based deep learning model is expected to provide more useful fire scene information, such as the fuel load, fuel type, potential fire spread path, and thus help predict the fire evolution and critical events occurrence accurately.

Physics-guided model training. The issue of interpretability and lacking large-scale fire test data is expected to be solved by physics-guided model training, which is proposed to address the data imbalance problem. The principle is to introduce physical knowledge, e.g., the scaling law, government equations, and fire dynamics in the training process. To be more specific, the closed-form and partial differential physical equations are served as a term in the loss function with a certain weight or embedded in each layer to impact the training process directly. Then the multi-source fire test data can be well understood by AI.

References

1. M. Ahrens, R. Maheshwari, *Home structure fires* (National Fire Protection Association, Quincy, MA 02169, 2021)
2. NFPA 72. *National fire alarm and signaling code* (National Fire Protection Association. Quincy, MA 02169, 2022)
3. NFPA 13. *Standard for the Installation of Sprinkler Systems* (National Fire Protection Association, Quincy, MA 02169, 2022)
4. NFPA 13D. *Standard for the Installation of Sprinkler Systems in One-and Two-Family Dwellings and Manufactured Homes* (National Fire Protection Association, Quincy, MA 02169, 2022)
5. NFPA 13R. *Standard for the Installation of Sprinkler Systems* (National Fire Protection Association, Quincy, MA 02169, 2022)
6. C.A. Wilkie, A.B. Morgan (Eds.), *Fire retardancy of polymeric materials* (CRC press, 2009)
7. V. Kodur, P. Kumar, M.M. Rafi, Fire hazard in buildings: review, assessment and strategies for improving fire safety. PSU Res. Review., (2019)
8. K.M. Chandler, E.M. Crown, S.A. Brown, Consumer information and education effects on knowledge and choice of fire resistant upholstery. J. Consum. Aff. **25**(2), 339–357 (1991)
9. FEMA. Fire-Safe Seniors Program, https://www.usfa.fema.gov/prevention/outreach/fire_safe_seniors.html. Last accessed 13 Dec 2022
10. NFPA 220. *Standard on Types of Building Construction* (National Fire Protection Association, Quincy, MA 02169, 2021)
11. NFPA 720. Standard for the Installation of Carbon Monoxide(CO) Detection and Warning Equipment (National Fire Protection Association, Quincy, MA 02169, 2015)
12. NFPA 70. National Electrical Code, (National Fire Protection Association, Quincy, MA 02169, 2023)

13. S. Kerber, Analysis of changing residential fire dynamics and its implications on firefighter operational timeframes. Fire Technol. **48**(4), 865–891 (2012)
14. R.D. Peacock, P.A. Reneke, R.W. Bukowski, V. Babrauskas, Defining flashover for fire hazard calculations. Fire Saf. J. **32**(4), 331–345 (1999)
15. D. Drysdale, *An introduction to fire dynamics* (John wiley & sons, 2011)
16. W.M. Croft, Fires involving explosions—A literature review. Fire Saf. J. **3**(1), 3–24 (1980)
17. H.J. Kim, D. Lilley, Comparison of theories for room flashover, in 37th Aerospace sciences meeting and exhibit (1999), p. 343
18. V. Babrauskas, Estimating room flashover potential. Fire Technol. **16**(2), 94–103 (1980)
19. B.J. McCaffrey, J.G. Quintiere, M.F. Harkleroad, Estimating room temperatures and the likelihood of flashover using fire test data correlations. Fire Technol. **17**(2), 98–119 (1981)
20. H.J. Kim, D.G. Lilley, Flashover: A study of parameter effects on time to reach flashover conditions. J. Propul. Power **18**(3), 669–673 (2002)
21. K.J. Overholt, O.A. Ezekoye, Characterizing heat release rates using an inverse fire modeling technique. Fire Technol. **48**(4), 893–909 (2012)
22. M.D. Price, Using inverse fire modeling with multiple input signals to obtain heat release rates in compartment fire scenarios (Doctoral dissertation, University of Maryland, College Park, 2014)
23. W. Jahn, Using suppression and detection devices to steer CFD fire forecast simulations. Fire Saf. J. **91**, 284–290 (2017)
24. S.H. Koo, J. Fraser-Mitchell, S. Welch, Sensor-steered fire simulation. Fire Saf. J. **45**(3), 193–205 (2010)
25. Y. Fu, H.V. Leong, G. Ngai, M.X. Huang, S.C. Chan, Physiological mouse: Toward an emotion-aware mouse. Univ. Access Inf. Soc. **16**(2), 365–379 (2017)
26. Y. Wang, S.C.F. Chan, H.V. Leong, G. Ngai, N. Au, Multi-dimension reviewer credibility quantification across diverse travel communities. Knowl. Inf. Syst. **49**(3), 1071–1096 (2016)
27. J. Li, C. Brown, D.J. Dzikowicz, M.G. Carey, W.C. Tam, M.X. Huang, Towards real-time heart health monitoring in firefighting using convolutional neural networks. Fire Saf. J. **140**, 103852 (2023)
28. S. Levine, A. Kumar, G. Tucker, J. Fu, Offline reinforcement learning: Tutorial, review, and perspectives on open problems. arXiv preprint arXiv:2005.01643
29. C. Li, Y. Wang, D. Li, D. Chu, M. Ma, An effective method of evaluating pension service quality using multi-dimension attention convolutional neural netwosrks. Int. J. Software Eng. Knowl. Eng. **31**(04), 533–543 (2021)
30. T.C. Kwok, E.Y. Fu, E.Y. Wu, M.X. Huang, G. Ngai, H.V. Leong, Every little movement has a meaning of its own: Using past mouse movements to predict the next interaction, in 23rd International conference on intelligent user interfaces (2018), pp. 397–401
31. M. Ngai, E. Fu, A. Tam, A. Yang, G. Ngai, Finding the signal from the smoke: a real-time, unattended fire prevention system using 3D CNNs. J. Stud. Res., 11 (3), (2022)
32. A.E. Mensch, A. Hamins, W.C. Tam, Z.J. Lu, K. Markell, C. You, M. Kupferschmid, Sensors and machine learning models to prevent cooktop ignition and ignore normal cooking. Fire Technol., 1–24 (2021)
33. W.C. Tam, E.Y. Fu, A. Mensch, A. Hamins, C. You, G. Ngai, H. Leong, Prevention of cooktop ignition using detection and multi-step machine learning algorithms. Fire Saf. J. **120**, 103043 (2021)
34. W.W. Yuen, W.C. Tam, W.K. Chow, Assessment of radiative heat transfer characteristics of a combustion mixture in a three-dimensional enclosure using RAD-NETT (with application to a fire resistance test furnace). Int. J. Heat Mass Transf. **68**, 383–390 (2014)
35. W.C. Tam, W.W. Yuen, W.K. Chow, Numerical study on the importance of radiative heat transfer in building energy simulation. Numer. Heat Transf., Part A: Appl. **69**(7), 694–709 (2016)
36. W.W. Yuen, C.L. Chow, W.C. Tam, Analysis of radiative heat transfer in inhomogeneous nonisothermal media using neural networks. J. Thermophys. Heat Transfer **30**(4), 897–911 (2016)

37. W.C. Tam, W.C. Tam, W.W. Yuen, *OpenSC: An open-source calculation tool for combustion mixture emissivity/absorptivity* (US Department of Commerce, National Institute of Standards and Technology, Gaithersburg, MD, USA, (2019)
38. W.W. Yuen, W.C. Tam, Point mean beam length, a new concept to enhance the computational efficiency of multidimensional, non-gray radiative heat transfer. Numer. Heat Transf., Part B: Fundam. **79**(3), 113–129 (2020)
39. W. Yuen, W.C. Tam, A fundamental assessment of the concept of mean beam length for application for multi-dimensional non-gray radiative heat transfer. Int. J. Therm. Sci. **179**, 107681 (2022)
40. A. Chenebert, T.P. Breckon, A. Gaszczak, A non-temporal texture driven approach to real-time fire detection, in 2011 18th IEEE international conference on image processing (IEEE, 2011), pp. 1741–1744
41. Z. Yin, B. Wan, F. Yuan, X. Xia, J. Shi, A deep normalization and convolutional neural network for image smoke detection. IEEE Access **5**, 18429–18438 (2017)
42. D. Shen, X. Chen, M. Nguyen, W.Q. Yan, Flame detection using deep learning, in 2018 4th International conference on control, automation and robotics (ICCAR) (IEEE, 2018), pp. 416–420
43. S. Aslan, U. Güdükbay, B.U. Töreyin, A.E. Cetin, Deep convolutional generative adversarial networks based flame detection in video (2019). arXiv preprint arXiv:1902.01824
44. M. Soleymani, M. Pantic, T. Pun, Multimodal emotion recognition in response to videos. IEEE Trans. Affect. Comput. **3**(2), 211–223 (2011)
45. Z.Q. Zhao, P. Zheng, S.T. Xu, X. Wu, Object detection with deep learning: A review. IEEE Trans. Neural Netw. Learn. Syst. **30**(11), 3212–3232 (2019)
46. R. Venkateswarlu, Eye gaze estimation from a single image of one eye, in Proceedings Ninth IEEE international conference on computer vision (IEEE, 2003), pp. 136–143
47. X. Geng, Z.H. Zhou, K. Smith-Miles, Automatic age estimation based on facial aging patterns. IEEE Trans. Pattern Anal. Mach. Intell., **29**(12), 2234–2240 (2007)
48. A. Asuncion, D. Newman, *UCI machine learning repository* (2007)
49. Y. Sugano, Y. Matsushita, Y. Sato, Learning-by-synthesis for appearance-based 3d gaze estimation, in Proceedings of the IEEE conference on computer vision and pattern recognition (2014), pp. 1821–1828
50. J. Wang, W.C. Tam, Y. Jia, R. Peacock, P. Reneke, E.Y. Fu, T. Cleary, P-Flash–A machine learning-based model for flashover prediction using recovered temperature data. Fire Saf. J. **122**, 103341 (2021)
51. E.Y. Fu, W.C. Tam, J. Wang, R. Peacock, P.A. Reneke, G. Ngai, H.V. Leong, T. Cleary, Predicting flashover occurrence using surrogate temperature data. in Proceedings of the AAAI Conference on Artificial Intelligence, vol. 35, no. 17 (2021), pp. 14785–14794
52. W.C. Tam, E.Y. Fu, J. Li, R. Peacock, P. Reneke, G. Ngai, H.V. Leong, T. Cleary, M.X. Huang, Real-time flashover prediction model for multi-compartment building structures using attention based recurrent neural networks. Expert Syst. Appl. **223**, 119899 (2023)
53. W.C. Tam, E.Y. Fu, R. Peacock, P. Reneke, J. Wang, J. Li, T. Cleary, Generating synthetic sensor data to facilitate machine learning paradigm for prediction of building fire hazard. Fire Technol., 1–22 (2020)
54. P.A. Reneke, R.D. Peacock, S.W. Gilbert, T. Cleary, *CFAST–Consolidated fire and smoke transport (Version 7) Volume 5: CFAST Fire Data Generator (CData)* (NIST TN 1889v5. National Institute of Standards and Technology, Gaitersburg, MD, 2021)
55. R.D. Peacock, P.A. Reneke, G.P. Forney, CFAST–Consolidated model of fire growth and smoke transport (Version 7) Volume 2: User's guide. NIST Technical Note 1889v2 (2017)
56. M. McKinnon, C. Weinschenk, D. Madrzykowski, *Modeling gas burner fires in ranch and colonial style structures* (Report. Underwriters Laboratories Inc., Columbia, MD, 2020)
57. W.C. Tam, E.Y. Fu, J. Li, X. Huang, J. Chen, M.X. Huang, A spatial temporal graph neural network model for predicting flashover in arbitrary building floorplans. Eng. Appl. Artif. Intell. **115**, 105258 (2022)

58. V. Babrauskas, Heat release rates, in SFPE handbook of fire protection engineering (Springer, New York, NY, 2016), pp. 799–904
59. V. Vapnik, The support vector method of function estimation, in Nonlinear Modeling (Springer, Boston, MA, 1998), pp. 55–85
60. D. Madrzykowski, C. Weinschenk, Impact of fixed ventilation on fire damage patterns in full-scale structures. Natl. Crim. Justice Ref. Serv., 252831 (2019)
61. P.A. Reneke, M. Bruns, S.W. Gilbert, C.P. MacLaren, R.D. Peacock, T.G. Cleary, D.T. Butry, *Towards a process to quantify the hazard of fire protection design alternatives* (2019)
62. V. Babrauskas, *Glass breakage in fires.* (Fire science and technology inc., 2011)
63. M.J. Hurley, D.T. Gottuk, J.R. Jr. Hall, K. Harada, E.D. Kuligowski, M.Puchovsky, C.J. Wieczorek (Eds.), *SFPE handbook of fire protection engineering* (Springer, 2015)
64. A. Graves, J. Schmidhuber, Framewise phoneme classification with bidirectional LSTM and other neural network architectures. Neural Netw. **18**(5–6), 602–610 (2005)
65. A. Persily, A. Musser, D.D. Leber, *A collection of homes to represent the US housing stock.* US NISTIR 7330 (National Institute of Standards and Technology, Gaithersburg, MD, 2006)
66. R.K. Yuen, E.W. Lee, S.M. Lo, G.H. Yeoh, Prediction of temperature and velocity profiles in a single compartment fire by an improved neural network analysis. Fire Saf. J. **41**(6), 478–485 (2006)
67. T. Zhang, Z. Wang, H.Y. Wong, W.C. Tam, X. Huang, F. Xiao, Real-time forecast of compartment fire and flashover based on deep learning. Fire Saf. J. **130**, 103579 (2022)
68. J.L. Hodges, B.Y. Lattimer, K.D. Luxbacher, Compartment fire predictions using transpose convolutional neural networks. Fire Saf. J. **108**, 102854 (2019)
69. L. Fan, W.C. Tam, Q. Tong, E.Y. Fu, T. Liang, An explainable machine learning based flashover prediction model using dimension-wise class activation map. Fire Saf. J., 103849 (2023)
70. Z. Wang, T. Zhang, X. Wu, X. Huang, Predicting transient building fire based on external smoke images and deep learning. J. Build. Eng. **47**, 103823 (2022)
71. Z. Wang, Y. Ding, T. Zhang, X. Huang, Automatic real-time fire distance, size and power measurement driven by stereo camera and deep learning. Fire Saf J, 103891 (2023)
72. Z. Wang, T. Zhang, X. Huang, Predicting real-time fire heat release rate by flame images and deep learning. Proc. Combust. Inst. **39**(3), 4115–4123 (2023)

Data-Driven Wireless Fire Hose Flow Rate Apparatus

Christopher U. Brown, Gregory W. Vogl, and Wai Cheong Tam

Abstract A wireless sensor network was created to measure water-flow rate in a fire hose. An integrated electronic piezoelectric (IEPE) accelerometer was chosen as the sensor to measure the flow rate based on the vibrations generated by water flowing through a fire hose. The flow apparatus, including the accelerometer, was lightweight, small, and easily attached and removed to any location along the fire hose, not obstructing the water's flow path. A relationship between the dominant-frequency metric and the flow rate was applied with a custom graphical user interface for quick, real-time, visual referencing of flow rate in a fire hose by fire personnel. The wireless flow apparatus was used with realistic firefighting hose conditions (i.e., holding the hose nozzle, nozzle motion during simulated fire suppression, simulated hose dragging) to evaluate the apparatus function and determine the influence of the hose motion, accelerometer location, and nozzle spray on the dominant-frequency metric. While more research is needed, such as enhancing the robustness of the dominant frequency metric, and physically hardening the apparatus, this research shows the potential of a "smart" fire hose for improved situational awareness during fire suppression.

C. U. Brown (✉) · W. C. Tam
Fire Research Division, Engineering Laboratory, National Institute of Standards and Technology, Gaithersburg, MD, USA
e-mail: christopher.brown@nist.gov

G. W. Vogl
Intelligent Systems Division, Engineering Laboratory, National Institute of Standards and Technology, Gaithersburg, MD, USA

X. Huang and W. C. Tam (eds.), *Intelligent Building Fire Safety and Smart Firefighting*, Digital Innovations in Architecture, Engineering and Construction,
https://doi.org/10.1007/978-3-031-48161-1_17

1 Introduction

1.1 Smart Fire Fighting—Water Flow at the Fire Hose Nozzle

Placing and flowing water from the initial hose lines at a fire is very important for the success of the fire attack. The water flowing from the hoses helps cool the environment which improves the chance of survival of trapped occupants, protects the fire fighters from excessive heat, and extinguishes the fire. Therefore, hoses are simultaneously a fire fighter's and occupant's lifeline [1]. Knowing the rate at which water is flowing through those hoses is a critical part of those lifelines; especially if that rate is zero, which unfortunately happens.

Applying 'smart' technology to a fire hose could improve the awareness of the hose's status and the chance of a successful fire attack. Harnessing the power of 'smart' technology to improve situational awareness of the hoses was a part of the vision of Smart Fire Fighting as documented in the Research Roadmap for Smart Fire Fighting [2]. A 'smart' system uses sensors to collect data, provides the data in an understandable format to a user, then allows the user to decide fire attack parameters like water flow rate. Today, human users alone make these decisions but in the future software may assist them.

Human fire fighters are the backbone of the fire service. The safety of fire fighters who risk their lives on the fireground could benefit from a smart sensor network that could perform two tasks: determine if water is flowing at the fire hose nozzle and communicate this information back to the pump operator at the fire engine or incident commander (IC). We envision that the smart fire hose system includes a sensor at the nozzle wirelessly transmitting the flow rate data to the IC and pump operator, and a way to present the data to the IC, as shown in Fig. 1. The IC must be able to make quick decisions regarding the water flow for suppression activities on the fireground, thus the data presented to the IC must be easily understood.

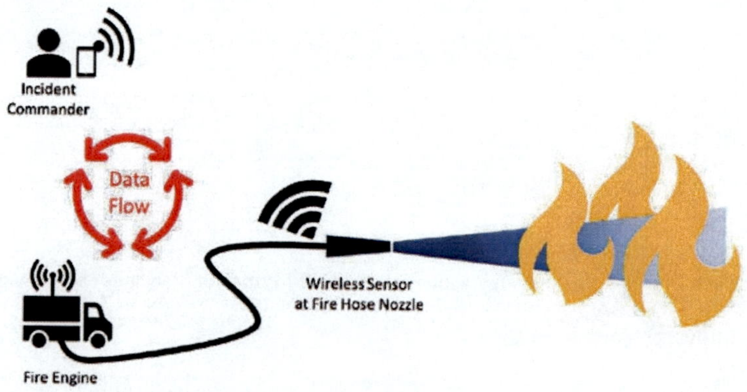

Fig. 1 A smart sensor network can measure flow rate at the fire hose nozzle and wirelessly transmit the flow rate to the IC and pump operator for decision making [3]

1.2 Current Technology

Fire engines have a gauge on the pump panel for the amount of water in their supply tank. The water level will decrease when the engine is operating using only its own water supply and water is flowing out of the engine. If the first arriving engine initially supplies water from its own supply tank, the decreasing level of tank water will indicate that water is flowing out of the engine, but not necessarily flowing from the nozzle (i.e., inside a building structure) as intended. If the engine has an external water supply, from a hydrant or tanker, then the volume of water will be replenished, and the tank volume will not indicate that water is leaving the engine. In both cases, the engine's water tank level will not clearly indicate if water is flowing from the nozzle.

Some engines will indicate pressure loss on their pump panel gauge when the hose nozzle is opened. Additional water pressure must then be added to increase the hose pressure after the nozzle is opened. The pump operator must be at the pump panel to increase the pressure to the target pressure once the nozzle is opened. By increasing pressure to the hose line, the pump operator assumes that water is flowing from the nozzle as intended.

On other engines, the hose pressure can be preset, and once the nozzle is opened and the pressure in the hose line drops, the engine pump will automatically compensate by increasing pressure to achieve the preset target pressure. The pump operator does not need to stay at the pump panel to wait for the fire fighters to open the nozzle so the pump operator can increase the pressure. The pump operator may, or may not, be able to hear the pump sounds while standing next to the pump panel as the engine's pump automatically increases water pressure. Unfortunately, this also means the pump operator may not know if water is flowing from the nozzle on the other end of the hose line that is out of sight.

1.3 On the Scene

Consider the following scenario describing a house fire with fire fighters using the current fire hose (Fig. 2). The first fire engine arrives at the front of a single-family house with fire showing from an upstairs window. The engine officer conducts the initial size up by walking completely around the house. Concurrently, the nozzle fire fighter stretches a 60 m (200 foot) preconnected 4.5 cm (1.75 inch) hose from the engine to the front door of the house while the backup fire fighter brings hand tools from the fire engine. The officer finishes the exterior assessment, and the two-person team prepares to enter the house through the front door. When the team is ready to enter the house, they signal to the pump engine operator to charge the hose line with water. The team enters the front door of the house and advances the pressurized hose into the house to the fire.

Fig. 2 Fire hose deployment preparing to enter a house on fire [4]

The two-person team entering the house, as shown in Fig. 2, is relying on the water from the hose to protect them from the heat and to extinguish the fire. Once the hose line has been charged, the pump operator may conduct other duties on the fire ground away from the pump panel, including placing ladders to windows of the house, connecting an intake hose from a hydrant to the engine, chocking wheels, etc. Once the suppression team disappears into the house, the pump operator may not know that the fire fighters are flowing water.

Radios are currently used to communicate between the fire fighters at the nozzle and the pump operator or IC to provide feedback about water flow. However, this is not always possible to do with competing radio traffic, while the team is advancing the hose, and conducting suppression activities that take two hands.

1.3.1 Low Nozzle Flow Due to a Rupture or Leak

Presently, water pressure measured at the fire engine's pump panel is often used by the pump operator to determine if water is flowing at the hose nozzle. Fireground threats to normal water flow such as hose damage and hose blockage can make reliable decisions based on water pressure misleading (Table 1). A pressure loss indicated on the fire engine's pump panel may occur when water flows from the nozzle as intended, or unintentionally through a ruptured hose (Fig. 3). A ruptured hose can occur as a result of wear and tear, a burn hole in the hose, a leaking coupling, or from being damaged by vehicle tires or structure debris. These threats could cause water to flow unintentionally from somewhere along the hose before the nozzle, decreasing pressure at the pump panel as if water was flowing properly from the nozzle. Fire

Table 1 Fireground threats to normal fire hose water flow and the results of reacting to the pump pressure [3]

Threats to normal water flow	Results to water flow	Result to water flow at nozzle	Pressure at pump	Reaction	Results to water flow at nozzle
Hose hole from normal wear and tear	Water leaks from hose	Flow decreases or stops	Decrease	Pump pressure increases	Water flow may or may not be adequate
Hose burn holes					
Leaking couplings between hose sections					
Hose rupture from car tire or structural collapse					
Hose crimped around corner	Water flow is partially or fully blocked	Flow decreases or stops	Remains the same	None	Water flow remains partially or fully blocked
Hose pinched under a door, furniture, car tire or debris					
In line valve is closed					
Closed nozzle due to hurt firefighter					

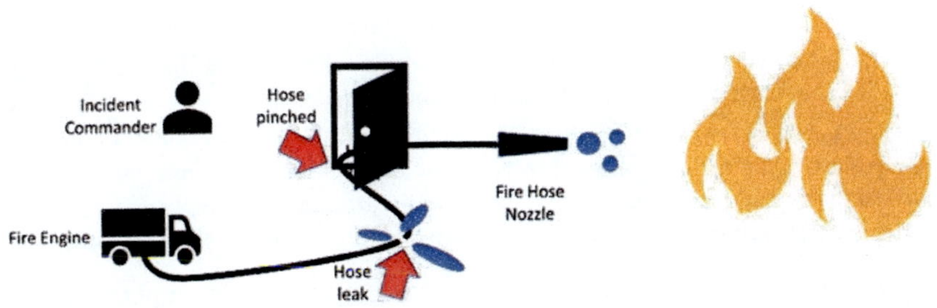

Fig. 3 Threats to normal nozzle flow include holes from wear and tear or burns, leaking couplings, a hose rupture from car tires or structural collapse, or a partially blocked hose from crimping or pinching, yet these dangerous conditions may not be obvious at the pump panel where water pressure may appear normal [3]

fighter fatalities have occurred because fire has burned through the hose and water did not reach the hose nozzle [5, 6].

The pressure loss would cause the engine's pump to increase the water pressure, automatically or manually, to account for the decrease in pressure assuming that water is flowing from the nozzle. But if there is a hose failure, less or no water may be flowing from the nozzle where it is needed. The safety of the fire fighters and occupants is threatened by these types of dangerous events; measuring only pressure at the pump panel is deceiving.

1.3.2 Low Nozzle Flow Due to Blockage

When the fire engine measures only water pressure at the fire engine's pump panel, the pump operator may not know if a hose is partially or fully blocked preventing water flow at the nozzle (Table 1). Sufficient water pressure may show at the pump panel even if the hose is partially or fully blocked preventing water from reaching the nozzle. Several threats on the fireground can partially or fully block a fire hose (Fig. 3). A charged hose line advanced inside a structure could become partially blocked as a result of being crimped around a sharp corner or past a piece of furniture. A hose pinched under a door, under a piece of furniture, under a vehicle tire, or under fallen debris could also reduce water flow at the nozzle. Water flow through the hose could be fully blocked by a closed in-line valve, debris in the hose, or by a closed nozzle bale that cannot be opened by an incapacitated fire fighter. The fully blocked hose would appear at the pump panel to have sufficient pressure.

1.4 The Goal and Objectives of This Chapter

The safest way for the IC to know that water is flowing from the hose nozzle is to have real-time water flow information sent to them. The goal of this chapter is to provide an overview of the development of a wireless-sensor network to measure water flow in a fire hose. Our approach is to collect vibration data, send it to the IC's laptop, and then convert it into a flow rate based on data-derived correlations. Flow rate will be determined in real-time to improve fireground situational awareness. We envision that a robust, wireless, flow apparatus includes the following features:

1. measures water flow through a fire hose or fire suppression water pipe
2. is portable, small, and durable
3. has a flexible, easy, external attachment at any location along a fire hose or pipe
4. is easy for the incident commander to determine water flow via a portable electronic device.

Such a device would improve fire fighting situational awareness and provide additional information for an incident commander to determine which fire hoses are supplying water, and how much, at a complex incident scene.

The objectives of this chapter are to highlight how the flow metric was determined [3, 7], highlight the wireless flow apparatus [8, 9], highlight improvements made to the system [10], and provide additional insights for system performance in realistic use scenarios [10].

2 Development of the Data-Driven Sensor Network

2.1 Background Literature on Sensing Technology

Different types of sensors have been attached to the outside surface of rigid pipes to measure fluid or gas flow. One type of sensor is an accelerometer. Accelerometers can be attached securely to the outside surface of a rigid pipe and detect the vibration of the fluid flow within the pipe. Different metrics corresponding to the vibrations can then be correlated with the flow rate.

Vibrations from fluid flow can occur in any relatively rigid pipe with fluid, such as water and oil flowing through utility or industrial pipes. Studying pipe vibration is important for utility pipelines because over a long period of time the vibrations can result in fatigue loading and eventually lead to pipe failure. Research has been conducted to measure fluid flow in utility pipelines using accelerometers [11–17]. The study of vibration in flexible pipe like a fire hose has received much less attention.

There are advantages to using external sensors such as accelerometers to measure flow inside a pipe. Accelerometers do not obstruct flow or become contaminated by coming into contact with the fluid inside the pipe. They are easy to install, relocate, maintain, or replace because they are accessible on the outside of a pipe. They are also small and lightweight. In the case of a fire hose, it is important that an external sensor not obstruct water flow and not be heavy or bulky.

Water flowing through a flexible fire hose is turbulent flow and consists of a flow field of eddies. The turbulent kinetic energy of these eddies is transferred from large eddies into smaller eddies whose energy then affects the pressure against the pipe wall. The pressure fluctuations induce pipe vibrations, and accelerometers are sensitive to these types of pipe vibrations.

Two groups of sensors used in the literature for measuring fluid flow in rigid pipes are piezoelectric pressure transducers and piezoelectric accelerometers. Clinch used a cluster of piezoelectric pressure transducers located flush along the inner surface of a water pipe to detect pressure fluctuations from turbulent water flow [18]. A relationship was determined between pressure fluctuations and flow rate. Flow velocities ranged from 5 m/s to 13 m/s.

A system of three accelerometers was used by Kim and Kim to measure the flow rate on the exterior of a pipe using an external excitor to generate vibrations [19]. Three accelerometers at different distances from the excitor measured the vibration and estimated flow rate based on changing wave number. Sources of error were considered, such as accelerometer spacing, relative accelerometer position, phase mismatch and magnitude mismatch between accelerometers, and measurement noise.

Their experimentally determined flow rate was within 12% of the 'true' flow rate. The internal flow rate was between 0 and 3.08 L/s.

A technique to measure the flow rate from a pipe exterior using accelerometers was developed by Evans et al. [20] and Pittard et al. [21]. A piezoelectric accelerometer attached to the outside surface of a pipe was used to detect turbulent flow-induced pipe vibrations. Data was collected during a 2 s time span at a sampling frequency of 2048 Hz. Fourteen flow rates were used between 400 L/min and 1500 L/min. The standard deviation of the time-series signals increased with flow rate.

Evans et al. also looked at peak frequency shifts with flow rate using accelerometers [20]. As the flow rate increased, there was a small decrease in peak frequency. The decrease in peak frequency with increasing flow rate in pipes was also noted by Blevins [22]. An increase in peak amplitude with increasing flow rate was found. Pipe material properties and diameter were shown to affect the vibration and flow rate relationship.

Accelerometers were used by Thompson et al. to determine a relationship between pipe wall vibration and average flow speed, pipe diameter, pipe thickness, and flow passing through baffle plates with various hole sizes [23]. The root mean square of the pipe wall acceleration increased nearly quadratically with the average fluid speed (3 m/s–11.5 m/s) and increased with decreasing pipe wall thickness. For the baffle plate experiments, the magnitude of the vibration decreased with increasing downstream distance from the turbulence source.

Dinardo et al. [24] and Campagna et al. [25] used a laser Doppler vibrometer (LDV) to measure pipe vibrations due to internal fluid flow. Dinardo et al. showed a linear relationship between the first harmonic amplitude of the transverse acceleration and the flow rate inside a pipe. Campagna et al. advanced this work and determined that sensitivity of the measurement is influenced by pipe material (steel and PVC) and pipe size. They also showed for a narrow range of increasing flow rates, the peak harmonic frequency decreased.

Medeiros et al. [26] and Medeiros et al. [27] used an accredited flow meter lab to evaluate Evans et al. [20] flow rate measurement technique using an accelerometer attached to the surface of a water pipe for application to the oil and water industry. The flow rate is estimated from the standard deviation of the accelerometer measurement. Three measurements were taken at each flow rate and the flow rate was also measured by a standard Coriolis flow meter for comparison. Signals caused by the effects of the pump, air compressor, and resonance frequency of the pipeline were discarded by filtering.

The standard deviation of the voltage from both pressure and acceleration sensors had an increasing relationship with flow rate, as shown by Lannes et al. [28] and similarly shown by others for accelerometers [20, 21, 23, 26, 27]. The flow rate was measured in both straight pipe and pipe bends. The piezoelectric sensors were embedded close to the inner surface of the pipe to increase their sensitivity to the pressure fluctuations from turbulent flow while an accelerometer was attached to the surface of the pipe to detect vibration in the vertical direction for comparison to the pressure sensors. The increase in turbulent flow at the pipe bends increased the sensitivity of the piezoelectric sensors and accelerometers to changes in flow rate [28].

2.2 *Determining Flow Metric with Wired Accelerometer Experiments*

The initial experiments were focused on using wired piezoelectric accelerometers as the sensors to determine flow rate near the nozzle of a fire hose (Fig. 4). The accelerometers were epoxied to the exterior of a nominal 4.5 cm (1.75 in) inner diameter, 15 m (50 ft), fire attack hose. The flow rate was lowered in 19 LPM (5 GPM) increments from 606 LPM (160 GPM) until 0 LPM as measured by a commercial flow meter (the abbreviation 'LPM' represents liters per minute (L/min) and 'GPM' represents gallons per minute included throughout this text because it is conventionally used by the United States fire service).

Raw accelerometer data was used to calculate the standard deviation at each flow rate (Fig. 5a). Previous literature has shown that the standard deviation of acceleration increased as flow rate increased in rigid pipes [20, 21, 23, 26–28]. This relationship was found to be similar for flexible fire hose at lower flow rates, from 0 LPM to at least about 300 LPM (80 GPM), however at higher flow rates, above 300 LPM (80 GPM), the standard deviation of acceleration became unpredictable, often peaking and decreasing [3, 8]. The bell-shaped curve therefore made the standard deviation of acceleration metric not suitable as the only metric for determining the flow rate in a flexible fire hose (Fig. 5b).

Another metric chosen was the dominant frequency at each flow rate. The time-series data was converted to the frequency-domain using a Fast Fourier Transform (FFT). A dominant frequency at each flow rate was determined based on the peak frequency from the power spectrum (Fig. 6). For relatively rigid pipe, previous research has shown a decreasing dominant frequency with increasing flow rate [20, 22, 25]. A similar relationship was also found for flexible fire hose [3, 8]. Dominant

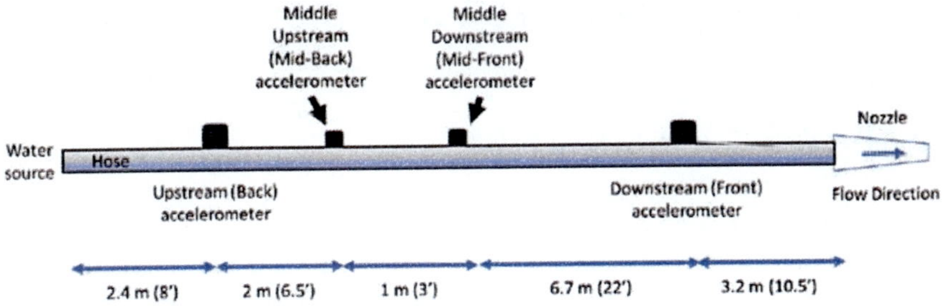

Fig. 4 The four accelerometer locations along the 15 m (50 ft) fire hose [3]

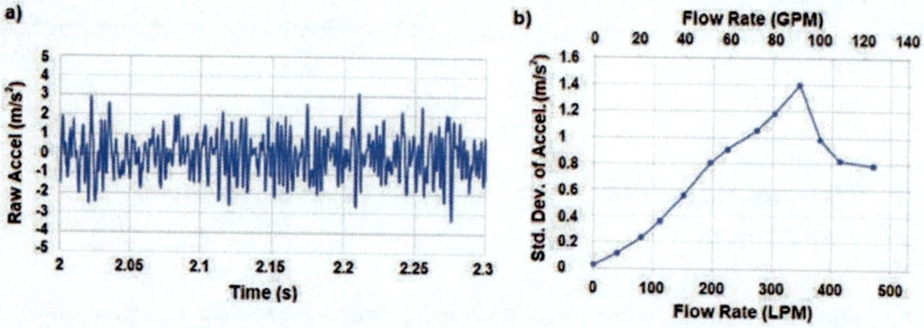

Fig. 5 **a** A sample of raw acceleration data in a time-series plot from an accelerometer during a flow test (379 LPM (100 GPM)) for a 0.3-s-long time period and **b** the standard deviation of the acceleration signal for the entire flow range

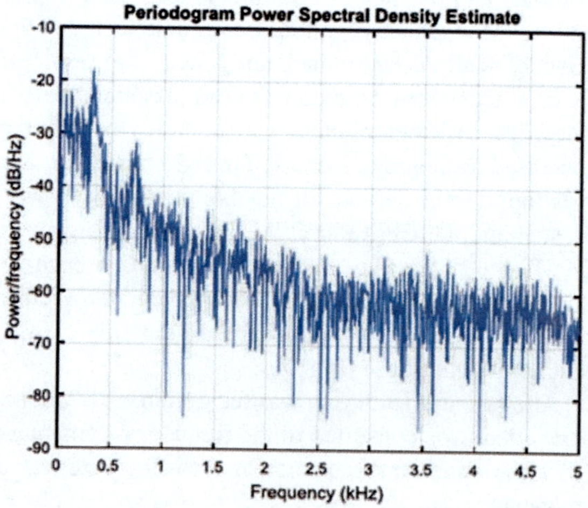

Fig. 6 A sample of the dominant frequency based on the power spectrum. A peak can be seen around 300 Hz for a flow rate of 606 LPM (160 GPM)

frequency determined by impact testing matched the dominant frequency as determined by flow tests. Dominant frequency was therefore chosen as the primary metric to determine flow rate for the fire hose in subsequent experiments.

2.3 *Wireless Sensor Network and Deployment*

A wireless sensor network was applied with the dominant frequency metric and other measures to robustly determine water flow rate in a fire hose in real-time to improve fireground situational awareness [8, 9]. The advantages of wireless sensor networks (Fig. 1) have resulted in their application to rigid pipes and utility pipelines to detect flow and assess pipeline health where long distances of pipe can be checked from fewer remote locations [12, 14, 15, 29, 30]. Wireless technology has been applied to fire hoses. An in-line, turbine, commercial wireless flow meter is available for fire hoses that transmits data to a hand-held device using Bluetooth[1] that reports the flow rate [31].

Wireless Sensor Network (WSN) Components. The wireless sensor network for this set of experiments had three main components: (1) the accelerometer (PCB model 352C33[1]) as the sensor attached to the exterior fabric of the fire hose (similar to that shown in Fig. 4), (2) the wireless node (LORD MicroStrain IEPE-LXRS[1]) (Fig. 7), and (3) the base station (Fig. 7) connected to, and powered by, a laptop via a USB port. The sensor collects data, the wireless node transmits the data, and the base station receives the data and passes it to a laptop for analysis.

Wireless Experiment Setup. A commercial fire attack hose (approximately 15 m (50 ft) long, nominal 4.5 cm (1.75 in) inner diameter) was used for the wireless experiments. Water was supplied to the hose from a building interior water supply. A commercial turbine flow meter was used to determine the actual flow rate for the experiments. The hose remained in a stationary position resting on the ground throughout the experiments.

A data sampling rate of 1 kHz was applied while data was sampled for 10 s every 30 s. The node takes time to transmit data depending on the amount of sampled data and the transmit rate [32]. The transmit rate for the nodes was automatically set based on the time increment between samples. The data received by the base station was passed to the software in nearly real-time acquisition.

The water flow rate was measured about every 19 LPM (5 GPM) from approximately 606 LPM (160 GPM) to 0 LPM (Fig. 8). Each flow rate was maintained for approximately 5 min for a minimum of 30 consecutive data sampling bursts at a single flow rate. The nozzle was closed for several data bursts between each flow rate. The standard deviation of acceleration can be seen to peak around 398 LPM (105 GPM) for all three sensors in the time-series plot from one experiment (Fig. 8).

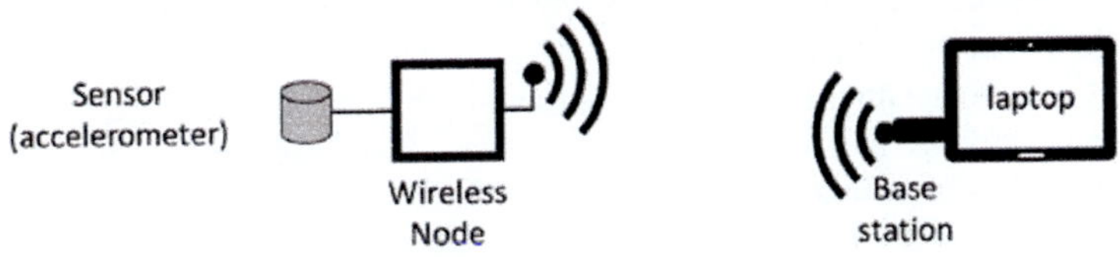

Fig. 7 The wireless sensor network components: (1) the accelerometer sensor, (2) the wireless node, and (3) the base station at the laptop [8]

[1] *Certain commercial entities, equipment, or materials may be identified in this document in order to describe an experimental procedure or concept adequately. Such identification is not intended to imply recommendation or endorsement by the National Institute of Standards and Technology, nor is it intended to imply that the entities, materials, or equipment are necessarily the best available for the purpose.*

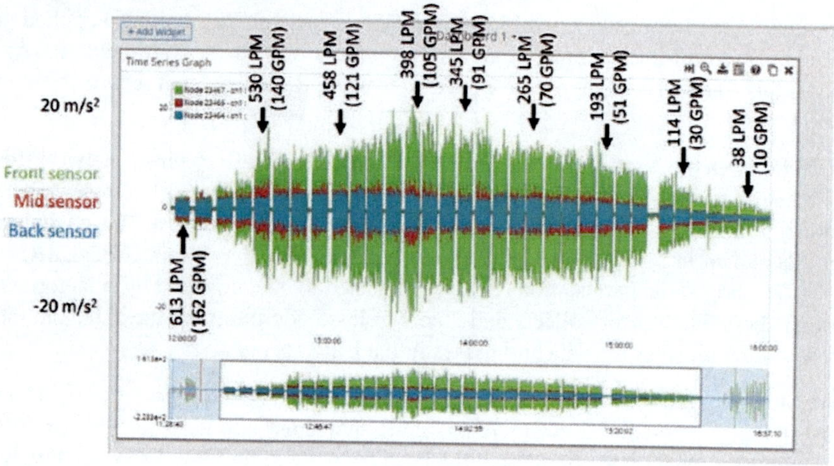

Fig. 8 Standard deviation of acceleration data collected in a time-series plot for the entire water flow range [8]

Dominant Frequency Curve. Based on the previous experiments using the wired accelerometers, the primary metric of interest to determine flow rate was the dominant frequency of the hose [3]. The time-domain accelerometer data (Fig. 5a) was converted to the frequency-domain using a Fast Fourier Transform (FFT) (Fig. 6). An average dominant frequency at each flow rate was determined based on the peak frequency from the power spectrum.

The raw accelerometer time-series data from the Front accelerometer was collected and processed in the custom data acquisition program. The software converted the time-series data to the frequency domain in real-time. A Savitzky-Golay filter [33] was used to smooth the frequency domain data in which the dominant frequency was identified for each flow rate.

The use of filtering techniques resulted in reduced data scatter and standard deviation for the dominant frequency at each flow rate. The standard deviation bar at zero flow is large because the signal noise instead of the flow in the hose dominated the signal at zero flow. Other limits were used within the software to identify the zero-flow condition since the dominant frequency was inconsistent due to signal noise. Figure 9 was used to determine the flow rate in real-time based on the dominant frequency of the water flowing through the hose during subsequent flow tests.

Human Perception. Due to the complexity on the fireground, communicating data driven fire hose flow to an IC using a hand-held device is critical. A two-light indicator was applied as a means of visually and quickly communicating the fire hose flow to the IC (Fig. 10). The IC can refer to the two blue lights and know that the attack hose has a flow rate above 140 GPM (530 LPM) at the nozzle (or any preset water flow goal), the water flow rate goal for a fire hose when attacking a single-family dwelling fire [34].

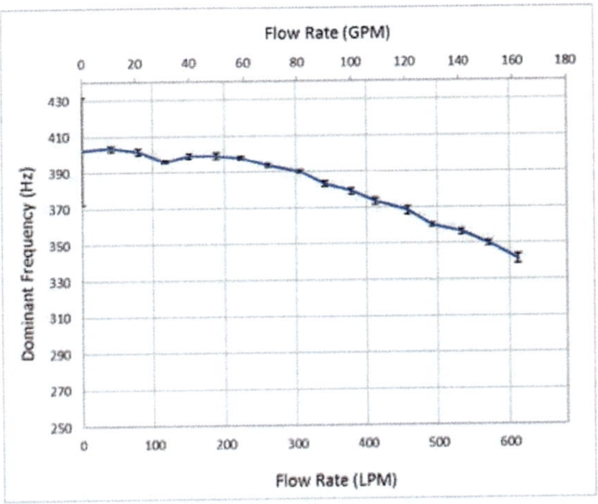

Fig. 9 Dominant frequency with standard deviation bars versus flow rate data after modifying the data sampling parameters resulting in reduced scatter, except at zero flow where the signal noise dominated causing scatter

Fig. 10 The two-light indicator for quick reference, no lights indicate no-flow, one blue light indicates flow less than 140 GPM (530 LPM), two lights indicate flow exceeds 140 GPM (530 LPM) (the United States fire service uses GPM for hose flow rate by convention however this display can be modified to report either GPM or LPM)

A colored flow bar displays the flow rate numerically if additional detail is needed by the IC (Fig. 11). The bar was red when the flow rate was less than 50 GPM (189 LPM) (dominant frequency between 399 and 403 Hz), yellow between 50 GPM (189 LPM) and 140 GPM (530 LPM) (dominant frequency between 399 and 356 Hz) and green when the flow rate was 140 GPM (530 LPM) or above (dominant frequency 356 Hz or less).

Real-Time Fire Hose Flow Rate. Data processing software was used to receive the raw time-series accelerometer data and convert the time-series to frequency-series using an FFT. The dominant frequency was then determined by the processing software to identify the flow rate using Fig. 9. The graphical user interface (GUI) for the software, includes the two-light indicator (Fig. 10) and the colored flow bar

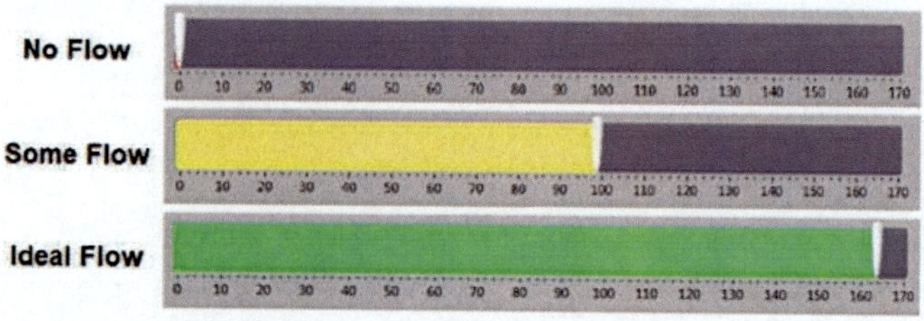

Fig. 11 The colored flow bar meter indicates the flow rate in GPM, red for flow less than 50 GPM, yellow for flow between 50 and 140 GPM, and green for flow above 140 GPM

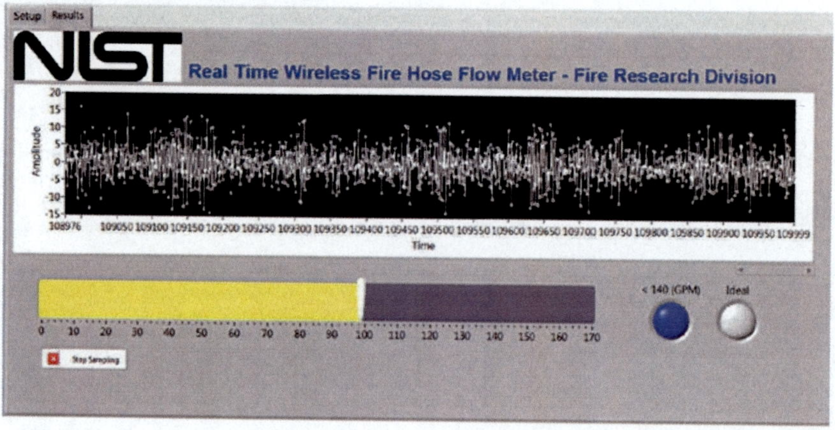

Fig. 12 The Flow dashboard for 100 GPM with yellow colored bar meter and with one blue light illuminated

meter (Fig. 11), and is shown in Fig. 12. The full GUI dashboard is intended for research purposes and not intended for the IC.

2.4 *Realistic Firefighting Conditions*

Additional experiments were conducted after making improvements to the flow apparatus hardware, such as using a smaller accelerometer, and modifying the apparatus so that it could be easily removed and attached to any location along the fire hose. Unlike the previous experiments where the hose remained stationary on the ground, the wireless flow apparatus was then applied to realistic firefighting hose conditions to evaluate the upgraded apparatus function and determine the influence of the hose motion, accelerometer location, and nozzle spray on the dominant-frequency metric [10].

Specifically, these additional experiments explored:

1. how the wireless flow apparatus functioned under realistic firefighting hose movements (holding the hose nozzle, nozzle motion during simulated fire suppression, hose motion during simulated hose dragging)
2. if the system responded differently at different locations along a fire hose (front, middle, back)
3. if the system was affected by the type of nozzle stream (straight stream versus a fog stream).

Upgraded Hardware Attachment. During previous experiments [3, 8] the accelerometer base was epoxied to the outer surface of the hose which limited the position of the accelerometer along the hose. For these current experiments, the accelerometer was epoxied to an aluminum alloy base to provide more flexibility with the accelerometer location on the hose (Fig. 13). The aluminum base was strapped tightly with Velcro to the hose. The accelerometer was protected with a plastic cover attached to the aluminum base. The removable sensor unit (aluminum base, accelerometer, and plastic protective cover) offered flexibility to attach and remove the accelerometer anywhere along the hose.

Upgraded Accelerometer. Previously, a larger accelerometer (PCB Piezotronics 352C33) was used to collect the vibration data. For these experiments, a smaller accelerometer (PCB Piezotronics 352A24) was applied to reduce weight and reduce the size of the accelerometer for practical field application. The profile of the sensor unit including the protective plastic cover was therefore reduced.

Hose Conditions. The same commercial fire-suppression hose with nominal 4.5 cm (1.75 in) inner diameter, and approximately 15 m (50 ft) long, was used as described in previous experiments (Sect. 2.2). The sensor unit was strapped to three different hose locations: front, middle, and back. The front location was approximately 2.4 m (8 ft) from the hose nozzle. The middle location was approximately the middle of the hose

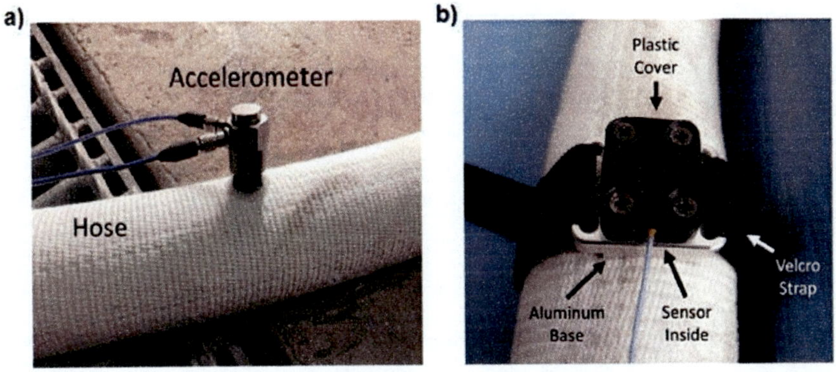

Fig. 13 **a** The accelerometer epoxied to the fire hose exterior and **b** the removable accelerometer strapped to the fire hose exterior. The sensor unit included the accelerometer epoxied to an aluminum base under a plastic protective cover. A Velcro strap secured the sensor unit to the fire hose exterior

length, or 7.6 m (25 ft) from the hose nozzle. The back location was approximately 2.4 m (8 ft) from the upstream hose attachment point to the water source. A typical fog-style hose nozzle was used to set either a straight stream or fog stream as well as to change the flow rate. A commercial turbine flow meter was attached between the nozzle and the fire hose to measure the actual water flow rate.

Previous experiments were conducted with the hose in a stationary position on the ground in a straight stream configuration. For these current experiments, data was collected with the nozzle set for a straight stream or a fog stream for several hose conditions: hose stationary on and above the ground, hose nozzle motion simulating fire suppression, and hose motion on the ground simulating dragging.

Test Matrix. The following table summarizes the number of experiments for each hose location, nozzle stream, and hose condition (Table 2).

Data Collection Settings. Besides the hose conditions, several other experimental parameters were modified compared to the previous experiments. The approximate flow rates used for these current experiments ranged from 0 LPM to approximately 568 LPM (150 GPM), in increments of approximately 114 LPM (30 GPM), instead of increments of 19 LPM (5 GPM) as was done previously. The fewer number of

Table 2 The number of experiments conducted for the various hose conditions

Accelerometer location on hose	Nozzle stream	Hose condition	Number of experiments
Front	Straight	Stationary—On ground	6
	Fog	Stationary—On ground	5
	Straight	Stationary—Holding	13
	Fog	Stationary—Holding	11
	Straight	Motion—T-Pattern	6
	Fog	Motion—T-Pattern	4
Middle	Straight	Stationary—On ground	4
	Fog	Stationary—On ground	3
	Straight	Motion—Left/Right	5
	Fog	Motion—Left/Right	3
	Straight	Motion—Dragging	3
	Fog	Motion—Dragging	3
Back	Straight	Stationary—On ground	4
	Fog	Stationary—On ground	4
	Straight	Motion—Left/Right	3
	Fog	Motion—Left/Right	3
	Straight	Motion—Dragging	3
	Fog	Motion—Dragging	3

flow rates helped streamline the experiments. In the future, if more flow rate data are needed, then more flow rates will be applied.

For these current experiments, one data-collection arrangement was used, a 10 kHz sampling frequency with a 1 s burst of data every 30 s. Initially 5 bursts of data were collected at each flow rate, but this was increased to 10 bursts to be able to discard outliers in the data and still have 5 data bursts.

As described previously, the time-domain, acceleration data collected from the accelerometers from the 10 data bursts were converted to the frequency-domain using a Fast Fourier Transform (FFT) (Sect. 2.2) (Fig. 6) and an average dominant frequency at each flow rate was calculated. A decreasing trend for dominant frequency was noted in the previous studies with increasing flow rate.

Stationary—Hose on Ground. For the stationary hose condition on the ground, the accelerometer was positioned at the front, middle, and back locations. The accelerometer at the front and back hose locations was attached to the hose on concrete, while the hose was on grass at the middle hose location. The nozzle was elevated by a cinder block, hanging off and not contacting the block, to prevent impacts between the nozzle and concrete that might have been detected by the accelerometer. For this hose condition, the water flow vibrations detected by the accelerometer may be affected by the ground, either concrete or grass. This hose configuration simulated a fire fighter holding the hose nozzle while the remaining hose including the accelerometer was on the ground.

Stationary—Holding. For the hose condition where the hose was held stationary, the accelerometer was only attached at the front hose location. The hose and nozzle were elevated off the ground, either held by a person, or the sensor was suspended between cinder blocks to simulate a person holding the hose and nozzle. When the nozzle was suspended on blocks, the nozzle was elevated above the concrete to reduce water flow vibrations between the nozzle and concrete that might be detected by the accelerometer. For this hose condition, the water flow vibrations should not be affected by the ground, as in the stationary hose on ground condition.

Motion—Moving Nozzle in T-Pattern. For this condition, the accelerometer was in the front location, elevated off the ground, with the nozzle held by a person, and a second person holding the hose upstream of the sensor such that the sensor was suspended between holding points. In most cases, the hose was supported by two blocks with the sensor suspended between the blocks to simulate two people holding the hose. For either case, the hose nozzle was moved in a T-pattern to simulate the lead fire fighter directing water on a room fire with a backup fire fighter supporting the hose. The nozzle was elevated above the concrete to reduce water flow vibrations between the nozzle and concrete that might be detected by the accelerometer. Motion of the hose was maintained during the data acquisition phase for the accelerometer. This hose condition added motion to the previous hose holding condition by introducing only the effect of the hose motion to the water flow vibrations.

Motion—Moving Hose Forward and Backward. For this condition, the accelerometer was in either the middle or back hose location, with the accelerometer attached to the hose resting on the grass or concrete. The hose was moved forward and backward in line with the flow direction to simulate the hose being dragged on the ground. Motion of the hose was maintained during each data acquisition phase for the accelerometer for approximately 1 s. This hose condition was similar to the stationary hose on ground condition but with the additional hose motion that may affect the vibrations detected by the accelerometer.

Motion—Moving Hose Left and Right. For this condition, the accelerometer was in either the middle or back hose location, with the accelerometer attached to the hose resting on the grass or concrete. The hose was moved left and right, perpendicular to the flow direction, to simulate the hose being moved or dragged on the ground. The left and right motion of the hose at various frequencies was maintained during the data acquisition phase for the accelerometer. This hose condition was similar to the forward and backward hose movement, but the motion was perpendicular to the water flow direction.

Data Collection with Sensor at Front Hose Location—Stationary Holding. For the stationary, hose-holding condition, there were 13 experiments with the straight stream (Fig. 14) and 11 experiments with the fog stream (Fig. 15). Each experiment included five (5), 1 s, data collection bursts for a total of 65 dominant frequencies for each flow rate for the straight stream and 55 dominant frequencies for the fog stream. For the holding condition, the hose was either held by a person, or suspended between two blocks, so the accelerometer was suspended off the ground to simulate the nozzle being held by a fire fighter. Depending on the holding configuration, the accelerometer was either approximately 1 m (3 ft) (person holding) or approximately 2.4 m (8 ft) (when suspended) from the hose nozzle.

Both the straight and fog stream experiments showed trends towards a decreasing dominant frequency with increasing flow rate, however because of the variability of

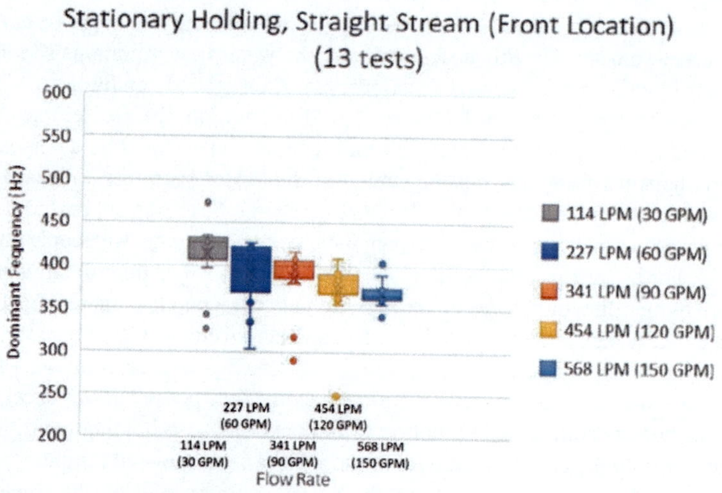

Fig. 14 The dominant frequency for each flow rate with the accelerometer at the front hose location while the hose was held with a straight stream

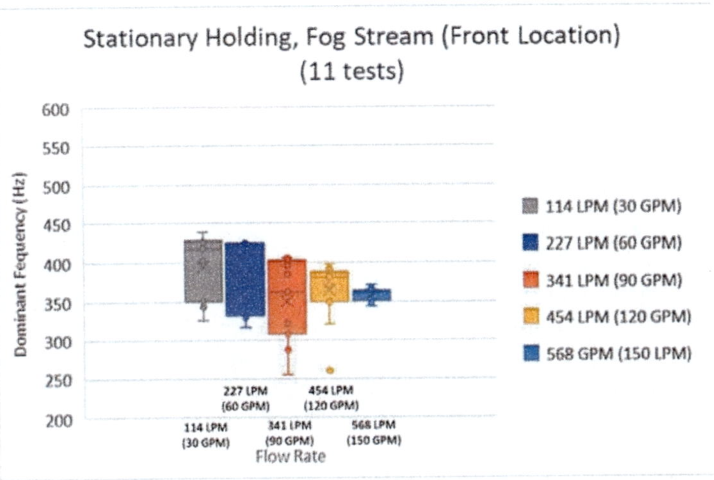

Fig. 15 The dominant frequency for each flow rate with the accelerometer at the front hose location while the hose was held with a fog stream

the data from one flow rate to the next, there was not a unique dominant frequency for each flow rate. The fog stream data showed more variability than the straight stream data.

Please refer to [10] for results and discussion for the other hose conditions and locations as noted in Table 2.

2.5 *Concluding Remarks*

The goal of these experiments was to develop a robust, wireless, flow apparatus to measure water flow through a fire hose and communicate the flow to an incident commander to improve firefighting situational awareness. The wireless flow apparatus was used with realistic firefighting hose movements (i.e., holding the hose nozzle, nozzle motion during simulated fire suppression, simulated hose dragging) to evaluate the upgraded apparatus function and determine the influence of the hose motion, sensor location, and hose spray, on the dominant frequency metric.

The sensor unit, including the accelerometer, aluminum alloy base, and plastic cover, was lightweight, easily portable, and small. The separate wireless node however was still large. Neither the sensor unit nor the wireless node is capable of withstanding high temperatures as would be expected on the fireground, or wet conditions that could compromise the electrical components of the sensor unit or the wireless node. Fireground hardening of the apparatus is still required.

The sensor unit was flexible and easily tightly attached and removed at any location along the fire hose using a Velcro strap. The strap also allows the sensor unit to be attached to different diameter fire hoses or rigid, water suppression, pipes. This was a significant advancement over the previous design where the accelerometer was epoxied to the exterior hose fabric. Based on the results, no loss of vibration signal was observed using the strap.

These experiments continued to show that the wireless flow apparatus on the hose exterior was capable of indicating that water was flowing through a flexible fire hose, however measuring the flow rate was still challenging. In the experiments, although a decreasing trend in dominant frequency with increasing flow rate was observed, a unique dominant frequency was not able to be determined for a given flow rate because of the data variability. One frequency would often correspond to multiple flow rates and therefore, a clear metric for measuring water flow is still undetermined. A GUI via a portable laptop computer was used during the experiments to observe that water was flowing through the fire hose, and the same, or similar, GUI could easily be used by an incident commander.

Some general observations were made from the experiments. The type of hose motion (dragging or fire suppression) did not appear to influence the dominant frequency. Experiments with the sensor elevated above the ground resulted in less variability than experiments with the hose resting on the ground. The hose nozzle stream type, either straight or fog, did not appear to influence the dominant frequency. Variability in the data at the back hose location was greatest and suggests there may be a location dependent difference in the dominant frequency metric for flexible fire hose. The results from these experiments showed improvement of the wireless, flow apparatus hardware, but additional research is still needed to understand the dominant frequency metric.

Further data analysis to evaluate other metrics, or combination of metrics, to determine flow rate may yield an improved relationship for flow rate. An effective field tool to assist the fire service may not require that the flow rate be determined as a value. Knowing that the flow rate is above or below a threshold is still valuable field data to distinguish if water flow rate is sufficient, or whether or not water is flowing at all. The results of these experiments suggest that comparing the dominant frequency of similar flow rates does not result in a unique flow rate. The same dominant frequency could often be interpreted as multiple flow rates. However, in some cases there was a difference between the dominant frequency at the lowest flow rate of 114 LPM (30 GPM) which was often greater than 400 Hz and the highest flow rate of 568 LPM (150 GPM) which was often less than 400 Hz. Although not ideal, a comparison between only the lowest and highest flow rates, using the extreme ends of the dominant frequency curve, could be used to give a no-flow (0 Hz), low-flow (< 400 Hz), or sufficient water flow (> 400 Hz) signal, to improve situational awareness.

3 Practical Considerations for a Nozzle Flow Meter and Future Research Needs

The following are considerations for applying a wireless sensor network to a fire hose.

1. Power—The sensor, wireless nodes, and base station required power. In this study, power was supplied to the accelerometers from the nodes, and from the laptop to the base station. For the fire hose application, the wireless nodes and sensor will need a source of power. Replaceable or rechargeable batteries may work.
2. Hostile environment—The wireless sensor network in this study was applied to the fire hose in a non-hostile, outside, environment. The wireless sensor network on a fire hose on the fireground will need to be capable of withstanding high temperature, water, and must be durable to survive being dragged on the ground while attached to the exterior fabric of the fire hose. The protective case used in this study was intended to provide only minimum protection during testing.
3. Equipment size—In this study, the accelerometer and the wireless node were separate, and the node was too large to be applied directly to a fire hose. Combining the accelerometer and the wireless node together and reducing the node size would be required for adoption to an actual fire hose. Although already sufficiently lightweight, a smaller single unit containing both the sensor and node with a lower profile would be less likely to be damaged during use on the fireground.
4. Signal strength and range—The wireless node in this study used radio frequency (RF) signals to transmit data to the base station. In order to maximize the signal range, physical barriers should be minimized, radio frequency interference and electromagnetic interference should be minimized, and metal near the node or base station antenna should be minimized [32, 35]. For an actual fire hose application, the node will not likely have a clear line of sight to the base station, either due to structural obstructions like interior and exterior walls, floors, ceilings, or doors. Ground terrain may prevent a clear line of site in a forest or if the node is underground in a basement. There may also be civilian sources of RF signals as well as radio interference from fire department sources that could interfere or weaken the RF signal from the flow meter. Preventing the loss of a signal due to these factors will be important.
5. Metric robustness—The results from these experiments showed that additional research is still needed to understand and improve the dominant frequency metric. The application of a machine learning (ML) model to predict the flow rate based on the dominant frequency metric, or combination of metrics, may help.
6. A flow meter at the fire hose nozzle could have additional benefits besides the application for structural firefighting. Increased situational awareness regarding nozzle water flow could likewise benefit the fire service in extremely hazardous environments such as chemical, radiation, or explosive environments, where humans do not have sufficient protection. A flow meter at the hose nozzle of a charged hose line with a robot in these environments could provide feedback

when beyond the IC's line of sight. A robot mounted camera may not be able to visually provide confirmation of water flow in smoky or dark conditions that block the camera from capturing images.

7. Situational awareness is also important for wildland fire fighting. Measuring water flow at many locations along a forestry hose would improve situational awareness where a forestry hose line may be stretched unattended and out of sight for hundreds of feet along a fire line. Typically, not all wildland fire fighters carry radios so communication with the engine pump operator could be difficult in rolling hills or forest. It is easy to consider that a wildland attack hose may develop a hose leak due to a puncture from ground debris or develop a burn hole from a spot fire or flying embers. A sensor at multiple nozzles or hose locations along the hose line could measure flow rate for the pump operator and improve situational awareness.

Two areas are considered for future research efforts to advance this proof-of-concept system by continuing to evaluate the relationship between the dominant frequency and the flow rate. A wired sensor system, as used previously (Sect. 2.2), would allow more vibration data to be collected without the constraints involved using data bursts with the wireless system. A larger volume of data would allow for an improved average acceleration determined for each flow rate. It is possible that this could improve the dominant frequency metric. The dominant frequency results from previous experiments (Sect. 2.3) had less variability than the results from the current experiments and additional research is needed to understand the differences. Once the dominant frequency relationship to flow rate is refined with a larger data set, the relationship could then be applied to the more practical smart wireless sensor system to evaluate if it is suitable.

The second area for future work is the application of a machine learning (ML) model to predict the flow rate based on the dominant frequency metric, or combination of metrics. Existing data would be used to train a ML model and could be supplemented with additional experimental data if more is required. The wired sensor system may be helpful to collect larger data sets for training the model. Lastly, the model's performance would be tested in the field and the robustness evaluated.

References

1. A.M. Petrillo, Hoses: Large and Small, from LDH to Forestry Lines. Fire Appar. Emerg. Equipment. **24**(6), (2019). https://digital.fireapparatusmagazine.com/fireapparatus/june_2019/MobilePagedReplica.action?pm=2&folio=16#pg16, Last accessed 6 May 2023
2. A. Hamins, C. Grant, N. Bryner, A. Jones, G. Koepke, Research roadmap for smart fire fighting (National Institute of Standards and Technology, Gaithersburg, MD, NIST Special Publication (SP) 1191, 2015). https://doi.org/10.6028/NIST.SP.1191
3. C.U. Brown, G.W. Vogl, W.C. Tam, Measuring flow rate for a fire hose using wired accelerometers for smart fire fighting (National Institute of Standards and Technology, Gaithersburg, MD, 2019), NIST Technical Note 2075. https://doi.org/10.6028/NIST.TN.2075

4. L. Moore-Merrell, S. Kerber, G.P. Horn, D.L. Smith, Effects of crew size on firefighter health and safety. Int. Fire Serv. J. Lead. & Manag., **15**, (2021)
5. K. Curran, *5 Investigates discovers key firefighting tool has failed in deadly fires across country* (WCVB5, 2014). https://www.wcvb.com/article/5-investigates-discovers-key-firefighting-tool-has-failed-in-deadly-fires-across-country-1/8211747. Last accessed 2023/4/12
6. Fire Engineering, *Hose failure in fatal boston fire raises concerns* (2014). https://www.fireengineering.com/firefighting/hose-failure-in-fatal-boston-fire-raises-concerns/#gref. Last accessed 12 April 2023
7. C.U. Brown, G. Vogl, W.C. Tam, Measuring Water flow rate in a flexible fire hose using an accelerometer. in Suppression, detection and signaling research and applications symposium proceedings (SUPDET), Sept 17–19 (Denver, CO, 2019). https://www.nfpa.org/-/media/Files/News-and-Research/Resources/Research-Foundation/Symposia/2019-SUPDET/Presentations/SUPDET19BrownVoglTam-paper.ashx
8. C.U. Brown, G.W. Vogl, W.C. Tam, Measuring water flow rate for a fire hose using a wireless sensor network for smart fire fighting (National Institute of Standards and Technology, Gaithersburg, MD, 2019), NIST Technical Note 2074. https://doi.org/10.6028/NIST.TN.2074
9. C.U. Brown, G.W. Vogl, W.C. Tam, Measuring water flow rate for a fire hose using a wireless sensor network for smart fire fighting, Fire Technology (Special Issue on Smart Systems), **57**, 3125–3250 (published online Jan 2, 2021). https://doi.org/10.1007/s10694-020-01054-1
10. C.U. Brown, G.W. Vogl, W.C. Tam, *Measuring water flow rate for a fire hose using a wireless sensor network for smart fire fighting, an update using various hose conditions* (National Institute of Standards and Technology, Gaithersburg, MD, NIST Technical Note 2245, 2023). https://doi.org/10.6028/NIST.TN.2245
11. G. Park, H.H. Cudney, D.J. Inman, Feasibility of using impedance-based damage assessment for pipeline structures. Earthquake Eng. Struct. Dynam. **30**, 1463–1474 (2001). https://doi.org/10.1002/eqe.72
12. Y. Kim, T. Schmid, Z.M. Charbiwala, J. Friedman, M.B. Srivastava, NAWMS: nonintrusive autonomous water monitoring system, in Proceedings of the 6th ACM conference on Embedded network sensor systems 2008 (Raleigh, NC, 2008), pp. 309–321
13. Y. Urthaler, E.M. Luther, L.E. Breaux, J. Austin, S. McNeill, M. Tognarelli, A methodology for assessment of internal flow-induced vibration (FIV) in subsea piping systems. in Proceedings of the ASME 2011 30th International Conf on ocean, offshore and Arctic engineering (Rotterdam, The Netherlands, 2011)
14. G. Owojaiye, Y. Sun, Focal design issues affecting the deployment of wireless sensor networks for pipeline monitoring. Ad Hoc Networks **11**, (2013). https://doi.org/10.1016/j.adhoc.2012.09.006
15. G.C. Nwalozie, A.C.O. Azubogu, Design and implementation of pipeline monitoring system using acceleration-based wireless sensor network. Int. J. Eng. Sci. **3**(9), 49–58 (2014)
16. S. Yazdekhasti, K.R. Pratla, S. Atamturktur, A.A. Khan, Novel vibration-based technique for detecting water pipeline leakage. Struct. Infrastruct. Eng. (2016). https://doi.org/10.1080/15732479.2016.1188318
17. L. Mistretta, G.C. Giaconia, A. Valenza, E. Napoli, C. Gianguzzi, M.L. Presti, F.D. Puma, Embedding monitoring systems for cured-in-place pipes, in Applications in electronics pervading industry, environment and society, lecture notes in electrical engineering 429, ed. by A. De Gloria (2018). https://doi.org/10.1007/978-3-319-55071-8_2
18. J.M. Clinch, Measurements of the wall pressure field at the surface of a smooth-walled pipe containing turbulent water flow. J. Sound Vib. **9**(3), 398–419 (1969)
19. Y.K. Kim, Y.H. Kim, A three accelerometer method for the measurement of flow rate in pipe. J. Acoustical Society of America **100**(2), 717–726 (1996). https://doi.org/10.1121/1.416211
20. R.P. Evans, J.D. Blotter, A.G. Stephens, Flow rate measurements using flow-induced pipe vibration. Trans. ASME **126**, 280–285 (2004). https://doi.org/10.1115/1.1667882
21. M.T. Pittard, R.P. Evans, R.D. Maynes, J.D. Blotter, Experimental and numerical investigation of turbulent flow induced pipe vibration in fully developed flow. Rev. Sci. Instrum. **75**(7), 2393–2401 (2004). https://doi.org/10.1063/1.1763256

22. R.D. Blevins, *Flow-Induced Vibration* (Litton Educational Publishing, NY, NY, Van Nostrand Reinhold Company, 1977), pp.287–311
23. A.S. Thompson, D. Maynes, J.D. Blotter, Internal turbulent flow induced pipe vibrations with and without baffle plates, in Proceedings of the ASME 2010 3rd Joint US-European fluids engineering summer meeting and 8th international conference on nanochannels, microchannels, and minichannels (Montreal, Canada, 2010)
24. G. Dinardo, L. Fabbiano, G. Vacca, Fluid flow rate estimation using acceleration sensors. in Seventh Int Conference on Sensing Technology (2013), pp. 221–225
25. M.M. Campagna, G. Dinardo, L. Fabbiano, G. Vacca, Fluid flow measurements by means of vibration monitoring. Meas. Sci. Technol. **26**, (2015). https://doi.org/10.1088/0957-0233/26/11/115306
26. K.A.R. Medeiros, C.R.H. Barbosa, E.C. Oliveira, Flow measurement of piezoelectric accelerometers: application in the oil industry. Pet. Sci. Technol. **33**, 1402–1409 (2015). https://doi.org/10.1080/10916466.2015.1044613
27. K.A.R. Medeiros, F.L.A. Oliveira, C.R.H. Barbosa, E.C. Oliveira, Optimization of flow rate measurement using piezoelectric accelerometers: application in water industry. Measurement **91**, 576–581 (2016)
28. D.P. Lannes, A.L. Gama, T.F.B. Bento, Measurement of flow rate using straight pipes and pipe bends with integrated piezoelectric sensors. Flow Meas. Instrum. **60**, 208–216 (2018)
29. J. Froehlich, E. Larson, T. Campbell, C. Haggerty, J. Fogarty, S.N. Patel, HydroSense: Infrastructure-mediated single-point sensing of whole-home water activity. in UbiComp 2009 (Orlando, FL, 2009), pp. 235–244
30. S. Saseendran, V. Nithya, Automated water usage monitoring system. in International conference on communication and signal processing (India, 2016), pp. 99–103
31. *Taskforce tips fire fighting equipment products website, about the sho-flow bluetooth series flow meter*, https://www.tft.com/Product-Series/SHO-FLOW-Series. Last accessed 6 May 2023
32. IEPE-Link-LXRS Wireless IEPE Sensor Node User Manual, *LORD sensing systems microStrain* (Williston, VT, 2014). http://files.microstrain.com.s3.amazonaws.com/IEPE-Link%20User%20Manual%20(8500-0043).pdf. Last accessed 6 May 2023
33. S.J. Orfanidis, *Introduction to signal processing* (Prentice-Hall, Englewood Cliffs, NJ, 1996)
34. *NFPA 1710—Standard for the organization and deployment of fire suppression operations, Emergency medical operations, and special operations to the public by career fire departments, 2016 Edition, Section 5.2.4.1* (National Fire Protection Association, Quincy, MA)
35. WSDA-200-USB Wireless USB base station user manual, LORD Sensing Systems MicroStrain, Williston, VT. https://www.microstrain.com/sites/default/files/wsda-200_usb_user_manual_8500-0077.pdf. Last accessed 6 May 2023

Digital Technologies for Fire Evacuations

R. Lovreglio, D. Paes, Z. Feng, and X. Zhao

1 Introduction

Evacuating buildings during fires is still one of the main strategies to reduce the impacts of such disasters on building occupants. Evacuations consist of a rapid relocation of humans from hazardous building areas to safer areas away from a threat. Safe areas can be either assembly points (in case of total evacuations) or safe areas within the building itself, such as adjacent fire compartments (in the case of partial evacuation of hospitals and retirement facilities) or refuge areas (in case of multistorey buildings) [1–3]. As such, measuring the evacuation performance of a building is an essential step in many performance-based regulations worldwide [4, 5]. Different solutions have been proposed in the literature and adopted in the fire industry to estimate the time required to evacuate a building. These solutions include manual engineering calculations and computer-based tools [6, 7]. The latest literature indicates that a total of 72 computer-based evacuation models are currently in use in the industry and academia [8]. On the other hand, it is also fundamental to assess if existing evacuation plans are effectively working for existing buildings and to train building occupants on how to behave during a fire disaster. In this case, fire evacuation drills represent the most used tool worldwide to assess evacuation plans and for training [9].

There has been great progresses in the evacuation field. In fact, thanks to the rapid increase in computational power, there has been a significant shift toward the development and use of computer-based evacuation simulation tools [7, 8]. However, there are still two key limitations in existing evacuation simulation models. In terms

R. Lovreglio (✉) · D. Paes · Z. Feng
School of Built Environment, Massey University, Auckland, New Zealand
e-mail: R.Lovreglio@massey.ac.nz

X. Zhao
Department of Civil and Coastal Engineering, University of Florida, Gainesville, United States

X. Huang and W. C. Tam (eds.), *Intelligent Building Fire Safety and Smart Firefighting*, Digital Innovations in Architecture, Engineering and Construction,
https://doi.org/10.1007/978-3-031-48161-1_18

of the decision-making capabilities of the simulated agents, most of the simulated evacuees' decisions are often an input of the model instead of an output of the model (see, for instance, the decision to start evacuating) [10, 11]. This represents a limitation for several reasons: it might give users the temptation to manage the scenario in order to reduce the overall evacuation time, and the simulated scenario might not capture all the dynamics that might reasonably be expected [11]. On the other hand, the navigation capabilities of simulated evacuees rely heavily on the original hydraulic equations proposed in 1950–1970, and do not account for ongoing changing population demographics and culture [12]. As such, there is a need for more studies on how to model and simulate evacuation decision-making as well as re-calibrating existing navigation models or–even better–the discovery of the *fundamental crowd behavioural rules* (i.e., rules describing how individuals make decisions about speed and direction, distance to others and obstacles, etc., in crowds). Finally, there are also well-known limitations associated with fire evacuation drills. Current drills might lack the pedagogical features required to effectively train building occupants [9]. New digital technologies could provide useful tools and solutions to overcome all these existing limitations in the evacuation field.

To date, many new technologies have been adopted to design, construct, and operate buildings [13, 14]. The technologies include mixed-reality solutions, such as Augmented Reality (AR) and Virtual Reality (VR), Building Information Modelling (BIM), Artificial Intelligence (AI) and Machine Learning (ML), among others. These technologies have been adopted in different fire engineering domains and provide promising results to enhance the safety of buildings [15]. The goal of this chapter is to provide an introduction to these technologies (i.e., AR, VR, BIM, AI and ML) and an overview of how they have been applied in the fire evacuation domain to enhance evacuation performance and achieve smart evacuation outcomes. Finally, this work discusses how some of these technologies might overlap with each other in existing evacuation applications.

This chapter provides definitions of the digital technologies under investigation in Sect. 2, while an overview of the latest application of these technologies is provided in Sect. 3. Final discussions and conclusions are provided in Sect. 4.

2 Technologies Background

This section provides definitions for the five digital technologies under investigation in this chapter. Section 2.1 illustrates the differences between Artificial Intelligence and Machine Learning, while Sect. 2.2 discusses the differences between Augmented Reality and Virtual Reality. Finally, Sect. 2.3 provides definitions related to Building Information Modelling.

2.1 Artificial Intelligence and Machine Learning

Artificial Intelligence (AI) refers to applying technologies to build smart machines, computers, and/or systems that can perform tasks that require human intelligence, e.g., being able to see, speak, analyse, understand, make decisions, etc. In turn, Machine Learning (ML) is a subfield of AI that enables machines, computers, and/or systems to automatically learn and improve from data. ML algorithms build mathematical models based on data, aiming to make predictions or decisions without being explicitly programmed to perform the task [16]. Moreover, Deep Learning (DL) is a subset of ML that is based on artificial neural networks (with at least three layers, including the input and output layers). The relationships between AI, ML, and DL are illustrated in Fig. 1.

With the rapid development of computational capabilities, AI has ushered in a new era of unprecedented technological innovations. AI has successfully been applied to nearly all the research fields, such as medical diagnoses, facial recognition, autonomous vehicles, power systems, and material discovery. For instance, ChatGPT, an AI chatbot developed by OpenAI and launched in November 2022, can engage in human-like conversations to answer questions and assist humans with tasks such as writing, coding, etc. It has attracted 100 million active users within just two months and has great potential to disrupt research and derive new research directions [17].

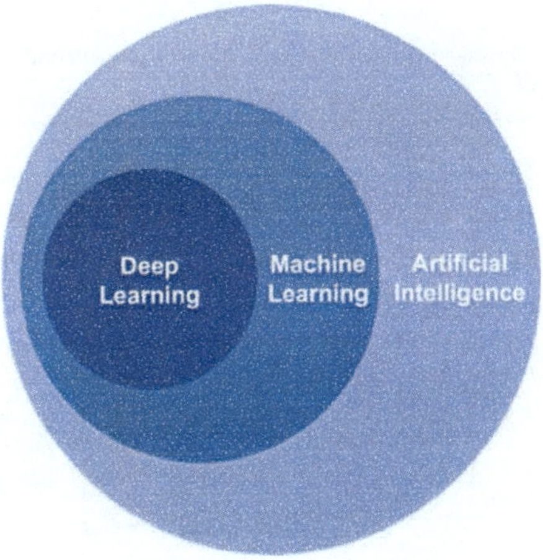

Fig. 1 The relationships between AI, ML, and DL

2.2 Virtual and Augmented Reality

Virtual Reality (VR) and Augmented Reality (AR) are advanced visualisation technologies that can display digital content across the user's surrounding space, providing them with intuitive and natural ways of navigating through, exploring, and interacting with the virtual or augmented environment. VR and AR technologies can also encompass haptic and auditory stimuli. Despite their differences, these technologies share three main components, namely: the contents (information being conveyed), the medium tools (software and hardware), and the users. The position of these two technologies in the virtuality continuum is shown in Fig. 2.

In AR, the digital content is blended with views of the real environment delivering an "augmented" experience of reality [21], while in VR, both the content and the surrounding environment are entirely computer-generated [22]. In other words, AR blends the virtual and real worlds by overlapping real-world views with virtual information, and VR is a completely computer-generated experience—there is nothing else in it other than virtual elements. For these reasons, while VR is often used whenever there is a need to replicate, replace or simulate the real world, AR would be suitable for users who perform tasks in the real world and could benefit from additional visual information [23]. In fact, in most AR applications, augmentations (the digital content displayed) must be distinguishable from real objects in a real-world view. Typically, augmentations are deliberately not realistic and do not intend to

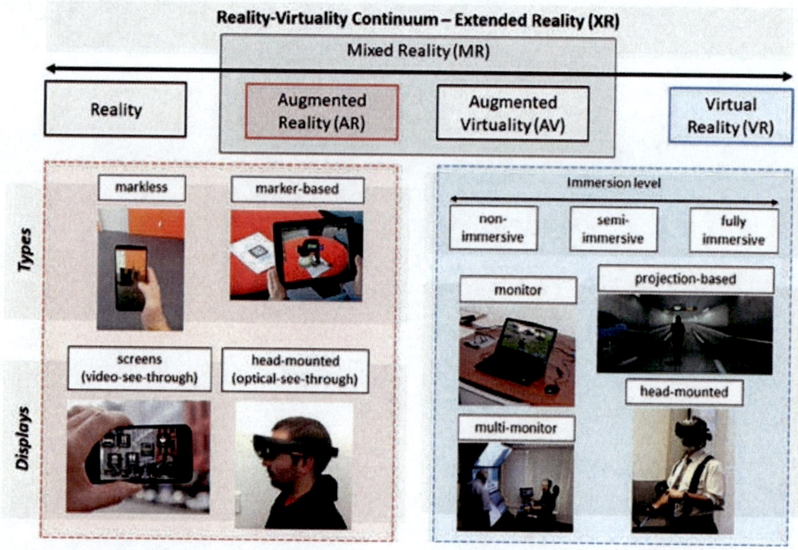

Fig. 2 VR and AR position in the virtuality continuum and their hardware solutions. This figure uses modified figures published in [18–20]

deceive the observer in any way so that they can distinguish virtual data from real objects.

Since late 2014, advances in computer processing power, display and sensing technologies have enabled the development and spread of affordable off-the-shelf VR and AR solutions. They have different setups and specifications; however, they generally encompass (besides digital content and user): a display, built-in infrared sensors or cameras, body movement and gesture tracking sensors, interaction devices, and software to stitch all these components together. These components can vary largely, especially regarding the display type. In AR, the displays can be optical-see-through (Microsoft® HoloLens™) or video-see-through (smartphones, tablets), handheld or head-mounted, as shown in Fig. 2. AR devices use built-in video cameras or special depth-sensing/infrared cameras to capture real-world data. Registration—the process of capturing real-world data in the format of images or infrared scans to overlay it with the digital content—is of major importance in AR to ensure that augmentations align properly with real-world objects [21]. The alignment of augmentations onto real-world scenes is achieved through different methods, depending on whether an AR system is marker-based or markerless.

While marker-based AR uses physical markers placed across the real environment as anchor points for the augmentations, markerless AR utilises other types of positioning technologies such as GPS (Global Positioning System) or SLAM (Simultaneous Localisation and Mapping). In VR, displays can be projection-based (conventional monitors, CAVE-like setups [24]) or head-mounted (Oculus Rift™ and the HTC Vive™), monoscopic or stereoscopic, as shown in Fig. 2. Also, VR systems can be non-immersive or immersive. Usually, in low-end non-immersive systems, the display mode offers monoscopic views of a digital model, and interaction devices are limited to a simple mouse and keyboard. High-end immersive VR (IVR) systems allow users to literally walk through the virtual space, manipulate virtual objects with hand gestures, and shift viewpoints with head movements—as if they were in the real world—while the display delivers stereoscopic views of the virtual world (see Fig. 2).

2.3 *Building Information Modelling*

Building Information Modelling (BIM) is a set of processes to create, analyse, and manage building models with physical and functional characteristics [14]. BIM is applied throughout a project's lifecycle, including planning, design, procurement, construction, operation, maintenance, renovation, and demolition. BIM provides a common data environment that facilitates the exchange and coordination of data and streamlines the communication and collaboration among different stakeholders of a building project, including clients, architects, engineers, contractors, facility managers, manufacturers, suppliers, and authorities (Fig. 3).

Unlike other Computer-Aided Design (CAD) systems, the operation and function of BIM are realised through parametric objects. A parametric object is an object

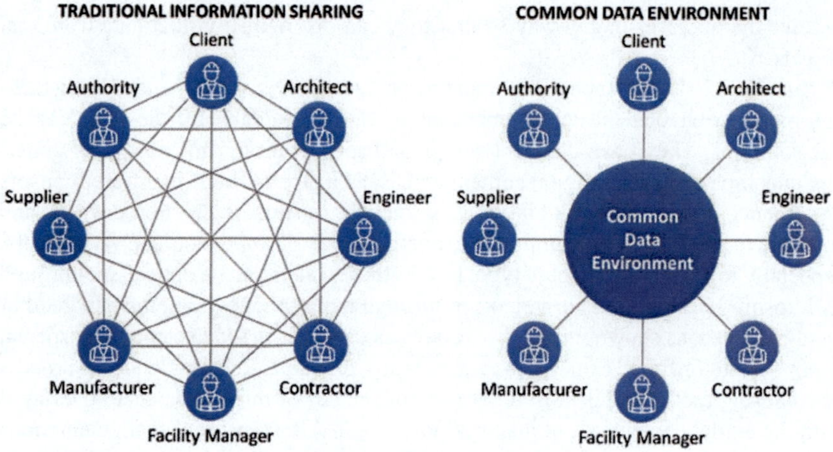

Fig. 3 Traditional information sharing versus common data environment

within a BIM model (e.g., a wall, floor, door, window, equipment, or fixture) defined by a set of parameters and rules [14]. The parameters and rules determine the associated properties and features of an object, such as geometry, appearance, materials, cost, physical properties, thermal properties, position, and the behavioural relationship with other objects. The properties and data of an object build up the core of BIM. The information is exchanged among stakeholders and utilised to enable various products and processes, such as visualisation, simulation, clash detection, cost estimation, structural analysis, and energy analysis. Further, scripts and algorithms can be applied to control and manipulate parameters. This transfers the abovementioned production and processes into an automated procedure. One of the examples is generative design, in which a software application generates design alternatives and creates iterations to provide optimised solutions and informs decision-making [25]. For instance, generative design helps to create thousands of floorplans and explore the optimised layout based on input parameters and design goals.

3 Evacuation Applications

This section provides an overview of how the five digital technologies being considered in this chapter have been applied in the evacuation domain. Section 3.1 illustrates the applications using AI and ML, while Sect. 3.2 discusses the different applications using both AR and VR. Finally, Sect. 3.3 provides examples related to BIM.

3.1 *Artificial Intelligence and Machine Learning for Evacuation*

In recent years, AI and ML have provided a new approach to improve the modelling and understanding of human behaviour in disasters, such as fires, hurricanes, and earthquakes [26–28]. With the capability of automatically learning and improving based on the input data, AI models often have flexible modelling structures to capture the complex relationships from the data and thus produce highly accurate predictions of human decision-making in disasters. For example, Zhao et al. [26] applied random forest (a random forest classifier is illustrated in Fig. 4) to model pre-evacuation decision-making in a building fire event and found that the random forest model produced high prediction accuracy and F1 score (a popular performance metric for machine learning classifiers, which is particularly useful when class imbalance problems exist). Sahebi et al. [29] also found random forest model had a better performance than other machine learning models when predicting the duration of hospital emergency evacuation during fire. More recently, Xu et al. [27] found that the decision tree has the highest capability in predicting the evacuation decisions of households when the input data (in this case, survey data) has a small sample size.

In addition to the increased model accuracy, the AI models can also produce new knowledge of the relationships between the features and targets (e.g., the relationships between sociodemographic factors and evacuation decisions) by leveraging the explainable AI (XAI) techniques [30]. For instance, Zhao et al. [26] found that the AI-based evacuation model can identify the important features affecting people's evacuation decisions in a building fire, reveal the interactions between the features, and capture the coupled nonlinear relationships. A new finding uncovered by XAI in [23] is that during a building fire event, a decision-maker is more sensitive to the percentage of responding occupants than the number of them; thereby, the people sitting in the back are often responding more slowly than the people in the front.

Based on the application scenarios, AI-based evacuation models may be divided into two main categories, i.e., planning and real-time models [31]. The planning models are typically trained once with static historical datasets and classical machine

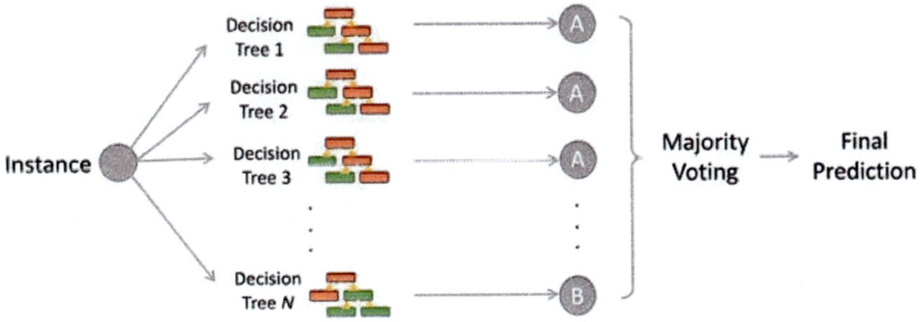

Fig. 4 Diagram of a random forest model predicting the choice to evacuate (**A**) or not evacuate (**B**)

learning methods such as tree-based models, support vector machines, and Bayesian networks. These models are mainly used for pre-event planning purposes [26, 32, 33]. In turn, the real-time models refer to the models trained/updated periodically with dynamic data streams and deep learning methods such as Long Short-Term Memory (LSTM) and Transformers. These models can generate highly accurate predictions in (near) real-time to assist decision-making in emergencies [34, 35].

Therefore, AI and ML have been demonstrated to have a high potential to fundamentally transform the research paradigm of modelling evacuation behaviours in fire evacuations and to significantly improve emergency management and policy-making to save more lives.

3.2 *Virtual and Augmented Reality for Evacuation*

Research institutions and industry practitioners worldwide have investigated the benefits and challenges of VR and AR implementation in various fields. For their inherent characteristics of combining (or not) real-world and virtual information, while AR systems have been mostly used to train/guide people on how to perform evacuations [36], VR solutions have been additionally used to study how building occupants behave during disasters (to advance or adjust safety protocols and evacuation procedures) [37]. To date, AR has not been used yet to investigate human behaviour during fires [38].

VR and AR solutions specifically designed to enhance fire evacuation readiness and efficacy are still at the proof-of-concept stages and are mostly found at universities. A few prototypes have been successfully developed and tested. For example, a mobile AR building evacuation training application through which users could learn how to reach the nearest exit in the event of a fire [39] (see Fig. 5a). There is also a prototype application combining BIM and marker-based video-see-through AR to provide occupants with interactive real-time guidance information so that they can find the best evacuation route [40]. The interactive visual information in this prototype included the occupant's current location, the building exit location, the shortest exit route, and the animations for evacuation guidance. A smart video-see-through AR navigation and emergency evacuation system has also been proposed as a user-friendly and intuitive guidance tool to provide building occupants with the optimal escape path to exit the building in the event of a fire. Combining the Internet of Things (IoT) and ML technologies, this innovative system can detect emergency events and deliver sensing data, predict disaster area, propagation, and occupants' location to then derive the optimum evacuation path for each person individually [41] (see Fig. 5b). Finally, an optical-see-through AR was proposed for fire warden training [42] (see Fig. 5c). In this case, the users learn how to contain a fire, activate the alarm, identify the location of a safe vest, evacuate building occupants and evacuate the building.

VR prototypes are slightly less challenging to be identified in the evacuation field. For example, a non-immersive VR simulation has been developed to train users on fire

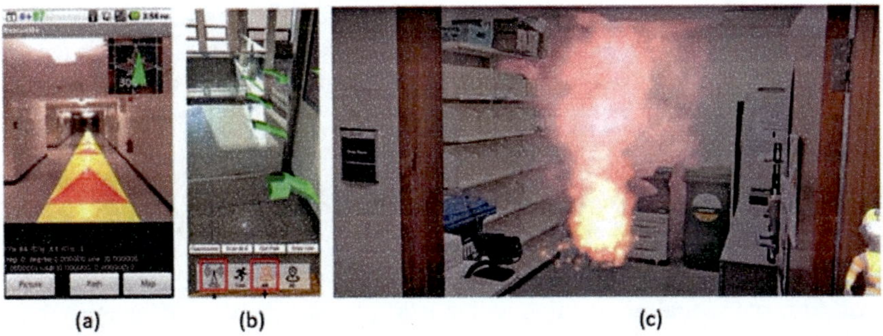

Fig. 5 AR applications for evacuation: **a** RescueMe [39]; **b** the smart evacuation system proposed in [41]; **c** the fire warden training solution proposed in [42]

safety and emergency evacuation procedures and has been shown to enable higher knowledge retention after a month compared to conventional slide-based training [43]. A similar project also used a non-immersive VR-based fire evacuation training tool for healthcare staff (see Fig. 6a) and compared it against a slide-based training mode. This project's results also indicated that VR-based training was more effective regarding knowledge acquisition and retention [44]. VR has also been shown to be a successful tool for training children on fire safety, who could successfully transfer to the real world the procedural skills learned in the simulation [45] (see Fig. 6b). Another study used a smartphone-based HMD VR system to study occupants' exit choice behaviour during evacuations. The validity of using VR to understand human behaviour was demonstrated by showing that occupants' exit choice behaviour during the evacuation in VR was similar to that in the field experiment [46]. When it comes to VR-based behavioural investigations (see Fig. 6c), their ecological validity—the extent to which they can trigger realistic behavioural responses, i.e., similar to the ones observed in the real world—remains a concern and is yet to be determined [47].

Before committing companies to invest in these new technologies, testing and ensuring their effectiveness is paramount. The literature is rich and divided on this

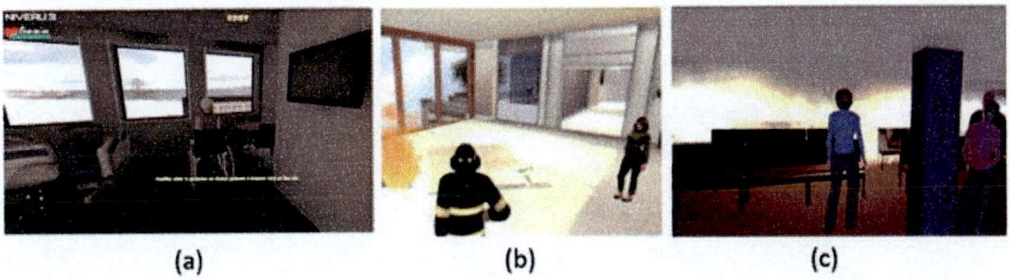

Fig. 6 VR applications for evacuation: **a** hospital fire safety training tool proposed in [44]; **b** safety training tool for children proposed in [45]; **c** behavioural experiment discussed in [47]

subject, and certain characteristics of the system and the users appear decisive. Findings are still preliminary [48] but consistently indicate that these technologies can provide safe, adaptable, and cost-effective fire evacuation training and research tools.

3.3 *Building Information Modelling for Evacuation*

There are different ways of applying BIM for fire evacuation research. First, the geometric and spatial information within BIM can be linked to computational fire modelling tools to simulate fire dynamics [49, 50]. Dimyadi et al. [51] compared a BIM Rule Language (BIMRL) approach and a Blender approach (with the IfcBlender add-on tool) to transfer geometric and spatial information from Industry Foundation Classes (IFC)-based BIM to Fire Dynamics Simulator (FDS) (see Fig. 7). They concluded that the BIMRL approach outperformed the Blender approach because it allows a customised query to extract additional semantical data for FDS, such as the physical properties and thermal properties of an object from an IFC model. Siddiqui et al. [52] proposed a strategy to facilitate the two-way data exchange between BIM and Fire Safety Engineering (FSE) tools. The idea is to feed data from BIM to perform fire and evacuation modelling and then send the modelling results back to BIM to inform design changes and optimisation. Further, Yakhou et al. [53] presented a working prototype to assess evacuation design and performance, which allows two-way data flow by integrating Autodesk Revit (a popular BIM tool) and Pathfinder (an egress simulator). Revit feeds the data of geometry, room usage, occupant loads and profiles to Pathfinder, and in turn, Pathfinder returns the simulated evacuation outcomes to Revit.

Second, BIM can be used as a visualisation platform to investigate evacuation performance. Sun and Turkan [54] integrated FDS and agent-based modelling with BIM to simulate fire growth and evacuation performance for different building layouts. BIM offers design inputs to modify building geometry and visualises simulation results that include fire-spreading zones and evacuation routes. The outcomes can be applied to optimise the building design and investigate the factors influencing fire evacuation. Wang et al. [55] demonstrated a BIM-based fire safety management tool that can assess evacuation design, plan escape routes, create visuals of

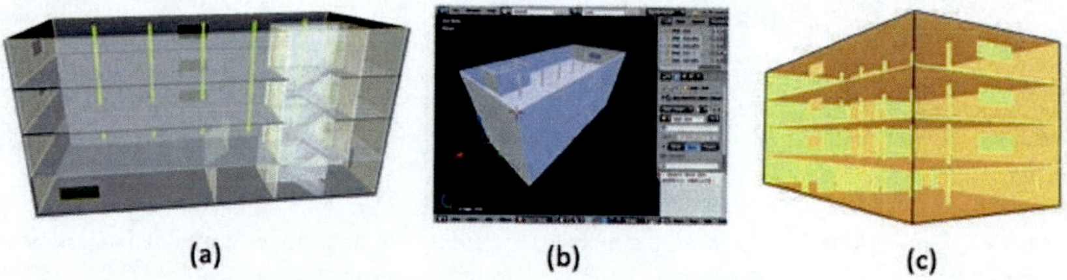

Fig. 7 Visualisation of the same building in **a** BIMRL, **b** Blender and **c** FDS SmokeView [51]

hazardous areas and escape routes, and check fire safety equipment maintenance records. All these BIM-based applications represent a step toward the possibility of having generative design tools for fire safety engineering [25].

Third, BIM can serve as a platform for real-time data monitoring and evacuation management. Cheng et al. [56] developed a BIM-based monitoring system with Bluetooth sensor networks. The monitoring system is visualised in a BIM environment with markers indicating the locations of fire, evacuators, and rescuers. This real-time information is obtained by sensors and transmitted via a Bluetooth network. The monitoring system can inform evacuation and rescue decision-making. Similarly, Chen et al. [57] applied the Internet of Things (IoT) with BIM to create a BIM-based visualisation and warning platform for fire rescue. This system can monitor the real-time visibility, temperature, and CO level data and display them in a BIM model. Also, the BIM model can control LED evacuation signages to direct evacuators.

To sum up, the literature demonstrates the capability of BIM for fire evacuation in the following aspects: (1) serving as a database to provide data and information (e.g., geometry, room usage, and occupancy load) for external fire engineering modelling tools to simulate fire dynamics; (2) serving as a visualisation platform to investigate evacuation performance and support decision-making, for instance, hazardous zones and evacuation routes can be visualised and scrutinised in a BIM model; (3) serving as an integrated platform for real-time data monitoring and evacuation management. In particular, sensors installed in a building can be connected in a BIM environment; therefore, real-time monitoring and control are allowed in case of fire.

4 Discussion and Conclusions

This chapter aims to provide an overview of how different emerging technologies can support fire evacuations. In this chapter, we focused on the following: Augmented Artificial Intelligence (AI), Machine Learning (ML), Reality (AR) and Virtual Reality (VR), and Building Information Modelling (BIM). However, there are many more technologies that can provide meaningful contributions to the fire evacuation domain not included in our analysis, such as the Internet of Things (IoT), drones, and quantum computing.

In this work, we provide an introduction to the selected technologies and how these have been applied in the fire evacuation domain to enhance evacuation performance and achieve smart evacuation outcomes. Sections 3.1, 3.2 and 3.3 show the potential of these technologies as they have been used for several purposes and in different stages of a building's lifecycle, from its design using BIM to assess evacuation performance to its operation phase using AR and VR to guide and train building occupants. Further, the analysis shows that AI and ML have a high potential to support research and shift the paradigm of investigations and modelling of evacuation behaviours in fire evacuations. However, one of the main obstacles in applying ML algorithms to investigate human behaviour is the lack of suitable data required to train decision-making models. Although several case studies have been published for

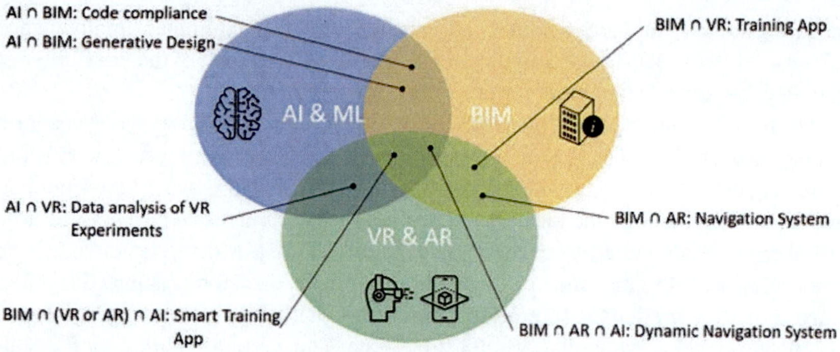

Fig. 8 Examples of overlapping among AR, VR, BIM, AI and ML (the ∩ symbol means intersection)

different evacuation decisions (see, for instance, the existing pre-evacuation database published in [58]), most of these data do not provide adequate data to assess the impact of several factors on the evacuation decision. On the other hand, new datasets on several evacuation decisions (see, for instance [59, 60]) are constantly being published, reducing this limitation year by year.

Sections 3.1, 3.2 and 3.3 also provide an overview of how the analysed technologies can be used in the fire evacuation domain to enhance evacuation performance. However, these technologies do not work in silos and can be integrated with each other or with other technologies not analysed in this work (e.g., IoT and quantum computing). Figure 8 provides examples of overlapping among AR, VR, BIM, AI and ML. For instance, AI can be used in combination with BIM to generate different building plans to optimise the evacuation performance of a building using the generative design framework proposed by Lovreglio et al. [25] or to assess the evacuation compliance of a building according to several standards and code by expanding the work done by Dimyadi et al. [51].

AI can also be combined with VR to enhance our understanding of how people behave during a variety of fire evacuation scenarios. In fact, VR provides a very useful tool for collecting evacuation data with reasonable ecological validity [61]. To date, most of the data generated using VR experiments are analysed using traditional statistical methods such as regression models and statistical tests. AI and ML can provide novel tools to better understand how evacuees behave. Furthermore, AR and VR can be combined with BIM to develop a new generation of training tools, as shown by Rüppel et al. [62] or to help to build occupants evacuate a building during a real emergency [40].

It is finally possible to derive more advanced integrations of these technologies if we refer to the intersections among the three bubbles in Fig. 8. There could be a dynamic AR-based navigation system. In this case, an AI tool can be implemented to identify the best evacuation route in real time, given the status of a fire emergency in a building [41]. Another possibility is using AI to enhance the capabilities of existing VR and AR-based evacuation training tools. AI could be used to enhance the capabilities of non-player characters, allowing them to have more natural interactions with the users and assisting them in the learning process by answering questions or adjusting teaching strategies depending on the mistakes made during their training. This approach has been investigated in other fields, for instance, [63, 64], and it could be applied in the evacuation training domain in the coming years.

In conclusion, many digital technologies can play a key role in supporting the research on fire evacuation dynamics and providing tools that can significantly improve the evacuation performance of new and existing buildings. As such, this chapter provides some examples that can inspire new research and applications in this domain.

References

1. N.C. McConnell, K.E. Boyce, Refuge areas and vertical evacuation of multistorey buildings: the end users' perspectives. Fire Mater. **39**, 396–406 (2015). https://doi.org/10.1002/fam.2205
2. A. Rahouti, R. Lovreglio, D. Nilsson, E. Kuligowski, P. Jackson, F. Rothas, Investigating evacuation behaviour in retirement facilities: case studies from New Zealand, Fire Technol. (2020). https://doi.org/10.1007/s10694-020-01058-x
3. E. Ronchi, D. Nilsson, Modelling total evacuation strategies for high-rise buildings. Build. Simul. **7**, 73–87 (2013). https://doi.org/10.1007/s12273-013-0132-9
4. G.V. Hadjisophocleous, N. Benichou, A.S. Tamim, Literature review of performance-based fire codes and design environment. J. Fire. Prot. Eng. **9**, 12–40 (1998). https://doi.org/10.1177/104239159800900102
5. A. Alvarez, B.J. Meacham, N.A. Dembsey, J.R. Thomas, Twenty years of performance-based fire protection design: Challenges faced and a look ahead. J. Fire. Prot. Eng. **23**, 249–276 (2013). https://doi.org/10.1177/1042391513484911/ASSET/IMAGES/LARGE/10.1177_1042391513484911-FIG1.JPEG
6. S.M. V. Gwynne, E.R. Rosenbaum, Employing the hydraulic model in assessing emergency movement, in SFPE Handb Fire Prot Eng (Springer New York, New York, NY, 2016), pp 2115–2151. https://doi.org/10.1007/978-1-4939-2565-0_59
7. E.D. Kuligowski, Computer Evacuation Models for Buildings, in SFPE Handb Fire Prot Eng (2016), pp. 2152–2180
8. R. Lovreglio, E. Ronchi, M.J. Kinsey, An online survey of pedestrian evacuation model usage and users. Fire Technol. **56**, 1133–1153 (2019). https://doi.org/10.1007/s10694-019-00923-8
9. S.M.V. Gwynne, E.D. Kuligowski, K.E. Boyce, D. Nilsson, A.P. Robbins, R. Lovreglio, J.R. Thomas, A. Roy-Poirier, Enhancing egress drills: Preparation and assessment of evacuee performance. Fire Mater. **43**, 613–631 (2017). https://doi.org/10.1002/fam.2448
10. R. Lovreglio, Modelling decision-making in fire evacuation based on the random utility theory—PhD Thesis, Politecnico of Bari, Milan and Turin (2016). https://doi.org/10.13140/RG.2.1.1695.5281/1
11. S.M. Gwynne, L.M. Hulse, M.J. Kinsey, Guidance for the Model Developer on Representing Human Behavior in Egress Models, Fire Technol., 1–26 (2015). https://doi.org/10.1007/s10694-015-0501-2
12. P. Thompson, D. Nilsson, K. Boyce, D. McGrath, Evacuation models are running out of time. Fire Saf. J. **78**, 251–261 (2015). https://doi.org/10.1016/j.firesaf.2015.09.004
13. A. Sawhney, M. Riley, J. Irizarry, *Construction 4.0—An innovation platform for the built environment* (2020)

14. R. Sacks, C. Eastman, G. Lee, P. Teicholz, *BIM handbook: A guide to building information modeling for owners, designers, engineers, contractors, and facility managers* (Wiley, 2018)
15. M.Z. Naser, C. Lautenberger, E. Kuligowski, Special Issue on "Smart Systems in Fire Engineering, Fire Technol. **57**, 2737–2740 (2021). https://doi.org/10.1007/S10694-021-01196-W/METRICS
16. C. Bishop, N. Nasrabadi, Pattern recognition and machine learning, Springer. N. Y. (2006). https://doi.org/10.1007/978-0-387-45528-0
17. E.A.M. van Dis, J. Bollen, W. Zuidema, R. van Rooij, C.L. Bockting, ChatGPT: five priorities for research, Nat. 2023 6147947. **614**, 224–226 (2023). https://doi.org/10.1038/d41586-023-00288-7
18. R. Lovreglio, D. Borri, L. Dell'Olio, A. Ibeas, A discrete choice model based on random utilities for exit choice in emergency evacuations. Saf. Sci. **62**, 418–426 (2014). https://doi.org/10.1016/j.ssci.2013.10.004
19. R. Lovreglio, M.T. Kinateder, Augmented reality for pedestrian evacuation research: promises and limitations, Saf. Sci. **128**, 104750 (2020). https://doi.org/10.1016/j.ssci.2020.104750
20. E. Ronchi, D. Nilsson, S. Kojić, J. Eriksson, R. Lovreglio, H. Modig, A.L. Walter, A virtual reality experiment on flashing lights at emergency exit portals for road tunnel evacuation, Fire Technol. **52**, (2016). https://doi.org/10.1007/s10694-015-0462-5
21. R. Azuma, Y. Baillot, R. Behringer, S. Feiner, S. Julier, B. MacIntyre, Recent advances in augmented reality. IEEE Comput. Graph. Appl. **21**, 34–47 (2001). https://doi.org/10.1109/38.963459
22. J. Steuer, Defining virtual reality: dimensions determining telepresence. J. Commun. **42**, 73–93 (1992). https://doi.org/10.1111/J.1460-2466.1992.TB00812.X
23. X. Li, W. Yi, H.-L. Chi, X. Wang, A.P.C. Chan, A critical review of virtual and augmented reality (VR/AR) applications in construction safety. Autom. Constr. **86**, 150–162 (2018). https://doi.org/10.1016/J.AUTCON.2017.11.003
24. C. Cruz-Neira, D.J. Sandin, T.A. DeFanti, Surround-screen projection-based virtual reality: The design and implementation of the CAVE, in Proc 20th Annual Conf Compute Graph Interact Tech SIGGRAPH 1993 (1993), pp. 135–142. https://doi.org/10.1145/166117.166134
25. R. Lovreglio, P. Thompson, Z. Feng, Automation in fire safety engineering using BIM and generative design. Fire Technol. **58**, 1–5 (2022). https://doi.org/10.1007/S10694-021-01153-7/FIGURES/1
26. X. Zhao, R. Lovreglio, D. Nilsson, Modelling and interpreting pre-evacuation decision-making using machine learning, Autom. Constr., (2020)
27. N. Xu, R. Lovreglio, E.D. Kuligowski, T.J. Cova, D. Nilsson, X. Zhao, Predicting and assessing wildfire evacuation decision-making using machine learning: findings from the 2019 Kincade Fire, Fire Technol. **59**, 793–825 (2023). https://doi.org/10.1007/S10694-023-01363-1/TABLES/7
28. X. Zhao, R. Lovreglio, E. Kuligowski, D. Nilsson, Using artificial intelligence for safe and effective wildfire evacuations, Fire Technol., 1–3 (2020). https://doi.org/10.1007/s10694-020-00979-x
29. A. Sahebi, K. Jahangiri, A. Alibabaei, D. Khorasani-Zavareh, Using artificial intelligence for predicting the duration of emergency evacuation during hospital fire. Disaster Med. Public Health Prep. **17**, e229 (2023). https://doi.org/10.1017/DMP.2022.187
30. C. Molnar, *Interpretable machine learning* (Lulu. com, 2020)
31. Y. Sun, X. Zhao, R. Lovreglio, E. Kuligowski, AI for large-scale evacuation modeling: promises and challenges, in Interpret Mach Learn Anal Des Assessment, Inf Decis Mak Civ Infrastruct (2023)
32. S. Grajdura, S. Borjigin, D. Niemeier, Fast-moving dire wildfire evacuation simulation. Transp. Res. Part D Transp. Environ. **104**, 103190 (2022). https://doi.org/10.1016/J.TRD.2022.103190
33. L. Hong, V. Frias-Martinez, Modeling and predicting evacuation flows during hurricane Irma. EPJ Data Sci. **9**, 29 (2020). https://doi.org/10.1140/EPJDS/S13688-020-00247-6
34. X. Li, S. Hasan, A. Culotta, Identifying hurricane evacuation intent on Twitter. Proc. Int. AAAI Conf. Web Soc. Media. **16**, 618–627 (2022). https://doi.org/10.1609/ICWSM.V16I1.19320

35. K.C. Roy, S. Hasan, A. Culotta, N. Eluru, Predicting traffic demand during hurricane evacuation using Real-time data from transportation systems and social media. Transp. Res. Part C Emerg. Technol. **131**, 103339 (2021). https://doi.org/10.1016/J.TRC.2021.103339
36. R. Lovreglio, M. Kinateder, Augmented reality for pedestrian evacuation research: Promises and limitations. Saf. Sci. **128**, 104750 (2020). https://doi.org/10.1016/J.SSCI.2020.104750
37. Z. Feng, V.A. González, R. Amor, R. Lovreglio, G. Cabrera-Guerrero, Immersive virtual reality serious games for evacuation training and research: A systematic literature review. Comput. Educ. **127**, 252–266 (2018). https://doi.org/10.1016/J.COMPEDU.2018.09.002
38. R. Lovreglio, Virtual and augmented reality for human behaviour in disasters: a review, in Fire evacuation model. Tech Conf (2020), p. 14. https://files.thunderheadeng.com/femtc/2020_d3-02-lovreglio-paper.pdf (accessed 21 Feb 2022)
39. J. Ahn, R. Han, An indoor augmented-reality evacuation system for the Smartphone using personalised Pedometry. Human-Centric Comput. Inf. Sci. **2**, 18 (2012). https://doi.org/10.1186/2192-1962-2-18
40. S. Kanangkaew, N. Jokkaw, T. Tongthong, A real-time fire evacuation system based on the integration of building information modeling and augmented reality. J. Build. Eng. **67**, 105883 (2023). https://doi.org/10.1016/J.JOBE.2023.105883
41. S.J. Yoo, S.H. Choi, Indoor AR navigation and emergency evacuation system based on machine learning and IoT technologies. IEEE Internet Things J. **9**, 20853–20868 (2022). https://doi.org/10.1109/JIOT.2022.3175677
42. D. Paes, Z. Feng, R. Lovreglio, M. King, H.K. Shad, P. Sasikumar, prototyping and testing an optical see-through augmented reality fire safety training system for building occupants, Under Rev., (2023)
43. S. Shiradkar, L. Rabelo, F. Alasim, K. Nagadi, Virtual world as an interactive safety training platform, Inf. **12**, 219 (2021). https://doi.org/10.3390/INFO12060219
44. A. Rahouti, R. Lovreglio, S. Datoussaïd, T. Descamps, Prototyping and validating a non-immersive virtual reality serious game for healthcare fire safety training. Fire Technol. **57**, 3041–3078 (2021). https://doi.org/10.1007/S10694-021-01098-X/FIGURES/15
45. Ü. Çakiroğlu, S. Gökoğlu, Development of fire safety behavioral skills via virtual reality. Comput. Educ. **133**, 56–68 (2019). https://doi.org/10.1016/J.COMPEDU.2019.01.014
46. Y. Feng, D.C. Duives, S.P. Hoogendoorn, Using virtual reality to study pedestrian exit choice behaviour during evacuations. Saf. Sci. **137**, 105158 (2021). https://doi.org/10.1016/J.SSCI.2021.105158
47. S. Arias, J. Wahlqvist, D. Nilsson, E. Ronchi, H. Frantzich, Pursuing behavioral realism in Virtual Reality for fire evacuation research. Fire Mater. **45**, 462–472 (2021). https://doi.org/10.1002/FAM.2922
48. L.W. Menzemer, E. Ronchi, M.M.V. Karsten, S. Gwynne, J. Frederiksen, A scoping review and bibliometric analysis of methods for fire evacuation training in buildings. Fire Saf. J. **136**, 103742 (2023). https://doi.org/10.1016/J.FIRESAF.2023.103742
49. M. Spearpoint, Transfer of architectural data from the IFC building product model to a fire simulation software tool. J. Fire. Prot. Eng. **17**, 271–292 (2007). https://doi.org/10.1177/1042391507074681
50. J.A.W. Dimyadi, M.J. Spearpoint, R. Amor, *Generating fire dynamics simulator geometrical input using an IFC-based building information model*, (2007). https://ir.canterbury.ac.nz/handle/10092/680 (Accessed 7 June 2021)
51. J. Dimyadi, W. Solihin, R. Amor, Using IFC to support enclosure fire dynamics simulation, in Lect Notes Comput Sci (Including Subser. Lect Notes Artif Intell Lect Notes Bioinformatics) (Springer Verlag, 2018), pp. 339–360. https://doi.org/10.1007/978-3-319-91638-5_19
52. A.A. Siddiqui, J.A. Ewer, P.J. Lawrence, E.R. Galea, I.R. Frost, Building information modelling for performance-based fire safety engineering analysis—a strategy for data sharing, J. Build. Eng., 102794 (2021). https://doi.org/10.1016/J.JOBE.2021.102794
53. N. Yakhou, P. Thompson, A. Siddiqui, J. Abualdenien, E. Ronchi, The integration of building information modelling and fire evacuation models. J. Build. Eng. **63**, 105557 (2023). https://doi.org/10.1016/J.JOBE.2022.105557

54. Q. Sun, Y. Turkan, A BIM-based simulation framework for fire safety management and investigation of the critical factors affecting human evacuation performance. Adv. Eng. Informatics. **44**, 101093 (2020). https://doi.org/10.1016/J.AEI.2020.101093
55. S.H. Wang, W.C. Wang, K.C. Wang, S.Y. Shih, Applying building information modeling to support fire safety management. Autom. Constr. **59**, 158–167 (2015). https://doi.org/10.1016/J.AUTCON.2015.02.001
56. M.Y. Cheng, K.C. Chiu, Y.M. Hsieh, I.T. Yang, J.S. Chou, Y.W. Wu, BIM integrated smart monitoring technique for building fire prevention and disaster relief. Autom. Constr. **84**, 14–30 (2017). https://doi.org/10.1016/J.AUTCON.2017.08.027
57. X.S. Chen, C.C. Liu, I.C. Wu, A BIM-based visualisation and warning system for fire rescue. Adv. Eng. Informatics. **37**, 42–53 (2018). https://doi.org/10.1016/J.AEI.2018.04.015
58. R. Lovreglio, E.D. Kuligowski, S.M.V. Gwynne, K. Boyce, A Pre-evacuation database for use in egress simulations, Fire Saf. J. (under Rev. (2018))
59. R. Lovreglio, E. Dillies, E. Kuligowski, A. Rahouti, M. Haghani, Exit choice in built environment evacuation combining immersive virtual reality and discrete choice modelling. Autom. Constr. **141**, 104452 (2022). https://doi.org/10.1016/J.AUTCON.2022.104452
60. R. Lovreglio, E. Kuligowski, A pre-evacuation study using data from evacuation drills and false alarm evacuations in a university library. Fire Saf. J. **131**, 103595 (2022). https://doi.org/10.1016/J.FIRESAF.2022.103595
61. M. Kinateder, E. Ronchi, D. Nilsson, M. Kobes, M. Müller, P. Pauli, A. Mühlberger, Virtual reality for fire evacuation research, in Comput Sci Inf Syst (IEEE, Warsaw, 2014), pp. 313–321. https://doi.org/10.15439/2014F94
62. U. Rüppel, K. Schatz, Designing a BIM-based serious game for fire safety evacuation simulations. Adv. Eng. Informatics. **25**, 600–611 (2011). https://doi.org/10.1016/j.aei.2011.08.001
63. K. Buchta, P. Wojcik, K. Nakonieczny, J. Janicka, M. Igras-Cybulska, NUX Characters-interaction with voice assistants in Virtual Reality, in Proc.—2022 IEEE Int Symp. Mix Augment Real Adjunct, ISMAR-Adjunct 2022 (2022), pp. 917–918. https://doi.org/10.1109/ISMAR-ADJUNCT57072.2022.00204
64. K. Buchta, P. Wojcik, K. Nakonieczny, J. Janicka, D. Galuszka, R. Sterna, M. Igras-Cybulska, Modeling and optimising the voice assistant behavior in virtual reality, in Proc.—2022 IEEE Int Symp Mix Augment Real Adjunct, ISMAR-Adjunct 2022. (2022), pp. 397–402. https://doi.org/10.1109/ISMAR-ADJUNCT57072.2022.00086

Cardiovascular Function and Deleterious Adaptations Among Firefighters: Implications for Smart Firefighting

Dillon J. Dzikowicz, Salah S. Al-Zaiti, and Mary G. Carey

1 Introduction

Firefighting is strenuous work that causes significant strain on the heart and cardiovascular system. As a result, firefighters are disproportionately affected by serious cardiac events when compared to other emergency personnel [1, 2]. Nearly 50% of all duty-related fatalities among firefighters since the 1970s are due to sudden cardiac death (SCD), a fatal tachy-arrhythmia that leads to the cessation of blood flow in the body [3–5]. Although fire suppression activities comprise approximately 5% of total annual working time in firefighters, SCD most likely occurs during or immediately after these activities [5, 6]. Exacerbating the elevated risk of SCD is the high prevalence of obesity, hypertension, and other chronic conditions among firefighters [8]. Thus, firefighting in combination with underlying chronic conditions strains the heart and precipitates serious cardiac events, including SCD (Fig. 1).

Herein, we first will explain the physiological mechanisms responsible for the strain on the heart and cardiovascular system precipitated by fire suppression activities, as well as the deleterious long-term cardiovascular adaptations in firefighters. We also discuss how these alterations play a role in the pathogenesis of serious cardiac events and SCD. Second, we will discuss how chronic conditions exacerbate the strain on the heart. Third, we propose new avenues of future research including wearables for real-time monitoring and interventions focused on risk factor modification as a means to reduce the strain on the heart and cardiovascular system, and ultimately reduce serious cardiac events among firefighters.

D. J. Dzikowicz (✉) · M. G. Carey
School of Nursing, University of Rochester, 255 Crittenden Blvd, Rochester, NY 14620, USA
e-mail: Dillon_Dzikowicz@URMC.Rochester.edu

S. S. Al-Zaiti
School of Nursing, University of Pittsburgh, 3500 Victoria St, Pittsburgh, PA 15261, USA

X. Huang and W. C. Tam (eds.), *Intelligent Building Fire Safety and Smart Firefighting*, Digital Innovations in Architecture, Engineering and Construction,
https://doi.org/10.1007/978-3-031-48161-1_19

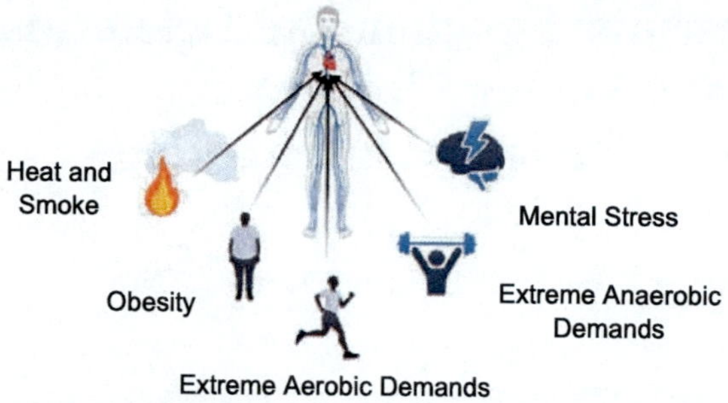

Fig. 1 Factors which influence firefighters' cardiovascular function and health. Heat and smoke from fires, obesity and chronic conditions, extreme aerobic and anaerobic demands, and mental stress cause significant strain on the heart. Figure created with BioRender; Credit to DJ Dzikowicz

2 Sudden Cardiac Death Among Firefighters

Due to the nature of their profession, firefighters suffer from considerable strain on the heart and cardiovascular system, and, consequently, are at a higher risk of serious cardiac events, including sudden cardiac death (SCD) [1, 2]. By SCD, we specifically mean an unexpected natural death due to an unresuscitable arrythmia that occurs within a short period of time in a person without any known fatal condition. When compared to other first responders, firefighters have double the odds of serious cardiac events including SCD compared to police officers and nearly four times the odds compared to emergency medical responders [2]. Of all cardiac events, SCD is the most concerning because it is fatal and has consistently accounted for the largest share of on-duty firefighter deaths [3–5]. Since the National Fire Protection Association began investigating causes of firefighter death in 1977, SCD was found to account for approximately 50% of all on-duty firefighter deaths [3–5]. Moreover, despite greater attention on SCD being such a significant cause of death among firefighters, data published by the National Fire Protection Association has reported a 64% relative increase in SCD among firefighters from 2005 (44%) to 2021 (72%), and differentially affecting younger firefighters (20–30 years of age) [3–5].

Most recent literature suggests 30–50% of all serious cardiac events including SCD events occur during fire suppression activities despite these activities accounting for less than 5% of total annual occupational time spent [5, 6]. During fire suppression activities, firefighters face extreme environmental conditions including noxious gas fumes and elevated temperatures. While fire suppression activities slightly differ between professional, volunteer, and wildland firefighters, they are all exposed to extreme environments. To protect themselves from these harsh conditions, firefighters wear heavy insulated personal protective equipment, which can weigh up to 55 pounds (25 kg), and performing aerobic and anaerobic activities such as ladder climbing, search and rescue, and moving heavy equipment [1, 9]. Collectively, this places considerable demands on the heart. For example, aerobic demands during fire suppression activities can exceed oxygen consumption greater than 40 ml/kg/min, approximately 12 times greater than at rest (reference: resting 3.5 ml/kg/min) [9, 10]. This exponential demand placed on the heart during fire suppression activities are met by heightened sympathetic arousal. However, this heightened sympathetic arousal can be maladaptive as it alters myocardial blood flow and induces myocardial ischemia. Subsequently, this pathophysiologic pathway is thought to be a major precursor to serious cardiac events including SCD [11].

3 Deleterious Cardiovascular Adaptations in Firefighters

In this section, we will discuss the three main mechanisms responsible for serious cardiac events including SCD among firefighters: heightened sympathetic arousal, alterations in blood flow, as well as myocardial ischemia. These three mechanisms are interlinked and contribute to the larger arena of deleterious cardiovascular adaptations that precipitate serious cardiac events including SCD.

Heightened Sympathetic Arousal

Sympathetic arousal, defined by the activation of the sympathetic nervous system, is an adaptive mechanism to increase the contractility of the heart to meet the exponential demands placed on the heart during fire suppression activities. Sympathetic arousal begins with the ringing of the fire bell in the firehouse, continues during transport to the fire scene, maximizes during fire suppression activities, and may persist even upon the completion of fire suppression activities as depicted in Fig. 2 below [12–14]. Most studies measure sympathetic arousal by evaluating heart rate but new evidence has observed a similar pattern with blood pressure [14]. Upon the ringing of the fire bell suddenly, previous studies have demonstrated a significant rise in sympathetic arousal likely due to a conditioned response among professional firefighters [12]. Marciniak et al. [13] reported that fire bells contributed to greater sympathetic arousal when compared to other bells such as for medical emergencies representing the conditioned response the fire bell has on sympathetic arousal [13]. Moreover, the fire bell may also induce mental stress during the anticipatory fire suppression activities; Liao and Carey (2015) demonstrated that mental stress may further exacerbate sympathetic arousal even before the actual firefighting activity [15]. While transporting to the scene, sympathetic arousal continues to escalate until it peaks during fire suppression activities [13].

This figure depicts the degree of sympathetic arousal measured during the entirety of fire suppression activities across three hypothetical scenarios: normal, elevated, and heightened. In the normal scenario, the firefighter's sympathetic arousal is triggered with the ringing of the fire bell, reaches a maximum during fire suppression activities, and has a rapid recovery to a resting baseline upon completion of the fire suppression activities. However, in elevated or heightened scenarios, this trajectory becomes maladaptive. In the elevated scenario, the firefighter's sympathetic arousal follows the same trajectory as the normal scenario but does not recover as drastically. Further, in the heightened scenario, the firefighter's sympathetic arousal follows the

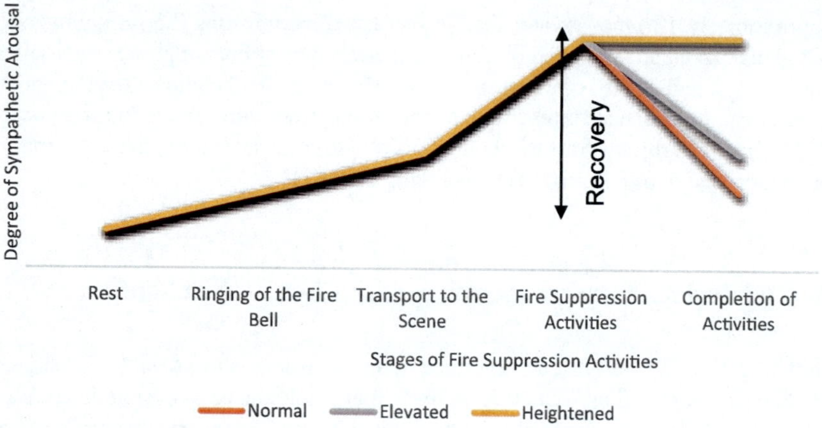

Fig. 2 Degree of sympathetic arousal during fire suppression activities

same trajectory as the normal scenario but does not recover at all. Recovery, defined as a withdrawal of sympathetic arousal and parasympathetic re-activation, appears to be critical in the pathophysiology of serious cardiac events including Sudden Cardiac Death among firefighters. Credit to DJ Dzikowicz.

Sympathetic arousal is necessary to meet extreme physical demands during fire suppression activities because it bolsters heart rate and blood flow [1, 9, 10]. During fire suppression activities, firefighters face extreme environmental conditions: temperatures ranging from a low of 98° F to a high of 106° F; noxious gas fumes; and, hazardous conditions [1, 9, 10]. Firefighters face these extreme environmental conditions all while wearing heavy insulated personal protective equipment, and must exert significant aerobic effort (stair and ladder climbing), anaerobic power (forcible entry, search, and rescue operations), and static exertion (heavy materials handling, cutting, and chopping to ventilate the building, and advancing charged hose lines) [1, 9, 10]. To provide an adequate flow of oxygenated blood, myocardial contractility, modulated by sympathetic arousal, must strengthen and occur faster [16, 17]. In addition to modulating the heart itself, sympathetic arousal also constricts the vasculature and retains fluid [16, 17]. Collectively, these effects of sympathetic arousal support adequate heart rate and blood flow to support the extreme physical demands during fire suppression activities. During the duration of fire suppression activities, these effects may become more pronounced. For example, the extreme physical environment in combination with the insulting gear and laborious work causes dehydration by losing fluids, e.g., sweating. This further exacerbates sympathetic arousal as a compensatory mechanism for declining intravascular volume [18, 19]. Therefore, sympathetic arousal is crucial for successfully meeting the demands during fire suppression activities, when they become excessive and persist during recovery after the completion of fire suppression activities ultimately leading to serious cardiac events and SCD [17].

Upon completion of fire suppression activities, sympathetic arousal may persist and cause delayed cardiovascular recovery, a known propagator of serious cardiac events and SCD [12, 17, 20]. Cardiovascular recovery, defined as a withdrawal of sympathetic arousal and parasympathetic re-activation, is a crucial physiological mechanism for preventing serious cardiac events including SCD [17]. More specifically, cardiovascular recovery is indexed within the first 20 s by parasympathetic re-activation solely mediated by changes in response to the activity in arterial baroreceptors as the later portion of recovery is indexed by a reduction in circulating catecholamines and withdrawal of sympathetic arousal [20, 21]. Thus, the most crucial part of recovery, may be within the first few moments of the cessation of fire suppression activities [21]. In a descriptional study, more than 50% of on-duty firefighters had exhibited reduced heart rate variability, a measure of persistent sympathetic arousal, after fire suppression activities [22]. Recently, Dzikowicz and Carey [23] aimed to investigate the physiological mechanism of cardiovascular recovery and reported that after exercise testing, recovery of heart rate was significantly associated with recovery of systolic and diastolic blood pressure meaning more rapid withdrawal of sympathetic arousal and parasympathetic re-activation may be necessary to promote a better and healthier hemodynamic condition [23].

Recovery of diastolic blood pressure to baseline levels, which corresponds to coronary perfusion, appears to be important to prevent myocardial damage (e.g. ischemia) during recovery from fire suppression activities [24, 25]. Animal models have demonstrated the linear relationship between coronary perfusion on myocardial properties and performance during diastole [24, 25]. Fernhall et al. [26] reported that among firefighters who underwent repeated bouts of training to heighten sympathetic arousal and then echocardiographic examination, significant reductions in left ventricular diastolic size and volume, mitral E/A (the ratio of passive left ventricular filling to the active atrial filling of the left ventricle), and left ventricular shortening fraction were observed during diastole, all suggesting poor relaxation of the myocardium [26]. Further study has suggested that these changes are sensitive to the physical environment [27]. For example, Lefferts et al. [27] reported that fire suppression activities involving structural wood fires exerted the greatest effect on hemodynamics, specifically blood pressure, due to the effects of heat in combination with environmental smoke [27]. While the clinical significance of the cardiac changes presented above is not fully understood, collectively they demonstrate significant cardiovascular workload during the diastolic phase in recovery, even among healthy, on-duty firefighters. Collectively, these changes effectively demonstrate how heightened sympathetic arousal is a main propagator of alterations in blood flow and subsequent ischemia which lead to serious cardiac events including SCD [17].

Alterations in Blood Flow

As prefaced above, the heightened sympathetic arousal which accompanies fire suppression activities in combination with the toxic environment in which fire suppression activities are performed significantly alters blood flow by shifting the hemostatic response to be pro-inflammatory and pro-coagulatory [1, 28]. During fire suppression activities, the extreme environmental conditions, exacerbated by the personal protective equipment and strenuous physical exertion, cause sympathetic arousal and subsequent vasoconstriction and pro-coagulatory properties of the arterial blood vessels as depicted in Fig. 3 [1, 9, 10]. Sympathetic arousal is indexed by the release of potent vasoconstrictors such as norephedrine as well as pro-coagulatory platelets from the spleen and reticular tissue [29, 30]. The vasoconstriction of the blood vessel causes greater damage to the delicate endothelial lining of the vessel wall [29, 30]. As a result, microdamage occurs causing a pro-inflammatory response, mediated by bradykinin and prostaglandin release, as a means to heal the tissue lining [31]. A pro-inflammatory response through these mediators is also pro-coagulatory. After fire suppression activities, an increase in platelet number and activity, coagulatory factors, and fibrin formation, all of which are the precursor to thrombogenicity, have been observed [31].

While some pro-inflammatory and pro-coagulatory measures such as platelet number return to baseline values within 2 h upon completion of fire suppression activities, other measures such as platelet activation remain significantly elevated into recovery [31–33]. This is an important consideration given the timing of most serious cardiac events including SCD occur duration or immediately after fire suppression activities [1, 10, 11]. A likely underlying mechanism is that pro-inflammatory and pro-coagulatory alterations in blood flow lead to other substrate including ischemia, infarction, or ventricular remodeling which are pro-arrhythmic and cause SCD. Remodeling is of specific interest as the phenomena of vasoconstriction, pro-inflammation, and pro-coagulation which occurs after one bout of strenuous

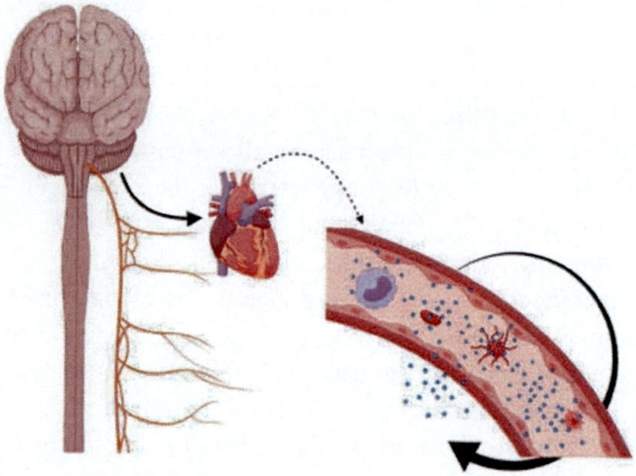

Fig. 3 Pathophysiological model linking sympathetic arousal to deleterious alterations in blood flow. This model explains how sympathetic arousal induces vasoconstriction, pro-inflammation, and pro-coagulation. First, sympathetic arousal as a response to the demands of fire suppression activities (1) directly increase the release of vasoconstricting catecholamines (e.g. norephedrine) depicted as blue particles and (2) directly increase heart rate. Combined, these two mechanisms induce shear stress on the vessel wall. Shear stress under high sympathetic activity is known to be a pro-inflammatory and pro-coagulatory environment, increasing the risk of thrombogenicity even in the absence of vascular disease. Figure created with BioRender; Credit to DJ Dzikowicz

exercise is repeated countless times over the career length of firefighters. Ultimately, these repeated bouts may place them at considerably higher risk [31].

Over time, the consequences of these repeated bouts of extreme demands which drive this combined phenomenon of vasoconstriction, pro-inflammation, and pro-coagulation induce arterial stiffness [34]. Arterial stiffness, described as the rigidity of the arterial wall, occurs and reflects gradual fragmentation and loss of elastin fibers and accumulation of stiffer collagen fibers in the arterial wall34. While this occurs transiently due to sympathetic arousal following acute perturbation among firefighters, increased arterial stiffness after firefighting activities may be particularly detrimental because it increases cardiac workload (e.g., increased afterload and compromised vasodilation) during diastole which is crucial for the restoration of coronary perfusion [35]. Moreover, arterial stiffness causes significant alterations in microvascular function and coronary perfusion, and novel markers of this alteration have been documented as predictors of serious cardiac events including SCD36,37. While arterial stiffness is important to understand, the alternations in microvascular function and coronary perfusion it causes include myocardial ischemia which affects a large proportion of firefighters [38].

Myocardial Ischemia

Myocardial ischemia, defined as a reduction in oxygenated blood flow to the myocardium, occurs among firefighters largely due to a mismatch between coronary artery supply and myocardial oxygen demands. When firefighters perform strenuous fire suppression activities, there is significantly greater myocardial oxygen demand as the heart attempts to supply oxygenated blood to the periphery. When these demands cannot be met by coronary supply, myocardial ischemia occurs [16]. Common causes of demand-mediated ischemia include coronary artery spasm, coronary embolism, arrhythmias, and alterations in blood pressure (e.g. hypertension, hypotension). In the setting of extreme environmental conditions including heat stress, dehydration, and carbon monoxide inhalation from noxious fumes, blood oxygenation reduces, blood viscosity rises and platelet aggregation becomes more common which can also induce myocardial ischemia [28, 75]. In some cases, platelet aggregation can also cause transient occlusion of the coronary artery. Such occlusion causes distal myocardial ischemia, and this model has been replicated among animals [39].

Myocardial ischemia is alarmingly prevalent among firefighters during the peak of fire suppression activities as well as during recovery, and is a robust predictor of serious cardiac events including SCD. Smith et al. [40] reported that 16% of a sample of firefighters had ischemic changes on the electrocardiogram and 20% had ventricular arrhythmias [40]. Al-Zaiti & Carey [22] reported that among a sample of firefighters undergoing exercise stress testing, 30% had changes on an electrocardiogram consistent with myocardial ischemia which persisted into the recovery period [22]. A known limitation of exercise stress testing with electrocardiography is that it is insensitive to myocardial ischemia of the smallest blood vessels or microvascular ischemia [41]; however, other more sensitive methods called hyperemic velocity based on myocardial blood flow suggest this prevalence be higher at approximately 45% [42, 43].

Physiologically, myocardial ischemia is concerning because it significantly alters depolarization, repolarization, and conduction velocity of the cardiomyocyte, and thereby acts as a potential trigger for life-threatening ventricular arrhythmias as seen in Fig. 4. Myocardial ischemia may last between 20–30 min before necrosis of tissue and subsequent infarct, even though firefighters may be completely asymptomatic or experience atypical symptoms (e.g. dyspnea, fatigue, etc.) [43]. If coronary flow restores and myocardial ischemia resolves, then the myocardium will recover but with a brief period of contractile dysfunction which still increases the risk of SCD [44, 45]. If coronary flow is not restored, the tissue becomes infarcted meaning that irreversible necrotic damage commences [44]. Either or, rest is essential for the firefighter to decrease the significantly elevated myocardial demands placed on the heart, and yet may be a propagator of serious cardiac events and SCD16. Premature ventricular contractions and other ectopic beats during recovery that happen to alterations in depolarization, repolarization, and conduction velocity of the cardiomyocyte are known to predict long-term cardiovascular mortality [22, 40, 46, 47]. There is an electrical instability between ischemic and nonischemic myocardium. The delayed repolarization velocity in one area but not the other can create a reentry circuit, which in turn leads to an arrhythmogenic foci for ventricular tachycardia/ventricular fibrillation.

Another cause of ischemia can be atherosclerotic heart disease. Atherosclerotic heart disease is characterized by the remodeling of the subendothelial accumulation of fatty deposits on the arterial wall which narrows the lumen [49]. These fatty deposits also promote inflammation, recruiting leukocytes, foam cells, and platelets

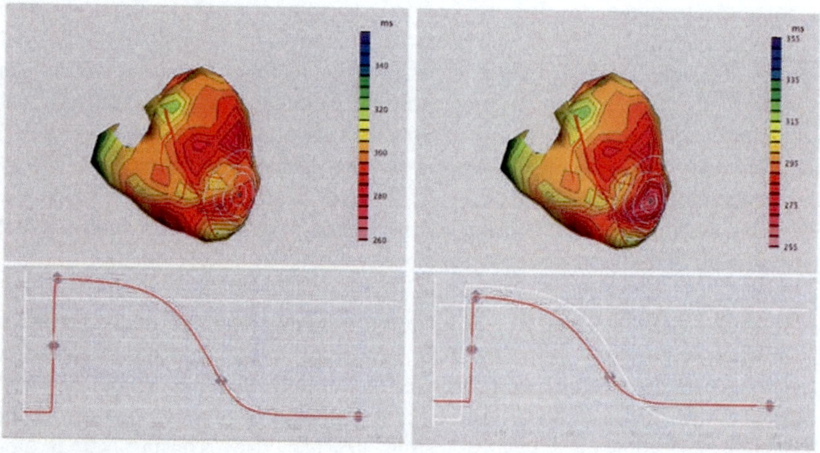

Fig. 4 Simulation of electrophysiological changes of the myocardium and cardiomyocyte due to myocardial ischemia. The normal electrophysiological potentials of the myocardium and normal cardiomyocyte action potential (left) compared to the ischemia-induced electrophysiological potentials of the myocardium and normal cardiomyocyte action potential (right). As can be appreciated, ischemia reduces the velocity at peak action potential but prolongs recovery (Phase 3). Figure created with ECGSim [48]; Credit to DJ Dzikowicz

Fig. 5 The Progression of atherosclerotic heart disease. As seen in the figure (from left to right), as atherosclerotic heart disease progresses, coronary lumen increasing narrows, increasing the risk of platelet aggregation and pro-inflammatory mediation which would subsequently occlude distal blood flow and induce myocardial ischemia

exacerbating the fatty deposits into a more formed atheroma which can rupture, leading to distal occlusion or causing a superimposed thrombus seen in Fig. 5 [49]. Regardless of the exact mechanism, this process will reduce the flow of oxygenated blood to the myocardium and induce myocardial ischemia [49]. In a study examining autopsy records between 1991 to 2014, 276 of 627 (44%) firefighter fatalities were cardiac-related [50]. Among the cardiac-related cases, atherosclerotic heart disease was present in 87% of cases [50]. Furthermore, among cardiac cases, coronary artery occlusion, that is, greater than 75% occlusion of the coronary lumen, was found in 60% of cases; and more than half of the cases reviewed had coronary artery occlusion of more than one of the three major coronary arteries [51, 52]. Interestingly, the extent of coronary disease among firefighters compared to the general population was lower. In the general population, nearly all (>95%) of SCD victims had at least one occluded coronary artery [51, 52]. The lower prevalence of coronary occlusion suggests that the preceding mechanisms of heightened sympathetic arousal and alterations in blood flow drive demand-mediated myocardial ischemia and not necessarily arterial occlusion [51, 52].

4 Comorbidities and Aging Augment Cardiovascular Workload Among Firefighters

While heightened sympathetic arousal, alterations in blood flow, and myocardial ischemia are deleterious to cardiovascular health among firefighters, underlying comorbidities compound and further exacerbate this damage. Among a sample of on-duty firefighters, a large proportion of firefighters were overweight or obese (90%), hypertensive (31%), and active or previous smokers (24%) [22]. Obesity, in particular, is dangerous because it is associated with an increased risk of job-related disability, lower job performance, and interacts with other cardiac conditions [53]. Obesity can also interact with other comorbidities such as hypertension and alter the electrical conduction of the heart, and reduce overall cardiorespiratory fitness, both important predictors of serious cardiac events and SCD [23, 43, 54, 55]. For example, Dzikowicz and Carey [54] demonstrated the interaction between obesity and hypertension significantly prolonged cardiac depolarization, increasing the risk of arrhythmia, and Smith et al., (2018) demonstrated obesity and hypertension significantly increased the mass of the heart (e.g. cardiomegaly) and led to the development of ventricular hypertrophy, a physiological substrate which contributes to lethal arrhythmias and SCD [43, 54].

In addition to metabolic comorbidities, Carey et al. (2012) reported that many firefighters reported sleep deprivation (59%), binge drinking behavior (58%), poor mental well-being (21%), depression (11%), poor physical well-being (8%), or poor social bonding (4%) all of which may also strain the cardiovascular system and taxing to overall health56. Chronic sleep deprivation reduces the ability to think, handle complex mental tasks, solve problems, form new memories, and has been associated with persistent sympathetic arousal [15, 56]. Furthermore, alcohol-use disorders are associated with depressive episodes, severe anxiety, insomnia, suicide, and abuse of other drugs [56]. Firefighters have consistently reported high rates of alcohol use and misuse; Carey et al. [56] reported that among 112 firefighters 58% reported binge drinking behavior and 14% reported hazardous drinking behavior. Evidence suggests firefighters had a have a higher occurrence of binge drinking than members of other emergency occupations (e.g. police officers, emergency medical services) as well as the general population and the relationship between employment type and binge drinking is mediated by gender [56, 76]. Specifically, in terms of poor mental well-being and depression, Liao et al. [57] reported these conditions were associated with persistent sympathetic arousal [57]. This is concerning because without adequate restoration of parasympathetic dominance, this significantly increases the risk of serious cardiac events including SCD.

Another critical comorbidity among firefighters is aging itself. The fire service has a high retention rate, meaning firefighters remain in the occupation [58]. In the US, there are nearly 25,000 firefighters, most are men (82%) and the average age is 47 years old; thus, the occupation is aging [58]. Since firefighting is a lifetime career with very little attrition, many firefighters continue to serve in their late 50 s significantly increasing their risk of serious cardiac events including SCD1,9,10. For example, one study demonstrated that firefighters aged 20–65 years were found to have similar cardiorespiratory health and physical fitness as the sedentary segment of the general population of the same age [59]. Furthermore, in the same study, firefighters between the ages 40–45-, 50–55-, and 60–65-years-old had low maximal aerobic capacity (31, 28, 26 ml/kg/min), high percent body fat (26%, 29%, 30%), high resting blood pressure (136/86, 140/90, 143/93 mm Hg), and the majority in each age range (66%, 83%, 93%) fell below the lowest published recommendations for maximal aerobic capacity [59]. More recently, Dzikowicz and Carey [60] examined the relationship between career length and age and noted that career length more so than age was the driving factor for myocardial ischemia severity which reflects the occupational burdens of firefighting on the heart [60]. This is an important consideration as cardiovascular risk stratification is largely age-dependent and does not take occupational considerations.

5 Future Directions for Smart Firefighting

In this chapter, we discussed three mechanisms responsible for serious cardiac events including SCD among firefighters: heightened sympathetic arousal, alterations in blood flow, as well as myocardial ischemia in addition to the compounding effects of comorbidities. In the following section, we will discuss a number of research directions to address these mechanisms in an effort to reduce serious cardiac events and SCD.

Starting Anti-Platelet Therapies

One potential intervention is the use of anti-platelet agents to prevent platelet aggregation and alterations in blood flow. A common anti-platelet drug is an aspirin which has been known to reduce serious cardiac events in individuals of the general population who have cardiovascular disease (CVD) risk factors. However, multiple studies using varying designs have demonstrated neither acute nor chronic low-dose aspirin therapy improves endothelial function and mitigates vascular inflammation [72–75]. Investigational use of other anti-platelet and anti-inflammatory agents may be necessary.

Improving Exercise Capacity

Given that the majority of time spent by firefighters is at rest, firefighters are plagued by physical inactivity defined as less than moderate to intense activity for at least 150 min per week. Routine physical activity has been demonstrated to reduce resting sympathetic arousal and is an obtainable goal for firefighters. Among a large cohort of 1,104 firefighters, Yang et al. (2019) demonstrated that higher baseline push-up capacity, a measure of more regular physical activity, was associated with lower systolic blood pressure and lower incidence of CVD events [70]. As means to increase physical activity among firefighters, Donahue et al. [71] implemented a telehealth-delivered 6-week functional circuit exercise intervention and Mediterranean Diet plan among 15 firefighters and demonstrated a significant improvement in firefighter physical activity and resting blood pressure [71]. Telehealth services are ideal for firefighters as their work requires them to be within the firehouse.

Reducing the Prevalence of Obesity and Metabolic Syndrome

Reductions in obesity, metabolic syndrome, blood pressure, and heart rates which all play an interacting role to alter blood flow may be essential to reducing serious cardiac events including SCD. Rates of obesity, metabolic syndrome, blood pressure, and resting heart rate among firefighters exceed Healthy People 2030 targets and are higher than the general population [66]. As prefaced, Donahue et al. (2022) implemented a telehealth-delivered intervention which included implementation of the Mediterranean Diet which emphasized vegetables, fruit, two to low/non-fat dairy, and fish and poultry, and reported a significant decrease in weight- 1.4 kg in just 6 weeks [71]. Moreover, this led to significant improvements in exercise performance [71]. In a different prospective interventional study recruiting firefighters from a single firehouse, a 12-week low-glycemic nutritional fitness program significantly lowered the prevalence of metabolic syndrome, and the average risk factors present for metabolic syndrome [64]. An important consideration for interventions aimed at addressing obesity and metabolic syndrome among firefighters is to also improve health knowledge. In a recent study on health knowledge among younger firefighters (30–39 years old), those with at least one risk factor had improved attitudes toward regular anaerobic exercise [63]. Incorporating holistic interventions which also address health education may be crucial for the longevity and sustainability of interventions.

The sense of community and brotherhood among firefighters is a unique aspect which can be a powerful motivator for reducing the prevalence of obesity and metabolic syndrome. While previous interventions have aimed at implementing more individual focused programs to promote positive behavioral changes and risk reduction, most results have been modest and short-lived because they do not conform to the firefighter's community and brotherhood. Firefighting, especially in the United States, is a lifelong job lasting 20–25 years and involves firefighters sleeping, eating, and working together for a series of days and repeated throughout their whole career. Carey et al. [56] wrote that due to this work schedule, firefighters develop close social bonds and connections with one another which is important in ameliorating distress and developing lifestyle patterns [56]. In their work, Carey et al. [56] demonstrated that firefighters who reported poor social bonding were more likely to report poor mental-health and be more likely obese and hypertensive [56]. With this in mind, we advocate that future research incorporate this strong sense of community and brotherhood into intervention designs. Instead of focusing on the individual, changing workplace norms and adopting policies that are supportive of healthy behaviors, the US Fire Administration may be able reach more firefighters and reduce the risk of serious cardiac events and SCD.

Adhering to Annual Physicals and Exercise Testing

By ensuring firefighters undergo annual physical exams including cardiopulmonary testing such as other professions in the military and transportation sectors, myocardial ischemia during physical activity can be easily identified which can help reduce the prevalence of serious cardiac events and SCD [65]. Cardiopulmonary testing refers to stressing the body by putting an exercise load on it and sometimes measuring the response using electrocardiography, the gold standard for myocardial ischemia, as well as blood pressure. Cardiopulmonary testing with electrocardiography is diagnostic for myocardial ischemia and indicative of atherosclerotic disease, and largely available across hospital systems. While physical exams for firefighters include review of the body systems, vital signs, and exercise stressing testing at the time of hire, there is no national mandate to maintain "good health" standards [53, 65]. When compared to other professions, this is a significant departure in the standard of practice, as well as a departure from the International Associate of Firefighters' (IAFF) own Fire Service Joint Labor/Management Wellness-Fitness Initiative which recommends cardiopulmonary testing for all firefighters at time of hire and annually thereafter.

However, current IAFF guidelines advise that firefighters under sub-maximal cardiopulmonary testing without electrocardiography. Under this protocol, firefighters exercise until reaching an 85 percent of the predicted maximum heart rate (maximum heart rate = 220-age) [53, 65]. For sub-maximal cardiopulmonary testing with electrocardiography, the focus is more so on heart rate, firefighter symptomology such as exertional dyspnea, and prediction of maximum exercise capacity. However, it does not routinely include electrocardiography. While cardiopulmonary testing without electrocardiography is relatively inexpensive, easier to administer by an exercise specialist, and does not require medical personnel, it does not accurately measure myocardial ischemia which is a significant limitation.

Addressing Mental Health

As we have discussed, firefighting is a demanding profession that can take a toll on the mental and physical health of firefighters. Firefighters are exposed to a wide range of physical and emotional challenges on a daily basis, which can lead to mental health issues. In particular, firefighters may be at risk of developing depression, anxiety, and post-traumatic stress disorder (PTSD). Carey et al. [56] reported that among a sample of 112 firefighters, 21% reported poor mental well-being, specifically 11% reported depression and 4% reported poor social bonding [56]. As discussed, firefighters are prone to using substances such as alcohol to cope with these mental health issues; Zegel et al. [77] reported empirical evidence of the compounding effect on alcohol use with underlying mental health problems including PTSD77. While mental health issues are common in all types of firefighters, volunteer firefighters report more severe psychiatric symptoms and more structural barriers to receiving care compared to career firefighters [78].

A number of programs and interventions have been developed to address mental health among the fire service with most prioritizing individual firefighters. Such programs include training on stress management and coping strategies, prioritizing self-care, and seeking support from peers. For example, a number of studies have looked at models to develop coping strategies to handle stressors related to firefighting and have reported encouraging results [79, 80]. However, these results have not been entirely reproducible. In one study which tested a four-hour resilience training intervention as part onboarding training failed to reduce symptoms of mental health issues after 6- and 12-months [80]. The researchers cited significant organizational culture, lack of resources, and poor awareness and stigma as possible reasons for the study's null findings [80]. Other studies have focused more on improving self-care, specifically sleep. In a recent study, most firefighters (69%) reported less than ideal sleep quality and 21% screened positive for clinical insomnia, with no significant difference between volunteer and career firefighters [81]. In the same study, clinical insomnia was significantly associated with PTSD, anxiety, and depression [81]. Innovatively, Carey et al. [82] reported a "Quiet Firehouse" intervention which restricted unnecessary fire alarms, reduced light levels, and regulated the temperature ideal for sleep [82]. While in pilot form, the results supported that reducing environmental stimuli in the firehouse improved sleep quality as well as actually lowering blood pressure

[82]. Thus, addressing mental health concerns can also have a positive impact on cardiovascular health.

To improve mental health among firefighters, fire departments and fire organizations must take action as well. The IAFF has responded to mental health concerns among firefighters by developing a peer support program for firefighters, developing and offering a one-day, eight-hour, in-person training focused on building individual and organizational resilience in the fire service, and evening providing mental health clinicians an intensive course to build best practices when treating professional firefighters,

Introducing Real-Time Physiological Monitoring

Other future work should focus on the development of better physiological monitoring during fire suppression activities. Newer wearable technologies such as electrocardiographic patches placed over the heart now make it possible for real-time monitoring of the heart and other vital signs during actual fire suppression activities [61]. The development of machine learning and other highly intelligent models may be predictive of serious cardiac events and relay this information quickly to field supervisors [62]. However, the development of such algorithms must be mindful and use data collected and applicable to firefighters. For example, Li et al. (2023) reported that a machine learning algorithm using electrocardiographic data collected among on-duty firefighters during fire suppression activities had excellent accuracy for the detection of arrhythmias. However, when the algorithm was trained on data collected from a sample of largely younger, male, White, and stationary subjects, accuracy declined by as much as 40%. While further work is needed before the dissemination of this particular model, this example underscores that any algorithms for real-time physiological monitoring must use training data collected from firefighters due to the unique nature of their work and their unique risks. Furthermore, further investigation as to the feasibility of wearing such wearable devices during fire suppression activities, the lead placement of the wearable devices, the signal quality collected from wearable devices, the transmission of data during fire suppression activities, and the development of multiparameter monitoring including measurements reflective of heightened sympathetic arousal, alterations in blood flow, and myocardial ischemia to be all-encompassing remains to be investigated.

6 Conclusion

Firefighters are among the most trusted professions and provide an essential service for public safety. In order to best protect our communities, it is a national priority to reduce the incidence of serious cardiac events including SCD among professional firefighters [68, 69]. Nearly half of the on-duty fatalities among firefighters are due to SCD and the potential mechanisms include heightened sympathetic arousal, alterations in blood flow resulting in myocardial ischemia. Confounding the risk of SCD in addition is the fact that the fire service is aging and suffers from a remarkably high prevalence of the metabolic disease which includes obesity, hypertension, diabetes mellitus, sleep apnea, and other cardiometabolic conditions. A number of interventions have been tested to address the elevated risk among firefighters including aspirin therapy, improving exercise capacity, reducing the impact of metabolic syndrome, mandating annual physicians with cardiopulmonary testing, and addressing mental health concerns. Moreover, some classical interventions like exercise and diet are being delivered in interaction and digital ways to increase uptake and maintain sustainability. With that said, these interventions largely remain in the developmental stage and are not nationally mandated. In addition, development of novel technologies for real-time monitoring of firefighters is being developed with significant potential at identifying precursor ischemia and preventing serious cardiac events including SCD in real-time.

References

1. S.N. Kales, D.L. Smith, Firefighting and the Heart: Implications for Prevention. Circulation **135**(14), 1296–1299 (2017). https://doi.org/10.1161/CIRCULATIONAHA.117.027018. (PMID: 28373524)
2. M. Han, S. Park, J.H. Park, S.S. Hwang, I. Kim, Do police officers and firefighters have a higher risk of disease than other public officers? A 13-year nationwide cohort study in South Korea. BMJ Open **8**(1), e019987 (2018). https://doi.org/10.1136/bmjopen-2017-019987.PMID:29391373;PMCID:PMC5878257
3. R. Fahy, *US firefights fatalities due to sudden cardiac death 1995–2004*, National Fire Protection Association. (National Institute for Occupational Safety and Health (NIOSH), 200)
4. R. Fahy, US firefights deaths in 2021. National Fire Protection Association. NFPA J., 2022
5. A. Farioli, J. Yang, D. Teehan, D.M. Baur, D.L. Smith, S.N. Kales, Duty-related risk of sudden cardiac death among young US firefighters. Occup Med (Lond). **64**(6), 428–435. https://doi.org/10.1093/occmed/kqu102. Epub 2014 Aug 7. PMID: 25104277; PMCID: PMC7811883
6. S.N. Kales, E.S. Soteriades, C.A. Christophi, D.C. Christiani, Emergency duties and deaths from heart disease among firefighters in the United States. N. Engl. J. Med. **356**(12), 1207–1215 (2007). https://doi.org/10.1056/NEJMoa060357. (PMID: 17377158)
7. D.M. Baur, C.A. Christophi, S.N. Kales, Metabolic syndrome is inversely related to cardiorespiratory fitness in male career firefighters. J. Strength Cond. Res. **26**(9), 2331–2337 (2012). https://doi.org/10.1519/JSC.0b013e31823e9b19. (PMID: 22067249)
8. S.M. Moffatt, D.F. Stewart, K. Jack, M.D. Dudar, E.D. Bode, K.C. Mathias, D.L. Smith, Cardiometabolic health among United States firefighters by age. Prev Med Rep. **14**(23), 101492 (2021). https://doi.org/10.1016/j.pmedr.2021.101492.PMID:34354906;PMCID:PMC8322457
9. S.N. Kales, D.L. Smith, Sudden cardiac death in the fire service. Occup. Med. (Lond.) **64**(4), 228–230 (2014). https://doi.org/10.1093/occmed/kqu057. (PMID: 24850812)
10. E.S. Soteriades, D.L. Smith, A.J. Tsismenakis, D.M. Baur, S.N. Kales, Cardiovascular disease in US firefighters: a systematic review. Cardiol Rev. **19**(4), 202–215. https://doi.org/10.1097/CRD.0b013e318215c105. PMID: 21646874
11. C.M. Albert, M.A. Mittleman, C.U. Chae, I.M. Lee, C.H. Hennekens, J.E. Manson, Triggering of sudden death from cardiac causes by vigorous exertion. N. Engl. J. Med. **343**(19), 1355–1361 (2000). https://doi.org/10.1056/NEJM200011093431902. (PMID: 11070099)
12. B. Choi, S. Ko, S. Kojaku, Resting heart rate, heart rate reserve, and metabolic syndrome in professional firefighters: A cross-sectional study. Am. J. Ind. Med. **60**(10), 900–910 (2017). https://doi.org/10.1002/ajim.22752. (Epub 2017 Sep 4 PMID: 28869309)

13. R.A. Marciniak, C.J. Tesch, K.T. Ebersole, Heart rate response to alarm tones in firefighters. Int. Arch. Occup. Environ. Health **94**(5), 783–789 (2021). https://doi.org/10.1007/s00420-020-01646-y. (Epub 2021 Jan 10 PMID: 33423092)
14. P.J. Rynne, C.C. Derella, C. McMorrow, R.L. Dickinson, S. Donahue, A.A. Almeida, M. Carty, D.L. Feairheller, Blood pressure responses are dependent on call type and related to hypertension status in firefighters. Blood Press. **32**(1), 2161997 (2023). https://doi.org/10.1080/08037051.2022.2161997. (PMID: 36597210)
15. L.M. Liao, M.G. Carey, Laboratory-induced mental stress, cardiovascular response, and psychological characteristics. Rev. Cardiovasc. Med. **16**(1), 28–35 (2015). https://doi.org/10.3909/ricm0730. (PMID: 25813794)
16. S. Michael, K.S. Graham, G.M. Davis, Cardiac autonomic responses during exercise and post-exercise recovery using heart rate variability and systolic time intervals—a review. Front. Physiol. **29**(8), 301 (2017)
17. K.T. Ebersole, D.J. Cornell, R.J. Flees, C.M. Shemelya, S.E. Noel, Contribution of the autonomic nervous system to recovery in firefighters. J. Athl. Train. **55**(9), 1001–1008 (2020). https://doi.org/10.4085/1062-6050-0426.19.PMID:32841323;PMCID:PMC7534942
18. P.C. Fehling, J.M. Haller, W.K. Lefferts, E.M. Hultquist, M. Wharton, T.W. Rowland, D.L. Smith, Effect of exercise, heat stress and dehydration on myocardial performance. Occup. Med. **65**(4), 317–323 (2015)
19. J. Gonzalez-Alonso, R. Mora-Rodríguez, E.F. Coyle, Stroke volume during exercise: interaction of environment and hydration. Am. J. Physiol.-Hear. Circ. Physiol. **278**(2), H321–H330 (2000)
20. C.R. Cole, E.H. Blackstone, F.J. Pashkow, C.E. Snader, M.S. Lauer, Heart-rate recovery immediately after exercise as a predictor of mortality. N. Engl. J. Med. **341**(18), 1351–1357 (1999)
21. Y.J. van de Vegte, P. van der Harst, N. Verweij, Heart rate recovery 10 seconds after cessation of exercise predicts death. J. Am. Heart Assoc. **7**(8), e008341 (2018). https://doi.org/10.1161/JAHA.117.008341.PMID:29622586;PMCID:PMC6015434
22. S.S. Al-Zaiti, M.G. Carey, The prevalence of clinical and electrocardiographic risk factors of cardiovascular death among on-duty professional firefighters. J Cardiovasc Nurs. **30**(5), 440–446 (2015). https://doi.org/10.1097/JCN.0000000000000165. PMID: 24874885; PMCID: PMC4247346
23. D.J. Dzikowicz, M.G. Carey, Correlates of autonomic function, hemodynamics, and physical activity performance during exercise stress testing among firefighters. Biol. Res. Nurs. **29**, 10998004221143508 (2022)
24. J.R. Resar, R.M. Judd, H.R. Halperin, V.P. Chacko, R.G. Weiss, F.C. Yin, Direct evidence that coronary perfusion affects diastolic myocardial mechanical properties in canine heart. Cardiovasc. Res. **27**(3), 403–410 (1993). https://doi.org/10.1093/cvr/27.3.403. (PMID: 8490939)
25. D. Merkus, F. Kajiya, H. Vink, I. Vergroesen, J. Dankelman, M. Goto, J.A. Spaan, Prolonged diastolic time fraction protects myocardial perfusion when coronary blood flow is reduced. Circulation **100**(1), 75–81 (1999). https://doi.org/10.1161/01.cir.100.1.75. (PMID: 10393684)
26. B. Fernhall, C.A. Fahs, G. Horn, T. Rowland, D. Smith, Acute effects of firefighting on cardiac performance. Eur. J. Appl. Physiol. **112**, 735–741 (2012)
27. E.C. Lefferts, A.J. Rosenberg, G. Grigoriadis, S.O. Wee, S. Kerber, K.W. Fent, G.P. Horn, D.L. Smith, B. Fernhall, Firefighter hemodynamic responses to different fire training environments. Vasc. Med. **26**(3), 240–246 (2021)
28. R.M. Bruno, L. Ghiadoni, G. Seravalle, R. Dell'Oro, S. Taddei, G. Grassi, Sympathetic regulation of vascular function in health and disease. Front. Physiol. **24**(3), 284 (2012)
29. N.K. Foster, J.B. Martyn, R.E. Rangno, J.C. Hogg, R.L. Pardy, Leukocytosis of exercise: role of cardiac output and catecholamines. J. Appl. Physiol. **61**(6), 2218–2223 (1986)
30. N. Li, N.H. Wallén, P. Hjemdahl, Evidence for prothrombotic effects of exercise and limited protection by aspirin. Circulation **100**(13), 1374–1379 (1999)
31. A.L. Hunter, A.S. Shah, J.P. Langrish, J.B. Raftis, A.J. Lucking, M. Brittan, S. Venkatasubramanian, C.L. Stables, D. Stelzle, J. Marshall, R. Graveling, Fire simulation and cardiovascular health in firefighters. Circulation **135**(14), 1284–1295 (2017)

32. S.S. Hegde, A.H. Goldfarb, S.E. Hegde, Clotting and fibrinolytic activity change during the 1 h after a submaximal run. Med. Sci. Sports Exerc. **33**(6), 887–892 (2001)
33. X. Lin, M.S. El-Sayed, J. Waterhouse, T. Reilly, Activation and disturbance of blood haemostasis following strenuous physical exercise. Int. J. Sports Med. **20**(03), 149–153 (1999)
34. T. Nagel, B. Melton, G.J. Grosicki, Fighting arterial stiffness: A case for the importance of cardiorespiratory fitness in firefighters. Med. Hypotheses **1**(162), 110831 (2022)
35. C.A. Fahs, H. Yan, S. Ranadive, L.M. Rossow, S. Agiovlasitis, G. Echols, D. Smith, G.P. Horn, T. Rowland, A. Lane, B. Fernhall, Acute effects of firefighting on arterial stiffness and blood flow. Vasc. Med. **16**(2), 113–118 (2011)
36. W.K. Lefferts, K.S. Heffernan, E.M. Hultquist, P.C. Fehling, D.L. Smith, Vascular and central hemodynamic changes following exercise-induced heat stress. Vasc. Med. **20**(3), 222–229 (2015)
37. T.R. Nagel, B.F. Melton, J.M. Keeler, B.L. Cross, W.T. Blumenburg, A.A. Flatt, L. Stoner, G.J. Grosicki, Cardiorespiratory fitness is inversely associated with aortic arterial stiffness in firefighters. J. Occup. Environ. Med. **64**(10), e641–e646 (2022)
38. S.J. Zieman, V. Melenovsky, D.A. Kass, Mechanisms, pathophysiology, and therapy of arterial stiffness. Arterioscler. Thromb. Vasc. Biol. **25**(5), 932–943 (2005). https://doi.org/10.1161/01.ATV.0000160548.78317.29. (Epub 2005 Feb 24 PMID: 15731494)
39. A. Schömig, G. Richardt, Cardiac sympathetic activity in myocardial ischemia: release and effects of noradrenaline. Basic Res. Cardiol. **85**(Suppl 1), 9–30 (1990). https://doi.org/10.1007/978-3-662-11038-6_2. (PMID: 2091611)
40. D.L. Smith, G.P. Horn, B. Fernhall, R.M. Kesler, K.W. Fent, S. Kerber, T.W. Rowland, Electrocardiographic responses following live-fire firefighting drills. J. Occup. Environ. Med. **61**(12), 1030–1035 (2019). https://doi.org/10.1097/JOM.0000000000001730.PMID:31599801;PMCID:PMC8630674
41. S.S. Al-Zaiti, K.N. Runco, M.G. Carey, Increased T wave complexity can indicate subclinical myocardial ischemia in asymptomatic adults. J Electrocardiol. **44**(6), 684–688 (2011). https://doi.org/10.1016/j.jelectrocard.2011.07.017. Epub 2011 Sep 14. PMID: 21924433; PMCID: PMC3200448
42. T.J. Anderson, F. Charbonneau, L.M. Title, J. Buithieu, M.S. Rose, H. Conradson, K. Hildebrand, M. Fung, S. Verma, E.M. Lonn, Microvascular function predicts cardiovascular events in primary prevention: long-term results from the Firefighters and their endothelium (FATE) study. Circulation **123**(2), 163–169 (2011). https://doi.org/10.1161/CIRCULATIONAHA.110.953653. (Epub 2011 Jan 3 PMID: 21200002)
43. D.L. Smith, E.L. Graham, J.A. Douglas, K. Jack, M.J. Conner, R. Arena, S. Chaudhry, Subclinical cardiac dysfunction is associated with reduced cardiorespiratory fitness and cardiometabolic risk factors in firefighters. Am. J. Med. **135**(6), 752–760 (2022)
44. D.J. Hausenloy, D.M. Yellon, Ischaemic conditioning and reperfusion injury. Nat. Rev. Cardiol. **13**(4), 193–209 (2016). https://doi.org/10.1038/nrcardio.2016.5. (Epub 2016 Feb 4 PMID: 26843289)
45. F. Tomai, F. Crea, L. Chiariello, P.A. Gioffrè, Ischemic preconditioning in humans: models, mediators, and clinical relevance. Circulation **100**(5), 559–563 (1999)
46. D.J. Dzikowicz, M.G. Carey, Exercise-induced premature ventricular contractions are associated with myocardial ischemia among asymptomatic adult male firefighters: Implications for enhanced risk stratification. Biol. Res. Nurs. **22**(3), 369–377 (2020)
47. M.M. Refaat, C. Gharios, M.V. Moorthy, F. Abdulhai, R.S. Blumenthal, M.A. Jaffa, S. Mora, Exercise-induced ventricular ectopy and cardiovascular mortality in asymptomatic individuals. J. Am. Coll. Cardiol. **78**(23), 2267–2277 (2021)
48. A. Van Oosterom, T.F. Oostendorp, ECGSIM: an interactive tool for studying the genesis of QRST waveforms. Heart **90**(2), 165–168 (2004)
49. M. Chen, Q. Liu, S. Zhou, The networks between the sympathetic nervous system and immune system in atherosclerosis. J. Am. Coll. Cardiol. **68**(4), 431–432 (2016)
50. D.L. Smith, J.M. Haller, M. Korre, P.C. Fehling, K. Sampani, L.G.G. Porto, C.A. Christophi, S.N. Kales, Pathoanatomic findings associated with duty-related cardiac death in US firefighters: A case-control study. J. Am. Heart Assoc. **7**, e009446 (2018)

51. C.A. Warnes, W.C. Roberts, Sudden coronary death: relation of amount and distribution of coronary narrowing at necropsy to previous symptoms of myocardial ischemia, left ventricular scarring and heart weight. Am. J. Cardiol. **54**(1), 65–73 (1984)
52. M.J. Davies, Anatomic features in victims of sudden coronary death. Coronary artery pathology. Circulation. **85**(1 Suppl), I19-24 (1992)
53. M.L. Wilkinson, A.L. Brown, W.S. Poston, C.K. Haddock, S.A. Jahnke, R.S. Day, Peer reviewed: Physician weight recommendations for overweight and obese firefighters, United States, 2011–2012. Prev. Chronic Disease. **11**, (2014)
54. D.J. Dzikowicz, M.G. Carey, Obesity and hypertension contribute to prolong QRS complex duration among middle-aged adults. Ann. Noninvasive Electrocardiol. **24**(6), e12665 (2019)
55. D.J. Dzikowicz, M.G. Carey, Widened QRS-T angle may be a measure of poor ventricular stretch during exercise among on-duty firefighters. J Cardiovasc Nurs. **34**(3), 201–207 (2019). https://doi.org/10.1097/JCN.0000000000000554. PMID: 30520777; PMCID: PMC6528825
56. M.G. Carey, S.S. Al-Zaiti, G.E. Dean, L. Sessanna, D.S. Finnell, Sleep problems, depression, substance use, social bonding, and quality of life in professional firefighters. J. Occup. Environ. Med. **53**(8), 928–933 (2011)
57. L.M. Liao, S.S. Al-Zaiti, M.G. Carey, Depression and heart rate variability in firefighters. SAGE open medicine **2**, 2050312114545530 (2014)
58. T.C.E. Zippia, *Fire fighter demographics and statistics in the US* (2023)
59. K. Saupe, M. Sothmann, D. Jasenof, Aging and the fitness of fire fighters: the complex issues involved in abolishing mandatory retirement ages. Am. J. Public Health **81**(9), 1192–1194 (1991). https://doi.org/10.2105/ajph.81.9.1192.PMID:1951832;PMCID:PMC1405652
60. D.J. Dzikowicz, M.G. Carey, Severity of myocardial ischemia is related to career length rather than age among professional firefighters. Workplace health & safety **69**(4), 168–173 (2021)
61. A. Coca, R.J. Roberge, W.J. Williams, D.P. Landsittel, J.B. Powell, A. Palmiero, Physiological monitoring in firefighter ensembles: wearable plethysmographic sensor vest versus standard equipment. J. Occup. Environ. Hyg. **7**(2), 109–114 (2010). https://doi.org/10.1080/15459620903455722. (PMID: 20017053)
62. U. Pluntke, S. Gerke, A. Sridhar, J. Weiss, B. Michel, Evaluation and classification of physical and psychological stress in firefighters using heart rate variability. Annu Int Conf IEEE Eng Med Biol Soc. **2019**, 2207–2212 (2019). https://doi.org/10.1109/EMBC.2019.8856596. (PMID: 31946339)
63. J. Ras, L. Leach, Firefighters' health knowledge, cardiovascular disease risk factors, and sociodemographic characteristics as predictors of firefighters attitudes toward health. J. Occup. Environ. Med. **64**, e705–e713 (2022)
64. M.G. Carey, S.S. Al-Zaiti, L.M. Liao, H.N. Martin, R.A. Butler, A low-glycemic nutritional fitness program to reverse metabolic syndrome in professional firefighters: results of a pilot study. J Cardiovasc Nurs. **26**(4), 298–304 (2011). https://doi.org/10.1097/JCN.0b013e31820344d7. PMID: 21263343; PMCID: PMC3114284
65. E. von Heimburg, J.I. Medbø, M. Sandsund, R.E. Reinertsen, Performance on a work-simulating firefighter test versus approved laboratory tests for firefighters and applicants. Int. J. Occup. Saf. Ergon.: JOSE **19**(2), 227–243 (2013)
66. C.A. Gómez, D.V. Kleinman, N. Pronk, G.L. Wrenn Gordon, E. Ochiai, C. Blakey, A. Johnson, K.H. Brewer, Addressing health equity and social determinants of health through healthy people 2030. J Public Health Manag Pract. **27**(Suppl 6), S249-S257 (2021). https://doi.org/10.1097/PHH.0000000000001297. PMID: 33729197; PMCID: PMC8478299
67. M. Carnethon, L.P. Whitsel, B.A. Franklin, P. Kris-Etherton, R. Milani, C.A. Pratt, G.R. Wagner, Worksite wellness programs for cardiovascular disease prevention: a policy statement from the American Heart Association. Circulation **120**(17), 1725–1741 (2009)
68. S.L. Tamers, R. Pana-Cryan, T. Ruff, J.M. Streit, M.A. Flynn, A. Childress, C.C. Chang, E. Novicki, T. Ray, D.E. Fosbroke, C.L. Geraci, *The NIOSH Future of Work Initiative Research Agenda*
69. W.A. Robison, D.F. Williams, P. Grandillo, *NIOSH Extramural Research and Training Program, annual report of fiscal year 2020*

70. J. Yang, C.A. Christophi, A. Farioli, D.M. Baur, S. Moffatt, T.W. Zollinger, S.N. Kales, Association between push-up exercise capacity and future cardiovascular events among active adult men. JAMA Netw. Open **2**(2), e188341 (2019). https://doi.org/10.1001/jamanetworkopen.2018.8341.PMID:30768197;PMCID:PMC6484614
71. S. Donahue, C. McMorrow, A.A. Almeida, D.L. Feairheller, Feasibility and perception of a diet and exercise intervention delivered via telehealth to firefighters. Int J Telerehabil. **14**(1), e6458 (2022). https://doi.org/10.5195/ijt.2022.6458.PMID:35734390;PMCID:PMC9187031
72. S.J. McEntire, S.E. Reis, O.E. Suman, D. Hostler, Effects of low-dose aspirin therapy on thermoregulation in firefighters. Saf Health Work. **6**(3), 256–262 (2015). https://doi.org/10.1016/j.shaw.2015.06.003. Epub 2015 Jun 26. PMID: 26929836; PMCID: PMC4674504
73. D.L. Smith, N.M.G. Friedman, S.I. Bloom, W.L. Armero, B.D. Pence, M.D. Cook, B. Fernhall, G.P. Horn, J. Woods, Firefighting induces acute inflammatory responses that are not relieved by aspirin in older firefighters. J. Occup. Environ. Med. **61**(7), 617–622 (2019). https://doi.org/10.1097/JOM.0000000000001626. (PMID: 31090673)
74. O. Olafiranye, D. Hostler, D.G. Winger, L. Wang, S.E. Reis, Effect of aspirin on acute changes in peripheral arterial stiffness and endothelial function following exertional heat stress in firefighters: The factorial group results of the Enhanced Firefighter Rehab Trial. Vasc. Med. **20**(3), 230–236 (2015). https://doi.org/10.1177/1358863X15571447. (Epub 2015 May 4 PMID: 25939657)
75. C.R. Henry, D. Satran, B. Lindgren, C. Adkinson, C.I. Nicholson, T.D. Henry, Myocardial injury and long-term mortality following moderate to severe carbon monoxide poisoning. JAMA **295**(4), 398–402 (2006)
76. H. Weir, D.M. Stewart, R.G. Morris, Problematic alcohol consumption by police officers and other protective service employees: A comparative analysis. J. Crim. Just. **40**(1), 72–82 (2012)
77. M. Zegel, A. Lebeaut, N. Healy, J.K. Tran, A.A. Vujanovic, Mental health correlates of probable posttraumatic stress disorder, probable alcohol use disorder, and their co-occurrence among firefighters. Behav. Modif. **46**(2), 395–421 (2022). https://doi.org/10.1177/01454455211033517. (Epub 2021 Jul 29 PMID: 34323099)
78. I.H. Stanley, J.W. Boffa, M.A. Hom, N.A. Kimbrel, T.E. Joiner, Differences in psychiatric symptoms and barriers to mental health care between volunteer and career firefighters. Psychiatry Res. **247**, 236–242 (2017). https://doi.org/10.1016/j.psychres.2016.11.037. (Epub 2016 Nov 28 PMID: 27930964)
79. S.K. Yang, Y. Ha, Predicting posttraumatic growth among firefighters: the role of deliberate rumination and problem-focused coping. Int. J. Environ. Res. Public Health **16**(20), 3879 (2019). https://doi.org/10.3390/ijerph16203879.PMID:31614945;PMCID:PMC6843524
80. P.M. Skeffington, C.S. Rees, T.G. Mazzucchelli, R.T. Kane, The primary prevention of PTSD in firefighters: Preliminary results of an RCT with 12-month follow-up. PLoS ONE **11**(7), e0155873 (2016). https://doi.org/10.1371/journal.pone.0155873.PMID:27382968;PMCID:PMC4934826
81. H. Cramm, R. Richmond, L. Jamshidi, M. Edgelow, D. Groll, R. Ricciardelli, J.C. MacDermid, M. Keiley, R.N. Carleton, Mental health of canadian firefighters: the impact of sleep. Int. J. Environ. Res. Public Health **18**(24), 13256 (2021). https://doi.org/10.3390/ijerph182413256.PMID:34948864;PMCID:PMC8701315
82. M.G. Carey, A.N. Baldzizhar, C.R. Miterko, B.W. Qualls, R.L. Vincent, R. Merrick, G.E. Dean, A quiet firehouse: reducing environmental stimuli among professional on-duty firefighters. J. Occup. Environ. Med. **60**(2), 186–190 (2018). https://doi.org/10.1097/JOM.0000000000001199. (PMID: 29049092)

Robotic Firefighting: A Review and Future Perspective

Meng Wang, Xinghao Chen, and Xinyan Huang

Abstract Firefighters are constantly exposed to the danger of hot, dark, and toxic fire environments during their firefighting operations and suffer injuries now and then. Today, firefighting robots are gradually deployed to support fire services and enter the explosive, toxic, and smoky fire scene for detection, mitigation, and rescue. Robotic firefighting aims to reduce the risk of casualties and improve disposal efficiency significantly. However, firefighting robots are still far from massive applications, because of the problematic remote control in complex fire incidents and their limited autonomy and small working range. There is still an urgent need for intelligent and autonomous firefighting robots, as well as research and development to expand their firefighting tasks and reliability. This chapter reviews the development of firefighting robots and UAVs over the last six decades, from early master-slave remote control to the latest sensor-driven semi-autonomous control. It also summarizes the classification of firefighting robots and their respective development directions. Finally, we point out the key technologies and challenges in the autonomous intelligent firefighting robots and predict the future application of firefighting robots and their interaction with fire services.

1 Introduction

1.1 History of Robot

A robot, in general, is any automatically operated machine that replaces human effort, so it may not resemble human beings in appearance or perform functions in a humanlike manner [1]. The first true robot, as we recognize it today, was created in the early twentieth century by Czech writer Karel Čapek in his play "R.U.R."

M. Wang · X. Chen · X. Huang (✉)
Research Centre for Fire Safety Engineering, Department of Building Environment and Energy Engineering, The Hong Kong Polytechnic University, Hong Kong, China
e-mail: xy.huang@polyu.edu.hk

X. Huang and W. C. Tam (eds.), *Intelligent Building Fire Safety and Smart Firefighting*, Digital Innovations in Architecture, Engineering and Construction,
https://doi.org/10.1007/978-3-031-48161-1_20

(Rossum's Universal Robots). The play introduced the term "robot" into the popular lexicon and told the story of a factory that created artificial humans to do manual labor. In the following decades, robots began to be used in industry and manufacturing, particularly in automotive production lines.

The first programmable robot was developed in 1954 by George Devol, called the "Unimate". This robot was used to perform repetitive tasks in a factory setting, such as welding and assembly. In 1971, The Stanford Research Institute built the first general-purpose mobile robot "Shakey", which had a far-reaching impact on the fields of robots and artificial intelligence (AI) for the sake of applying the A* search algorithm, the Hough transform, and the visibility graph method [2]. In the 1990s, with the improvement of control technology and communication technology, people gradually tried to use robots in precision operations and extreme environments, such as medicine and space exploration [3]. Figure 1 shows the historical milestones for the development of robots.

1.2 History of Firefighting Robot

Firefighting robots, as the name implies, are specialized robots that serve in firefighting operations. In a narrow sense, it refers only to robots and unmanned aerial vehicles (UAVs) that perform firefighting operations, such as those carrying water cannons and other firefighting equipment. In a broader sense, it also includes inspection robots that conduct daily inspections to eliminate fire hazards (i.e., the fire detection and safety management), logistics robots that assist in transporting heavy equipment, rescue robots that carry out the transfer of trapped people, and other robots and UAVs [4] related to firefighting work. Firefighting robots can assist or even replace firefighters in dangerous firefighting tasks, and can improve the firefighting and rescue capabilities while safeguarding the health and safety of firefighters.

Since 1986, the Tokyo Fire Department has first used the "Rainbow 5" firefighting robot [5] in the fire rescue. Thus, robotic technologies have been in the firefighting field for more than 30 years. The "Rainbow 5" (see Fig. 2a) has a crawler chassis as its walking mechanism and is driven by an electric motor. Equipped with a high-flow fire cannon, gas detection sensors and monitoring cameras, and other components. In 1987, two Sweden companies, PE Holmgren and Rivteknik, collaborated to launch the Brokk 100 demolition robot (see Fig. 2b). As a small demolition robot, it can be radio-controlled from a long distance, with functions such as breaking, digging, and carrying heavy objects. It can move through a standard doorway and climb stairs (up or down), with a working time of 500 h. By 2000, the German company LUF had developed the LUF 60 firefighting robot and achieved a commercial success (see Fig. 2c). LUF 60 included rail carriers, high-powered tracked vehicles, and high-pressure firefighting guns, which was mostly used in tunnels, underground, petrochemical, and other places [6]. The first commercial firefighting robot in the United States was developed by Maine-based Howe & Howe Technologies, named the RS1-T2 Thermite (see Fig. 2d). The prototype was unveiled in 2012, and its

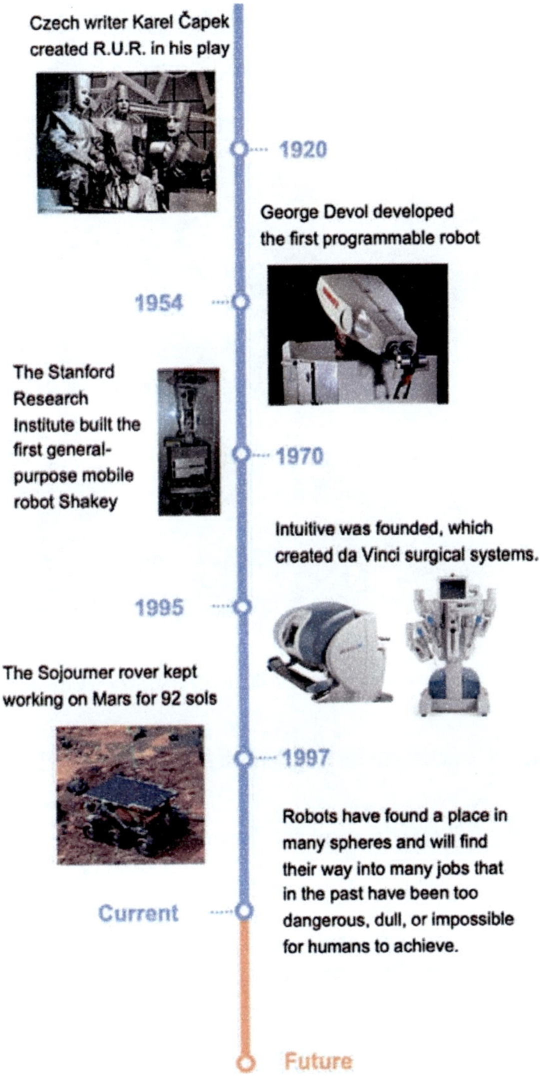

Fig. 1 The historical milestones in the development of robots

technology came from the military robots used by the US Army [7]. More recently, the French company Shark Robotics developed the COLOSSUS firefighting robot in 2017. It can eject the water up to distances of 250 m. In the case of the Notre Dame fire in 2019, the COLOSSUS led the way into the most dangerous areas of the cathedral, equipped with a heat-proof and water-resistant chassis and powerful all-terrain treads. It successfully extinguished flames and clearing away the debris that could have harmed its human operators [8].

Fig. 2 **a** *Rainbow 5* firefighting robot, Tokyo Fire Department, **b** *Brook 100* for firefighting, PE Holmgren and Rivteknik, **c** *LUF 60* firefighting robot, LUF, and **d** *Thermite RS-1*, HOWE & HOWE

2 Classification of Firefighting Robots

2.1 Classification by Function

Firefighting robots can be divided into detection robots, extinguishing robots, smoke extraction robots, explosive disposal robots, lifesaving robots, and logistics robots according to their main functions as shown in Table 1.

a. Detection Robot

The detection robot is composed of mobile mechanism, integrated sensor system, communication unit, remote controller, and other components. It has the functions of temperature detection, toxic and harmful gas monitoring, fire source tracking, real-time scene transmission, real-time mapping, and trapped person search and positioning. It can detect the interior of the fire and identify and determine the accident environment before the firefighters enter the fire. It can provide real-time fire scene information for the incident commanders. Figure 3a shows the Detection robot.

b. Fire Suppression Robot

The fire suppression robot is composed of mobile mechanism, fire extinguishing unit, communication unit, remote controller, and other components. The cameras it

Table 1 Six types of firefighting robots classified by their functions

Type	Function and features
Detection robot	• Temperature detection • Toxic and harmful gas monitoring • Fire source tracking • Real-time scene transmission and mapping • Trapped person search and positioning
Fire suppression robot	• Remotely controlling the robot's movements • Reaching high-risk areas • Targeting and suppressing fires
Smoke extraction robot	• Smoke and dust removal • Thermal radiation barrier • Washing fumes and exhaust gases
Explosive disposal robot	• Providing tactile feedback to the operator • Hazardous materials cleanup • Explosive ordnance disposal
Lifesaving robot	• Finding and transporting people • Deliver materials for assistance
Logistics robot	• Transporting heavy supplies and equipment

Fig. 3 **a** Detection robot, **b** Fire suppression robot, and **c** Smoke extraction robot

carries can assist firefighters in remotely manipulating the robot's movement and targeting the fire. It has the function of fire extinguishing and can enter the high-risk area where firefighters cannot approach for operation. The fire extinguishing unit is usually a water cannon, fog cannon, or fire foam cannon. Figure 3b shows the Fire suppression robot.

c. Smoke Extraction Robot

The smoke extraction robot is composed of mobile mechanism, exhaust fan, communication unit, remote controller, and other parts (see Fig. 3c). Some robots are also equipped with water tanks and fogging devices for spraying water mist. In firefighting and rescue scenarios, they are capable of smoke and dust removal, blocking thermal radiation, and washing smoke and exhaust gas, which is of great significance to improve rescue safety.

d. Explosive Disposal Robot

The explosive disposal robot is composed of a sturdy chassis and a manipulator arm with a range of tools and sensors (see Fig. 4a). The robot is usually controlled remotely by an operator, who uses a combination of cameras and other sensors to navigate and perform tasks. The key function of the firefighting explosive disposal robot is to safely move explosive materials to a safe place or detonate them in a controlled manner. The value of the firefighting explosive disposal robot lies in its ability to perform tasks that are too dangerous for human operators to perform. By using a robot to handle explosive materials, firefighters can reduce the risk of injury or death from accidental detonations [9].

e. Lifesaving Robot

Lifesaving robots often have a variety of structures. Some robots are focused on finding and transporting unconscious trapped people. For example, a bucket with a conveyor belt is used to transfer the trapped person to the robot's internal compartment and then take the trapped person away from the hazardous environment. Other robots focus on delivering protective equipment and supplies to trapped people to assist them in rescuing themselves [10]. The RoboCue, made by the Tokyo Fire Department, is a lifesaving robot already in use (see Fig. 4b).

f. F. Logistics Robot

Logistics robots usually consist of a mobile mechanism and a shelf platform (see Fig. 4c). They are used to assist in transporting oxygen cylinders, water belts, breaking equipment, medical equipment and other heavy supplies and equipment to front-line firefighters. Load capacity and efficiency are often the first practical specifications to be considered for logistics robots. For firefighters, climbing stairs with heavy loads is the most desired robot function. However, no robot has yet emerged that can climb stairs at high speed with sufficient carrying capacity.

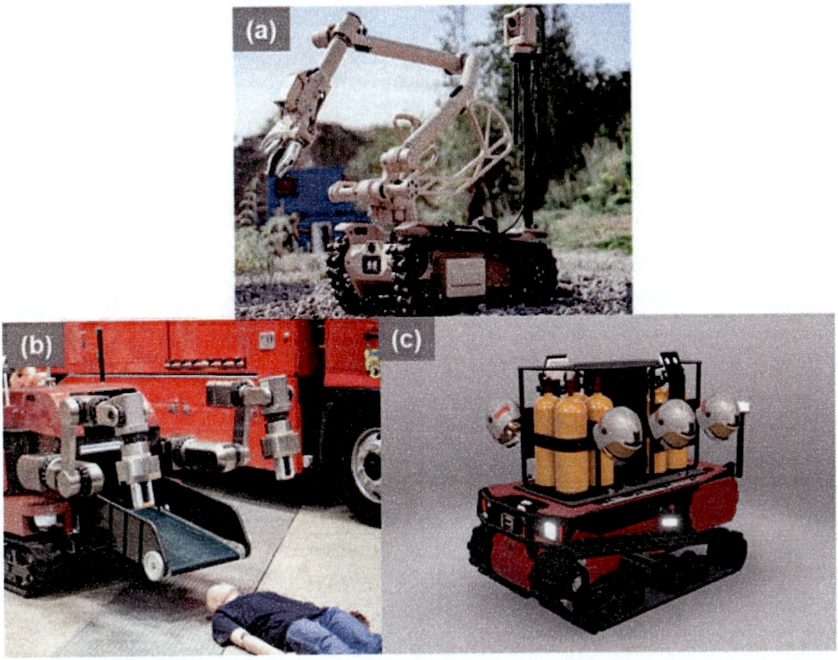

Fig. 4 **a** *T7* Multi-Mission Robotic System made by L3HARRIS, **b** *RoboCue* rescue robot, Tokyo Fire Department, and **c** *TEC800*, ANGATEC

2.2 Classification by Intelligence

Level 1. Remote (Manual) Control Robot

Since the emergence of firefighting robots, most products are robots that require manual remote control. The operator controls every movement of the robot through the operator's console or remote control [11]. Controlling the robot's movement is similar to that of a toy car. Through the monitoring, control the pointing of the water cannon like playing a video game. This is one of the most mature and easy-to-implement control methods, however, it also greatly limits the role of firefighting robots because of various drawbacks [7].

The first obvious drawback is the learning cost associated with robot control. Firefighters often need long periods of training and practice to become proficient at handling the robot. And with fewer real fire scenarios suitable for firefighting robots, it is difficult for firefighters to gain enough experience. Failure in initial attempts can lead to serious injuries and fatalities. The pressure of heavy responsibility is not conducive to the progress of novice operators too.

The second disadvantage is that remote control operations often require visual guidance. In most use scenarios, firefighting robots tend to operate within the visual range of the operator. The operator is next to the robot, observing the robot's path of movement and operational targets in real time for control. In very rare cases, such as in training, firefighters will use the surveillance camera carried by the robot for out-of-sight remote control operations. However, due to the narrow view, it is often difficult to respond to unexpected situations. Cameras also often fail in smoke scenarios caused by fire [12].

Level 2. Partially Autonomous Robot

With the development of sensor technology and information technology, program-assisted control has also been gradually introduced into the research field of fire-fighting robots. Constant speed cruise allows the robot to move quickly on a flat and obstacle-free field. Automatic obstacle avoidance can avoid robot collision accidents caused by operation errors. Monitoring of temperature and gas concentration can be achieved with the help of various sensors. Part of it can do the detection of fire sources and people.

In terms of control, some products have introduced features such as one-touch deployment and one-touch hill climbing. Partially autonomous robots are already advanced firefighting robots at this stage. Technologies such as sensors and program control are introduced as auxiliaries. But it does not effectively combine and solve the problem of remote-control type. Except for a very few simple operations being integrated into a one-button type, it does not reduce the burden of the operator. Large amounts of environmental data were simply collected, adding to the information overload of the fire commander.

Some researchers are currently working on the development of AI-assisted one-click operate functions. The goal and type of operations will be decided by the human. The action of the operation will be done by a robot guided by AI technology. For example, the operator selects a flame and the fire extinguishing task through an interactive interface, and the robot autonomously completes the operational tasks of aiming and spraying extinguishing agent.

Level 3. Autonomous Robot

Autonomous firefighting robots are the most cutting-edge research direction for firefighting robots. Researchers aim to achieve full autonomy in robot movement, reconnaissance, and operations.

In terms of robot movement, autonomous navigation is achieved with the help of synchronous positioning map building, path planning, dynamic obstacle avoidance and other technologies. During the detection, robots complete the analysis of the fire environment deduction and prediction with deep learning using the real-time collection of temperature and gas data and multi-spectrum image information. They directly return intuitive information such as the location of the fire source, the location of trapped people, dangerous areas and fire development to help fire commanders judge the fire situation and command operations. During the operation, robots can achieve the positioning of the target object and the autonomy of the robot operation with the help of traditional computer vision, deep learning, visual servo control and other technologies. The differences among the three levels are shown in Fig. 5.

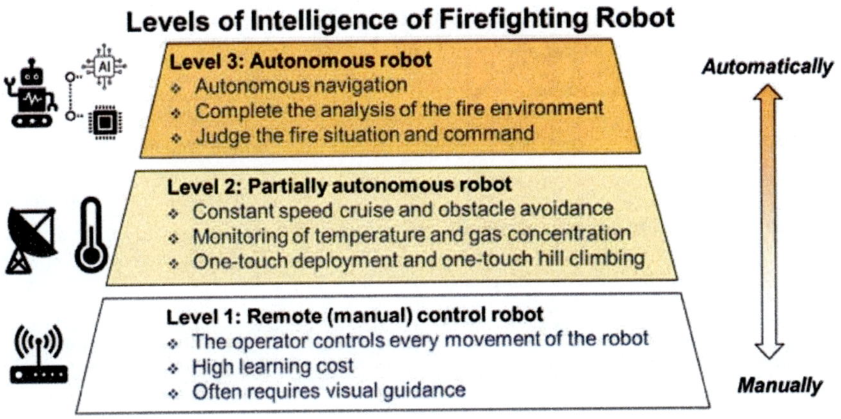

Fig. 5 Function differences among the three levels of firefighting robots

2.3 *Classification by Movement Mode*

According to the movement mode, firefighting robots can be divided into five types as shown in Table 2.

Wheeled chassis: Like an automobile, a firefighting robot using a wheeled chassis has high mobility. However, wheel-based firefighting robots have several disadvantages. Firstly, they can be easily damaged or destroyed by flames or hot gases, which can cause them to lose function or even explode. Secondly, due to the limitation of robot size, the wheels cannot be too big, and the robot's mobility in uneven fields will be significantly reduced. Finally, the use of wheels can cause traction issues in sandy or muddy environments, making it difficult for the robot to move effectively. Therefore, wheeled firefighting robots are often used for daily inspections in sites with level road environments.

Tracked chassis: Firefighting robots that use tracked chassis are the most common type of firefighting robots today. They are designed to operate in more challenging environments, such as high temperatures and ground covered with sharp debris. Tracked chassis provides greater stability and traction than wheeled chassis, making them more suitable for operation in rough terrain. They can achieve a large load and carry large equipment into the fire operation. The crawler has a large contact area and high friction, so it has a strong towing or propulsion force and can perform clearing and towing operations. At the same time, in the erection of high-flow water cannon for extinguishing operations, it is not easy to be affected by recoil. However, they require a significant amount of power to operate, and may suffer from a lack of range in emergency situations. Another disadvantage of the tracked chassis is its slow speed. In situations where emergency deployment is required, waiting for a tracked firefighting robot to reach its designated location takes a lot of time and will seriously affect the efficiency of firefighting and rescue.

Table 2 Five types of firefighting robots classified by movement mode

Type	Function and features	Working environment	Issues
Wheeled chassis	Daily inspections with high mobility	Level road	Can be easily damaged or destroyed in high temperature; Cannot operate complex landforms
Tracked chassis	Achieve a large load and carry large equipment into the fire operation with greater stability and traction	Rough terrain	Slow speed; Require a significant amount of power
Quadruple chassis	Perform detection tasks with quick speed and high efficiency	Challenging terrain	A small load capacity; Difficult to control and navigate; Fragile and require more maintenance
Hybrid chassis	Providing improved versatility, adaptability and maneuverability	All terrain except high obstacles	More fragile and require more maintenance
Unmanned aerial vehicle	Navigate tight spaces; Reach difficult locations; Provide aerial views of the fire scene; Operate at a safe distance from the fire; Reduce the risk of injury to firefighters	In air	Low weight capacity; susceptibility to recoil; Not able to approach hot areas; Difficult to control and stabilize in the air, especially with high temperatures and dense smoke

Quadrupedal chassis: Quadrupedal chassis firefighting robots, also known as four-legged firefighting robots (see Fig. 6a), are the most challenging to build and operate. However, they are the most effective ones. They are designed to move quickly and efficiently through challenging environments, such as steep inclines and dense smoke [13]. However, quadrupedal chassis firefighting robots also have several disadvantages. Firstly, they can be more difficult to control and navigate through complex environments, such as buildings or forests. Secondly, they require more maintenance and repair than other types of firefighting robots, which can add to the cost and complexity of operations. Thirdly, quadruped chassis firefighting robots have small load capacities, making it difficult to carry equipment for operations. Often, they can only perform detection tasks. Finally, quadrupedal chassis firefighting robots may be more fragile than other types of firefighting robots, making them more susceptible to damage or failure in challenging environments.

Hybrid chassis: Due to the above several land mobile way chassis have their own advantages and disadvantages. Several research teams have developed several hybrid structure chassis. The wheel-track hybrid chassis and wheel-foot hybrid chassis are two examples of combining the benefits of different locomotion modes, providing robots with improved versatility, adaptability, and maneuverability in various terrains.

Fig. 6 **a** *Spot* robot dog, Boston Dynamics, **b** Multimodal robot

The wheel-track hybrid chassis integrates both wheels and tracks into a single system, allowing a robot to seamlessly switch between these two locomotion modes. This design often features wheels for efficient movement on hard, flat surfaces and tracks for enhanced traction and stability on rough, uneven terrain. The chassis typically consists of a robust frame that supports the robot's body. It also houses the drive units for the wheels and tracks. When the robot transitions to track mode, the wheels retract or become part of the track system, distributing the robot's weight over a larger surface area. This increases traction and reduces the chances of getting stuck in muddy or sandy conditions.

The wheel-foot hybrid chassis combines the characteristics of wheels and feet, providing robots with the ability to traverse a wide variety of surfaces while maintaining high maneuverability (see Fig. 6b). With this design, each wheel is equipped with multiple foot-like appendages that can articulate independently, allowing the robot to adapt its locomotion to different terrains. These appendages, or "feet," can

Fig. 7 Fire monitoring UAV, DJI ENTERPRISE

be passively compliant or actively controlled, enabling the robot to grip or step on uneven surfaces, climb over obstacles, or maintain stability on slippery terrain. The wheel-foot hybrid chassis offers a balance between the speed and efficiency of wheeled locomotion and the adaptability and terrain negotiation capabilities of legged robots [14].

Unmanned Aerial Vehicle (UAV): The UAV, as a type of flying robot, is also increasingly being used in the development of firefighting applications [15–19]. Figure 7 shows a kind of fire monitoring UAV made by DJI ENTERPRISE. UAV-based firefighting robots have several advantages and disadvantages in their use during fires. The primary advantages include their high maneuverability, ability to pass over obstacles, good field of view, and remote operation capabilities. These features allow them to navigate tight spaces, reach difficult locations, provide aerial views of the fire scene, and operate at a safe distance from the fire, reducing the risk of injury to firefighters. However, UAVs also have several disadvantages. These include their low weight capacity, which may limit their ability to carry large equipment or fire suppressants, and their susceptibility to recoil when dispensing suppressants, which can affect their stability and aim precision. In addition, UAVs may not be able to approach hot areas as well as well-protected ground robots, which may limit their effectiveness in certain situations. Finally, UAVs may be subject to updrafts and hot, low-density air, which may make them difficult to control and stabilize in the air. Despite these limitations, UAV-based firefighting robots offer a promising tool for firefighting operations that can help firefighters respond to fires more effectively and safely [20].

Other innovative modes: In addition to traditional movement methods, researchers have also come up with some very innovative designs for firefighting robots. The SnakeFighter robot Anna Konda [21] designed by SINTEF, shown in Fig. 8a, is

Fig. 8 **a** The SnakeFighter Anna Konda, **b** The SAFFiR humanoid firefighting robot, and **c** The Dragon Firefighter aerial hose type robot

based on the idea of enabling a fire hose to move like a biological snake and perform fire intervention on its own. Combined water utilization is realized for hydraulic joint actuation, fire extinguishment, and robot cooling under high temperatures. A humanoid firefighting robot, SAFFiR [22] shown in Fig. 8b, is a bio-inspired bipedal robot utilized for finding and extinguishing fires aboard naval ships. It has parallel actuated biped using linear actuators. It is capable of omnidirectional walking, balancing in sea state conditions, traversing obstacles, and manipulating fire suppressors. As shown in Fig. 8c, it is an "aerial hose type robot" called the "Dragon Firefighter (DFF)" [23] that can fly along with the water jet. It can avoid obstacles and enter the fire spot from a distance, thus achieving direct fire extinguishing.

3 Key Technologies and Challenges

Firefighting robots are advanced machines designed to assist or replace human firefighters in hazardous environments. These robots employ a combination of cutting-edge technologies to perform their tasks, ensuring the safety of human lives and reducing property damage. This subsection will briefly describe the key technologies for firefighting robots, including mobile chassis, perception sensors, communication, control, navigation, and protection.

3.1 Mobile Chassis

In terms of the power source of the chassis, early firefighting robots were generally powered by internal combustion engines. In recent years, the development of the electric vehicle industry has been driven by the advancement of battery technology, for which benefits are improved efficiency, reduced emissions, and quieter operation. Many firefighting robots equipped with electric chassis are starting to enter the market. However, the challenging working conditions faced by firefighting robots, characterized by high temperature, high pressure, and the constant threat of collision and extrusion, present unique challenges for their power systems. Lithium batteries in this case have potential risks of spontaneous combustion and explosion [24]. Consequently, the development of an explosion-proof battery compartment has become a crucial technological advancement in the design of firefighting robot chassis. This specialized compartment is designed to mitigate the risks associated with the use of lithium batteries, ensuring the safe operation of the robot in hazardous conditions.

Regarding the structure of the chassis, various modes such as wheeled, tracked, and quadrupedal have been introduced and utilized in firefighting robots. When designing the chassis, the selection of an appropriate mode is of paramount importance, as it should align with the specific environment. Additionally, careful consideration must be given to the overall structure of the chassis to fulfill several key requirements. These include achieving a large load capacity to carry necessary equipment and tools, ensuring stability to operate effectively in challenging terrains, facilitating strong mobility to navigate obstacles and debris, and enabling robust obstacle-crossing capabilities to overcome barriers encountered during firefighting operations.

The design of an efficient and reliable chassis is crucial for the overall performance of firefighting robots. By incorporating the appropriate power source and selecting the optimal chassis structure, these robots can effectively and safely carry out their tasks in high-risk environments, assisting firefighters and enhancing their capabilities in combating fires and other emergencies.

3.2 Perception Sensors

To perform complex detection and operational tasks, firefighting robots often use a variety of sensors to sense and collect information from the outside world (see Fig. 9), some of which are specifically designed to work in harsh and hazardous conditions. Here are some of the sensors commonly being used in firefighting robots:

Camera. Cameras are crucial sensors for firefighting robots, providing visual information about the environment [25]. They can be used to detect smoke, flames, and other signs of fire [26]. Cameras provide real-time video footage of the environment, allowing firefighters to make informed decisions about how to proceed. However, they can be obstructed by smoke and vape oils, limiting their effectiveness in some

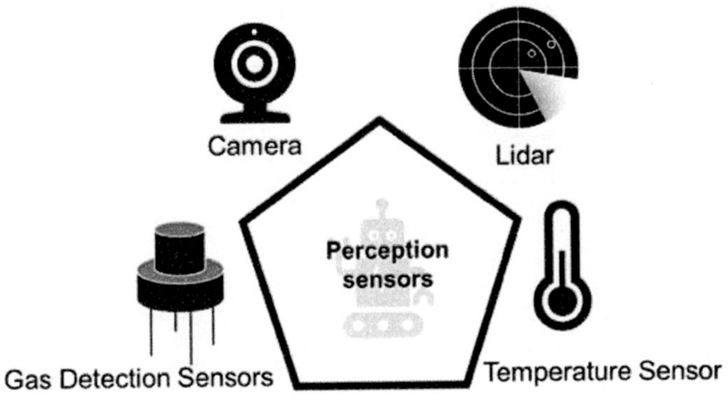

Fig. 9 Common Sensors used in firefighting robots

cases. High temperature also poses a significant threat to their safety. Some improvements that could be made to the camera sensors include using thermal imaging cameras that can detect heat signatures even in smoke, adding a toggle brush to clean the lens oil in front of the lens, and adding a low-absorption housing, and a water-cooled structure to the camera to control the temperature.

IR camera. Infrared (IR) cameras can detect heat signatures through smoke and can provide valuable information about the location and intensity of a fire. As with normal cameras, computer vision algorithms still need to be developed to recognise objects such as fire, smoke and their heat reflections in infrared images [27–32]. The Infrared image from IR cameras is shown in Fig. 10. However, IR cameras can be expensive, and when the smoke temperature is too high, it can also affect the observation of IR cameras. Improved sensitivity, measurement range, and resolution can make IR cameras more effective in detecting heat signatures in a firefighting environment. However, this may further increase the cost.

LiDAR. LiDAR sensors use lasers to detect objects in their path. They can be used to create detailed 3D maps of the environment (see Fig. 11a), which can be used for navigation and detecting obstacles. The principles of Lidar system are shown in Fig. 11b. However, high-resolution LiDARs can be expensive. And because they use shorter wavelength lasers for detection, LiDARs may fail in environments with large amounts of smoke. Switching to longer wavelength radar is a solution to deal with smoky environments even though the resolution of radar is far worse than LIDAR [27]. Recently, Ding et al. [33] proposed a novel approach to 4D radar-based scene flow estimation via cross-modal learning. The approach is motivated by the co-located sensing of radar and camera, which provides various forms of supervision cues to the radar scene flow estimation. This method is also useful in motion segmentation and ego-motion estimation.

Fig. 10 Infrared imaging in chemical fire detection applications

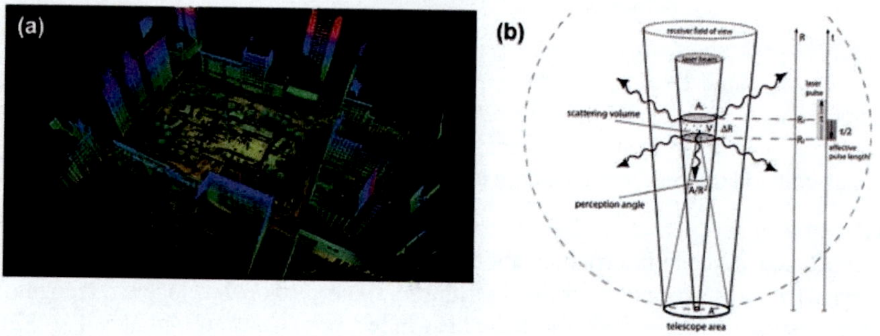

Fig. 11 **a** Cloud point map obtained by LiDAR scanning, and **b** Principle setup of a lidar system [34]

Temperature Sensor. Temperature sensors can be used to detect temperature changes, which can be used as an indication of the location and intensity of a fire, allowing firefighters to make informative decisions about how to proceed. However, the temperature field varies considerably in the vertical direction in the fire environment. The height at which the temperature sensor is set has a significant effect on temperature detection. Mounting the temperature sensor on a retractable mechanism may be a solution.

Gas Detection Sensors. Gas detection sensors identify toxic gases and hazardous materials, enabling firefighters to take appropriate safety measures. While these sensors can be expensive and less effective for certain gases, improved sensitivity and accuracy can enhance their detection capabilities. However, small gas detection sensors that can be mounted on robots are somewhat reliable in detecting the presence of gases but are often very inaccurate when it comes to detecting the concentration of certain gases [35]. Special temperatures, humidity and air pressure can also cause large errors in the sensors. Designing small and accurate gas detection sensors is a valuable research direction.

3.3 AI-Driven Data Analytics

During firefighting and rescue operations, firefighters and commanders often face a large amount of information. The robots used in these operations carry numerous sensors, resulting in a large amount of data that, if transmitted directly to firefighting commanders without processing, can cause information overload. AI can be used to process and analyse this data, summarizing it into more intuitive information that helps firefighters quickly understand the fire environment. For example, based on radar and visual information, using SLAM technology to directly generate 3D maps and mark the location of the fire source, using temperature information to directly generate temperature fields, and using different colors to distinguish the temperature range [36] can allow firefighters to quickly grasp the current situation. AI models also have predictive capabilities, which can provide important guidance for incident commanders to effectively perform their duties on the scene.

3.4 Communication

Wireless communication is critical for robots operating in hazardous or remote environments, where wired communication may not be feasible or safe [37]. The technical challenges in robot wireless communication include signal interference from other wireless devices, signal attenuation due to obstacles, and multi-robot interference when multiple robots are operating in the same area. To overcome these challenges, various techniques can be used, such as frequency hopping, beamforming, and distributed communication systems. Advances in wireless communication technology, such as 5G, are expanding the capabilities of robot wireless communication and enabling new applications. These technologies provide higher data transfer rates, lower latency, and improved reliability, which are essential for real-time control and mission-critical applications. In practice, the antenna pointing is required to be adapted to local conditions. Therefore, the design of an antenna-pointing automatic calibration system is beneficial to maintain the stability of communication.

3.5 Control and Navigation

For autonomous firefighting robots, they are required to use a variety of technologies to navigate in complex and hazardous environments, such as burning buildings or industrial facilities [38]. Some of the key technologies used include:

Self-locating. This technology allows the robot to determine its own position in space without relying on external beacons. Self-location typically involves the use of sensors such as cameras and radars to detect features in the environment that can be used to locate the robots. In addition to cameras and radars, inertial measurement units (IMUs) can also be used for self-location. An IMU is a sensor package that measures the robot's acceleration and rotation in three dimensions. By integrating this data over time, the robot can determine its own position and orientation in space. Combining internal and external data sources, extended Kalman filter [39] can be used to improve the accuracy of robot position estimation [40].

Mapping. Mapping technology enables the robot to create a map of the environment. This can be useful for identifying areas of interest, such as fire hotspots or hazardous materials, as well as for planning safe and efficient routes in the fire environment. Using radars or depth cameras, robots can intuitively acquire point cloud data of the environment. Based on the generated point cloud data, the distance to obstacles and obstacles themselves can be calculated. However, visual SLAM solutions based on monocular cameras cannot directly obtain point cloud data of the environment [41], but form grayscale or colour images. It is necessary to continuously move the robot's position and extract and match feature points to calculate the distance to obstacles. This method often only obtains relative distances, and the scale of the obtained map is distorted [42]. Only by combining absolute distances obtained from other detection methods for matching and calibration can the map be scaled to the true proportion. Figure 12a shows a result of a SLAM algorithm.

Path Planning. Distinguished from autonomous navigation, autonomous exploration does not contain explicit goals. Exploring unknown areas requires finding boundary conditions and optimizing the robot's walking path to complete the exploration of unknown environments at minimal cost and gathering as much information as possible. Truncated Signed Distance Function (TSDF) is a classical algorithm for real-time 3D reconstruction (see Fig. 12b).

Obstacle avoidance. Obstacle avoidance technology allows the robot to navigate around obstacles in its path, such as debris or furniture. This technology typically involves the use of sensors such as radar or ultrasound to detect obstacles and plan alternative routes. Obstacle avoidance is often included in the algorithm for path planning [44].

Together, these navigation technologies enable firefighting robots to explore autonomously in complex and hazardous environments, helping to keep human firefighters safe while also improving the speed and effectiveness of firefighting operations.

For non-autonomous firefighting robots, operational control is relatively simple. For example, for a firefighting robot equipped with a water cannon, only a program needs to be written to control the water cannon to change the horizontal and pitch angles, the opening and closing of the water cannon, and the pattern of the water column according to remote control commands.

For autonomous firefighting robots, they require a combination of hardware and software components to automate tasks such as fire detection and suppression, smoke control, and obstacle detection and removal. For fire detection, the robot is equipped with sensors such as thermal cameras to detect fires. To suppress the fires, it is equipped with suppression systems such as water cannons or foam dispensers to

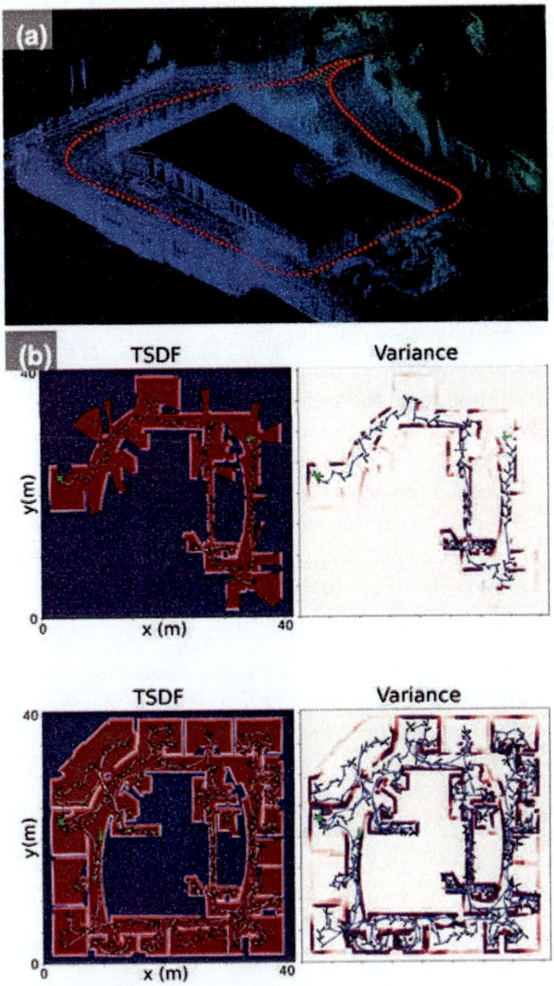

Fig. 12 **a** A result of a SLAM algorithm, red dots indicate keyframe positions, and **b** The robot's exploration of the Intel Map [43]

extinguish them. Smoke control involves using sensors and ventilation systems to detect and clear smoke from the environment, while obstacle detection and removal require sensors like LIDAR or ultrasound to detect obstacles and manipulators or other tools to remove them. These components are integrated into a robust and reliable system with control algorithms and user interfaces, allowing the robot to operate safely and effectively in complex and dangerous environments and display its operating status in real time [45].

In the field of autonomous operations, there are a large number of scientific and engineering problems that need to be solved. This is one of the most cutting-edge and promising directions in the field of firefighting.

3.6 Protective Measures

Protective measures are crucial considerations in the design of firefighting robots to ensure their reliable and safe operation in challenging environments. Two significant aspects of protective measures are ingress protection and thermal protection.

Ingress protection focuses on the robot's ability to withstand water and dust exposure, particularly important as firefighting robots often operate alongside water spray operations. To ensure reliable operation even in the presence of water, the entire robot system should be designed to be fully waterproof. This means that all components and connections should be adequately sealed to prevent water ingress. Additionally, firefighting robots should possess a certain level of dustproof performance due to the presence of smoke and dust particles at fire scenes. An ideal waterproof and dustproof rating for firefighting robots is recommended to be at least IP67 [46]. This level of protection not only safeguards the internal components of the robot but also enhances its durability and reliability in harsh environments.

Thermal protection is a critical consideration for firefighting robot design, especially in high-temperature environments [47]. Figure 13a shows a delivery robot caught fire by accident. All firefighting robots should go through high temperature tests (see Fig. 13b). To ensure effective thermal protection, several measures can be taken which include:

- **Reducing thermal absorption rate.** The outer surface areas of the robot can be made of light-colored or reflective materials or coated with low-absorption coatings to reduce the amount of heat absorbed.

Fig. 13 **a** Delivery robot catches fire on UC Berkeley campus [48], and **b** The high temperature protection tests of the fire-fighting robot [49]

- **Using fire-resistant materials.** The robot's exterior can be constructed using fire-resistant materials such as ceramics or metals that can withstand high temperature.
- **Incorporating insulation.** A middle thermal insulation layer can be added to the robot's exterior to reduce heat transfer to internal components.
- **Adding cooling systems.** Water cooling systems can be included to dissipate heat and maintain the robot's temperature at a safe level.
- **Battery thermal protection.** Battery-powered driven firefighting robots are at risk of thermal runaway under the threat of high temperature and possible mechanical damage [50]. To prevent overheating and potential explosion, thermal management systems can be implemented, such as heat spreaders, cooling systems, and explosion-proof containers [51–53]. The battery can also be placed in a separate compartment with thermal insulation to provide additional protection.

In summary, protective measures such as ingress protection and thermal protection play vital roles in ensuring the safe and reliable operation of firefighting robots. By implementing waterproof and dustproof designs, the robot's internal components are protected from environmental hazards. Effective thermal protection measures, including reducing heat absorption, using fire-resistant materials, incorporating insulation, and implementing cooling systems, safeguard the robot's integrity and functionality in high-temperature environments. These protective measures enhance the durability, efficiency, and overall performance of firefighting robots, enabling them to effectively assist in fire suppression and rescue operations.

4 Current Applications and Future Prospective

Figure 14 shows that Hong Kong Fire Services Department used fire-fighting robot to tackle cable bridge fire. As per the current technological requirements and industrial demands, modularization, autonomy, intelligence, and multi-robot collaboration will be the major trends for the future development of firefighting robots.

a. **Modularization.** Modularization refers to the ability to configure the robot's modules according to the specific needs of the task at hand. This approach allows for flexibility and adaptability, enabling the robot to perform a wide variety of tasks [54].
b. **Autonomy.** Autonomy refers to the robot's ability to operate independently without human intervention. Autonomous firefighting robots can make decisions based on sensor data and algorithms, allowing them to perform complex tasks in hazardous environments without risking human lives.
c. **Intelligence.** Intelligence refers to the robot's ability to process and analyze data from sensors and other sources, allowing it to make informative decisions and take appropriate action. This can include the ability to detect and locate fire hotspots, smoke, and other hazards, as well as the ability to adapt to changing environmental conditions.

Fig. 14 Hong Kong fire services department used fire-fighting robot to tackle cable bridge fire

d. **Multi-robot collaboration.** Multi-robot collaboration refers to the ability of multiple robots to work together to achieve a common goal. This approach can improve efficiency and effectiveness, as robots can divide up tasks and work in parallel, or collaborate to tackle more complex tasks [55, 56]. Overall, the development of firefighting robots that are modular, autonomous, intelligent, and capable of multi-robot collaboration will enable them to perform a wide range of tasks in hazardous environments with increased efficiency, effectiveness, and safety.

5 Summary

Firefighters are constantly exposed to the danger of hot, dark, and toxic fire environments during their firefighting operations and suffer injuries now and then. Today, firefighting robots are gradually deployed to support fire services and enter the explosive, toxic, and smoky fire scene for detection, mitigation, and rescue. Robotic firefighting aims to reduce the risk of casualties and improve disposal efficiency significantly. This chapter reviews the development of firefighting robots and UAVs over the last six decades, from early manual remote control to the latest sensor-driven semi-autonomous control. It also summarizes the classification of firefighting robots and their respective development directions. However, firefighting robots are still far from massive applications, because of the problematic remote control in complex fire incidents and their limited autonomy and small working range. There is still an urgent need for intelligent and autonomous firefighting robots, as well as research and development to expand their firefighting tasks and reliability.

Acknowledgements This work is funded by the Hong Kong Research Grants Council Theme-based Research Scheme (T22-505/19-N) and The Hong Kong Polytechnic University (P0045918).

References

1. H.P. Moravec, *Robot*, Encycl. Br (2023). https://www.britannica.com/technology/robot-technology
2. G. Steinbauer, *Expert Systems (Graz) Heuristic Search (Klagenfurt) Shakey, the Robot office environment* (n.d.), pp. 1–58.
3. N.G. Hockstein, C.G. Gourin, R.A. Faust, D.J. Terris, A history of robots: From science fiction to surgical robotics. J. Robot. Surg. **1**, 113–118 (2007). https://doi.org/10.1007/s11701-007-0021-2
4. B.Y. Lattimer, X. Huang, M.A. Delichatsios, Y.A. Levendis, K. Kochersberger, S. Manzello, P. Frank, T. Jones, J. Salvador, C. Delgado, E. Angelats, M.E. Parés, D. Martín, S. McAllister, S. Suzuki, Use of unmanned aerial systems in outdoor firefighting. Fire Technol. (2023). https://doi.org/10.1007/s10694-023-01437-0
5. K. Miyazawa, Fire robots developed by the Tokyo fire department. Adv. Robot. **16**, 553–556 (2002). https://doi.org/10.1163/156855302320535953
6. A. Kumar, A. Gaur, A. Singh, A. Kumar, K.S. Kulkarni, S. Lala, K. Kapoor, V. Srivastava, S.C. Mukhopadhyay, Fire sensing technologies: A review. IEEE Sens. J. **19**, 3191–3202 (2019). https://doi.org/10.1109/JSEN.2019.2894665
7. P. Liu, H. Yu, S. Cang, L. Vladareanu, Robot-assisted smart firefighting and interdisciplinary perspectives, in 2016 22nd int conf autom comput icac 2016 tackling new challenges autom comput (2016), pp. 395–401. https://doi.org/10.1109/IConAC.2016.7604952
8. R. Bogue, The role of robots in firefighting. Ind. Rob. **48**, 174–178 (2020). https://doi.org/10.1108/IR-10-2020-0222
9. A.O. Oke, A. Afolabi, Development of a robotic arm for dangerous object disposal, in 2014 6th Int Conf Comput Sci Inf Technol CSIT 2014—Proc (2014), pp. 153–160. https://doi.org/10.1109/CSIT.2014.6805994
10. R. Edlinger, C. Fols, A. Nuchter, An innovative pick-up and transport robot system for casualty evacuation, in SSRR 2022—IEEE int symp safety, secur rescue robot (2022), pp. 67–73. https://doi.org/10.1109/SSRR56537.2022.10018818
55. Y. Tamura, H. Amano, J. Ota, Analysis of firefighting skill with a teleoperated robot. Abs. Rob. J. **7**(1) (2020). https://doi.org/10.1186/s40648-020-00177-y
12. Y. Tamura, H. Amano, J. Ota, Analysis of firefighting skill with a teleoperated robot, ROBOMECH J. **7**, (2020). https://doi.org/10.1186/s40648-020-00177-y
13. P. Biswal, P.K. Mohanty, Development of quadruped walking robots: A review. Ain Shams Eng. J. **12**, 2017–2031 (2021). https://doi.org/10.1016/j.asej.2020.11.005
14. M. Bjelonic, C.D. Bellicoso, Y. De Viragh, D. Sako, F.D. Tresoldi, F. Jenelten, M. Hutter, Keep rollin'-whole-body motion control and planning for wheeled quadrupedal robots. IEEE Robot. Autom. Lett. **4**, 2116–2123 (2019). https://doi.org/10.1109/LRA.2019.2899750
15. Á. Madridano, A. Al-Kaff, P. Flores, D. Martín, A. de la Escalera, Software architecture for autonomous and coordinated navigation of uav swarms in forest and urban firefighting. Appl. Sci. **11**, 1–36 (2021). https://doi.org/10.3390/app11031258
16. J.J. Roldán-Gómez, E. González-Gironda, A. Barrientos, A survey on robotic technologies for forest firefighting: Applying drone swarms to improve firefighters' efficiency and safety. Appl. Sci. **11**, 1–18 (2021). https://doi.org/10.3390/app11010363
17. M.M. Valero, O. Rios, E. Planas, E. Pastor, Automated location of active fire perimeters in aerial infrared imaging using unsupervised edge detectors. Int. J. Wildl. Fire. **27**, 241–256 (2018). https://doi.org/10.1071/WF17093
18. P. Pecho, P. Magdolenová, M. Bugaj, Unmanned aerial vehicle technology in the process of early fire localization of buildings. Transp. Res. Procedia. **40**, 461–468 (2019). https://doi.org/10.1016/j.trpro.2019.07.067
19. H. Qin, J.Q. Cui, J. Li, Y. Bi, M. Lan, M. Shan, W. Liu, K. Wang, F. Lin, Y.F. Zhang, B.M. Chen, Design and implementation of an unmanned aerial vehicle for autonomous firefighting missions, in IEEE Int Conf Control Autom ICCA 2016-July (2016), pp. 62–67. https://doi.org/10.1109/ICCA.2016.7505253

20. C. Viegas, B. Chehreh, J. Andrade, J. Lourenço, Tethered UAV with combined multi-rotor and water jet propulsion for forest fire fighting. J. Intell. Robot. Syst. Theory Appl. **104**, 1–13 (2022). https://doi.org/10.1007/s10846-021-01532-w
21. J. Whitman, N. Zevallos, M. Travers, H. Choset, P. Liljebäck, Ø. Stavdahl, A. Beitnes, Snake robot urban search after the 2017 Mexico City Earthquake, in 2018 IEEE Int symp safety, Secur rescue robot SSRR 2018 (2006), pp. 7–12. https://doi.org/10.1109/SSRR.2018.8468633
22. *Shipboard Autonomous Firefighting Robot (SAFFiR)—Office of Naval Research* (2016). http://www.onr.navy.mil/en/MediaCenter/Fact-Sheets/Shipboard-Robot-Saffir.aspx
23. H. Ando, Y. Ambe, T. Yamaguchi, M. Konyo, K. Tadakuma, S. Maruyama, S. Tadokoro, Fire fighting tactics with aerial hose-type robot "dragon firefighter," in Proc IEEE work adv robot its soc impacts, ARSO 2019-Octob (2019), pp. 291–297. https://doi.org/10.1109/ARSO46408.2019.8948716
24. P. Sun, R. Bisschop, H. Niu, X. Huang, *A review of battery fires in electric vehicles* (Springer US, 2020). https://doi.org/10.1007/s10694-019-00944-3
25. K.A.P. Perumal, M.A.M. Ali, Z.H. Yahya, Fire fighter robot with night vision camera, in Proc—019 IEEE 15th Int Colloq Signal Process. Its Appl CSPA 2019 (2019), pp. 270–274. https://doi.org/10.1109/CSPA.2019.8696077
26. A.E. Çetin, K. Dimitropoulos, B. Gouverneur, N. Grammalidis, O. Günay, Y.H. Habiboğlu, B.U. Töreyin, S. Verstockt, Video fire detection—Review. Digit. Signal Process. A Rev. J. **23**, 1827–1843 (2013). https://doi.org/10.1016/j.dsp.2013.07.003
27. J.W. Starr, B.Y. Lattimer, Evaluation of navigation sensors in fire smoke environments. Fire Technol. **50**, 1459–1481 (2014). https://doi.org/10.1007/s10694-013-0356-3
28. J. Zhu, W. Li, D. Lin, H. Cheng, G. Zhao, Intelligent fire monitor for fire robot based on infrared image feedback control. Fire Technol. **56**, 2089–2109 (2020). https://doi.org/10.1007/s10694-020-00964-4
29. M. Bhattarai, M. Martinez-Ramon, A deep learning framework for detection of targets in thermal images to improve firefighting. IEEE Access. **8**, 88308–88321 (2020). https://doi.org/10.1109/ACCESS.2020.2993767
30. J.H. Kim, S. Jo, B.Y. Lattimer, Feature selection for intelligent firefighting robot classification of fire, smoke, and thermal reflections using thermal infrared images, J. Sensors. **2016**, (2016). https://doi.org/10.1155/2016/8410731
31. J.H. Kim, J.W. Starr, B.Y. Lattimer, Firefighting robot stereo infrared vision and radar sensor fusion for imaging through smoke. Fire Technol. **51**, 823–845 (2015). https://doi.org/10.1007/s10694-014-0413-6
32. J.H. Kim, B.Y. Lattimer, Real-time probabilistic classification of fire and smoke using thermal imagery for intelligent firefighting robot. Fire Safety. J. 7240–7249 (2015). https://doi.org/10.1016/j.firesaf.2015.02.007
33. F. Ding, A. Palffy, D.M. Gavrila, C.X. Lu, *Hidden Gems: 4D radar scene flow learning using cross-modal supervision* (2023)
34. P.F. McManamon, Introduction to LiDAR, LiDAR Technol. Syst., 1–18 (2019). https://doi.org/10.1117/3.2518254.ch1
35. D. Kohl, J. Kelleter, H. Petig, Detection of fires by gas sensors. Sensors Updat. **9**, 161–223 (2001). https://doi.org/10.1002/1616-8984(200105)9:1%3c161::aid-seup161%3e3.0.co;2-a
36. X. Zhang, X. Wu, X. Huang, Smart real-time forecast of transient tunnel fires by a dual-agent deep learning model. Tunn. Undergr. Sp. Technol. **129**, 104631 (2022). https://doi.org/10.1016/j.tust.2022.104631
37. J.-H. Park, B.-W. Kim, D.-J. Park, M.-J. Kim, A system architecture of wireless communication for fire-fighting robots. IFAC (2008). https://doi.org/10.3182/20080706-5-kr-1001.00892
38. G.N. DeSouza, A.C. Kak, Vision for mobile robot navigation: A survey. IEEE Trans. Pattern Anal. Mach. Intell. **24**, 237–267 (2002). https://doi.org/10.1109/34.982903
39. K. Dhwaj, J.M. Kovitz, H. Tian, L.J. Jiang, T. Itoh, Half-mode cavity-based planar filtering antenna with controllable transmission zeroes. IEEE Antennas Wirel. Propag. Lett. **17**, 833–836 (2018). https://doi.org/10.1109/LAWP.2018.2818058

40. S. Thrun, Probabilistic robotics. Commun. ACM **45**, 52–57 (2002). https://doi.org/10.1145/504729.504754
41. K. Qian, Z. He, X. Zhang, 3D Point cloud generation with millimeter-wave radar, Proc. ACM Interactive, Mobile, Wearable Ubiquitous Technol. 4 (2020). https://doi.org/10.1145/3432221.
42. D. Scaramuzza, F. Fraundorfer, Tutorial: Visual odometry. IEEE Robot. Autom. Mag. **18**, 80–92 (2011). https://doi.org/10.1109/MRA.2011.943233
43. K. Saulnier, N. Atanasov, G.J. Pappas, V. Kumar, Information theoretic active exploration in signed distance fields, in Proc—IEEE International Conference Robot Autom (2020), pp. 4080–4085. https://doi.org/10.1109/ICRA40945.2020.9196882
44. J. Borenstein, Y. Koren, Real-time obstacle avoidance for fast mobile robots. IEEE Trans. Syst. Man Cybern. **19**, 1179–1187 (1989). https://doi.org/10.1109/21.44033
45. J. Gancet, E. Motard, A. Naghsh, C. Roast, M.M. Arancon, L. Marques, User interfaces for human robot interactions with a swarm of robots in support to firefighters, in Proc—IEEE International Conference Robot Autom (2010), pp. 2846–2851. https://doi.org/10.1109/ROBOT.2010.5509890
46. International Electrotechnical Commission(IEC), IEC 60529:1989+AMD1:1999+AMD2:2013 CSV—Degrees of protection provided by enclosures (IP Code), (2013). https://webstore.iec.ch/publication/2452
47. T. AlHaza, A. Alsadoon, Z. Alhusinan, M. Jarwali, K. Alsaif, New concept for indoor fire fighting robot, procedia—Soc. Behav. Sci. **195**, 2343–2352 (2015). https://doi.org/10.1016/j.sbspro.2015.06.191
48. B. Sophia, *KiwiBot catches fire outside Martin Luther King Jr. Student Union, Dly* (Californian, 2023). https://www.dailycal.org/2018/12/14/kiwibot-catches-fire-outside-mlk-student-union
49. W. Wang, W. Gao, S. Zhao, W. Cao, Z. Du, Robot protection in the hazardous environments, in Robot Oper Hazard Environment (InTech, 2017). https://doi.org/10.5772/intechopen.69619.
50. Y. Liu, H. Niu, J. Liu, X. Huang, Layer-to-layer thermal runaway propagation of open-circuit cylindrical li-ion batteries: Effect of ambient pressure, J. Energy Storage. **55**, 105709 (2022). https://doi.org/10.1016/j.est.2022.105709.
51. J. Sun, B. Mao, Q. Wang, Progress on the research of fire behavior and fire protection of lithium ion battery. Fire Saf. J. **120**, 103119 (2021). https://doi.org/10.1016/j.firesaf.2020.103119
52. J. Weng, D. Ouyang, Y. Liu, M. Chen, Y. Li, X. Huang, J. Wang, Alleviation on battery thermal runaway propagation: Effects of oxygen level and dilution gas. J. Power. Sources **509**, 230340 (2021). https://doi.org/10.1016/j.jpowsour.2021.230340
53. Q. Wang, B. Mao, S.I. Stoliarov, J. Sun, A review of lithium ion battery failure mechanisms and fire prevention strategies. Prog. Energy Combust. Sci. **73**, 95–131 (2019). https://doi.org/10.1016/j.pecs.2019.03.002
54. X. Liu, M. Zhang, W. Liu, Design method to modular robot system, in Proc. 2009 ASME/IFToMM Int Conf Reconfigurable Mech Robot ReMAR 2009 (2009), pp. 521–528
55. M.S. Innocente, P. Grasso, Self-organising swarms of firefighting drones: Harnessing the power of collective intelligence in decentralised multi-robot systems. J. Comput. Sci. **34**, 80–101 (2019). https://doi.org/10.1016/j.jocs.2019.04.009
56. J. Saez-Pons, L. Alboul, J. Penders, L. Nomdedeu, Multi-robot team formation control in the GUARDIANS project. Ind. Rob. **37**, 372–383 (2010). https://doi.org/10.1108/01439911011044831